TWO-DIMENSIONAL PHASE UNWRAPPING

TWO-DIMENSIONAL PHASE UNWRAPPING

THEORY, ALGORITHMS, AND SOFTWARE

Dennis C. Ghiglia

Sandia National Laboratories
Albuquerque, New Mexico

Mark D. Pritt

Lockheed Martin Corporation
Gaithersburg, Maryland

A WILEY-INTERSCIENCE PUBLICATION

JOHN WILEY & SONS, INC.

New York / Chichester / Weinheim / Brisbane / Singapore / Toronto

COVER

The image shown on the cover is a synthetic aperture radar interferogram generated by combining two passes over Death Valley, California by the Canadian Radarsat satellite. The fringes represent the wrapped phased differences between the complex-valued images from each pass. The terrain elevations can be computed from the phase differences, but the phase differences must first be unwrapped. This is a difficult phase unwrapping problem. In regions of steep terrain, the phase wraps very quickly and becomes corrupted where there are radar shadow and "layover" effects. Surface disturbances that occurred between the two passes, which were 24 days apart, also corrupt the phase.

This image was acquired as part of a program for the Terrain Modeling Project Office funded by the Defense Modeling and Simulation Office and the U.S. Army Topographic Engineering Center. The SAR data was provided by Radarsat International. The interferogram was generated and provided by Vexcel Corporation, Boulder, Colorado.

This text is printed on acid-free paper.

Published simultaneously in Canada.

Library of Congress Cataloging in Publication Data:

Ghiglia, Dennis C.
Two-dimensional phase unwrapping : theory, algorithms, and software / Dennis C. Ghiglia and Mark D. Pritt.
p. cm.
"A Wiley-Interscience publication."
Includes bibliographical references and index.
ISBN 0-471-24935-1 (cloth : alk. paper)
1. Synthetic aperture radar. 2. Signal processing—Mathematics. 3. Interferometry. I. Pritt, Mark D. II. Title.
TK6592.S95G45 1998
621.36'7—dc21 97-38034

10 9 8 7 6 5 4 3

CONTENTS

FOREWORD

Two-dimensional phase unwrapping is the type of problem that is typically the domain of the mathematician. It is both complex and abstract. However, phase unwrapping is also the core technology that enables radar interferometry.

Over the past decade interferometry has changed the way that we use radar data. Radar data are now used for precise measurement of surface topography in clouded regions. Additionally, spaceborne radar systems have proved effective for measuring surface changes from earthquakes and volcanic eruptions. These applications have created a new class of radar data users primarily involved in mapping and remote sensing applications.

In *Two-Dimensional Phase Unwrapping: Theory, Algorithms, and Software*, the authors unlock the mystery of phase unwrapping in interferometric data processing. This text provides a clear, concise treatment of phase unwrapping that cannot be found in any other source. It presents for the first time the relationship between theory and application. Its uniform treatment of the various phase unwrapping techniques makes it a valuable resource for any engineer or scientist involved in processing or exploitation of interferometric data.

I expect that radar interferometry will increase in importance over the coming decade with the development of airborne and spaceborne sensor systems designed to optimally exploit this technology. *Two-Dimensional Phase Unwrapping: Theory, Algorithms, and Software* is an important contribution to our understanding of radar interferometry that will benefit both research into advanced techniques and the design of these future sensor systems.

JOHN C. CURLANDER

President and CEO
Vexcel Corporation

PREFACE

Two-dimensional phase unwrapping arises most naturally in, but is not restricted to, interferometric applications. Measured or calculated phase values from two or more mutually coherent multidimensional signals are related in a nonlinear manner to a desired physical quantity of interest. The nonlinearity is in the form of "wraps" or cycle discontinuities where an underlying two-dimensional phase is wrapped into the interval $(-\pi, \pi]$.[1] The wrapped phase must somehow be unwrapped in order to provide an estimate of the underlying physical quantity. Estimation of surface topography from interferometric synthetic aperture radar (SAR) or extremely accurate profiling of mechanical parts by optical interferometers are two such examples.

Originally developed for military reconnaissance, SAR is now experiencing new life in civil applications. In fact civilian and commercial interests are rapidly becoming the drivers of technology. Clever utilization of the coherent SAR imagery in interferometric configurations makes possible the measurement of surface topography to accuracies much better than the spatial resolution (~ 0.3 meters to several meters) of the SAR images themselves. Indeed, as is commonplace with interferometers, measurement sensitivities are on the order of the operating wavelength, which is typically a few centimeters for SAR. Imaging geometries, noise, and other operational factors degrade performance somewhat from centimeter-scale accuracies, but nevertheless SAR interferometry makes possible global topographic mapping in a timely fashion, in daylight or at night, in all weather conditions, and with unprecedented accuracy. SAR interferometry also can detect deformations of the earth's crust on the order of millimeters, a capability that shows promise for the timely detection of earthquakes or volcanic eruptions.

These exciting possibilities have led to an explosive growth in the field of phase unwrapping as indicated by the increasing number of journal publications. Newcomers to SAR interferometry and related disciplines will eventually confront the phase unwrapping problem and, undoubtedly, will encounter a rather bewildering variety of ideas and algorithms, including those based on neural networks, simulated annealing, cellular automata, genetic algorithms, and other unusual constructs. Which of these are good? Which are not? We do

[1] Throughout this book we use the notation $(-\pi, \pi]$ to represent the interval $-\pi < x \leq \pi$.

not have complete answers to these questions, and in all likelihood there never will be a complete answer. However, we can help researchers approach the problem in the context of their own applications with some guidance from those who have been through the "school of hard knocks."

We have greatly simplified the investigative process by categorizing phase unwrapping approaches into two general classes: path-following and minimum-norm methods. We have provided detailed descriptions and analyses of four path-following approaches and four minimum-norm approaches. As is the case with politics, there are plenty of partisan participants firmly entrenched in their own camp with nary a nod of recognition to an opposing view. We ourselves are guilty of this to some degree. However, we provide enough theory, philosophy, algorithms, and references to allow researchers to make up their own minds in a much more timely fashion than if some sifting had not been done. Fortunately, unlike politics, sound scientific investigation makes it easier to switch "party affiliations" if necessary.

While this book is slanted toward SAR interferometry applications, it has much in common with other disciplines in which phase unwrapping plays an important role. In fact it is safe to say that the phase unwrapping problems encountered in SAR are about "as hard as they come." Practical solutions for SAR interferometry should easily find a place in related applications outside the field of SAR. We believe readers of this book will have no trouble bridging the gap.

This book provides another useful function. Most researchers are "quick studies" and soon learn which algorithms they would like to try. Newcomers usually lack the basic building blocks of operational codes from which timely results spring forth. As we all know, it takes time to develop and successfully implement a working algorithm. To this end, we have supplied a suite of C language programs and routines comprising over 6000 lines of source code that implement the ideas and algorithms discussed in this book. We have already done the laborious part of coding the algorithms, fixing the bugs, and contacting the authors of the algorithms to discover the "tricks" necessary to get them to work correctly. We encourage the reader to understand how the algorithms work, use them properly, modify them as appropriate, and improve them. These programs, which are listed in Appendix A, are available for downloading on the Wiley ftp server, along with a number of phase unwrapping datasets. See Appendix C for details.

The supplied software comprises only a subset of the vast suite of published and unpublished algorithms. We have selected and implemented those that we believe are representative of the current "state-of-the-art" in phase unwrapping, such as Flynn's minimum discontinuity algorithm, while not neglecting the older and more "classic" algorithms, such as the residue and branch cut algorithm of Goldstein, Zebker, and Werner. Omissions of certain algorithms are not a suggestion of inadequacy. We just could not include them all. By contrast, inclusion is not a claim of superiority. Instead, they should be viewed for the insight and perspective they provide. Indeed new and potentially powerful and useful algorithms are being developed all the time. This is the sign of a vital and dynamic field, and we should be encouraged rather than befuddled.

Many of the ideas, concepts, philosophies, and algorithms discussed are not, of course, our own. We have been fortunate to work with many talented colleagues who have helped us immensely for many years in many diverse ways. Assistance has come in the form of developing challenging problems, providing experimental data, offering first-rate mathematical and algorithmic assistance, supplying hardware and software support, and providing a stimulating and supportive work environment. For all of these reasons and more we wish to extend sincere thanks to Charles V. Jakowatz, Jr., Louis A. Romero, Gary A. Mastin, Daniel E. Wahl, Paul A. Thompson, Bruce C. Walker, William H. Hensley, David K. Johnson, Terry M. Calloway, Terry A. Bacon, Michael J. Roth, and Edward J. Nowicki.

We especially wish to thank Thomas J. Flynn, Paul H. Eichel, and Doug L. Bickel of Sandia National Laboratories for their thorough and insightful reviews of this book. Tom Flynn was especially helpful with his collaboration during the preparation of this work and the actual research investigations. He developed one of the phase unwrapping codes, and he graciously allowed us to present and distribute it to the research community. Michael J. Roth of the Johns Hopkins University Applied Physics Laboratory supplied us with two phase unwrapping examples that were generated by his IFSAR simulator. In addition Richard M. Goldstein of the Jet Propulsion Laboratory provided invaluable assistance with our understanding of the subtleties of the widely used algorithm credited to him and his colleagues, Howard A. Zebker and Charles L. Werner. Finally we wish to thank John C. Curlander of Vexcel Corporation for writing the Foreword.

This work was performed in part at Sandia National Laboratories, Albuquerque, NM. Sandia is a multiprogram laboratory operated by Sandia Corporation, a Lockheed Martin company, for the United States Department of Energy under Contract DE-AC04-94AL85000.

Dennis C. Ghiglia would like to express his thanks and appreciation to the management of Sandia National Laboratories and the U.S. Department of Energy for their support throughout his career. He also wishes to thank his son, Daniel C. Ghiglia, for his expertise and invaluable assistance with the artwork development and LaTeX typesetting, and his wife Ruth Ann and daughter Carla for their understanding and support during the time-consuming process of book writing.

Mark D. Pritt would like to express his thanks and appreciation to Hal Garner, a vice president of Lockheed Martin Corporation, for funding the research and writing of this book. He also wishes to thank his former manager, Sharon L. Smith, for her invaluable support of this effort, and his current manager, Rick W. Duke, for helping bring it to completion. Last, but certainly not least, he wishes to thank his wife Christine and his children Anna, Jacob, Emily, Samuel, and Noah for their enthusiastic support and encouragement.

DENNIS C. GHIGLIA
Sandia National Laboratories
MARK D. PRITT
Lockheed Martin Corporation

April 1997

TWO-DIMENSIONAL PHASE UNWRAPPING

1

INTRODUCTION TO PHASE UNWRAPPING

1.1 THE IMPORTANCE OF PHASE

Over the past several years there has been a marked increase in the development and application of sophisticated techniques for coherent signal and image processing. Coherent processing is characterized by a single quantity known as *phase* [1]. Transmission or reception of coherent signals requires accommodating and utilizing both temporal and spatial signal magnitudes and phase. By contrast, incoherent processes utilize only signal magnitudes or *intensities*.[1] Examples where coherent processing is required include synthetic aperture radar (SAR), synthetic aperture sonar, adaptive beamforming, acoustic imaging, projection and diffraction tomography, adaptive optics, magnetic resonance imaging (MRI), optical and microwave interferometry, seismic processing, X-ray crystallography, and aperture synthesis radio astronomy. In fact, in *any* process requiring Fourier synthesis or analysis, phase is important.

In 1981 Oppenheim and Lim published a now-classic paper illustrating the importance of phase in signals [2]. Even though this paper has nothing to do with phase unwrapping per se, we choose to repeat their experiment here to illustrate the importance of phase. But first, as an aid to the forthcoming discussion, we define the one-dimensional continuous Fourier transform by the following pair of relationships:

$$H(\omega) = \int_{-\infty}^{\infty} h(t)e^{-j\omega t}dt, \tag{1.1}$$

$$h(t) = \frac{1}{2\pi}\int_{-\infty}^{\infty} H(\omega)e^{j\omega t}d\omega, \tag{1.2}$$

[1] To be consistent with the optics literature, signal magnitude is referred to as *amplitude*, whereas squared magnitude is referred to as *intensity*.

where t typically represents time and ω denotes angular frequency (in radians per second). However, t could, for example, refer to spatial position, and ω would then denote spatial frequency (e.g., in $2\pi\times$cycles per meter). Similarly the one-dimensional discrete Fourier transform (DFT) is represented by

$$H_k = \sum_{n=0}^{N-1} h_n e^{-j2\pi nk/N}\,, \qquad k = 0, 1, ..., N-1, \tag{1.3}$$

$$h_n = \frac{1}{N}\sum_{k=0}^{N-1} H_k e^{j2\pi nk/N}\,, \qquad n = 0, 1, ..., N-1\,. \tag{1.4}$$

The quantities n and k here denote discrete sample indexes in the time and frequency domains, respectively, while N denotes the number of samples in both domains. The quantity h_n will be regarded as a sample of the continuous function $h(t)$ at time $t = nT$, and H_k will be interpreted as an approximation of the continuous spectrum $H(\omega)$ at the frequency $\omega = \omega_k = 2\pi k/(NT)$. The generalization of the above equations to two (or more) dimensions is straightforward.

Figure 1.1 shows two 512×512-pixel digitized photographs of Albert Einstein and the Mona Lisa, respectively. If we compute the two-dimensional DFT of each image and separate the resulting complex spectrum into a *magnitude* and *phase* spectrum, the results are as shown in Figures 1.2 and 1.3, respectively. (In reference to Equations 1.1 through 1.4, the magnitude spectrum is $|H|$ and the phase spectrum is $\psi = \arctan[\Im(H), \Re(H)]$, where $\Re$ and $\Im$ refer to the real and imaginary parts, and arctan is the four-quadrant arc tangent operator where $-\pi < \psi \leq \pi$.)

It is readily observed that each magnitude spectrum shows the distribution of the magnitudes of the spatial frequency components of the corresponding image,

(a) (b)

Figure 1.1 512×512-pixel digital photographs. (a) Albert Einstein. (b) Mona Lisa.

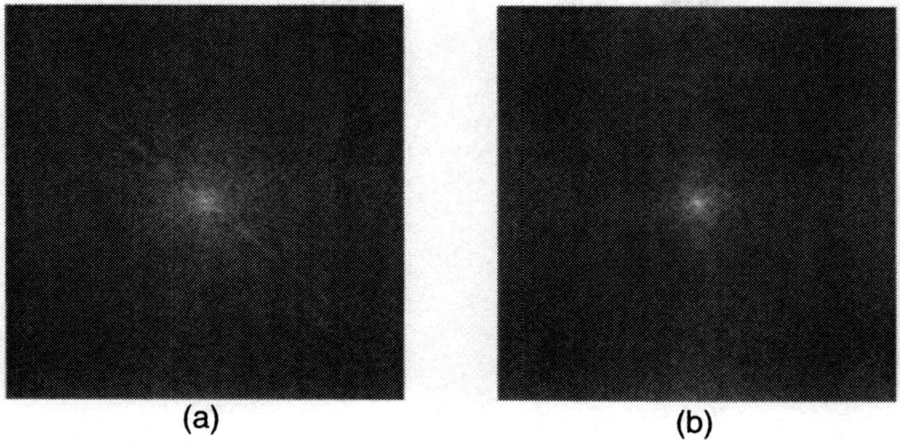

Figure 1.2 Two-dimensional Fourier transform magnitude spectra. (a) Log-scaled magnitude spectrum of Albert Einstein. (b) Log-scaled magnitude spectrum of the Mona Lisa. The spatial frequency origin is at the image center. The horizontal and vertical directions refer to spatial frequencies. The point of zero spatial frequency is in the center of each spectrum.

Figure 1.3 Two-dimensional Fourier transform phase spectra. (a) Phase spectrum of Albert Einstein. (b) Phase spectrum of the Mona Lisa. Note that the phase appears random with no apparent structure.

while the phase spectrum appears random with no apparent connection to the original image. However, it should be emphasized that the phase spectrum defines the relative *position* of each two-dimensional complex exponential comprising the Fourier decomposition. The importance of phase is thus illustrated by reconstructing or synthesizing not the original images by Fourier inversion (as we all know can be done exactly) but two new images produced by

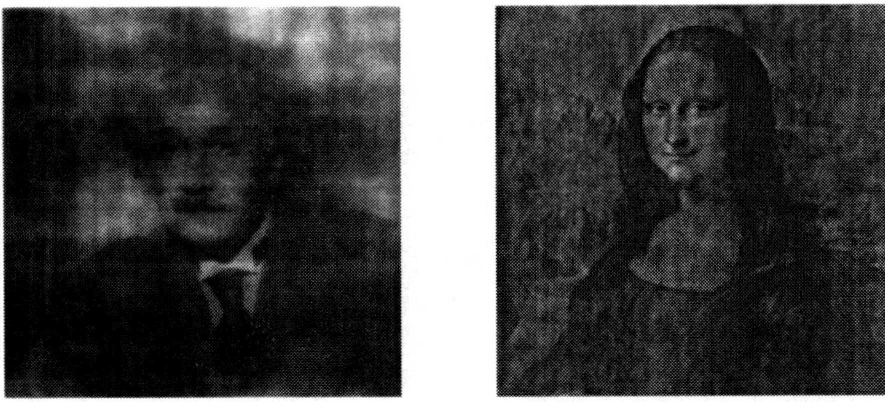

(a) (b)

Figure 1.4 Reconstruction of images by Fourier inversion with swapped phase spectra. (a) Synthesized image obtained by combining the magnitude spectrum of the Mona Lisa with the phase spectrum of Albert Einstein. (b) Synthesized image obtained by combining the magnitude spectrum of Albert Einstein with the phase spectrum of the Mona Lisa.

first swapping the respective phase spectra and then performing a two-dimensional inverse DFT. In other words, we assign the magnitude spectrum of the Mona Lisa to the phase spectrum of Albert Einstein and then perform the inverse DFT. Similarly we assign the magnitude spectrum of Albert Einstein to the phase spectrum of the Mona Lisa and then inverse transform.

The results of this apparently bizarre synthesis are shown in Figure 1.4. It is quite obvious that the interpretability of each image is due more to the content of the phase spectrum than the magnitude spectrum. A highly recognizable Albert Einstein was synthesized from the magnitude spectrum of the Mona Lisa coupled with the *phase* spectrum of Albert Einstein, for instance. Sharp edges, fine structure, and overall interpretability are maintained in the phase spectrum reconstruction. Incorrect magnitude spectral components simply add an overall cloudy or noiselike perturbation to the reconstruction. It is easily imagined that slight perturbation of the correct phase values could destroy completely the fidelity of the synthesized image. This simple experiment dramatically demonstrates the importance of phase in the reconstruction. Any coherent processing requires extreme care in order to maintain correct phase information.

1.2 THE NEED FOR PHASE UNWRAPPING

One subtle but very important point hinted at in the previous phase-swapping experiment is that the phase must be *extracted* from an actual signal by some mathematical operation. Even though we may treat phase as if it were a real signal, it is actually a *property* of a real signal. It is a property related to the temporal and/or spatial wavelength and thus influences the signal only through phase principal values, that is, only those values that lie between $\pm\pi$ radians. These principal values are also called *wrapped* phase values because the absolute phase (whatever it happens to be) is wrapped into the interval $(-\pi, \pi]$.

We know of no instance where phase itself is detected as an actual signal. Instead, we obtain wrapped phase values through mathematical operations on actual signals. So what's the point? The point is that the wrapping operation is a nonlinear process. Mathematically some unknown phase function φ is wrapped into the interval $(-\pi, \pi]$ by the following operation:

$$\psi(t) = \varphi(t) + 2\pi k(t), \tag{1.5}$$

where $k(t)$ is an integer function that forces $-\pi < \psi \leq \pi$. Thus $\psi(t)$ is a nonlinear function of $\varphi(t)$. In many applications $\varphi(t)$ relates to some physical quantity such as surface topography in interferometry [3]–[10], wavefront distortion in adaptive optics [11]–[13], the degree of magnetic field inhomogeneity in the water/fat separation problem of magnetic resonance imaging (MRI) [14]–[16] or the relationship between the object phase and its bispectrum phase in astronomical imaging [17], [18]. The highly nonlinear $\psi(t)$ that can be extracted from physical signals does, in general, provide useful information. However, the phase must be processed (unwrapped) before further use.

Simply stated, the phase unwrapping problem is concerned with obtaining an estimate $\phi(t)$ for the phase $\varphi(t)$ from the wrapped function $\psi(t)$. Thus the wrapped values must be *unwrapped* somehow to estimate $\varphi(t)$, the quantity relating to the physical property of interest.

We will now introduce the phase unwrapping problem in the context of synthetic aperture radar (SAR) interferometry, the subject with which we have the most experience. The following brief overview of SAR interferometry will be closely linked to classical optics and the fundamental properties of wave interference. It captures the essence of why two-dimensional phase unwrapping arises naturally in interferometric applications.

1.2.1 Synthetic Aperture Radar

Synthetic aperture radar is a valuable imaging technique with wide-ranging applications in reconnaissance and remote sensing. It allows high-resolution imaging at great distances (e.g., submeter resolution at a few kilometers range, or several meter resolution at hundreds of kilometers range) in all weather conditions, day or night. It works by transmitting coherent broadband microwave radio signals from an aircraft or spacecraft platform, receiving the signals reflected from the terrain, storing and processing the returns to synthesize a large aperture, and then focusing the data to form a high-resolution image of the terrain response to these radio signals [19], [20]. Depending on the collection mode (e.g., "stripmap" or "spotlight"), image formation is accomplished by various stages of time-domain correlation, data interpolation, and/or Fourier inversion. The bulk of this processing is accomplished digitally in a modern SAR. State-of-the-art SARs even perform such functions as waveform generation and motion compensation by digital means [21], [22].

Because a SAR provides its own illumination (microwaves) that penetrates virtually all weather disturbances, imagery can be obtained under almost any conditions. See Figure 1.5 for a typical imaging modality and relevant terrain illumination geometry. SAR imagery in many ways resembles optical imagery, as can be seen in the side-by-side comparison shown in Figure 1.6. The primary difference is that optical imagery shows the terrain reflectivity response to visible

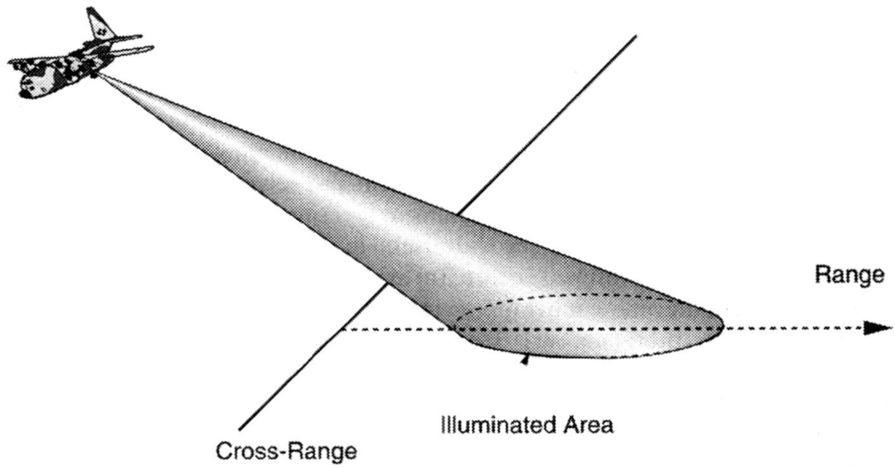

Figure 1.5 Typical SAR imaging modality and terrain illumination geometry.

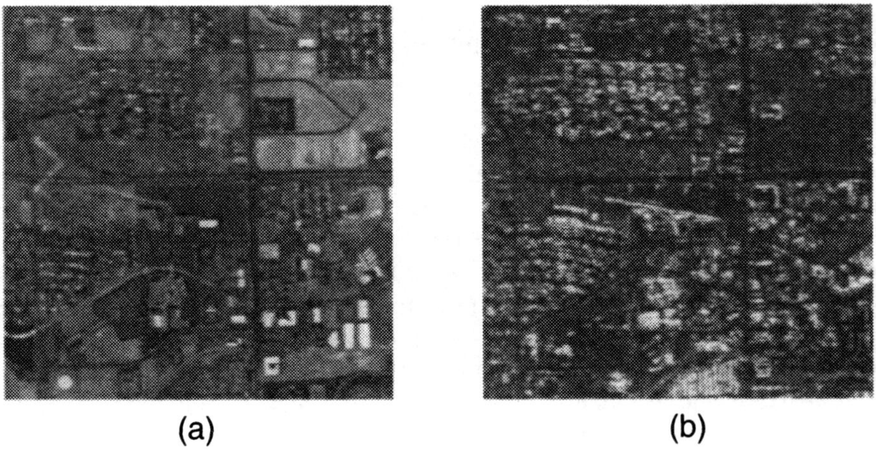

Figure 1.6 Side-by-side comparison of (a) optical and (b) radar imagery.

light, whereas radar imagery shows the terrain response to radio waves. While it is not possible to discern fine detail in this side-by-side comparison, SAR can indeed produce very high-resolution imagery. An example of the current state-of-the-art is shown in Figure 1.7. This spotlight-mode image of the U.S. Capitol building in Washington, DC, was acquired and processed to one-foot resolution by the Sandia National Laboratories Twin-Otter SAR system. Fine detail is strikingly clear. In many ways high-resolution SAR images resemble those from optical systems. However, the response of objects and terrain to the impinging microwave radiation can be vastly different than from optical waves. These differences can be both useful and frustrating depending on one's ability to utilize the relevant information.

There is a very valuable piece of information contained in SAR imagery that is

Figure 1.7 One-foot-resolution spotlight-mode image of the U.S. Capitol building in Washington, DC, acquired by the Sandia National Laboratories Twin-Otter SAR system.

not readily available to visual interpretation, however. This "hidden" information, known as *phase*, is what makes SAR imagery possible in the first place, and it provides the means for new and exciting applications based on interferometry [23].

The fundamental key to the success of SAR imaging is that the signals transmitted and received are coherent. This means that the transmitted signals are generated from a stable local oscillator in the radar and can be referenced in time and space to a common point. The received signals return after a precisely measurable time delay (depending on the distance to the terrain being illuminated) and they have a precisely measurable phase (or starting point on the radio wave) relative to the local oscillator reference. Coherence provides the means for synthesizing the effect of a large aperture (or antenna) which permits imaging at a much higher resolution than could be provided with the physically small antenna carried onboard the aircraft or spacecraft.

An important by-product of this coherent imaging is that phase information is preserved in the SAR imagery as well, although it cannot be seen when displayed as in Figure 1.6(b). Nevertheless, this phase information is available and can be exploited in remarkable ways, as we will see. Optical imagery, on the other hand, cannot be exploited in the same way as SAR imagery because the sun is an incoherent source of illumination and phase is not preserved. (However, coherent optical imagery acquired from appropriate laser illumination is what makes three-dimensional holography possible [25].)

1.2.2 Interference

We have talked about the fundamentally important concept of phase that allows subtle information to be obtained through interference effects. Phase refers to the relative position along the wave with respect to a reference point or the relative displacement between two (or more) waves, as shown in Figure 1.8.

Waves can and do interact with one another to give rise to varied and useful effects. The result of their interaction depends on the individual wave properties prior to their interaction. The addition of two waves with slightly different frequencies is shown in Figure 1.9. It is easy to see that where the waves are in phase (the crests or troughs line up), they add *constructively*; that is, they add up in amplitude. Where the waves are out of phase (the crests line up with the troughs), they add *destructively*. This constructive and destructive interaction among waves is known as *interference*. Notice that interference takes place between two or more waves. A single wave exhibits no interference properties unless it is divided into two or more waves that subsequently can be recombined. This is why *all interference exploitation experiments require at least two mutually coherent observations*. We will now describe a classic optical interference experiment and then later show how the same principles apply to SAR interferometry.

In 1801, in a simple but now classic experiment, Thomas Young first studied how light waves interfere. He devised an experiment that allowed him to observe

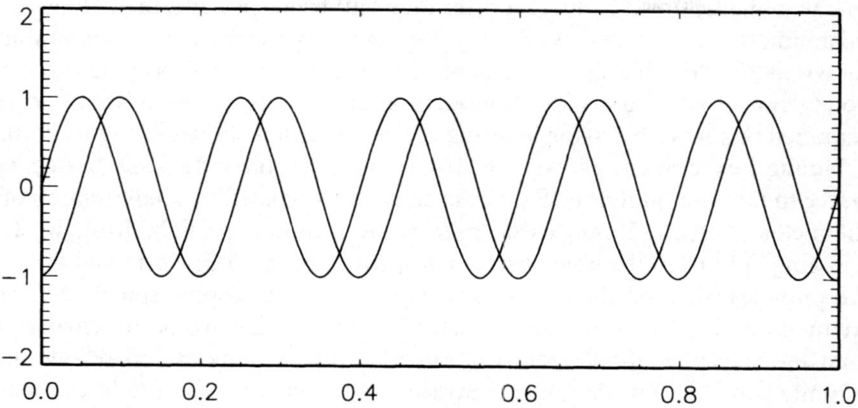

Figure 1.8 Phase difference between two identical waves.

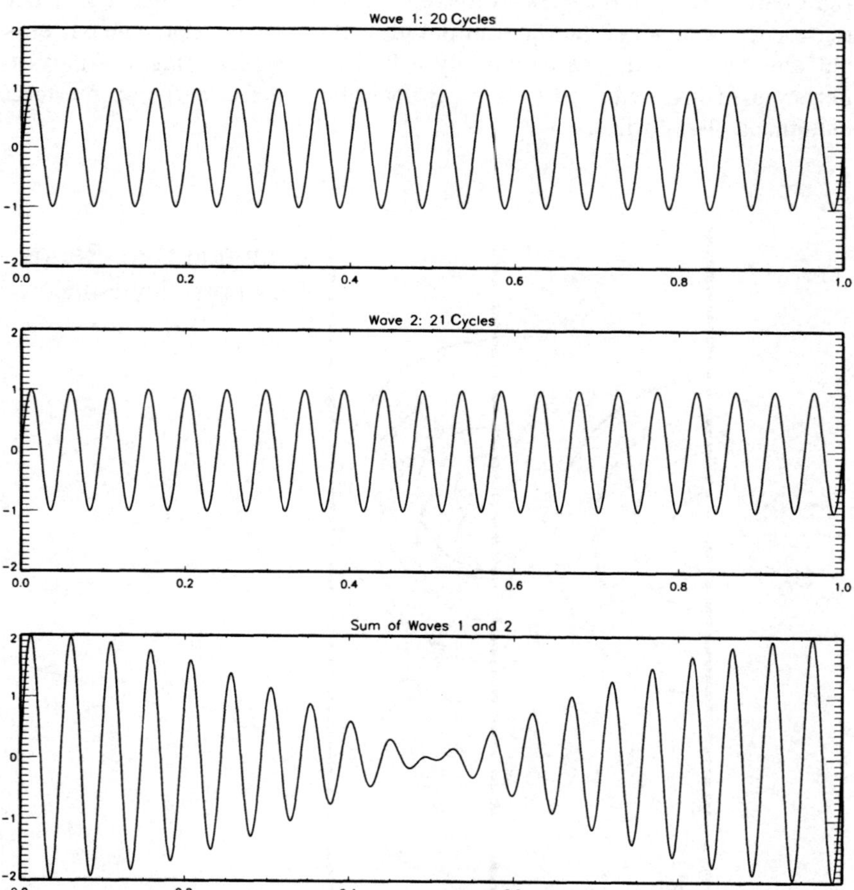

Figure 1.9 Addition of two waves with slightly different frequencies.

how two identical waves interact after they had been split and then allowed to recombine to form a new wave [26]. Young's experiment is more commonly known as the "double-slit" experiment for reasons that will become obvious. Young's work is the basis for all modern understanding of wave behavior. He examined this wave behavior by using the construction shown in Figure 1.10.

Young's experiment allows light from a monochromatic (single frequency) source to fall on a pinhole at S_0, whose diameter is typically a small fraction of a millimeter. (Actually Young used a narrow slit to allow more light through.) The emerging light from the hole in screen **A** spreads out by diffraction and falls on two pinholes (slits) on the next screen. Again two overlapping spherical waves expand into the space to the right of screen **B**. Because the two beams emanating from the screen originate from the same source S_0, they are correlated and spatially coherent. On the viewing screen **C** are seen a series of light and dark bands. The dark lines represent the minima of the combined wave disturbance, where destructive interference occurs and the waves cancel each other, and the bright bands between the dark bands represent the maxima, where constructive interference occurs and the wave amplitudes add to each other. The patterns of light and dark bands are commonly called *interference fringes*. Analysis of interference fringes can tell us a great deal about the underlying interference phenomena, as we will see.

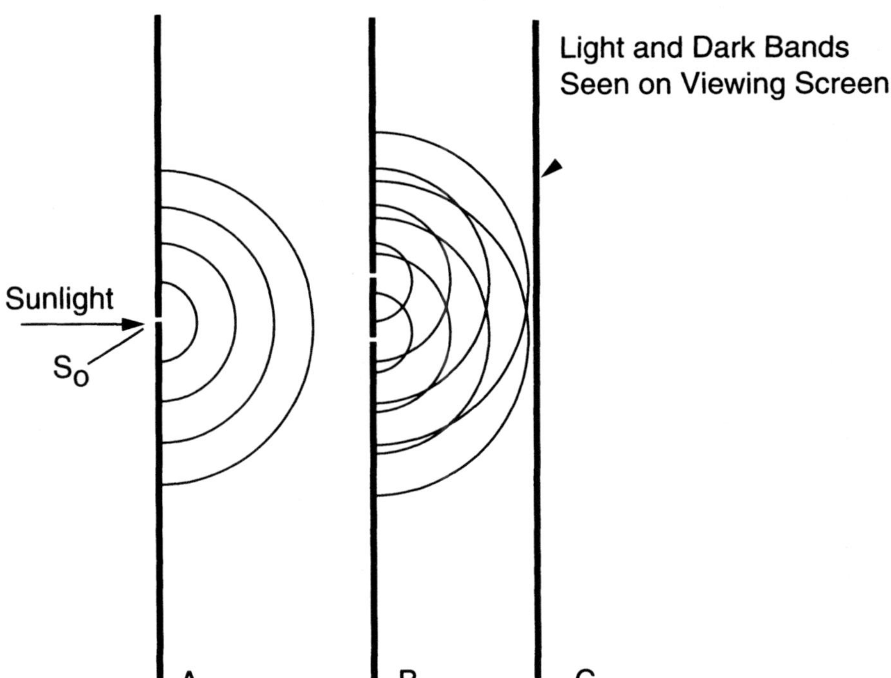

Figure 1.10 Young's two-slit experiment.

Now consider the diagram shown in Figure 1.11, which is based on Young's experimental setup. The distance **d** is the center-to-center distance between the two sources of light waves. The range r_1 is the distance from S_1 to the point **P** on the screen, and r_2 is the range from S_2 to the same point **P**. In this example r_2 is larger than r_1. The difference between these distances is called the *path difference*. Because r_2 is longer than r_1, it takes the light wave traveling along r_2 a little longer to get to point **P** than along r_1. Because the waves emanating from the two slits are spatially coherent (they come from a common wave impinging on screen **B**), the nonzero path difference implies that at point **P** there will be a *phase difference* between the light wave along r_1 and the light wave along r_2.

Now, if point **P** is moved down the viewing screen, the path difference will decrease until point **P** coincides with point **O**. At point **O** the path difference, and hence the phase difference, is zero and the waves are "in phase." We have already seen that when waves are in phase, they add constructively, and when they are completely out of phase, they add destructively. Thus it is easy to see that as the point **P** moves up and down the viewing screen, there will be a continuously varying intensity or brightness as the two waves add constructively or destructively depending on the path difference and various other parameters of the experiment (the wavelength of the light, the distance between the two slits, etc.). In particular, wherever the path length difference is equal to an integral

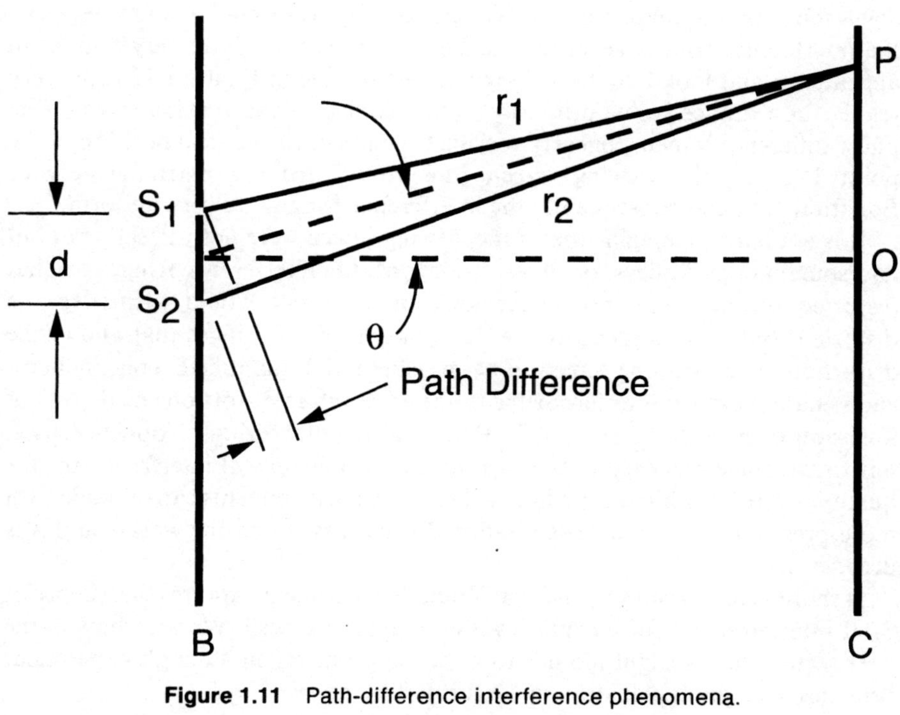

Figure 1.11 Path-difference interference phenomena.

number of wavelengths, a bright band will be observed. Mathematically constructive interference occurs where

$$|r_2 - r_1| = k\lambda, \tag{1.6}$$

where k is an integer and λ is the wavelength. Conversely, destructive interference produces dark bands where

$$|r_2 - r_1| = \left(k + \frac{1}{2}\right)\lambda. \tag{1.7}$$

Young determined the mathematical relationships between all of the relevant quantities shown and discovered that the phase difference divided by 2π is equal to the path difference $\Delta\phi$ divided by the wavelength. A little shuffling of these equations (based on the geometry) allows us to calculate the distance between the two sources of light as

$$d = \frac{(\Delta\phi)\lambda}{2\pi \sin\theta}, \tag{1.8}$$

where θ is the angle shown in the figure.

If the point **P** is confined to move along the viewing screen so that the angular deviation from the sources is small (in practice, θ is restricted to a few degrees), the interference fringes generated at the viewing screen **C** are very regular in appearance and look like those shown in perspective in Figure 1.12. This very regular appearance is due primarily to the flatness of the viewing screen. The phase difference is then a linearly varying function of the vertical position of the point **P** along the viewing screen (this results from a rearrangement of Equation 1.8), and consequently the interference fringes are very uniform.

It is not hard to imagine that if the viewing screen were not perfectly flat but had some bumps, valleys, or other distortions, the interference fringes on that distorted screen would no longer look uniform but would themselves be distorted. Indeed it is precisely the deformation of the fringes that allows the distortion of the viewing screen to be measured and quantified. This phenomenon enables optical interferometry to measure subtle distortions or defects of precision-manufactured parts [27]. Precision manufacturing of optical lenses, mirrors, turbine blades, and the like, routinely uses optical interferometry for quality control. Within certain limits, distortion measurements can be made with more precision if the source separation **d** is increased and the wavelength λ is decreased.

It should come as no surprise that Young's experiment captures the essence of SAR interferometry for terrain elevation mapping as well. We will show in the next section that a slight modification of the geometry in Young's experiment provides a simple explanation of SAR interferometry.

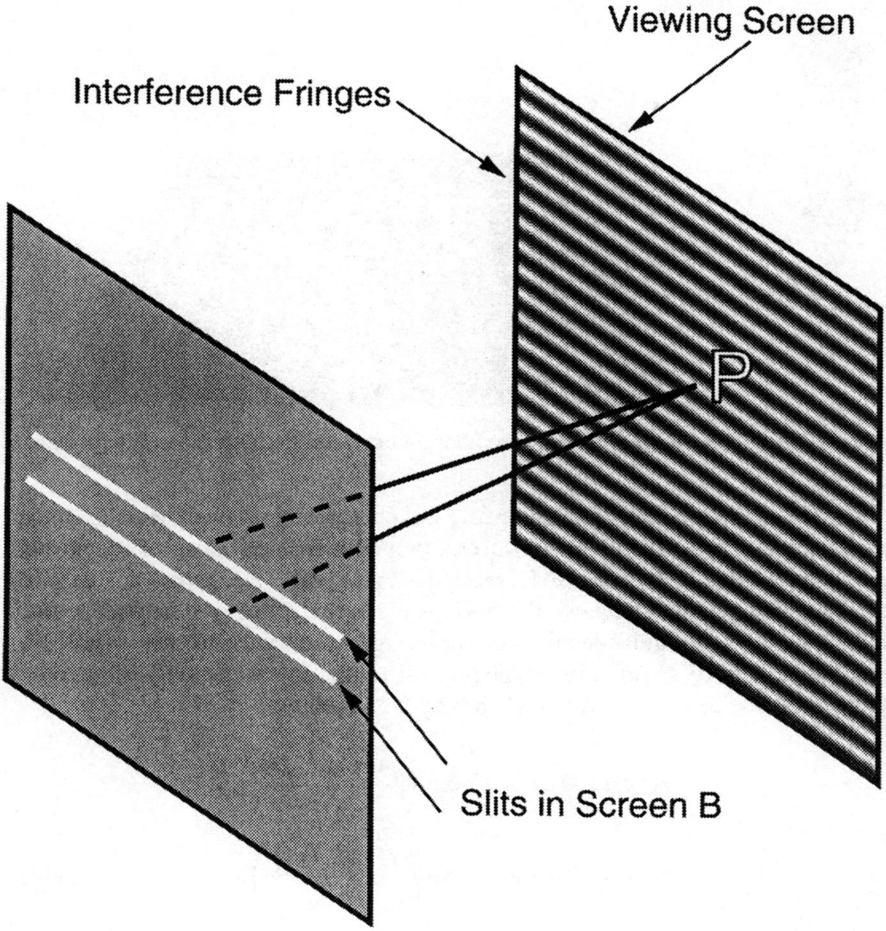

Figure 1.12 Typical interference fringes from Young's two-slit experiment.

1.2.3 SAR Interferometry for Terrain Elevation Mapping

Let us imagine replacing the two slits of Figure 1.11 with two radar antennas, the emerging energy now being coherent radar waves instead of light. The viewing screen becomes the ground, and we will tilt the whole geometry on its side, mimicking a side-looking SAR. Our new experiment looks like Figure 1.13. We again see an interference pattern on the viewing screen (ground), because, as before, the path difference from the two antennas to any point varies with position. Notice in this figure that the fringes possess a slight curvature. This is because the path difference varies in both the cross range as well as range directions. The exact spacing and curvature of the fringes are a function of parameters such as radar wavelength, distance from antenna to ground, field of view, and the separation of the two antennas.

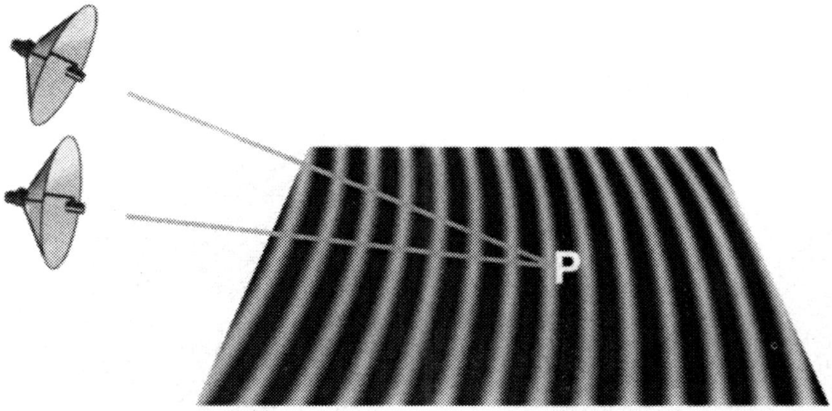

Figure 1.13 Two radar antennas producing interference patterns on the ground.

The viewpoint observed in Figure 1.13 is a perspective view of the experiment. Previously we saw that a SAR produces a two-dimensional image of the ground being illuminated. The two SAR receivers in Figure 1.13 produce two such images of the ground. Because the SAR is coherent, these two images possess phase as well as magnitude. In fact the image functions are comprised of a complex reflectivity (a function of the ground/radar interaction) modulated by a phase term due to the round trip path length as follows:

$$g_1(x,y) \cong a_1(x,y)\exp\left(\frac{j4\pi R_1(x,y)}{\lambda}\right),$$

$$g_2(x,y) \cong a_2(x,y)\exp\left(\frac{j4\pi R_2(x,y)}{\lambda}\right), \tag{1.9}$$

where a_i is the complex terrain reflectivity, R_i is the range from antenna i to the point (x,y), and λ is the wavelength.

If we arrange the imaging geometry precisely (this is key) and perform the signal processing correctly, the reflectivity terms in the image equations will be highly *correlated* (i.e., $a_1(x,y) = a_2(x,y) = a(x,y)$).[2] Then the two images can be interfered with each other by conjugate multiplication,

$$g_1(x,y)g_2^*(x,y) \cong |a(x,y)|^2 \exp\left\{j\frac{4\pi}{\lambda}[R_1(x,y) - R_2(x,y)]\right\}, \tag{1.10}$$

the real (or imaginary) part of which produces the two-dimensional fringe

[2] The concept of spatial correlation is important to much of the material that follows. The spatial correlation coefficient [20], [23] is a useful quality measure. It can be used to guide a phase unwrapping process [24] and provide consistent cross checks (see Chapter 3).

(a) (b)

Figure 1.14 (a) Two-dimensional fringe pattern from radar data. This image represents the real or imaginary part of the complex product of Equation 1.10, assuming the amplitude modulation is uniform. (b) Two-dimensional wrapped phase pattern from the hypothetical flat ground SAR experiment.

pattern of Figure 1.14(a). In this figure, for simplicity, we assume that the amplitude modulation $|a(x, y)|^2$ of the fringe pattern is constant. In other words, Figure 1.14(a) is a picture of $\cos\{(4\pi/\lambda)[R_1(x, y) - R_2(x, y)]\}$. In reality $|a(x, y)|^2$ is most definitely *not* constant but represents the spatially varying intensity of the conjugate product. The wrapped phase, extracted as the argument of the complex quantity of Equation 1.10, is shown in Figure 1.14(b). This figure is a picture of $\mathcal{W}\{(4\pi/\lambda)[R_1(x, y) - R_2(x, y)]\}$. The wrapped phase is, of course, not modulated by $|a(x, y)|^2$.

Because our hypothetical ground is perfectly flat, this fringe pattern is very uniform in appearance, save for the slight curvature as noted previously. Similarly the wrapped phase has a more or less uniformly repetitive appearance. Note that the hypothetically flat ground is not represented directly by the wrapped phase. The wrapping discontinuities must be removed to yield the predominantly planar terrain. If one were to remove the curvature, wrapping discontinuities, and spacing effects due to the (known) imaging geometry and wavelength, it would be possible to deduce that the imaged ground is indeed flat.

What happens if we now image actual undulating terrain? As the point **P** roams not only in the range and cross-range directions, but up and down as well, the path difference varies continuously in all three directions. We obtain a result similar to that in Figure 1.15.

After interfering the two images acquired in an actual SAR experiment, we produce the wrapped phase shown in Figure 1.16(a). The phase pattern contains the effects of geometry and wavelength for flat terrain as well as the influence of terrain height. Armed with precise knowledge of the imaging geometry, it is possible to remove the pattern due to flat terrain, leaving only the height-induced fringes.

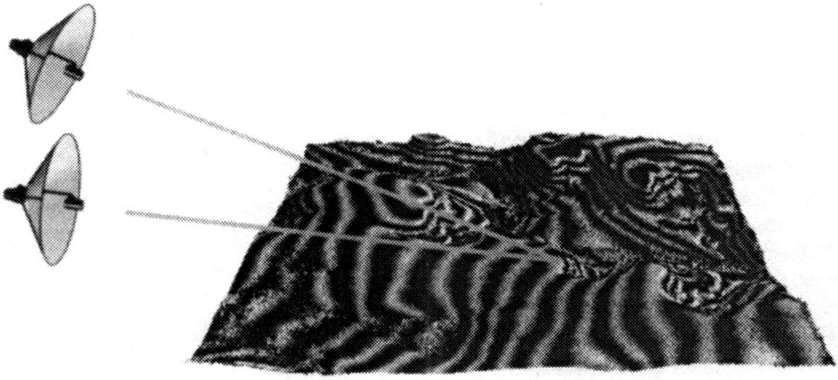

Figure 1.15 Typical interference pattern due to undulating terrain.

Because only wrapped values of phase are available from the conjugate product image (Equation 1.10), these values must be unwrapped to remove the phase discontinuities present. After unwrapping, the terrain height can be deduced using a scale factor relating phase to height [8], [20],

$$\frac{d\phi}{dh} = -\frac{4\pi}{\lambda}\frac{\Delta\Psi}{\cos\Psi}, \tag{1.11}$$

where Ψ is the angle between the propagation path and the ground and $\Delta\Psi$ is the angular separation of the two antennas as seen from the ground. (If $\Delta\Psi = 0$ and the two SAR images are separated in time, *millimeter* changes in the topography of the earth's surface due to earthquakes or volcanic activity can be detected [28].) In the final processing steps it is necessary to unwrap the principal (wrapped) values of phase and mathematically invert the terrain height-to-fringe relationship, arriving at an estimate of the terrain height at every point in the image (shown in perspective in Figure 1.16b). Chapter 3 provides additional details of SAR interferometry for terrain mapping.

The success or failure of this mapping application hinges on unwrapping the phase correctly. SAR interferometry produces two-dimensional phase maps that can, at times, be very difficult to unwrap. Extremely rough terrain, height-induced layover, superposition, shadows, low signal returns, poor image correlation, and signal aliasing can produce phase data containing a number of undesirable properties. It is safe to say that the phase unwrapping problems encountered in SAR are "about as hard as they come," and they encompass virtually all of the difficult issues encountered in applications outside of SAR.

1.3 ASPECTS OF ONE-DIMENSIONAL PHASE UNWRAPPING

It will be easiest to understand some of the difficulties encountered in two-dimensional phase unwrapping by illustrating several important aspects of phase

(a)

(b)

Figure 1.16 (a) SAR wrapped phase pattern due to natural terrain. (b) Perspective view of terrain height derived from SAR interferometry.

unwrapping with simple one-dimensional signals. Let us assume that we have obtained the complex signal,[3]

$$s(t) = e^{j5\pi t}, \qquad 0 \leq t \leq 1, \tag{1.12}$$

[3] The complex signal $s(t)$ could have been obtained, for example, by a one-dimensional slice in the y direction of the two-dimensional complex signal represented by Equation 1.10, assuming the signal magnitude was constant and the SAR was illuminating perfectly flat terrain. The one-dimensional wrapped phase would then be nothing more than the equivalent slice through the two-dimensional wrapped phase, similar to that shown in Figure 1.14(b).

and we wish to reconstruct from $s(t)$ the continuous phase $\varphi(t) = 5\pi t$. We can obtain only the wrapped phase $\psi(t)$ from the phase-extraction operation,

$$\psi(t) = \arctan[\Im(s(t)), \Re(s(t))], \tag{1.13}$$

where $\Re$ and $\Im$ refer to the real and imaginary parts, and arctan is the four-quadrant arc tangent operator where $-\pi < \psi(t) \leq \pi$. This arctangent operator also can be considered as a wrapping operator $\mathcal{W}$, where

$$\psi(t) = \mathcal{W}[\varphi(t)] \,. \tag{1.14}$$

Figure 1.17 plots the true and wrapped phase of Equation 1.12. It is seen that while $\varphi(t)$ is a linear function of t, the wrapped phase $\psi(t)$ is plotted as the sawtoothed pattern shown by the dashed line. Thus, for a given value of t, we obtain $\psi(t)$ on the dashed line. An estimate of $\varphi(t)$, denoted $\phi(t)$, can be obtained by adding an appropriate multiple of 2π radians to $\psi(t)$. Visually it is easily determined that nothing need be added to $\psi(t)$ for $0 \leq t < 0.2$ to give a correct estimate of the unwrapped phase within that same interval. Similarly, for $0.2 \leq t < 0.6$, one multiple of 2π must be added, whereas $2 \times 2\pi$ must be added in the interval $0.6 \leq t < 1.0$. We could determine the correct unwrapping

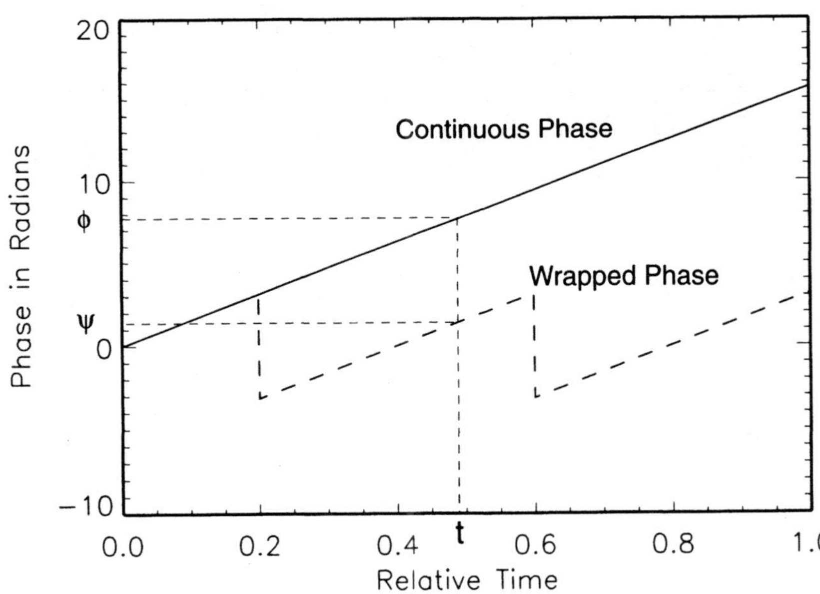

Figure 1.17 Plots of the true and wrapped phase of Equation 1.12. The true phase is a linear function of t (solid line), whereas the wrapped phase $\psi(t)$ is the sawtooth function (dashed line). Phase unwrapping is accomplished by detecting the wrapped phase jumps and adding an appropriate multiple of 2π radians.

from the graph by visually detecting the 2π jumps in the wrapped phase and adding the appropriate multiple of 2π.

While this simple example captures the spirit of one-dimensional phase unwrapping, implementation is problematic for several reasons. To work with actual signals in a computer requires that they be sampled and converted to digital signals. Thus continuous functions become a sequence or array of samples. If the sampled data are to represent the particular continuous signal in some respect, we know from sampling theory that they constitute an incomplete representation in the following ways: First, the sampled data measure the continuous signal at discrete points in time. Second, they may span a limited extent of the total signal duration. An insufficient sampling rate or duration may preclude reconstruction of the continuous signal to a satisfactory level of fidelity. Furthermore the wrapped phase that is extracted from the sampled signal may no longer allow unwrapping to *any* level of fidelity.

In summary, most phase unwrapping problems arise when one goes from the continuous to the discrete domain. In the continuous domain, in the absence of singularities,[4] there is only one way to unwrap phase, and the unwrapped solution is unique. In the discrete domain, however, certain assumptions must be introduced (e.g., the assumption of nonaliasing). In a sense phase unwrapping is equivalent to the problem of recovering the underlying *continuous* wrapped phase from the *discrete* wrapped phase that has been sampled.

1.3.1 Effect of Phase Aliasing on Phase Unwrapping

We can illustrate the effect of phase aliasing (i.e., insufficient sampling rate) with manipulation and processing of the following continuous phase function:[5]

$$\varphi(t) = 10 \sin 10t, \qquad 0 \le t \le 1 . \tag{1.15}$$

Sampling $\varphi(t)$ by placing N equally spaced samples across the time interval [0, 1] defines the sample interval $\Delta t = 1/N$. According to sampling theory, more than two samples per period of the highest frequency component must be obtained. The Nyquist rate defines this minimum sampling rate [29]. Alternatively, sampling at the Nyquist rate is equivalent to constraining the phase change to less than π radians per sample (in magnitude) everywhere. From Equation 1.15 we compute the phase change as follows:

$$\Delta\varphi = \dot{\varphi}\,\Delta t, \tag{1.16}$$

where $\dot{\varphi} = d\varphi/dt$. It is readily verified that $\dot{\varphi} = 100 \cos 10t$ and the phase gradient maxima occur at $t = n\pi/10, n = 0, 1, \ldots, N-1$. Thus imposing the

[4] The phase of a continuous complex function cannot be unwrapped if it has singularities (i.e., poles or zeros). This issue is discussed further in Section 2.3.

[5] Although the actual sampling process applies to the complex signal $A(t)\exp(j\varphi(t))$, it is sometimes simpler and more informative to illustrate the effect of sampling on $\varphi(t)$.

constraint

$$|\Delta\varphi| < \pi \tag{1.17}$$

implies that

$$\left|\frac{100}{N}\cos 10t\right| < \pi\,. \tag{1.18}$$

Thus

$$N > \frac{100}{\pi} \approx 31.83\,. \tag{1.19}$$

Equation 1.19 states that for $N \geq 32$ samples across the time interval, phase aliasing will not occur. Conversely, phase aliasing will occur in the vicinity of maximum (in magnitude) phase gradients for $N = 31$ or less. We hypothesize that simple phase unwrapping will fail (i.e., the unwrapped phase will bear little resemblance to the true phase) for aliased signals.

Before we illustrate the affect of phase aliasing on one-dimensional phase unwrapping, we must replace our "visual determination of wrapping discontinuities" mentioned in the previous section with a quantitative mathematical procedure. In 1982 Itoh analyzed the one-dimensional phase unwrapping problem and showed that the unwrapped phase can be obtained by integrating (summing) wrapped phase differences [30]. The unwrapped result will equal the true phase provided the true phase differences are less than π radians in magnitude everywhere.[6] In other words, if the true phase is not aliased, the unwrapping is correct.

Itoh's method is easily summarized as follows: Let $\mathcal{W}$ be the wrapping operator (defined by Equations 1.14 and 1.5) that wraps the phase into the interval $(-\pi, \pi]$. Thus

$$\mathcal{W}\{\varphi(n)\} = \psi(n) = \varphi(n) + 2\pi k(n), \qquad n = 0, 1, \ldots, N-1, \tag{1.20}$$

where $k(n)$ is an array of integers chosen so that

$$-\pi < \psi(n) \leq \pi. \tag{1.21}$$

The difference operator Δ is defined as

$$\begin{aligned} \Delta\{\varphi(n)\} &= \varphi(n+1) - \varphi(n), \\ \Delta\{k(n)\} &= k(n+1) - k(n), \qquad n = 0, 1, \ldots, N-2. \end{aligned} \tag{1.22}$$

Computing the difference of wrapped phases using Equations 1.20 and 1.22

[6] A *true* phase change of more than π radians in magnitude cannot be distinguished from a wrapping discontinuity.

yields the equation

$$\Delta\{\mathcal{W}\{\varphi(n)\}\} = \Delta\{\varphi(n)\} + 2\pi\Delta\{k_1(n)\}. \tag{1.23}$$

We again apply the wrapping operator to the above result to obtain

$$\mathcal{W}\{\Delta\{\mathcal{W}\{\varphi(n)\}\}\} = \mathcal{W}\{\Delta\{\psi(n)\}\} = \Delta\{\varphi(n)\} + 2\pi[\Delta\{k_1(n)\} + k_2(n)]. \tag{1.24}$$

The subscripts on k_1 and k_2 are used to distinguish the integer arrays found by the two wrapping operations. The result of Equation 1.24 is the wrapped difference of wrapped phases. Because the wrapping operator $\mathcal{W}$ produces values in the interval $(-\pi, \pi]$, the requirement

$$-\pi < \Delta\{\varphi(n)\} \leq \pi \tag{1.25}$$

implies that the term $2\pi[\Delta\{k_1(n)\} + k_2(n)]$ in Equation 1.24 must equal zero. Thus

$$\Delta\{\varphi(n)\} = \mathcal{W}\{\Delta\{\mathcal{W}\{\varphi(n)\}\}\} = \mathcal{W}\{\Delta\{\psi(n)\}\}, \tag{1.26}$$

which can be manipulated to yield the equation

$$\varphi(m) = \varphi(0) + \sum_{n=0}^{m-1} \mathcal{W}\{\Delta\{\mathcal{W}\{\varphi(n)\}\}\}. \tag{1.27}$$

Equation 1.27 states that the phase can be unwrapped by integrating the wrapped differences of wrapped phases. However, if Equation 1.25 is not satisfied, the true phase cannot be recovered. We can always obtain an unwrapped result from Equation 1.27 whether or not Equation 1.25 is satisfied. If Equation 1.25 is not satisfied, we will not know this *a priori*. Therefore the blind application of Equation 1.27 should be considered a means for obtaining an unwrapped *estimate* ϕ and not the *true* phase φ.

A simple one-dimensional phase unwrapping procedure based on Itoh's analysis is presented as the following algorithm. This procedure unwraps the phase in the array $\psi(i)$, $0 \leq i \leq N-1$.

Itoh's Method for One-Dimensional Phase Unwrapping

Step 1. Compute the phase differences: $D(i) = \psi(i+1) - \psi(i)$ for $i = 0, \ldots, N-2$.

Step 2. Compute the wrapped phase differences: $\Delta(i) = \arctan\{\sin D(i), \cos D(i)\}$ for $i = 0, \ldots, N-2$.

Step 3. Initialize the first unwrapped value: $\phi(0) = \psi(0)$.

Step 4. Unwrap by summing the wrapped phase differences:
$\phi(i) = \phi(i-1) + \Delta(i-1)$, for $i = 1, \ldots, N-1$.

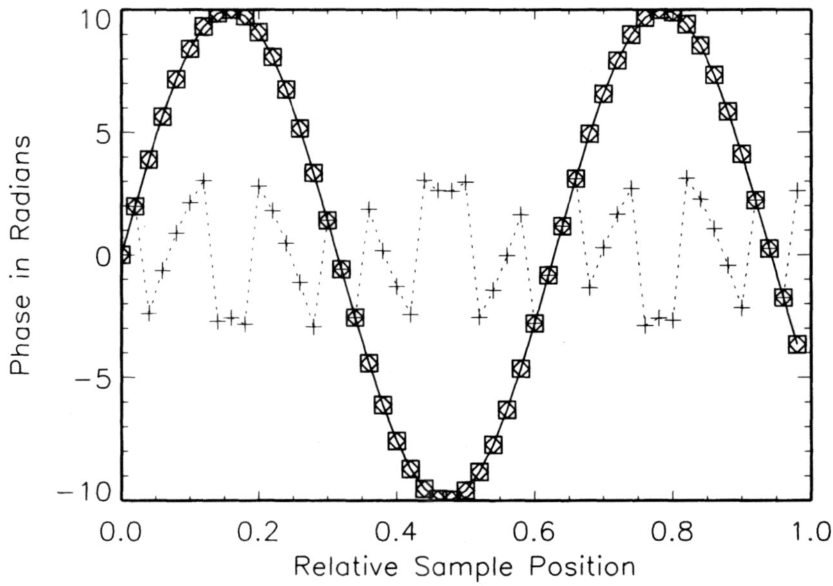

Figure 1.18 A sinusoidal phase function with 50 samples spanning the time interval is designated by the diamonds (◇). Samples of the wrapped phase are designated by +. The unwrapped result, designated by the squares (□), matches the true phase.

Figures 1.18 through 1.22 illustrate the affect of phase aliasing on Itoh's one-dimensional unwrapping procedure. In Figure 1.18 the sinusoidal phase function of Equation 1.15 is plotted with 50 samples spanning the time interval. The original phase samples are represented by the diamonds (◇). Samples of the wrapped phase are designated by +. The unwrapped phase samples, obtained from Itoh's method, are designated by the squares (□). Because the number of samples exceeds the critical number given by Equation 1.19, no phase aliasing occurs, and the unwrapped phase exactly matches the true phase. As the sampling rate is reduced, but kept greater than the Nyquist rate, the unwrapped phase exactly matches the true phase (see Figures 1.19 and 1.20).

However, when $N = 31$, the Nyquist sampling rate is violated, and the true phase changes by more than π radians in magnitude in the vicinity of the steepest phase gradients. Consequently the wrapped phase values no longer encode the true phase gradient, and the unwrapping process fails to reconstruct the true phase (Figures 1.21 and 1.22). Phase aliasing occurs abruptly and catastrophically as the number of samples is reduced below the critical value.

1.3.2 Effect of Noise on Phase Unwrapping

Aliasing is not the only concern with sampled data. Noise can also cause catastrophic failure in phase unwrapping. The combination of noise and

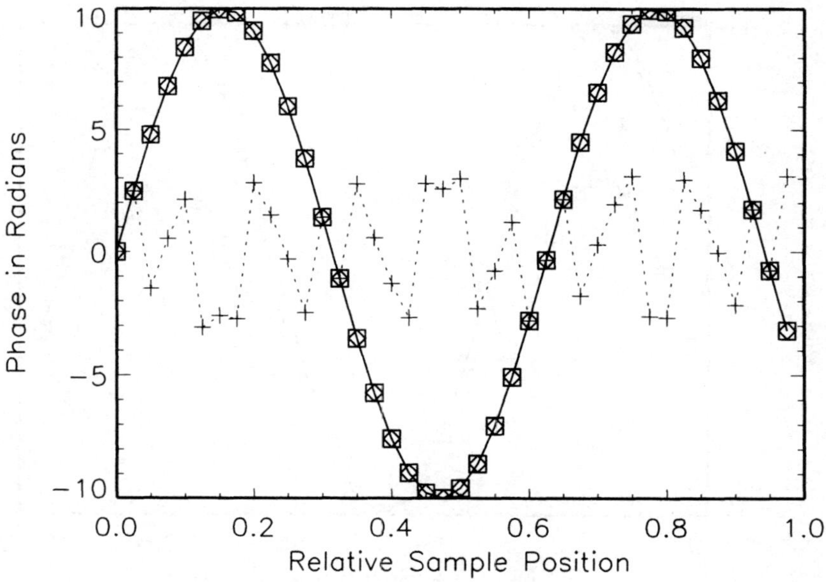

Figure 1.19 A sinusoidal phase function with 40 samples spanning the time interval is designated by the diamonds (◇). Samples of the wrapped phase are designated by +. The unwrapped result, designated by the squares (□), matches the true phase.

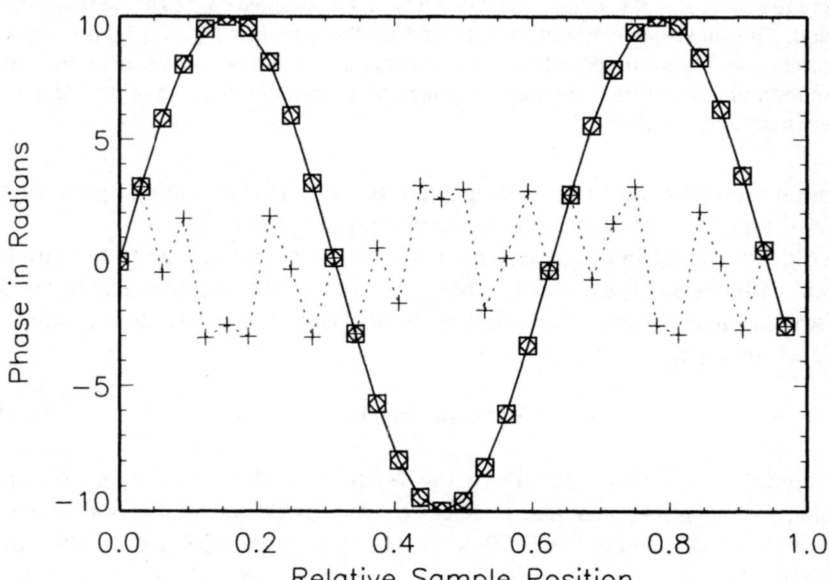

Figure 1.20 A sinusoidal phase function with 32 samples spanning the time interval is designated by the diamonds (◇). Samples of the wrapped phase are designated by +. The unwrapped result, designated by the squares (□), matches the true phase, but an aliasing condition is very near.

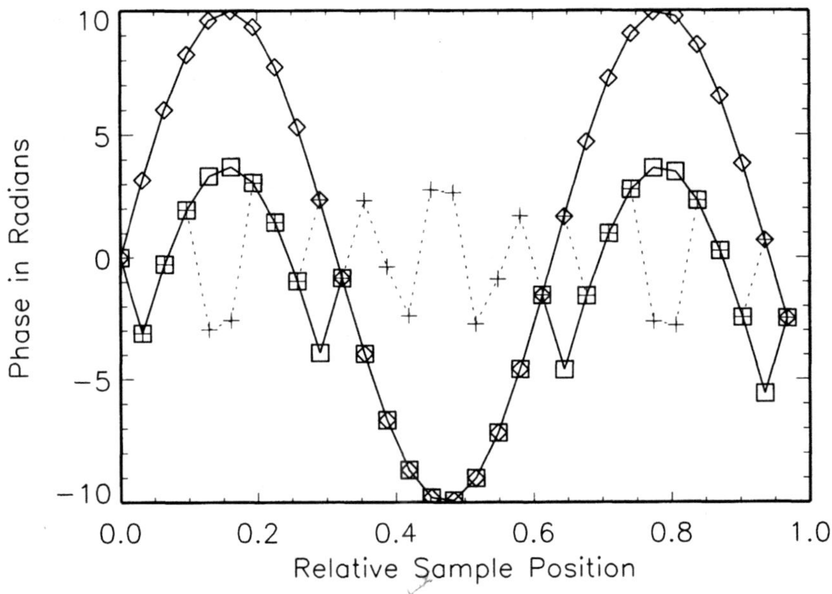

Figure 1.21 A catastrophic failure for phase unwrapping occurs when the phase becomes aliased. A sinusoidal phase function with 31 samples spanning the time interval is designated by the diamonds (◇). Samples of the wrapped phase are designated by +. Aliasing occurs in the vicinity of the steepest slopes of the true phase function. The unwrapped result, designated by the squares (□), no longer equals the true phase. Phase jumps of 2π multiples occur at the aliased points. In addition the unwrapped phase differs from the true phase by unknown multiples of 2π radians at unknown sample locations.

marginal sampling work together through the nonlinear wrapping operation to confound the most sophisticated phase unwrapping methods.

We illustrate the effect of noise by generating a unity-magnitude complex phasor with the same sinusoidal phase we used in the previous examples. In addition circular complex Gaussian noise $n(t)$ is added to produce a complex signal of the form

$$s(t) = \exp(j10 \sin 10t) + n(t). \tag{1.28}$$

This signal is sampled well above the Nyquist limit to avoid any aliasing problems. The noise variance is selected to provide various signal-to-noise ratios (SNR). We define the SNR to be the peak signal divided by the noise variance. As before, we extract the wrapped phase values by means of Equation 1.13. Phase unwrapping is attempted with Itoh's method (see Section 1.3.1) and the results are plotted and shown in Figure 1.23. The lower dashed curve is the original sinusoidal phase. The curves above (offset vertically to avoid confusion) represent the unwrapped results with decreasing SNRs.

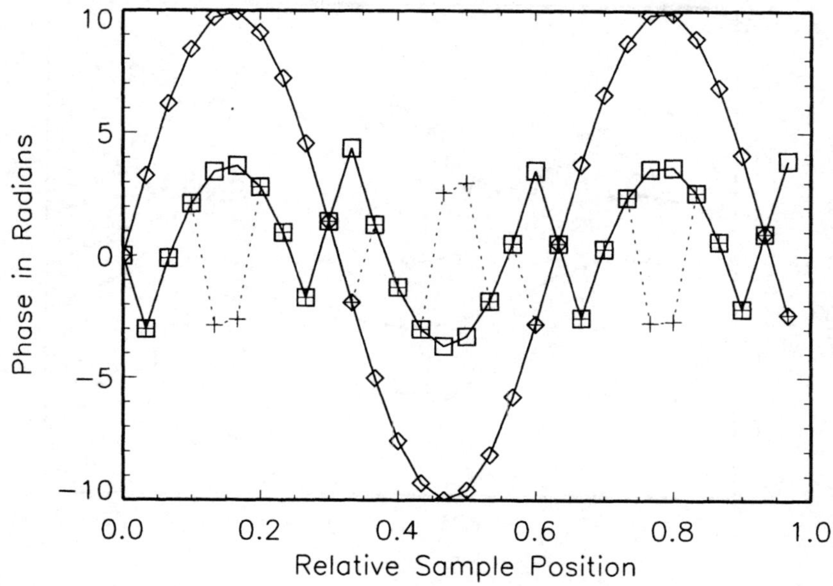

Figure 1.22 A second example of phase aliasing. A sinusoidal phase function with 30 samples spanning the time interval is designated by the diamonds (◇). Samples of the wrapped phase are designated by +. Again, aliasing is evident in the regions of steep slope. The unwrapped result, designated by the squares (□), no longer equals the true phase. Phase jumps of 2π multiples occur at the aliased points. In addition the unwrapped phase differs from the true phase by unknown multiples of 2π radians at unknown sample locations. Slight changes in sampling cause drastic changes in the unwrapped phase when aliasing is present. Compare this unwrapped phase with the unwrapped phase shown in Figure 1.21.

Notice that the unwrapping captures a noisy estimate of the true sinusoidal phase until the SNR drops too low. (The first instance of an incorrectly unwrapped phase is evident for SNR = 2.0). Once the SNR drops to near unity, catastrophic results occur.[7] The unwrapped phase for SNR = 1.0 bears little resemblance to the true signal. In this case the additive noise causes the complex phasor to swing wildly in phase, which in turn causes an effect analogous to phase aliasing. Correct unwrapping becomes impossible.

This effect is perhaps easiest to understand when we view the complex signal as a phasor perturbed by a noise cloud as shown in Figure 1.24. When the variance of the noise increases to the point where the noise cloud has a high probability of encompassing the origin, the phasor can swing wildly in both magnitude and phase from one instance in time to the next. In fact, when the phasor magnitude goes through zero, the phase changes by π radians in magnitude. These abrupt phase changes cannot be distinguished from wrapping

[7] This phenomenon is the same as the "threshold effect" in angle modulation [31].

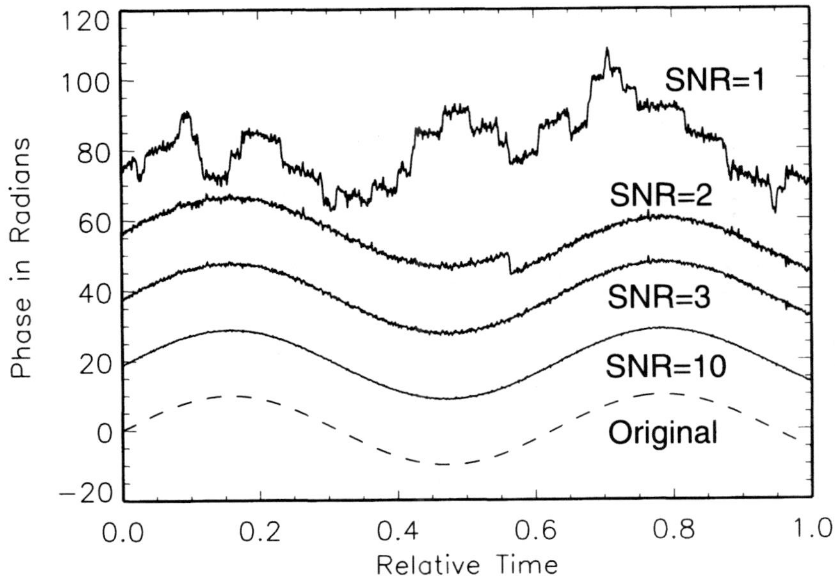

Figure 1.23 The effect of noise on phase unwrapping. The lower dashed curve is the original sinusoidal phase. The next higher curve (offset to avoid confusion) is the unwrapped result for SNR = 10. The next higher curves represent SNR = 3 and SNR = 2, respectively. Finally the upper curve represents the unwrapped result for SNR = 1. Catastrophic results occur when the SNR drops too low. The effect is analogous to phase aliasing.

discontinuities, and thus the subsequent phase unwrapping procedure fails to reproduce the true phase or even a reasonable estimate of the true phase.

As we will see, two-dimensional phase unwrapping amounts to integrating phase gradients over some valid domain. For two-dimensional complex signals, zero-magnitudes, phase discontinuities, noise, inconsistent phase gradients, and aliasing can be the source of *residues* which greatly complicate the phase unwrapping process. We will define residues in Chapter 2.

1.4 SUMMARY AND BRIEF LOOK AHEAD

Itoh's method is a specific implementation of the notion that the unwrapped phase can be obtained by integration of the phase gradient. This concept can be extended to N-dimensional signals with the following representation. Assuming the phase gradients are known along with the phase at some initial point $\mathbf{r}_0$, the phase at some other point $\mathbf{r}$ is obtained from the following path integral:

$$\varphi(\mathbf{r}) = \int_C \nabla\varphi \cdot \mathbf{dr} + \varphi(\mathbf{r}_0), \tag{1.29}$$

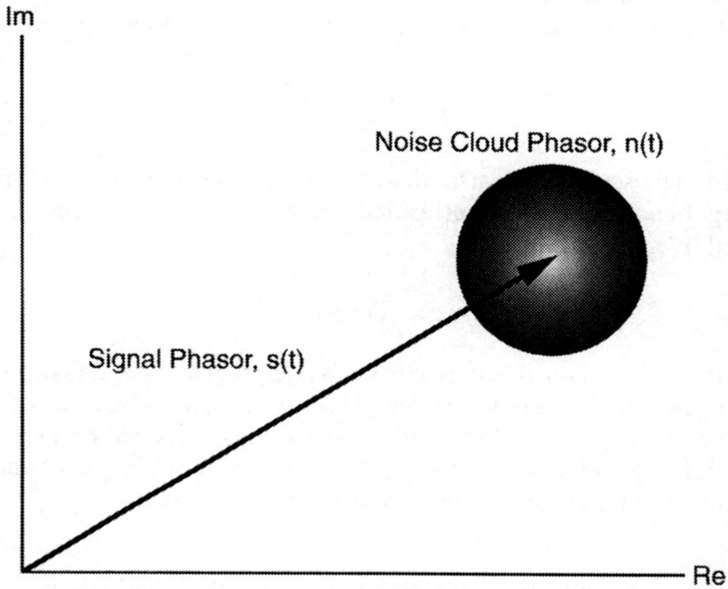

Figure 1.24 The effect of noise on phase unwrapping is easily understood when the complex phasor of interest is perturbed statistically and represented by a noise cloud. If the noise cloud has a high probability of encompassing the origin (or equivalently, the SNR is too low), the phasor can swing wildly in both magnitude and phase from one time sample to the next. The phase changes by π radians in magnitude whenever the phasor magnitude goes through zero. These abrupt changes cannot be distinguished from wrapping discontinuities, and subsequent phase unwrapping will fail.

where C is *any* path in N-dimensional space connecting the points $\mathbf{r_0}$ and $\mathbf{r}$, and $\nabla\varphi$ is the phase gradient.

We have shown previously that aliasing or noise can destroy our ability to unwrap phase correctly for one-dimensional signals. In two or more dimensions, aliasing, singularities, or noise can make the phase computed by Equation 1.29 *dependent on the integration path.* This can greatly complicate the problem. Indeed it is safe to say that *all* two- (or more) dimensional path-following[8] phase unwrapping algorithms are much more concerned with selecting an appropriate integration path than actually integrating gradients.

The theoretical requirements for the unwrapped phase to be independent of the integration path will be discussed in detail in Chapter 2. In addition the important concept of "residues" as applied to the phase unwrapping problem will be developed at length. Possible means for helping with the path-selection problem are addressed in Chapter 3, and the implementation of practical algorithms based on the evaluation of Equation 1.29 is the subject of Chapter 4.

Alternatively, if we accept the notion that the measured (or calculated) phase

[8] See Chapter 4.

gradients are corrupted by noise, we can solve the following equation to minimize the squared error ϵ^2 of the phase gradients in a least-squares sense:

$$\epsilon^2 = \int W(\nabla\phi - \mathbf{g})^2 dA\ , \tag{1.30}$$

where dA represents an elemental area. We assume that W is a specified weighting function, and the measured gradients $\mathbf{g}$ are the true gradients corrupted by noise as follows:

$$\mathbf{g} = \nabla\varphi + \mathbf{n}, \tag{1.31}$$

where $\mathbf{n}$ is a noise vector that corrupts the phase gradient $\nabla\varphi$. Equation 1.30 is the motivation for path-independent phase unwrapping algorithms that are based on a minimum-norm formulation. These techniques will be addressed in Chapter 5. In Chapter 6 we will compare the techniques of Chapters 4 and 5 and discuss methods for evaluating phase unwrapping algorithms.

REFERENCES

[1] J. M. Blackledge, *Quantitative Coherent Imaging*, Academic Press, London, 1989.

[2] A. V. Oppenheim and J. S. Lim, "The importance of phase in signals," *Proceedings of the IEEE*, Vol. 69, No. 5, pp. 529–541, May 1981.

[3] L. C. Graham, "Synthetic interferometer radar for topographic mapping," *Proceedings of the IEEE*, Vol. 62, No. 6, pp. 763–768, June 1974.

[4] R. M. Goldstein, H. A. Zebker, and C. L. Werner, "Satellite radar interferometry: Two-dimensional phase unwrapping," *Radio Science*, Vol. 23, No. 4, pp. 713–720, July–August 1988.

[5] H. A. Zebker and R. M. Goldstein, "Topographic mapping from interferometric synthetic aperture radar observations," *Journal of Geophysical Research*, Vol. 91, No. B5, pp. 4993–4999, April 10, 1986.

[6] F. K. Li and R. M. Goldstein, "Studies of multibaseline spaceborne interferometric synthetic aperture radars," *IEEE Transactions on Geoscience and Remote Sensing*, Vol. 28, No. 1, pp. 88–97, January 1990.

[7] A. L. Gray and P. J. Farris-Manning, "Repeat-pass interferometry with airborne synthetic aperture radar," *IEEE Transactions on Geoscience and Remote Sensing*, Vol. 31, No. 1, pp. 180–191, January 1993.

[8] P. H. Eichel, D. C. Ghiglia, C. V. Jakowatz, Jr., P. A. Thompson, and D. E. Wahl, "Spotlight SAR interferometry for terrain elevation mapping and interferometric change detection," Sandia National Laboratories Report, SAND93-2072, September 1993.

[9] D. Massonnet, M. Rossi, C. Carmona, F. Adragna, G. Peltzer, K. Feigl, and T. Rabaute, "The displacement field of the Landers earthquake mapped by radar interferometry," *Nature*, Vol. 364, pp. 138–142, July 8, 1993.

[10] D. C. Ghiglia and P. H. Eichel, "High-resolution synthetic aperture radar interferometry: Technology for precise terrain elevation mapping," *DSP and Multimedia Technology*, Vol. 3, No. 10, October 1994.

[11] D. L. Fried, "Least-squares fitting a wave-front distortion estimate to an array of phase-difference measurements," *Journal of the Optical Society of America*, Vol. 67, No. 3, pp. 370–375, March 1977.

[12] R. H. Hudgin, "Wave-front reconstruction for compensated imaging," *Journal of the Optical Society of America*, Vol. 67, No. 3, pp. 375–378, March 1977.

[13] R. J. Noll, "Phase estimates from slope-type wave-front sensors," *Journal of the Optical Society of America*, Vol. 68, No. 1, pp. 139–140, January 1978.

[14] G. H. Glover and E. Schneider, "Three-point Dixon technique for true water/fat decomposition with B_0 inhomogeneity correction," *Magnetic Resonance Medicine*, Vol. 18, pp. 371–383, 1991.

[15] J. Szumowski, W. R. Coshow, F. Li, and S. F. Quinn, "Phase unwrapping in the three-point Dixon method for fat suppression MR imaging," *Radiology*, Vol. 192, No. 2, pp. 555–561, August 1994.

[16] S. M. Song, S. Napel, N. J. Pelc, and G. H. Glover, "Phase unwrapping of MR phase images using Poisson equation," *IEEE Transactions on Image Processing*, Vol. 4, No. 5, pp. 667–676, May 1995.

[17] P. Negrete-Regagnon, "Practical aspects of image recovery by means of the bispectrum," *Journal of the Optical Society of America A*, Vol. 13, No. 7, pp. 1557–1576, July 1996.

[18] J. C. Marron, P. P. Sanchez, and R. C. Sullivan, "Unwrapping algorithm for least-squares phase recovery from the modulo 2π bispectrum phase," *Journal of the Optical Society of America A*, Vol. 7, No. 1, pp. 14–20, January 1990.

[19] J. Curlander and R. McDonough, *Synthetic Aperture Radar, Systems and Signal Processing*, Wiley, New York, 1991.

[20] C. V. Jakowatz, Jr., D. E. Wahl, P. H. Eichel, D. C. Ghiglia, and P. A. Thompson, *Spotlight-Mode Synthetic Aperture Radar: A Signal Processing Approach*, Kluwer, Boston, 1996.

[21] B. Burns and J. Cordaro, "A SAR image formation algorithm that compensates for the spatially-variant effects of antenna motion," *Proceedings of the SPIE*, Vol. 2230, Society of Photo-Optical Instrumentation Engineers, Bellingham, WA, pp. 14–24, 1994.

[22] B. Remund and C. Srivatsa, "A 500 MHz phase generator for synthetic aperture radar waveform synthesizers," *1991 IEEE GaAs IC Symposium*, pp. 349–352, 1991.

[23] D. L. Bickel and W. H. Hensley, "Design, theory, and applications of interferometric synthetic aperture radar for topographic mapping," Sandia National Laboratories Report, SAND96-1092, May 1996.

[24] D. Derauw, "Phase unwrapping using coherence measurements," *Proceedings of the SPIE*, Vol. 2584, Society of Photo-Optical Instrumentation Engineers, Bellingham, WA, pp. 319–324, 1995.

[25] R. Collier, C. Burckhardt, and L. Lin, *Optical Holography*, Academic Press, New York, 1971.

[26] M. Born and E. Wolf, *Principles of Optics*, Pergamon Press, Oxford, 1970, pp. 260–261.

[27] H. Smith, *Principles of Holography*, Wiley-Interscience, New York, 1969, ch. 8.

[28] D. Massonnet, "Satellite radar interferometry," *Scientific American*, Vol. 276, No. 2, pp. 46–53, February 1997.

[29] A. V. Oppenheim and R. W. Schafer, *Digital Signal Processing*, Prentice-Hall, New Jersey, 1975, p. 29.

[30] K. Itoh, "Analysis of the phase unwrapping problem," *Applied Optics*, Vol. 21, No. 14, p. 2470, July 15, 1982.

[31] P. Peebles, Jr. *Communication System Principles*, Addison-Wesley, Reading, MA, 1976, pp. 273–276.

2

LINE INTEGRALS, RESIDUES, AND 2-D PHASE UNWRAPPING

2.1 INTRODUCTION

As we observed in Chapter 1, two-dimensional phase unwrapping is embodied in the evaluation of the line integral

$$\varphi(\mathbf{r}) = \int_C \nabla\varphi \cdot \mathbf{dr} + \varphi(\mathbf{r_0}), \tag{2.1}$$

where C is *any* path in a domain D connecting the points $\mathbf{r_0}$ and $\mathbf{r}$, and $\nabla\varphi$ is the *gradient* of φ. From calculus, the value of a line integral of the form

$$I = \int_C \mathbf{F(r)} \cdot \mathbf{dr} \tag{2.2}$$

depends, in general, not only on the beginning and end points of the path C but on C itself [1]. We wish to show under what conditions the line integral in Equation 2.2 depends *only* on the end points of the path C and not on the specific path C selected. In other words, what special properties (if any) must Equation 2.2 have to allow a path-independent solution? We believe that the theory describing the properties of line integrals encompasses *all* the problems encountered in phase unwrapping. It is important to understand how these issues are addressed in algorithm designs, and how alternative approaches could be conceived.

We will discuss differential forms and conditions for path independence in Section 2.2 and summarize the four equivalent conditions under which a line integral is independent of the path. The extremely important concept of *residues* will be addressed in considerable detail in Section 2.3. It will be shown that the existence of residues in the wrapped phase is the sole measure of path dependence in the two-dimensional phase unwrapping problem. We will provide a thorough analysis that shows how phase residues differ from the residues found in complex variable theory and how the local phase behaves in the vicinity of phase residues. In addition we will show what properties a complex function must have in order to generate phase residues. These analyses and examples will allow the reader to look at images of wrapped phase functions and make some valuable assessments of the nature of the complex function from which the phase was extracted.

In Section 2.4 we discuss the theory of decomposing phase into two components. One component is residue-free and presents no problem for any phase unwrapping algorithm. The second component contains all the residues and captures all the difficulties of phase unwrapping. This phase decomposition provides additional insight into the phase unwrapping problem and suggests unwrapping strategies or useful postprocessing techniques.

In practice, phase unwrapping must be performed by a computer. In Section 2.5 we examine the two-dimensional phase unwrapping problem in terms of discrete samples and define the relevant terms used in algorithm development. We also present a simple means for detecting and locating residues in actual algorithm implementations in Section 2.6.

In Section 2.7 we show several examples of wrapped phase that contain residues and some of their interesting properties. We examine closely the local behavior of the wrapped phase in the vicinity of the residues, and we see that the phase behaves as predicted by theory. Section 2.8 concludes with an example motivating the need for sophisticated phase unwrapping methods, the subject of the chapters that follow.

2.2 DIFFERENTIAL FORMS AND CONDITIONS FOR PATH INDEPENDENCE

With the substitutions $\mathbf{F}(\mathbf{r}) = f(x, y)\,\mathbf{i} + g(x, y)\,\mathbf{j}$ and $\mathbf{dr} = dx\,\mathbf{i} + dy\,\mathbf{j}$, where $\mathbf{i}$ and $\mathbf{j}$ are the $x-$ and $y-$direction unit vectors, respectively, Equation 2.2 becomes

$$I = \int_C (f\,dx + g\,dy). \tag{2.3}$$

The expression

$$f\,dx + g\,dy \tag{2.4}$$

is called a first-order differential form in the two variables x and y in some domain D. This form is *exact*, or an *exact differential* in D, if it is the differential of a single-valued differentiable function $\varphi(x, y)$. That is, the differential form in Expression 2.4 is exact in D if, and only if, there is a single-valued differentiable function $\varphi(x, y)$ such that

$$f(x, y) = \frac{\partial \varphi}{\partial x}, \quad g(x, y) = \frac{\partial \varphi}{\partial y}, \tag{2.5}$$

everywhere in D.

In vector notation, Expression 2.4 is exact in D if, and only if, the vector function

$$\mathbf{F}(\mathbf{r}) = f(x,y)\,\mathbf{i} + g(x,y)\,\mathbf{j} \tag{2.6}$$

is the *gradient* of a single-valued scalar function $\varphi(x,y)$ in D:

$$\mathbf{F}(\mathbf{r}) = \nabla\varphi = \frac{\partial\varphi}{\partial x}\,\mathbf{i} + \frac{\partial\varphi}{\partial y}\,\mathbf{j}. \tag{2.7}$$

We know from vector calculus that the curl of the gradient is identically zero *if* certain conditions are imposed on the cross derivatives. Specifically,

$$\nabla \times (\nabla\varphi) = \nabla \times \{f(x,y)\,\mathbf{i} + g(x,y)\,\mathbf{j}\} = 0 \tag{2.8}$$

if, and only if,

$$\frac{\partial g}{\partial x} = \frac{\partial f}{\partial y}, \tag{2.9}$$

or (by substituting Equation 2.5 for f and g),

$$\frac{\partial^2\varphi}{\partial x\partial y} = \frac{\partial^2\varphi}{\partial y\partial x}. \tag{2.10}$$

Equation 2.10 holds for any single-valued differentiable function φ. However, as we will see in much more detail later, the equation can be violated at isolated points and then path independence will no longer hold.

Another condition for path independence can be derived by means of Green's theorem. If C is a simple closed path in D, then Green's theorem states that Equation 2.3 can be written

$$I = \iint_G \left(\frac{dg}{dx} - \frac{df}{dy}\right) dx\,dy, \tag{2.11}$$

where G is the region enclosed by the closed path C. It can be shown that this integral is zero for *every* closed path C if, and only if, Equation 2.9 holds. Thus the new condition for path independence is

$$\oint (f dx + g dy) = 0 \tag{2.12}$$

for every closed path C in the domain D.

In summary, the line integral

$$I = \int_C \mathbf{F}(\mathbf{r}) \cdot \mathbf{dr} = \int_C (f\, dx + g\, dy) \tag{2.13}$$

is independent of the path C if any of the following conditions are true. (Any one condition implies the other [2].)

SUMMARY OF CONDITIONS FOR PATH INDEPENDENCE

1. $\mathbf{F}(\mathbf{r}) \cdot \mathbf{dr} = f\, dx + g\, dy$ is an exact differential.
2. $\mathbf{F}(\mathbf{r}) = \nabla\varphi(\mathbf{r})$, which means that $\mathbf{F}$ is the gradient of a single-valued scalar function φ.
3. $\oint \mathbf{F}(\mathbf{r}) \cdot \mathbf{dr} = 0$, which means that the integral around every simple closed path is zero.
4. $\nabla \times \mathbf{F} \equiv 0$, which means that the curl of $\mathbf{F}$ is identically zero if, and only if, the cross derivatives are equal (Equations 2.9 and 2.10).

As we will demonstrate in Section 2.6, Condition 3 is used to detect path dependence in two-dimensional arrays. The other conditions are equivalent but they are usually not implemented.

2.3 THEORY OF RESIDUES AND APPLICATIONS TO PHASE UNWRAPPING

Armed with the conditions for path independence, how do we apply them to the two-dimensional phase unwrapping problem? If we associate φ with the phase we wish to recover through unwrapping, and associate $\mathbf{F}(\mathbf{r})$ with the gradient of the phase $\nabla\varphi$, we immediately see the connection between phase unwrapping and path integration as embodied in Equation 2.1 and similar equations encountered previously. The requirements on φ and $\mathbf{F}$ for path independence were summarized in Conditions 1 through 4 in Section 2.2. Thus, if the phase φ or its gradient $\nabla\varphi$ satisfy any of those conditions, phase unwrapping is path independent and the task is trivially simple. Simply evaluate Equation 2.1 sequentially along a path that covers the domain D, and the job is done.

In practice, two-dimensional phase unwrapping is *not* trivial. Measured data, such as wrapped phase or phase gradients, will violate the conditions for path independence given in Section 2.2. The "fun" then begins by selecting a path that covers the domain D and satisfies certain properties. Alternatively, various minimization criteria methods can be developed to provide unwrapped phase estimates without explicitly specifying an integration path.[1] In all cases, however, phase unwrapping amounts to "integrating gradients" in some fashion.

[1] See Chapter 5.

All of this is well and good, but we have not yet answered the following questions. Why does the wrapped phase sometimes contain certain properties that make phase unwrapping path dependent? Furthermore, how does the phase acquire these undesirable properties? We will answer these questions after setting the stage with a brief historical sketch.

The notion of phase "inconsistencies" in the context of two-dimensional unwrapping first appeared in the literature in 1987 in a paper by Ghiglia, Mastin, and Romero [3]. The authors found that the integration (summation) of wrapped phase differences over a path encompassing the two-dimensional domain of interest yielded inconsistent values. The unwrapped phase depended on the path followed. Furthermore it was observed that this inconsistent behavior was restricted to certain isolated points or regions. Simply stated, inconsistencies were detected by summing (integrating) wrapped phase differences (gradients) around every 2×2-sample path in the entire two-dimensional array of wrapped phases. Inconsistencies were flagged where these closed-path integrals yielded nonzero results. The signs (which are now called polarities or charges in the phase unwrapping literature) and the magnitudes of the local integrals were not utilized in any unwrapping strategy. Thus their true significance went unrecognized. In retrospect, these inconsistencies were detected by applying the discrete counterpart of Condition 3 of Section 2.2 as a test of path independence. The 2×2-sample path integral is the smallest closed loop that can be defined in a sampled image. This small loop helps localize the source of each phase inconsistency.

Later, in 1988, Goldstein, Zebker, and Werner coined the term "residues" to describe these inconsistencies [4]. (The inconsistencies were also called discontinuity sources by Huntley [5], [6].) Making the analogy with the residues encountered in complex variable contour integration, Goldstein, Zebker, and Werner determined that each residue "charge" must be balanced by imposing "branch cuts" or barriers that must not be crossed during the unwrapping process.

2.3.1 Complex Functions, Residues, and Local Phase Behavior

How close is the analogy between the residues of functions of a complex variable and the *phase residues* associated with the phase unwrapping problem? It may seem that phase residues are the same as the residues of complex functions, but they are not. Although the differences may be slight, it is important to note them.

From the theory of functions of a complex variable, we know that the problem of evaluating line integrals around closed paths (contour integrals) is equivalent to the *algebraic* problem of evaluating the enclosed residues. The "residue theorem" for functions of a complex variable states:

$$\oint f(z)dz = 2\pi j \times \text{sum of residues enclosed by the contour,} \qquad (2.14)$$

where z is the complex variable $z = x + jy$ and $j = \sqrt{-1}$. The residues are the *coefficients* of the $(z - z_i)^{-1}$ terms in the Laurent series expansion of $f(z)$ about the various singular points z_i [7]. The function $f(z)$ is defined to be analytic in some neighborhood of a singular point ($z = z_i$) except at the point itself. The term *residue* stems from the fact that the various $(z - z_i)^{-1}$ terms in the Laurent series expansions are the only terms that contribute to the contour integral.

The function $f(z) = 1/z$ serves as a simple example of a function that is analytic everywhere except at the one singular point $z_0 = 0$. By design, this function is its own Laurent series expansion. The coefficient of the $1/z$ term is the residue. Thus

$$\oint \frac{1}{z} dz = 2\pi j, \tag{2.15}$$

where it is assumed that the contour integral encircles the origin, which is the singular point of this function. In contrast, the simple function $f(z) = z$ is analytic everywhere, so $\oint f(z)dz = 0$ for any simple closed contour.

We emphasized in Chapter 1 that phase must be *extracted* from a complex function. Equivalently, if phase *gradients* are estimated or measured, some operation on a complex function is implied. There is no escaping this fact. What is it then about the complex function that determines whether the phase can be unwrapped in a path-independent manner? In other words, what properties of the complex function are responsible for generating phase residues?

It will be instructive to look at the extracted phase in the vicinity of certain points of interest. Let us assume that we intend to extract the wrapped phase $\psi(x, y)$ from some complex function $s(x, y)$ relating to the real-world experiment of interest. This complex function could have come from SAR interferometry, MRI, or a host of other applications mentioned earlier. Let us define

$$s(x, y) = |s(x, y)| \exp\{j\varphi(x, y)\}, \tag{2.16}$$

where $|s(x, y)|$ is the magnitude of the complex function and $\varphi(x, y)$ is the phase. Let us also assume that the magnitude of $s(x, y)$ is bounded; that is, $s(x, y)$ does not contain any poles.

Even though $|s(x, y)|$ is bounded, it is not necessarily an *analytic* function of the complex variable $z = x + jy$. For example, $s(x, y) = x - jy = \bar{z}$ is not analytic *anywhere*. ($\bar{z}$ is the complex conjugate of z). In fact $s(x, y)$ cannot be expressed solely as a function of z. We must also include functions of $\bar{z}$. Therefore, if we wish to expand $s(x, y)$ about an *arbitrary* point $z = z_0$, we must include powers of $z - z_0$ *and* powers of $\overline{z - z_0}$. Because we have assumed that $|s(x, y)|$ is bounded, it is not necessary to include negative powers in the expansion. Therefore we can write our expansion as

$$f(z, \bar{z}) = \sum_{n=0}^{\infty} \sum_{m=0}^{\infty} a_{mn} (z - z_0)^m \overline{(z - z_0)}^n. \tag{2.17}$$

It is likely that at some point $z = z_0$, we can expect $f(z_0) = 0$. This implies that in the vicinity of $z = z_0$, we have

$$f(z, \bar{z}) \approx a_{1,0}(z - z_0) + a_{0,1}\overline{(z - z_0)}, \tag{2.18}$$

where $a_{1,0}$ and $a_{0,1}$ are the appropriate expansion coefficients in Equation 2.17. (Note that the coefficient $a_{0,0}$ also must equal zero if $f(z_0) = 0$.) Thus, in the vicinity of the point where the complex function is zero, the function "looks like" $a_{1,0}(z - z_0) + a_{0,1}\overline{(z - z_0)}$.

We can now look at the *phase* of the complex function in the vicinity of this point. We can write

$$f(z, \bar{z}) = (a_{1,0} + a_{0,1})r \cos\theta + j(a_{1,0} - a_{0,1})r \sin\theta, \tag{2.19}$$

where $r = |z - z_0|$, and $\theta = \angle(z - z_0)$. The coefficients $a_{1,0}$ and $a_{0,1}$ need not be real. Indeed we can assume that they are arbitrary complex numbers. Without loss of generality, we can normalize the summation in Equation 2.19 by letting $|a_{1,0}| + |a_{0,1}| = 1$. Therefore we obtain

$$\begin{aligned} a_{1,0} &= \alpha \exp(j\gamma_1) \\ a_{0,1} &= (1 - \alpha) \exp(j\gamma_2), \end{aligned} \tag{2.20}$$

where $0 \leq \alpha \leq 1$ and γ_1 and γ_2 are arbitrary angles. After substituting Equation 2.20 into Equation 2.19 and simplifying, we obtain

$$f(z, \bar{z}) = f(r, \vartheta) = r \exp(j\beta)[\alpha \exp(j\vartheta) + (1 - \alpha) \exp(-j\vartheta)], \tag{2.21}$$

where $\vartheta = \theta + (\gamma_1 - \gamma_2)/2$ and $\beta = (\gamma_1 + \gamma_2)/2$.

If we now let

$$g(\alpha, \theta) = \alpha \exp(j\vartheta) + (1 - \alpha) \exp(-j\vartheta), \tag{2.22}$$

we see from Equations 2.21 and 2.22 that

$$\varphi = \angle f(r, \vartheta) = \beta + \angle g(\alpha, \vartheta). \tag{2.23}$$

According to the above equation the phase of the function $f(r, \vartheta)$ in the vicinity of z_0 is composed of a constant phase β plus the phase of the complex function g. It now suffices to find the phase of the function g. Some simple manipulations allow us to write

$$\psi = \angle g(\alpha, \vartheta) = \arctan\{(2\alpha - 1) \tan\vartheta\}. \tag{2.24}$$

Without loss of generality we can ignore the effect of the constant phase terms, β and $(\gamma_1 - \gamma_2)/2$, and concentrate on the behavior of the wrapped phase ψ as given by Equation 2.24. It is important to note that for an arbitrary value of α, as

ϑ sweeps out 2π radians, ψ sweeps out 2π radians also. Therefore the line integral of the phase gradient around the point $z = z_0$ accumulates $\pm 2\pi$ radians. The sign of the term $2\alpha - 1$ determines the sign of ψ. Two limiting conditions are worthy of note. (1) If $\alpha = 1$, then $\psi = \vartheta$. This result implies that the expansion given in Equation 2.18 is made up of only the $z - z_0$ term, and the phase is in the positive sense. (2) If $\alpha = 0$, then $\psi = -\vartheta$. This result implies that the expansion given in Equation 2.18 is made up of only the $\overline{z - z_0}$ term, and the phase is in the negative sense. For arbitrary α the local phase behavior results from a mixture of the $z = z_0$ and $\overline{z - z_0}$ terms as implied by Equation 2.24. Figure 2.1 shows typical local phase behavior in the vicinity of a residue for various values of α. Note that the phase changes by $\pm 2\pi$ as the origin is encircled.

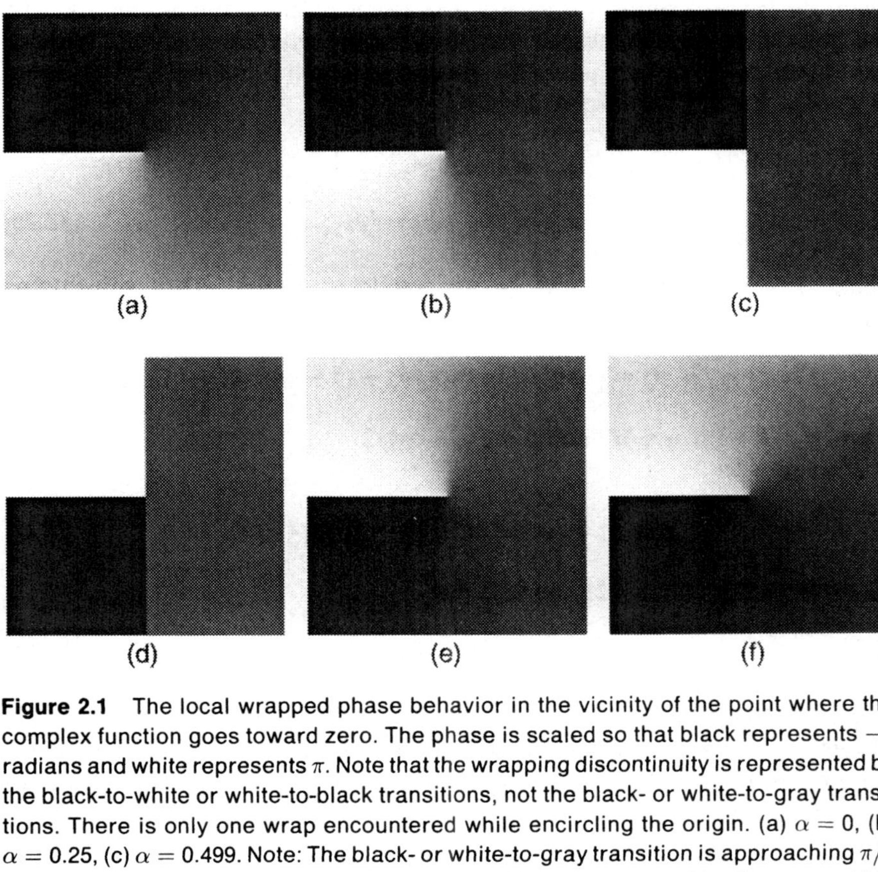

Figure 2.1 The local wrapped phase behavior in the vicinity of the point where the complex function goes toward zero. The phase is scaled so that black represents $-\pi$ radians and white represents π. Note that the wrapping discontinuity is represented by the black-to-white or white-to-black transitions, not the black- or white-to-gray transitions. There is only one wrap encountered while encircling the origin. (a) $\alpha = 0$, (b) $\alpha = 0.25$, (c) $\alpha = 0.499$. Note: The black- or white-to-gray transition is approaching $\pi/2$ radians. (d) $\alpha = 0.501$. As before, the black- or white-to-gray transition is approaching $\pi/2$ radians. (e) $\alpha = 0.75$, (f) $\alpha = 1$. Note: Any arbitrary constant added to the phase simply rotates the local wrapped phase about the origin. The sign of the wrapped phase is determined by the sign of $2\alpha - 1$. The variable α determines the mix of the $z - z_0$ and $\overline{z - z_0}$ terms in the expansion given in Equation 2.18.

While it is likely that a complex function can be zero at some isolated point, it is *not* likely (except in pathological cases) that the function *and* any of its partial derivatives are zero at the *same* point. Nevertheless, if it were true that the function and certain partial derivatives are zero at the same point, the series expansion (Equation 2.17) could be used to predict the behavior of the local phase. For example, in the highly contrived situation where the function and all partial derivatives up to order M are zero at the same point, the local phase can change by $\pm 2\pi(M+1)$ radians as the point is encircled (i.e., multiple wraps will be encountered). Figure 2.2 shows some of these special, but highly unlikely cases. The details of the local phase behavior analysis for more general cases are tedious and complicated and will not be discussed further. Suffice it to say that in almost all practical cases, the phase in the vicinity of a zero will resemble one of

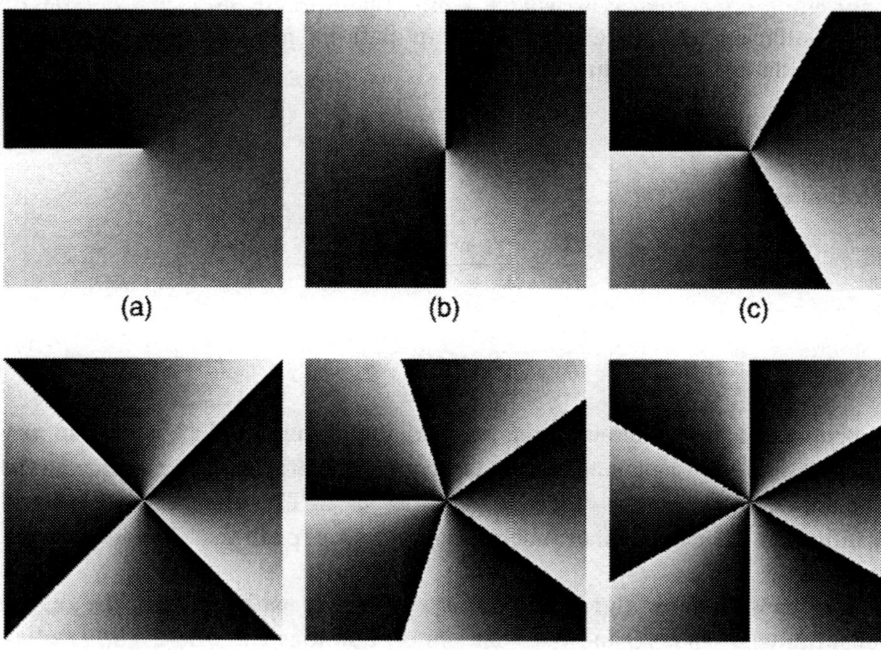

Figure 2.2 The local wrapped phase behavior in the vicinity of the point where the complex function, and all the partial derivatives up to order M go toward zero at the same point z_0. In addition these examples assume that the series expansion of the complex function $f(z, \bar{z})$ contains only powers of $z - z_0$. Therefore the local phase behavior is governed by the term $(z - z_0)^{M+1}$, whose phase "looks like" $(M+1)\vartheta$. (a) $M = 0$. The function goes toward zero but not its first partial derivatives. This is the typical behavior expected. (b) $M = 1$, (c) $M = 2$, (d) $M = 3$, (e) $M = 4$, (f) $M = 5$. Note: An arbitrary constant added to the phase simply rotates the local wrapped phase about the center. The sign of the wrapped phase is determined by the expansion coefficients. We are showing the positive sense in this figure.

the examples shown in Figure 2.1. Typically only a single wrap will be encountered in a path around the zero point.

In order to make further analysis simple, let us investigate the typical "local" phase function

$$\varphi(\mathbf{r}) = \theta(\mathbf{r}), \qquad -\pi < \theta \le \pi, \tag{2.25}$$

shown in Figure 2.3(a). $\theta(\mathbf{r})$ is the angle of the vector $\mathbf{r}$ with respect to the positive x axis. It is obvious that a phase discontinuity exists along the negative x axis. (Assume that the origin is in the center of the image and corresponds to the point $z = z_0$ where $|f(z)| = 0$.) Is this merely a simple wrapping discontinuity, or is there more to this than first meets the eye? Because the discontinuity appears to originate "out of nowhere," we might suspect something unusual is happening. Indeed φ is discontinuous, it is not differentiable everywhere, and the domain D is not simply connected, as we will see. Any of these aforementioned properties of φ are sufficient to violate conditions for path-independent integration. For example, let us rewrite Equation 2.25 as

$$\varphi(x, y) = \arctan\frac{y}{x}. \tag{2.26}$$

We find that

$$\begin{aligned} \frac{\partial\varphi}{\partial x} &= \frac{-y}{x^2 + y^2} = \frac{-\sin\theta}{r}, \\ \frac{\partial\varphi}{\partial y} &= \frac{x}{x^2 + y^2} = \frac{\cos\theta}{r}, \end{aligned} \tag{2.27}$$

and upon further differentiation, we find that Equation 2.10 is true in any domain of the $x - y$ plane that does not contain the origin. (For example, in the domain $D : \epsilon \le \sqrt{x^2 + y^2} < |\mathbf{r}|; \quad \epsilon > 0, \quad |\mathbf{r}| < \infty$.) Therefore the path integral of Equation 2.3 evaluated around any simple closed path that does *not* contain the origin is zero.

It is easy to show that trouble arises when the path encircles the origin. Substituting $x = r\cos\varphi$ and $y = r\sin\varphi$ into Equations 2.27 and 2.3 and noting that a closed path around the origin is represented by $r = 1$, we obtain

$$I = \int_0^{2\pi} d\varphi = 2\pi, \tag{2.28}$$

where the integral is evaluated once around the circle in the counterclockwise direction. The nonzero result indicates a positive phase residue, which in turn indicates path-dependent phase unwrapping. Therefore, even though our complex function is well-behaved in the vicinity of a zero, the phase function contains a *phase residue*. In practical applications, zeros are generators of residues in the extracted phase.

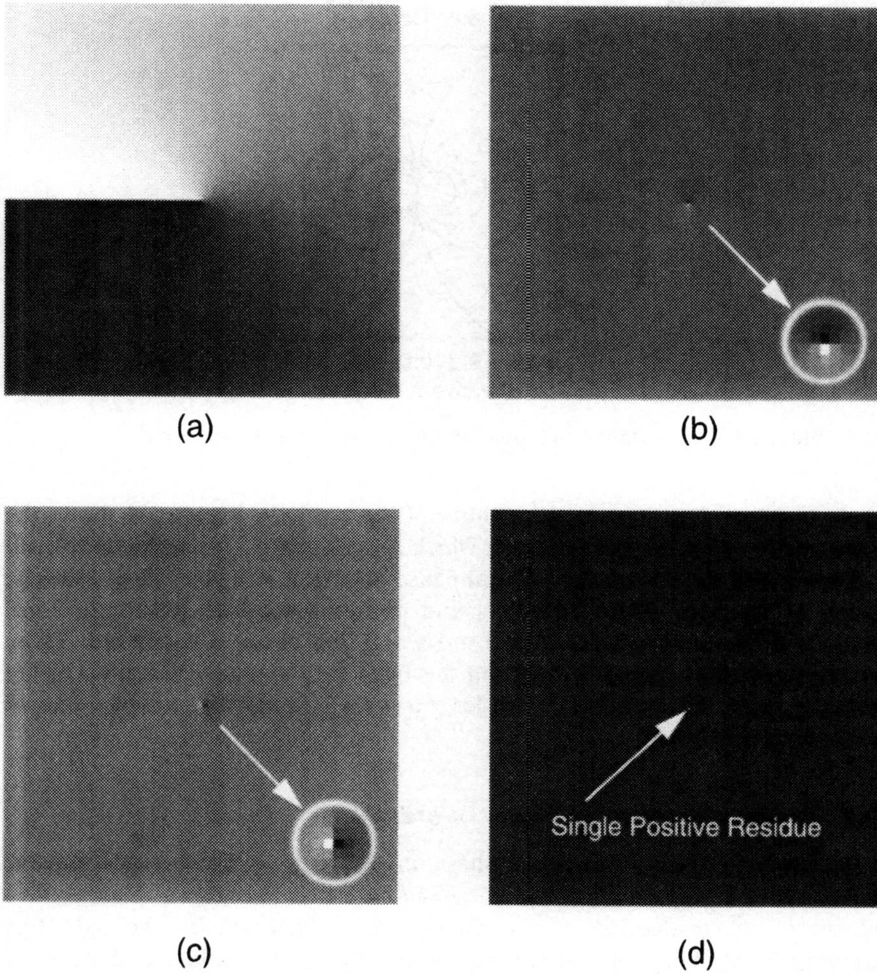

Figure 2.3 256 × 256-pixel digital images. (a) A phase function described by $\varphi(\mathbf{r}) = \theta(\mathbf{r})$, $-\pi \le \theta < \pi$, scaled for display. φ is scaled so that black represents $-\pi$ radians and white represents π radians. (b) The x-gradient component $\varphi_x(x, y)$. A magnified version is shown in the inset. (c) The y-gradient component $\varphi_y(x, y)$. A magnified version is shown in the inset. (d) A single positive residue is detected at the origin.

The gradient components of Equation 2.27 are shown in Figures 2.3(b) and (c), respectively, and graphically indicate potential trouble in the vicinity of the origin where they "blow up." The single positive residue at the origin is shown by the white dot in Figure 2.3(d). The rightmost terms in Equation 2.27 are mathematically equivalent to *dipoles* in electromagnetic field theory. Thus, where phase residues are present, the phase gradient components "look like"

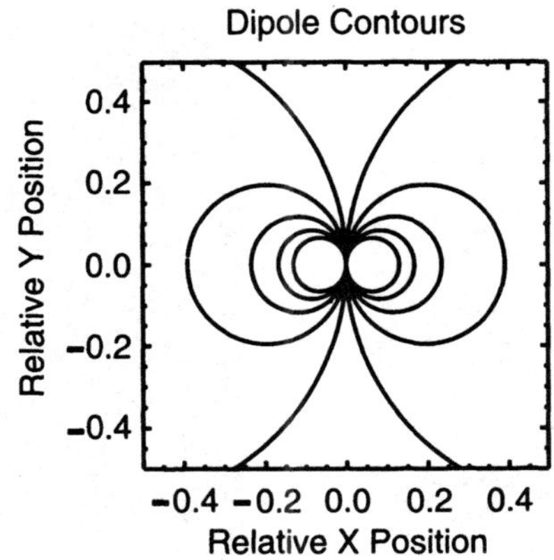

Figure 2.4 Constant amplitude contours of a typical dipole response.

dipole response functions in the vicinity of the residues. Figure 2.4 shows the contours of a typical dipole response, which illustrates the $1/r$ amplitude roll off.

Armed with the knowledge of local phase behavior, it is possible to examine images of wrapped phase functions and deduce something about the local behavior of the complex function from which the phase was derived. There are many examples spread throughout this book for the reader to test his or her deductive skills. We trust that the reader will never look at wrapped phase images in the same way again!

2.3.2 Residue Theorem for Phase Unwrapping

In the context of two-dimensional phase unwrapping, we have shown that the evaluation of Equation 2.2 around a *closed* path that encircles a *single* residue reduces to the evaluation of the phase residue enclosed by the path (see Equation 2.28). This concept can be simply extended to allow evaluation of closed-path integrals that contain several phase residues within the contour. Evaluation of the large closed-path integral reduces to the sum of small closed-path integrals around the individual residues, just as in complex variable contour integration [8]. Therefore

$$\oint \nabla\varphi(\mathbf{r}) \cdot \mathbf{dr} = 2\pi \times (\text{sum of enclosed phase residue charges}). \tag{2.29}$$

This is the *residue theorem* for two-dimensional phase unwrapping. From the previous discussion and analysis we can assign an integer to the residue "charge." In other words, the closed-path integral around a phase residue will equal some *integer* multiple of 2π radians. We emphasized, however, that it is unlikely to encounter residue charges other than ± 1 except for pathological phase constructions. This is in contrast to the residue theorem of complex functions (Equation 2.14). Because complex function residues are not restricted to integers, contour integrals are not restricted to the summation of 2π multiples.

It is easy to see that when the residue charges are balanced in a region, the line integral around that region is zero for *any* simple path chosen. Thus consistent unwrapping is possible *if, and only if,* all integration paths do not encircle unbalanced residue charges. Once the residue charges are balanced by connecting the residues of opposite polarity with branch cuts, unwrapping can occur along any path that does not cross branch cuts.

The residue concept forms the basis for virtually all path-following unwrapping methods.[2] Unwrapping barriers (branch cuts) must be defined explicitly or implicitly to balance residue charges. Unwrapping then proceeds along a path that does not cross these forbidden barriers but does visit all of the array locations.

Obviously there are many ways to perform charge balancing and path selection. Two-dimensional path-following algorithm development is still an active area of research and probably will remain so for some time, especially as additional information from the particular application provides clues as to how the unwrapping barriers should be defined. It should be emphasized that the wrapped phase values themselves cannot determine (in the absence of ancillary information or assumptions) any preferred set of branch cuts. It is not possible to tell where the unwrapping will run into trouble. *Trouble will occur simply by encircling an unbalanced residue charge.*

2.4 PHASE DECOMPOSITION INTO "IRROTATIONAL" AND "ROTATIONAL" COMPONENTS

Borrowing from electromagnetic field theory, it is possible to specify a vector field in terms of its divergence and curl [9], [10]. A vector function $\mathbf{F}(\mathbf{r})$ (as well as the field described by it) is called *irrotational* or *rotation free* throughout a domain D if, and only if,

$$\nabla \times \mathbf{F}(\mathbf{r}) = 0. \tag{2.30}$$

Thus, if Equation 2.30, which is also Condition 4 of Section 2.2 for path independence, is true, then $\mathbf{F}(\mathbf{r}) = \nabla\varphi(\mathbf{r})$, and hence $\varphi(\mathbf{r})$ is rotation free. In other words, $\varphi(\mathbf{r})$ has no curl components anywhere in D.

On the other hand, a vector function $\mathbf{F}(\mathbf{r})$, as well as the field described by it, is called *solenoidal* or *rotational* in D if, and only if,

$$\nabla \cdot \mathbf{F}(\mathbf{r}) = 0. \tag{2.31}$$

Equation 2.31 is true if, and only if, $\mathbf{F}(\mathbf{r})$ is the curl of some other vector function

[2] See Chapter 4.

$\mathbf{A}(\mathbf{r})$. Mathematically

$$\mathbf{F}(\mathbf{r}) = \nabla \times \mathbf{A}(\mathbf{r}), \tag{2.32}$$

where $\mathbf{A}(\mathbf{r})$ is the *vector potential* of the vector field described by $\mathbf{F}(\mathbf{r})$. Note that Equation 2.31 is *not* a condition for path independence. If $\varphi(\mathbf{r})$, which is the field described by $\mathbf{F}(\mathbf{r})$, contains solenoidal or rotational components, path-*dependent* unwrapping is to be expected. Fields of this type are called *nonconservative*.

If the divergence and curl of $\mathbf{F}(\mathbf{r})$ are specified everywhere in D, $\mathbf{F}(\mathbf{r})$ can be expressed as the sum of an *irrotational* vector function $\mathbf{F_I}(\mathbf{r})$, and a rotational vector function $\mathbf{F_R}(\mathbf{r})$. This result, which is known as the *Helmholtz decomposition theorem*, is expressed mathematically as

$$\begin{aligned} \mathbf{F}(\mathbf{r}) &= \mathbf{F_I}(\mathbf{r}) + \mathbf{F_R}(\mathbf{r}) \\ &= \nabla\varphi(\mathbf{r}) + \nabla \times \mathbf{A}(\mathbf{r}), \end{aligned} \tag{2.33}$$

with

$$\nabla \times \mathbf{F_I}(\mathbf{r}) = \nabla \cdot \mathbf{F_R}(\mathbf{r}) = 0. \tag{2.34}$$

The significance of this decomposition when applied to phase unwrapping is as follows: If the true phase function $\varphi(\mathbf{r})$ consists solely of irrotational components and $\nabla\varphi$ is known everywhere and exactly in D, path-independent unwrapping is possible. On the other hand, if $\varphi(\mathbf{r})$ contains rotational components *anywhere* in D, unwrapping is path dependent.

In practice, true phase discontinuities or dislocations in φ, aliasing [11], or noise-corrupted phase gradients $\nabla\varphi$ can generate the equivalent of *localized* rotational components. These perturbations are usually equivalent to generating zeros in the complex function. When residues are present, phase unwrapping becomes path dependent and nontrivial.

2.5 DISCRETE FORMULATION OF TWO-DIMENSIONAL PHASE UNWRAPPING

We have indicated that two-dimensional phase unwrapping amounts to integrating the phase gradient over a path that covers the domain of interest. In practice we must, of course, deal with sampled data. In order to do this, however, we must define the two-dimensional phase unwrapping problem in the discrete domain and specify the relevant terms that will be used throughout the book.

In the most general and least-restrictive sense, two-dimensional phase unwrapping is an impossible problem. For example, an unknown phase function φ corrupted by noise and wrapped into the interval $(-\pi, \pi]$ is impossible to recover unambiguously. However, certain assumptions of the underlying process or nature of the desired solution can make the phase unwrapping problem tractable, if not easy. More importantly, though, useful solutions can be obtained from rather simple assumptions.

The most common assumption is that the desired unwrapped phase has local phase derivatives (differences) that are less than π radians in magnitude everywhere.[3] We assume that we know the phase ϕ, modulo 2π, on a discrete grid of points:

$$\psi_{i,j} = \phi_{i,j} + 2\pi k, \quad k \text{ an integer,}$$
$$-\pi < \psi_{i,j} \le \pi, \quad i = 0, \ldots, M-1, \quad j = 0, \ldots, N-1. \tag{2.35}$$

Given the wrapped phase values $\psi_{i,j}$, we wish to determine the unwrapped phase values $\phi_{i,j}$ at the same grid locations. We assume that the phase differences of the $\phi_{i,j}$ are less than π radians in magnitude everywhere. In this case ϕ includes the effect of noise on the unknown phase φ. We hope only to obtain ϕ, which is assumed to be a useful embodiment of the unknown and unobtainable φ. This is the discrete formulation of the two-dimensional phase unwrapping problem. In practice, we must relax the local gradient assumption in some locations to meet other equally important constraints in the context of particular applications. We relax this assumption in a controlled manner by means of branch cuts, quality maps, and weights. We will elucidate this in detail in the later chapters.

We will utilize discrete gradient components as approximations to the continuous phase gradient. Specifically, we will utilize two sets of phase differences: those differences with respect to the i index and those with respect to the j index. From the wrapped phase $\psi_{i,j}$, we can compute the following wrapped phase differences:[4]

$$\Delta^x_{i,j} = \mathcal{W}\{\psi_{i+1,j} - \psi_{i,j}\}, \quad i = 0, \ldots, M-2, \quad j = 0, \ldots, N-1,$$
$$\Delta^x_{i,j} = 0 \quad \text{otherwise,} \tag{2.36}$$

and

$$\Delta^y_{i,j} = \mathcal{W}\{\psi_{i,j+1} - \psi_{i,j}\}, \quad i = 0, \ldots, M-1, \quad j = 0, \ldots, N-2,$$
$$\Delta^y_{i,j} = 0 \quad \text{otherwise,} \tag{2.37}$$

where the x and y superscripts refer to wrapped differences in the i and j indexes, respectively. Note that $\mathcal{W}$ defines a wrapping operator that wraps all values of its argument into the range $(-\pi, \pi]$ by adding or subtracting an integral multiple of

[3] This is equivalent to the requirement that the sampled signal from which ψ is *extracted* is not aliased.

[4] The wrapping operation is necessary to obtain a correct estimate for the phase gradient. See Itoh's analysis in Section 1.3.1.

2π radians from its argument. Equations 2.36 and 2.37 form the definition of wrapped phase differences used throughout this book.[5] We will use the definitions above extensively in all aspects of algorithm development, problem formulations, and implementation.

2.6 SIMPLE TESTS FOR PATH DEPENDENCE

If we substitute Equation 2.33 into Equation 2.2, we obtain

$$I = \int_C \mathbf{F}(\mathbf{r}) \cdot \mathbf{dr} = \int_C \nabla\varphi(\mathbf{r}) \cdot \mathbf{dr} + \int_C [\nabla \times \mathbf{A}(\mathbf{r})] \cdot \mathbf{dr}, \tag{2.38}$$

where the integral

$$\int_C \nabla\varphi(\mathbf{r}) \cdot \mathbf{dr} \tag{2.39}$$

represents the *path-independent* solution that captures only the *irrotational* components of φ, and

$$\int_C [\nabla \times \mathbf{A}(\mathbf{r})] \cdot \mathbf{dr} \tag{2.40}$$

represents the *path-dependent* components containing the *rotational* part of φ.

It should be emphasized that the *rotational components in* $\mathbf{F}(\mathbf{r})$ *(or in* $\varphi(\mathbf{r})$*) are the sole source of path dependence in two-dimensional phase unwrapping*. The rotational components in the phase are responsible for the residues. Simple tests based on Conditions 1 through 4 of Section 2.2 can be developed to detect rotational components in discrete arrays of wrapped phase values (or in arrays of x- and y-gradient components).

2.6.1 Detecting Residues in Two-Dimensional Arrays

In order to utilize the residues in the process of phase unwrapping, we must be able to determine their locations. We will now investigate a means for detecting and locating residues in two-dimensional phase arrays. Let us assume that we have an array that contains samples of a wrapped phase function $\psi(m, n)$. A small portion of such an array might consist of the samples shown in Figure 2.5. (Note that the samples shown in the figure represent the phase in *cycles*. They should be multiplied by 2π to obtain the wrapped phase in radians.) Let us now consider the integration of the gradient around the path shown in the figure. We

[5] We assume that the phase derivatives are computed discretely as local two-point phase *differences*. This is, of course, not a restriction. Any other functional approximation to derivatives is a viable candidate.

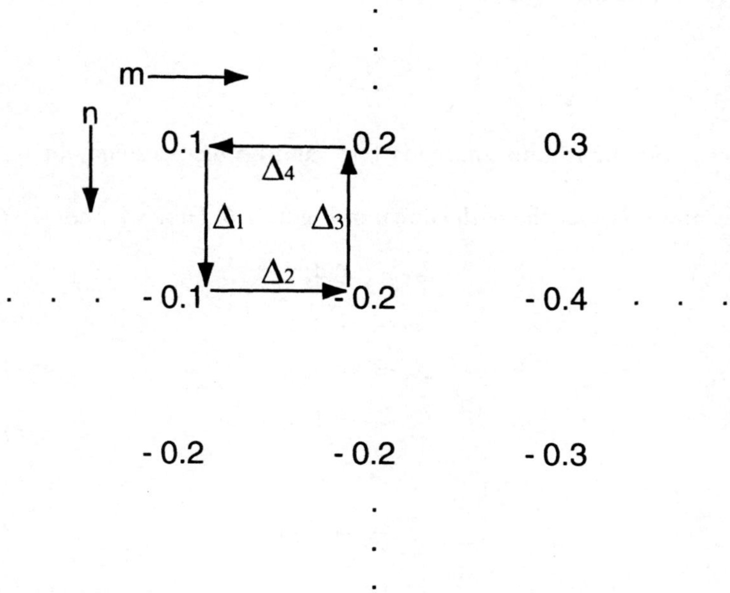

Figure 2.5 A small portion of a typical array of wrapped phase values. For simplicity the samples represent the phase in cycles and should be multiplied by 2π to obtain the wrapped phase in radians. The m and n index directions are shown along with a closed unwrapping path for detecting a possible residue.

recall that an estimate of the true gradient (i.e., phase derivative) is obtained by wrapping the differences of wrapped phase. For example, referring to Figure 2.5,

$$\begin{aligned}
\Delta_1 &= \mathcal{W}\{\psi(m, n+1) - \psi(m, n)\}, \\
\Delta_2 &= \mathcal{W}\{\psi(m+1, n+1) - \psi(m, n+1)\}, \\
\Delta_3 &= \mathcal{W}\{\psi(m+1, n) - \psi(m+1, n+1)\}, \\
\Delta_4 &= \mathcal{W}\{\psi(m, n) - \psi(m+1, n)\}.
\end{aligned} \tag{2.41}$$

Substituting the actual sample values from Figure 2.5 into Equation 2.41 we find

$$\begin{aligned}
\Delta_1 &= -0.2, \\
\Delta_2 &= -0.1, \\
\Delta_3 &= +0.4, \\
\Delta_4 &= -0.1\,.
\end{aligned} \tag{2.42}$$

We now compute the residue charge by summing the phase differences around the closed path as indicated in Figure 2.5:

$$q = \sum_{i=1}^{4} \Delta_i. \tag{2.43}$$

In this example the residue charge is zero, and no inconsistency or residue is present.

Let us now consider the path shown in Figure 2.6. Here we find

$$\begin{aligned} \Delta_1 &= -0.4, \\ \Delta_2 &= -0.2, \\ \Delta_3 &= -0.3, \\ \Delta_4 &= -0.1, \end{aligned} \tag{2.44}$$

and

$$q = \sum_{i=1}^{4} \Delta_i = -1.0\,. \tag{2.45}$$

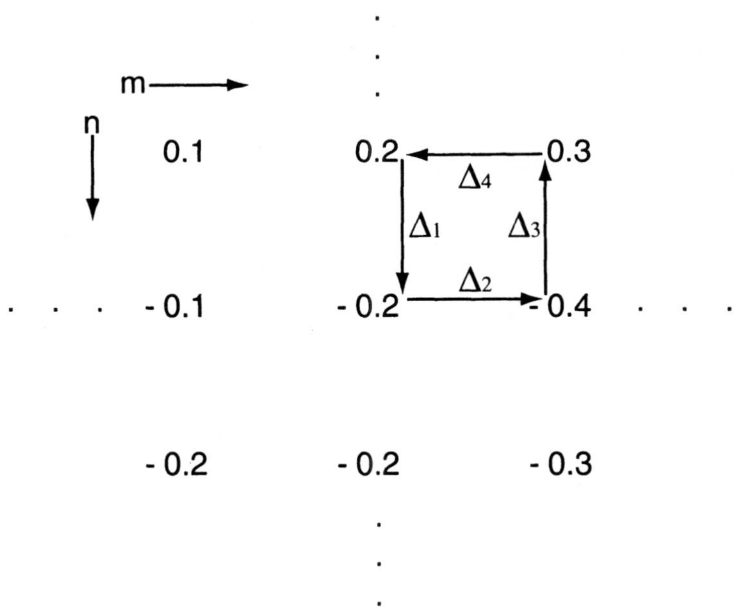

Figure 2.6 A small portion of a typical array of wrapped phase values. For visual simplicity the samples represent the phase in cycles and should be multiplied by 2π to obtain the wrapped phase in radians. The *m* and *n* index directions are shown along with another particular closed unwrapping path for detecting a possible residue.

Note that $\Delta_3 = \mathcal{W}\{0.3 - (-0.4)\} = -0.3$. Here we have found a nonzero path integral, so we label it a residue with a negative charge or polarity. If we perform similar path integrations over the remaining paths in the figure, we will find no more nonzero integrals or residues. What is going on here?

We now associate the Δs with an actual (not wrapped) phase difference of the underlying continuous phase function:

$$\begin{aligned}
\Delta_1 &= \phi(m, n+1) - \phi(m, n), \\
\Delta_2 &= \phi(m+1, n+1) - \phi(m, n+1), \\
\Delta_3 &= \phi(m+1, n) - \phi(m+1, n+1), \\
\Delta_4 &= \phi(m, n) - \phi(m+1, n).
\end{aligned} \tag{2.46}$$

Further insight is obtained by manipulating Equation 2.46. We obtain

$$\begin{aligned}
\sum_{i=1}^{4} \Delta_i = [\phi(m, n+1) &- \phi(m, n)] \\
&+ [\phi(m+1, n+1) - \phi(m, n+1)] \\
&- [\phi(m+1, n+1) - \phi(m+1, n)] \\
&- [\phi(m+1, n) - \phi(m, n)].
\end{aligned} \tag{2.47}$$

The four terms in the summation on the right side of this equation can be categorized in the following way: Associate the first term with the discrete counterpart to a partial derivative in the y direction. Thus

$$\phi(m, n+1) - \phi(m, n) \Rightarrow \phi_y(m, n), \tag{2.48}$$

where the subscript y refers to partial differentiation in the y direction. Similarly we associate the remaining terms in Equation 2.47 with the other partial derivatives as follows:

$$\begin{aligned}
\phi(m+1, n+1) - \phi(m, n+1) &\Rightarrow \phi_x(m, n+1), \\
\phi(m+1, n+1) - \phi(m+1, n) &\Rightarrow \phi_y(m+1, n), \\
\phi(m+1, n) - \phi(m, n) &\Rightarrow \phi_x(m, n).
\end{aligned} \tag{2.49}$$

Thus

$$q = \sum_{i=1}^{4} \Delta_i = [\phi_x(m, n+1) - \phi_x(m, n)] - [\phi_y(m+1, n) - \phi_y(m, n)]. \tag{2.50}$$

If, by similar reasoning, we make the following associations,

$$\phi_x(m, n+1) - \phi_x(m, n) \Rightarrow \phi_{xy} = \frac{\partial \phi_x}{\partial y},$$

$$\phi_y(m+1, n) - \phi_y(m, n) \Rightarrow \phi_{yx} = \frac{\partial \phi_y}{\partial x}, \tag{2.51}$$

we find that

$$q = \sum_{i=1}^{4} \Delta_i = \phi_{yx} - \phi_{xy}. \tag{2.52}$$

It is now obvious from Equation 2.52 that the closed-path integral is zero if, and only if, the cross derivatives are the same! This is precisely Condition 4 of Section 2.2[6] and is the discrete counterpart to Green's theorem in the plane:

$$\oint \nabla\phi \cdot d\mathbf{r} = \int\int \left(\frac{\partial^2 \phi}{\partial x \partial y} - \frac{\partial^2 \phi}{\partial y \partial x} \right) dx dy. \tag{2.53}$$

In other words, at the points where residues are detected, local curl components exist, the field (or wrapped phase) is nonconservative and path-independent integration (unwrapping) is not possible. *The existence of residues is a necessary and sufficient condition for path dependence in two-dimensional phase unwrapping.* The simple test described above should always be performed in practice. The locations and densities of residues are indicative of the severity of the phase unwrapping problem and graphically indicate where problems in the unwrapping process are likely to occur. In addition the residues provide clues as to how the branch cuts might be constructed or how the integration path might be selected. Ironically residues are both the problem *and* the solution.

2.7 RESIDUES, GRADIENTS, AND PHASE DECOMPOSITION: EXAMPLES

It is instructive to observe how the local phase properties developed in Section 2.3 carry over into the real world. We will do this with the help of the Helmholtz decomposition theorem of Section 2.4. While we are not yet in a position to explain in detail *how* the decomposition is performed, let us assume that it can be done.[7] Figure 2.7(a) depicts a 512 × 512-pixel wrapped phase function typical of

[6] Note that if phase *gradients* are provided everywhere (instead of wrapped phases), Equation 2.50 can be used directly to compute the locations and charges of all the residues.

[7] We use our ability to solve the least-squares phase unwrapping problem (Chapter 5) to obtain the irrotational-component (conservative) solution. The residual then contains the rotational components, and hence, the residues.

those that might be encountered in practice. Notice that several of the fringe lines seem to originate or terminate "out of nowhere" like the fringe line of Figure 2.2(a) or Figure 2.3(a). Positive and negative residues are marked as 4×4-pixel white and black squares, respectively, as shown in Figure 2.7(b). Residues are located near the origination/termination points of the disconnected

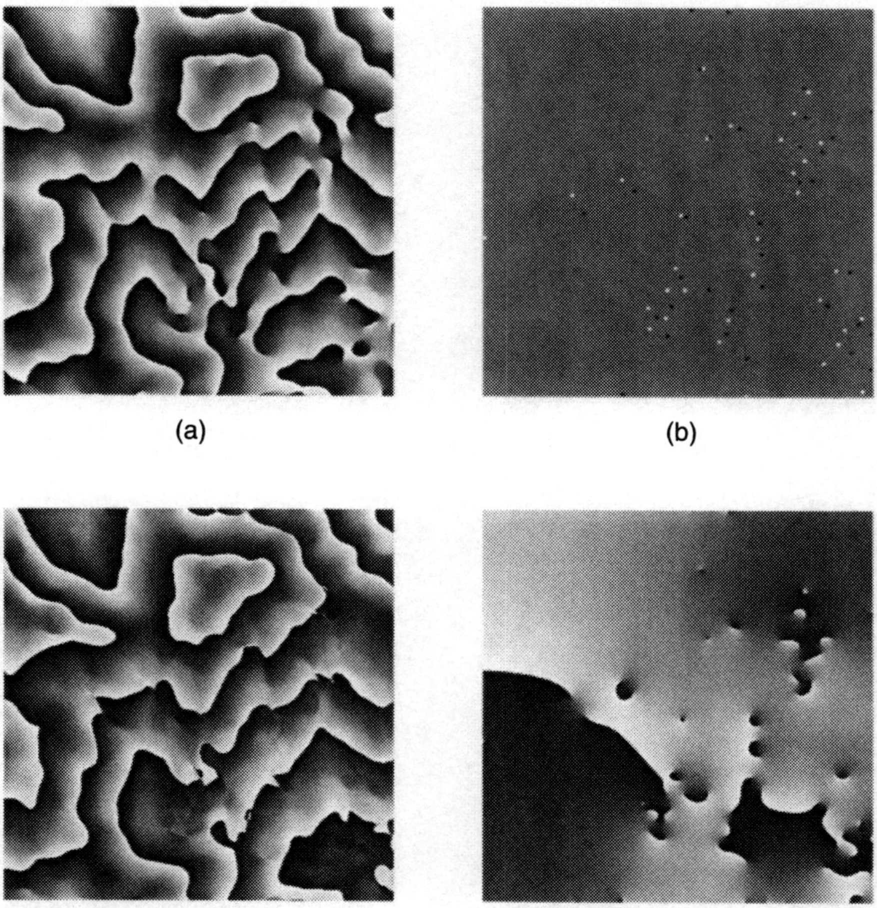

(a) (b) (c) (d)

Figure 2.7 The Helmholtz decomposition theorem applied to a 512×512-pixel array of wrapped phase. (a) A typical wrapped phase function scaled for display. Note that several fringe lines appear disconnected. (b) The residue map shows the positive and negative residues as white and black squares, respectively. (c) The *irrotational* or *rotation-free* wrapped phase component. (d) The *rotational* component. The decomposition is such that the wrapped sum of the components (c) and (d) yields the original data in (a).

fringe lines. The local phase behavior indicates the presence of zeros in the complex function from which this phase was derived

The wrapped phase of Figure 2.7(a) is decomposed into the *irrotational* or *rotation-free* wrapped phase component shown in Figure 2.7(c) and the *rotational* component shown in Figure 2.7(d). Except for an arbitrary additive

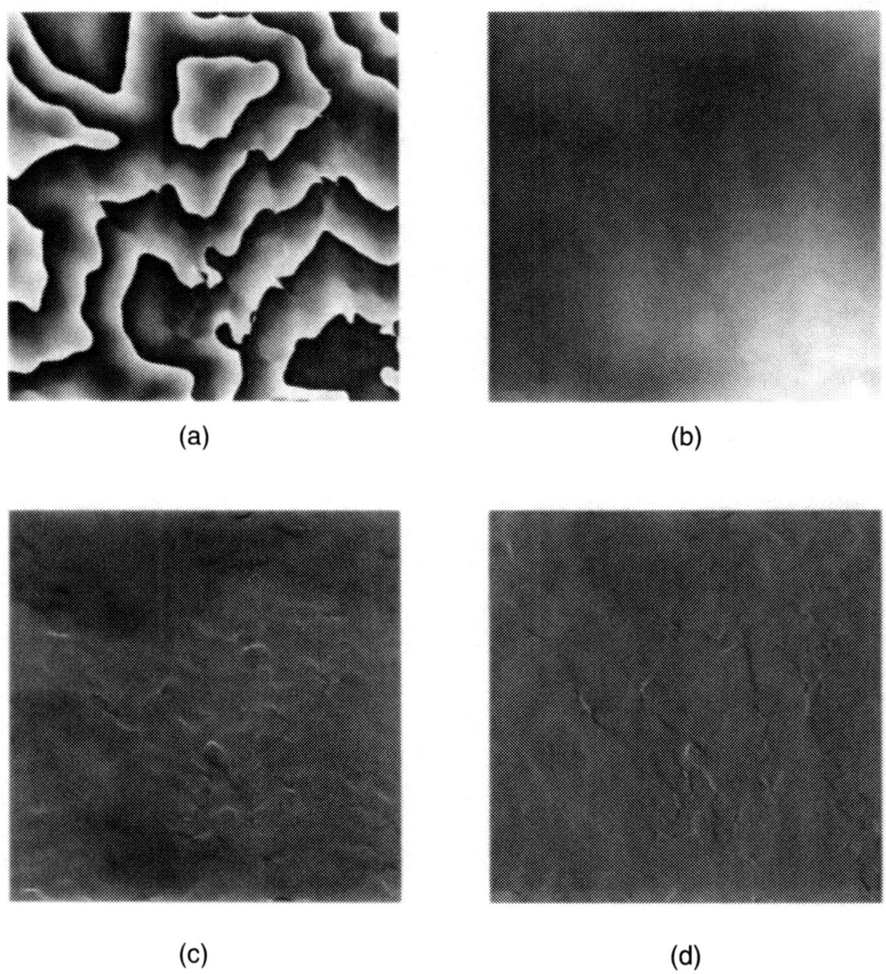

Figure 2.8 The wrapped phase that corresponds to the irrotational component of Figure 2.7(a). Note that all of the fringe lines remain connected in closed loops or begin and end at the array boundaries. This phase is uniquely unwrappable by any path-following algorithm. (b) The unwrapped phase of (a). (c) The *x*-gradient component φ_x of the irrotational phase. (d) The *y*-gradient component φ_y of the irrotational phase. Notice that the gradient components highlight some of the original phase topography of Figure 2.7(a).

phase, these components are unique. Collectively they capture *all* of the properties embodied in the original wrapped phase of Figure 2.7(a).

The irrotational component shown in Figure 2.8(a) is residue-free and can be uniquely unwrapped by any path-following algorithm. The unwrapped phase is

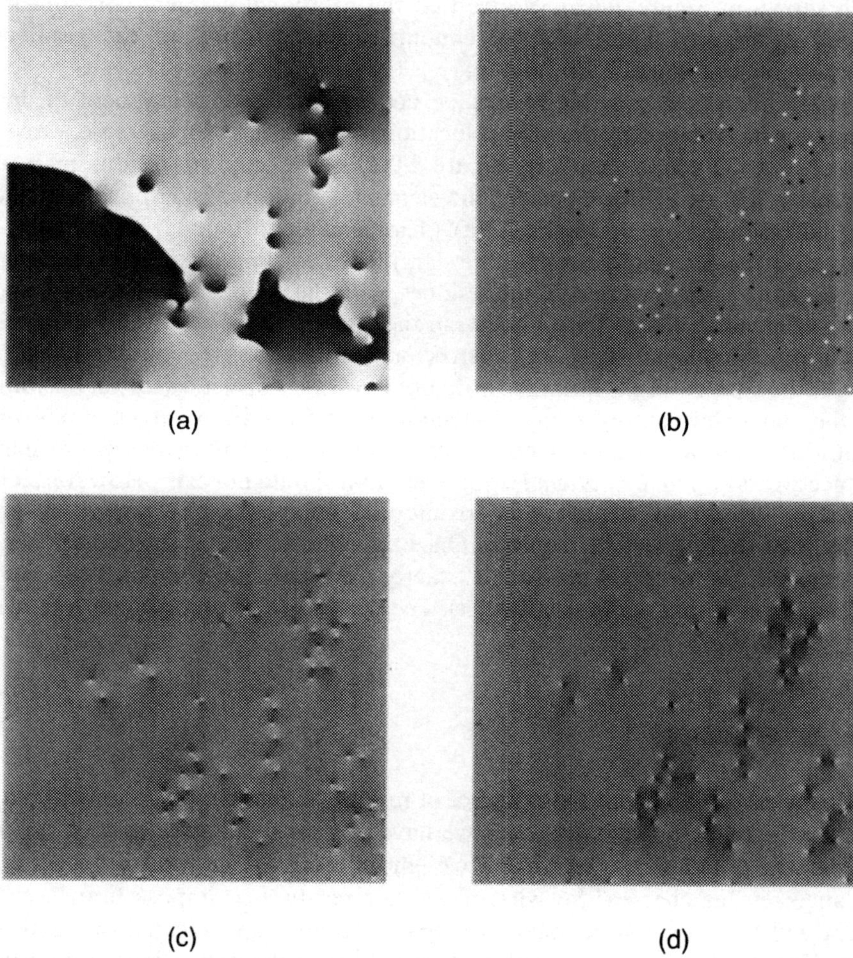

Figure 2.9 (a) The wrapped phase that corresponds to the rotational component of Figure 2.7(a). Note that all of the fringe lines are disconnected. This phase cannot be unwrapped uniquely. (b) The residue map shows the positive and negative residues as white and black squares, respectively. This residue map is identical to that shown in Figure 2.7(b), which indicates that the rotational phase component retains all the residues of the original wrapped phase. (c) The *x*-gradient component φ_x of the rotational phase. (d) The *y*-gradient component φ_y of the rotational phase. Notice that the rotational gradient components retain none of the original phase topography of Figure 2.7(a).

shown in Figure 2.8(b). For completeness, the x and y phase gradient components are shown in Figures 2.8(c) and (d). Notice that the gradients reflect the "topography" embodied in the phase

In contrast, the rotational component shown in Figure 2.9(a) contains all the residues of the original phase function of Figure 2.7(a). The residue map of the rotational component is shown in Figure 2.9(b) and is identical to the residue map shown in Figure 2.7(b). Notice that the rotational gradient components shown in Figures 2.9(c) and (d) contain a superposition of the gradient components of Figures 2.3(b) and (c).

Interestingly it is possible to reconstruct the rotational component of any wrapped phase from only the residue locations and charges. For example, we can reconstruct the phase shown in Figure 2.9(a) given only the residue map of Figures 2.9(b) or 2.7(b). Consider the elemental rotational component of the positive residue shown in Figure 2.3(a). (The elemental rotational component of a negative residue simply reverses the sign.) If we position elemental rotational components at the locations of the residues, with the proper signs imposed, add up the elemental phases for all points in the domain of interest,[8] and wrap the result into the range $(-\pi, \pi]$, we construct a rotational phase component that is *almost* the correct result. Because we do not know the correct *fractional* position of the elemental rotational components (i.e., the residue is at an unknown location within a 2×2-pixel box), our simple superposition procedure may induce an additional *irrotational* component that should not exist. However, it is a simple matter to remove this irrotational component by means of the Helmholtz decomposition theorem. Once the induced irrotational component is removed, the wrapped residual is exactly the rotational component desired (within an arbitrary additive constant). The results of this procedure are shown in Figure 2.10.

2.8 CONCLUSION

We have emphasized that the existence of residues causes the phase unwrapping process to be path dependent, but we have not yet demonstrated that this is necessarily a bad thing. So what if we ignore residues? So what if we do not balance residue charges? So what if we pair residues or impose branch cuts incorrectly? We do not yet have available to us any sophisticated two-dimensional phase unwrapping algorithms to help answer these questions, but do we really need them? The answer is yes!

The only algorithm for phase unwrapping we have described thus far is Itoh's method (see Section 1.3.1 in Chapter 1). Why not extend this algorithm to two dimensions by integrating the wrapped phase differences over a path that covers

[8] Superposition of elemental components is equivalent to summing the angles of the vectors from each positive residue to a point **P** in the domain of interest, then subtracting the sum of the angles of the vectors from the negative residues to the same point. The point **P** scans the domain, and the phases are computed for each sample.

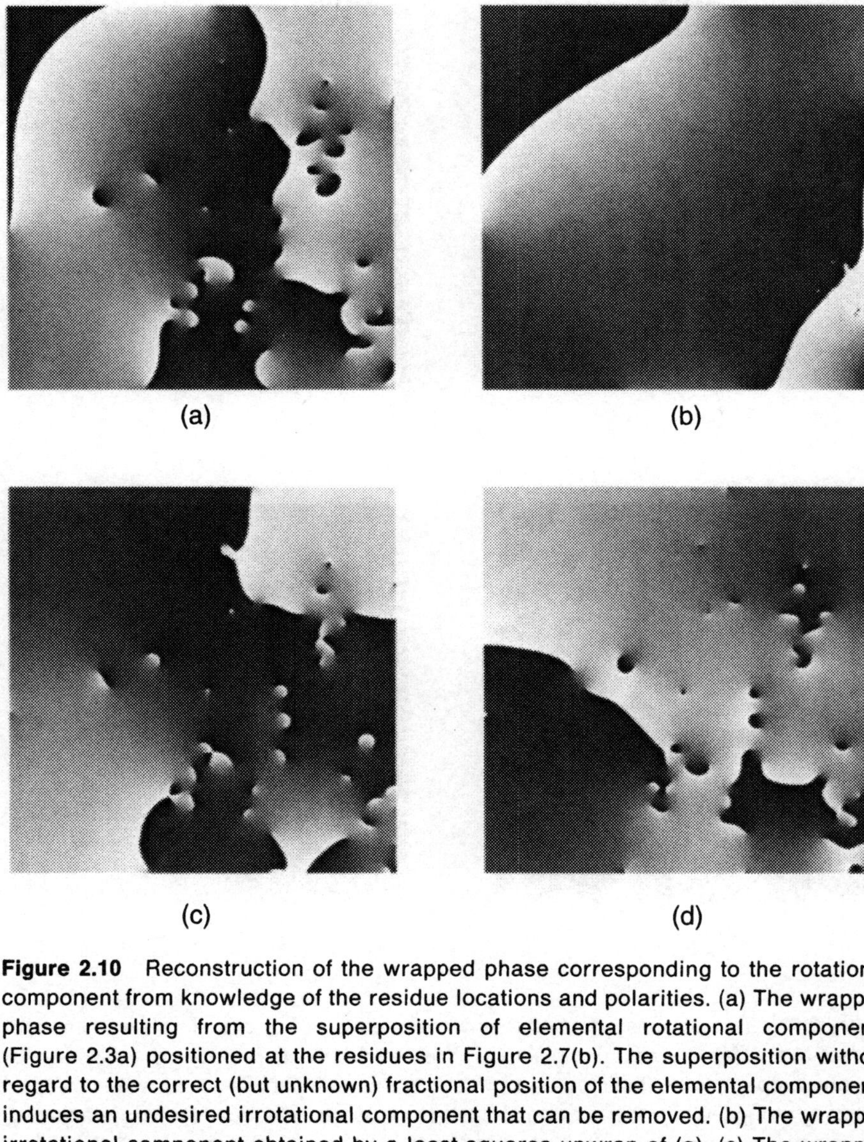

Figure 2.10 Reconstruction of the wrapped phase corresponding to the rotational component from knowledge of the residue locations and polarities. (a) The wrapped phase resulting from the superposition of elemental rotational components (Figure 2.3a) positioned at the residues in Figure 2.7(b). The superposition without regard to the correct (but unknown) fractional position of the elemental components induces an undesired irrotational component that can be removed. (b) The wrapped irrotational component obtained by a least-squares unwrap of (a). (c) The wrapped residual after subtracting (b) from (a). This result is the correct rotational component with the exception of an arbitrary additive constant. (d) The wrapped rotational component after adjusting the bias. This is identical to that shown in Figure 2.7(d).

the entire two-dimensional array and see what happens? Figure 2.11 illustrates the failure of this simple approach when residues are present. Figure 2.11(a), which is identical to Figure 2.8(a), shows a wrapped phase array of consistent data (i.e, no residues are present). Figure 2.11(b) shows the unwrapped phase

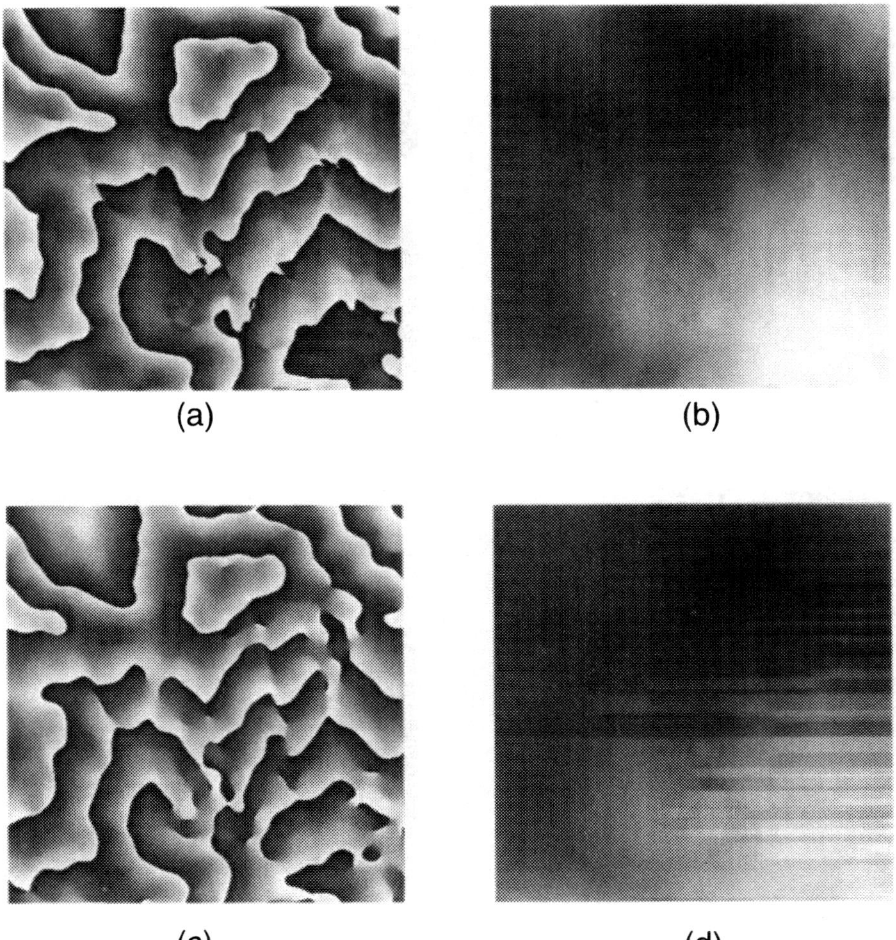

Figure 2.11 Two-dimensional phase unwrapping by means of a simple extension of Itoh's method. (a) The wrapped phase of consistent data. No residues are present. (b) The unwrapped phase depicted as an intensity map. The phase was first unwrapped vertically along the leftmost column, then the rows were individually unwrapped using the respective leftmost unwrapped phase as the initial value for that row. This simple unwrapping was successful. (c) Wrapped phase containing residues. (d) The simple unwrapping technique produced streaks emanating from the residues in the direction of the integration paths.

obtained by extending Itoh's one-dimensional method as follows: First, we unwrap the phase vertically along the leftmost column of the array. Next, we unwrap each row using the respective leftmost unwrapped phase as the initial value for that row. (The C language software for this simple phase unwrapping algorithm can be found in Section A.2 of Appendix A.) With data that are completely consistent, the unwrapping is successful. No unusual artifacts are visible in the unwrapped result. In addition the unwrapped phase has local x- and y-phase differences that are less than π radians in magnitude everywhere. This is exactly what is required. In fact Itoh's method yields exactly the same result as that shown in Figure 2.8(b).

Figure 2.11(c), which is the same as Figure 2.7(a), shows an array of wrapped phase that contains residues. The unwrapped result shown in Figure 2.11(d) shows streaks emanating from the locations of the residues. In addition the unwrapped phase no longer contains local phase differences that are less than π radians in magnitude everywhere. Some local phase differences have jumps of magnitude greater than π. These jumps are seen as the phase streaks. Because the wrapped phase "topography" gives no indication that large phase discontinuities should be present in the unwrapped phase, the result of Itoh's method is undesirable. Even if we used a different path to cover the array, streaks would emanate from the residues in the new direction traveled. It appears that more sophisticated unwrapping methods are required. But what conditions need to be imposed to ensure "good" results? That is the topic of the rest of this book.

We conclude this chapter with an important observation. The rotational gradient components shown in Figures 2.9(c) and (d) seem devoid of "topography." Indeed they retain *no* essence of the original wrapped phase gradients as the Helmholtz decomposition theorem demands. Therefore how can the residue charges be balanced in a manner that reflects the phase topography? The answer is that they cannot. Simply stated, methods for charge balancing, whether residue pairing, branch cuts, or something else, have no a priori rationale. The wrapped phase itself supports *no* unambiguous notion of how to place the unwrapping barriers.[9] In practice, one must inject a preferred path-following strategy or assumption based on the application at hand *and* some feel for the desired outcome. The true difficulty of two-dimensional phase unwrapping will become evident as we discuss the various strategies and assumptions of successful phase unwrapping algorithms.

REFERENCES

[1] E. Kreysig, *Advanced Engineering Mathematics*, Wiley, Inc., New York, 1966, ch. 6.

[2] C. R. Wylie, Jr., *Advanced Engineering Mathematics*, McGraw-Hill, New York, 1966, ch. 12.

[3] D. C. Ghiglia, G. A. Mastin, and L. A. Romero, "Cellular automata method for phase unwrapping," *Journal of the Optical Society of America*, Vol. 4, No. 1, pp. 267–280, January 1987.

[9] While some algorithms may attempt to obtain unwrapping-path guidance from the wrapped phase, there is no unique way to accomplish this. See Chapter 3.

[4] R. M. Goldstein, H. A. Zebker, and C. L. Werner, "Satellite radar interferometry: Two-dimensional phase unwrapping," *Radio Science*, Vol. 23, No. 4, pp. 713–720, July–August 1988.

[5] J. M. Huntley and J. R. Buckland, "Characterization of sources of 2π phase discontinuity in speckle interferograms," *Journal of the Optical Society of America A*, Vol. 12, No. 9, pp. 1990–1996, September 1995.

[6] J. M. Huntley, "Noise-immune phase unwrapping algorithm," *Applied Optics*, Vol. 28, No. 15, pp. 3268–3270, August 15, 1989.

[7] R. V. Churchill, *Complex Variables and Applications*, 2nd ed., McGraw-Hill, New York, 1960, ch. 7.

[8] E. Kreysig, *Advanced Engineering Mathematics*, Wiley, New York, 1966, pp. 700–701.

[9] G. B. Arfken and Hans J. Weber, *Mathematical Methods for Physicists*, 4th ed., Academic Press, San Diego, 1995.

[10] G. A. Korn and T. M. Korn, *Mathematical Handbook for Scientists and Engineers*, end ed., McGraw-Hill, New York, 1968, sec. 5.7.

[11] U. Spagnolini, "2-D phase unwrapping and phase aliasing," *Geophysics*, Vol. 58, No. 9, pp. 1324–1334, September 1993.

3

PHASE DATA, QUALITY MAPS, MASKS, AND FILTERS

3.1 INTRODUCTION

In this chapter we will present seven examples that illustrate the nature and challenges of phase unwrapping. These examples will be used in Chapters 4 and 5 to evaluate a number of phase unwrapping algorithms. In this chapter we will also examine the information (other than the phase derivatives and residues) that can be extracted from the phase[1] data to improve the phase unwrapping process. This information will be used to define "quality maps," which are arrays of values that indicate the quality or goodness of the phase data. Quality maps are required for some of the path-following algorithms to be presented in Chapter 4. We will also examine methods for extracting "masks" from the quality maps by means of thresholds. Masks are used for masking out or zero-weighting the low-quality phase values for some of the weighted L^p-norm algorithms to be presented in Chapter 5. They can also be used for some of the path-following algorithms of Chapter 4. We will conclude the chapter with a brief introduction to the noise filtering of phase data.

3.2 PHASE DATA EXAMPLES

In this section we will present seven examples of (wrapped) phase data that are representative of practical problems encountered in phase unwrapping. The examples include data from interferometric synthetic aperture radar (IFSAR) and magnetic resonance imaging (MRI), as well as simulated data for which the unwrapped surface is known. For each example we will show the wrapped phase, the residues, and, where available, the actual unwrapped surface. We will also

[1] We will use the term *phase* interchangeably with the term *wrapped phase* when there is no danger of confusion.

point out the features in each problem that present a challenge for phase unwrapping algorithms.

3.2.1 Synthetic Aperture Radar Interferometry

The first example consists of interferometric synthetic aperture radar (IFSAR) data. IFSAR is one of the most widely used applications of phase unwrapping. In Chapter 1 we presented a detailed overview of IFSAR. In this section we will review the basic concepts briefly and explain why the terrain elevations are proportional to the phase, drawing on References [1], [2], and [3].

The purpose of IFSAR is to generate a three-dimensional model in the form of digital elevation data of the imaged terrain. A SAR sensor generates a two-dimensional projection of the radar reflectivity of the three-dimensional terrain, much as an optical camera generates a two-dimensional projection of the optical reflectivity. The SAR sensor is typically carried aboard an airborne platform such as the AIRSAR system [4] or a spaceborne platform such as the ERS-1 satellite, the Space Shuttle Imaging Radar, or the *Magellan* spacecraft to Venus [4]. Unlike optical imaging, radar imaging generates a complex-valued function. Each pixel of the image represents the magnitude and the phase of the radar signal reflected from a particular scene point. The two-dimensional projection of the scene is obtained by measuring the range and an angle of each scene point relative to the SAR sensor. By acquiring a second SAR image from a sensor that has been displaced in elevation from the first one, and by co-registering the two images into spatial alignment with one another, we can estimate a second angle of each scene point. This allows us to compute the location of each scene point in three-dimensional space. The second SAR image can be obtained by means of two SAR antennas on the same platform as in the TOPSAR system [5] or by means of two separate passes of the same sensor as in the Shuttle Imaging Radar [6].

The second angle that is obtained from the pair of SAR images is labeled θ in Figure 3.1. It is derived from the path difference Δr as follows. The geometry of the sensors in the figure has been simplified for clarity, but the results generalize to any orientation of the sensors. When the sensors are far from the scene compared to the distance between the sensors, which is typically the case in an IFSAR system, the paths from the sensors to the scene are approximately parallel. Elementary geometry then shows that the path difference Δr is related to the angle θ by the relation

$$\Delta r \approx B \cdot \sin\theta = \left(\frac{B}{D}\right)h, \tag{3.1}$$

where B is the "baseline" distance between the two sensors, D is the path length to the scene point, and h is the terrain height of the scene point. It can be shown that path difference is proportional to the phase difference φ between the two SAR images: $\varphi = k \cdot \Delta r$, where k is the wavenumber $2\pi/\lambda$ and λ is the radar

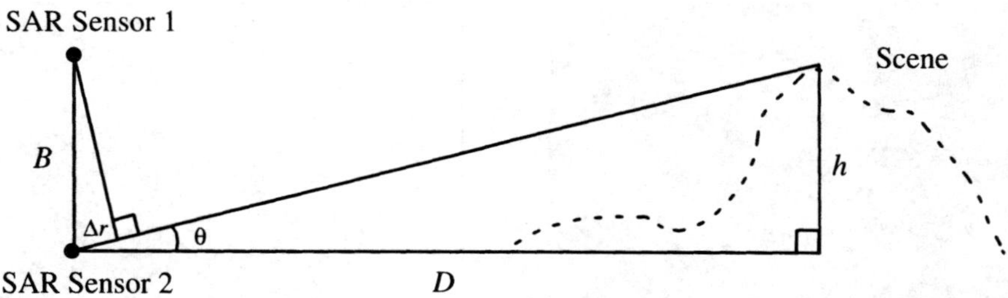

Figure 3.1 The geometry of SAR interferometry. Under the assumption that the distance D to the scene is large compared to the baseline distance B, it can be shown that the path difference Δr is proportional to the terrain elevation h.

wavelength. The phase difference is thus proportional to the terrain height:

$$\varphi \approx \left(\frac{2\pi B}{\lambda D}\right) h. \tag{3.2}$$

Because the phase difference φ is only known modulo 2π, it must be unwrapped to yield the actual phase, from which the terrain heights can be calculated by Equation 3.2 to yield the terrain elevation model.

Figure 3.2(a) shows a 512 × 512-pixel example of wrapped phase data from an IFSAR experiment.[2] The phase fringes, which are indicative of terrain height, are analogous to a contour map. There are patches of noise where the phase fringes are broken up and the terrain height cannot be determined. Within these patches the phase of the two SAR images were not correlated. Such patches of decorrelation can be caused by any one of a number of effects, including radar shadow and layover (which will be explained in Section 3.2.3), multi-path effects, temporal changes that occurred between the times at which the two images were acquired, and image misregistration. In this case the patches were caused by radar shadow. One could attempt to unwrap the phase data in these decorrelated regions, but the result would be meaningless, since one essentially would be unwrapping noise. A phase unwrapping algorithm must be able to unwrap the valid phase data around and between the decorrelated patches without being corrupted by the patches. The resulting terrain elevation model will then be correct between the patches but undefined within them.

Figure 3.2(b) shows the residues of the phase data of Figure 3.2(a). Most of the 5877 residues are concentrated within the decorrelated patches. Figures 3.2(c) and (d) show enlarged views of the phase data and the residues within the marked squares.

3.2.2 Magnetic Resonance Imaging

The water/fat separation problem [7]–[10], which arises in the field of magnetic resonance imaging (MRI), is another application where the need for phase

[2] Many of the images, including this one, have been reduced to sizes appropriate for the figures.

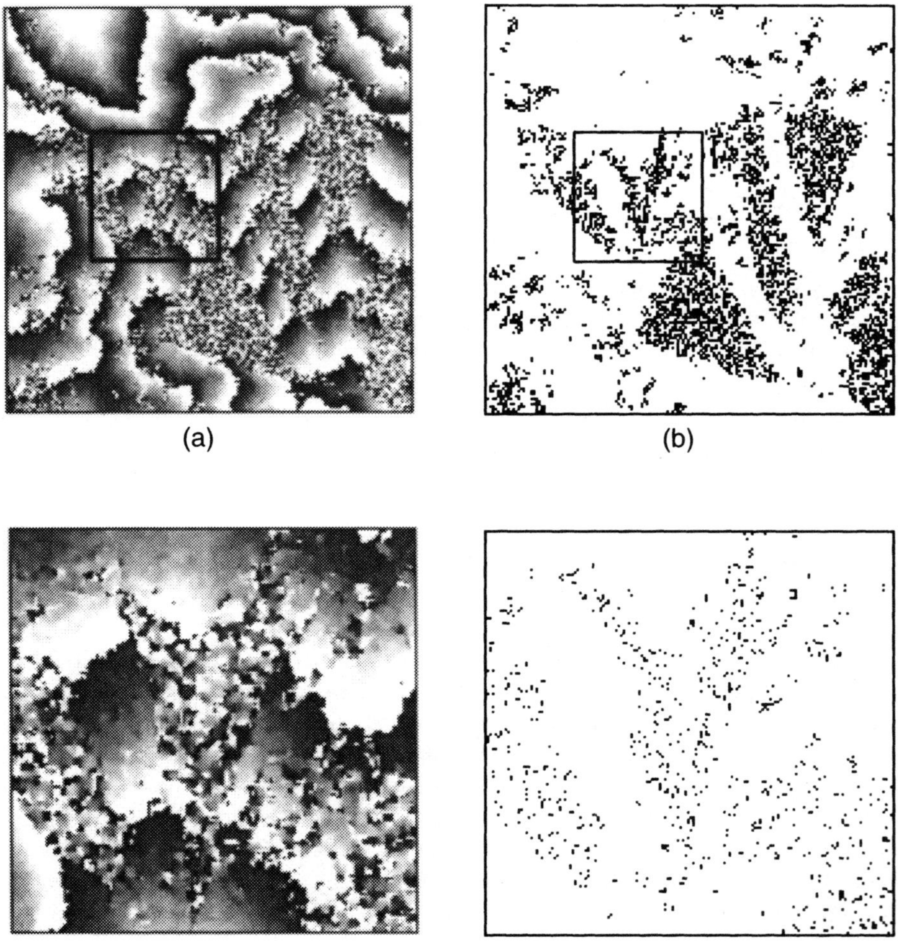

(a) (b) (c) (d)

Figure 3.2 The IFSAR data. (a) The wrapped phase differences. The phase "fringes," like a contour map, are indicative of terrain elevation. The patches where the fringes are noisy and broken up are regions of undefined phase. (b) The residues are concentrated in the noisy patches. (c) An enlarged view of the phase data within the rectangle. (d) The residues within the rectangle.

unwrapping arises. The problem can be understood as follows: A signal of the form

$$s = [W + F\exp(j\Phi)]\exp[j(\theta_0 + \varphi)] \tag{3.3}$$

is obtained by means of conventional MRI procedures. The terms W and F are the water and fat signal intensities, respectively. The term Φ is a water/fat chemical phase shift, θ_0 is a constant phase value for a given pixel, and φ is a phase shift that is due to inhomogeneities within the magnetic field. The parameter Φ can be manipulated by the MRI operator to obtain the following two signals:

$$s_0 = (W + F)\exp(j\theta_0), \tag{3.4}$$

$$s_1 = (W - F)\exp[j(\theta_0 + \varphi)]. \tag{3.5}$$

It follows that

$$\frac{\overline{s_0} s_1}{s_0 \overline{s_1}} = \exp(j2\varphi), \tag{3.6}$$

where $\overline{s_0}$ and $\overline{s_1}$ are the complex conjugates of s_0 and s_1, respectively.

Extracting the phase from Equation 3.6 by means of the four-quadrant arc tangent operator (see Equation 1.13) yields the wrapped values of the phase 2φ. These wrapped values must be unwrapped and divided by 2 to yield an estimate ϕ for the phase shift φ. The water and fat images can then be computed by the formulas

$$I_0 = \frac{|s_0 + s_1 \exp(-j\phi)|}{2}, \tag{3.7}$$

$$I_1 = \frac{|s_0 - s_1 \exp(-j\phi)|}{2}. \tag{3.8}$$

Phase unwrapping is therefore necessary to solve this water/fat separation problem.

Figure 3.3(a) shows the phase data for an MRI water/fat separation problem. [11]. The data are defined on a 256 × 256-pixel grid. Figure 3.3(b) shows the 1926 residues, which are concentrated where the phase is noisy. This problem is difficult to unwrap correctly because various regions inside the head appear to be completely isolated from one another by the noisy phase.

Figure 3.4(a) shows a second MRI water/fat separation problem [11], in this case involving a knee. Like the preceding MRI example, the data are defined on a 256 × 256-pixel grid. The 11,240 residues are shown in Figure 3.4(b). Most of these numerous residues lie in the background, which consists entirely of noise. The challenge for a phase unwrapping algorithm is to unwrap the phase data within the knee correctly and without corruption from the background noise.

3.2.3 Simulated IFSAR

While the preceding IFSAR and MRI problems are derived from actual data, they share the disadvantage that the correct unwrapped phase is unknown.

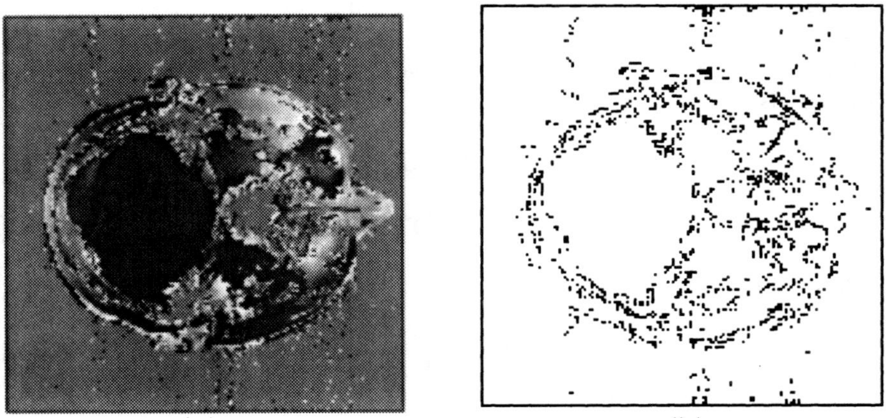

(a) (b)

Figure 3.3 The data for a water/fat separation problem from an MRI experiment. (a) The wrapped phase data. These data are defined by 2φ in Equation 3.6. (b) The residues.

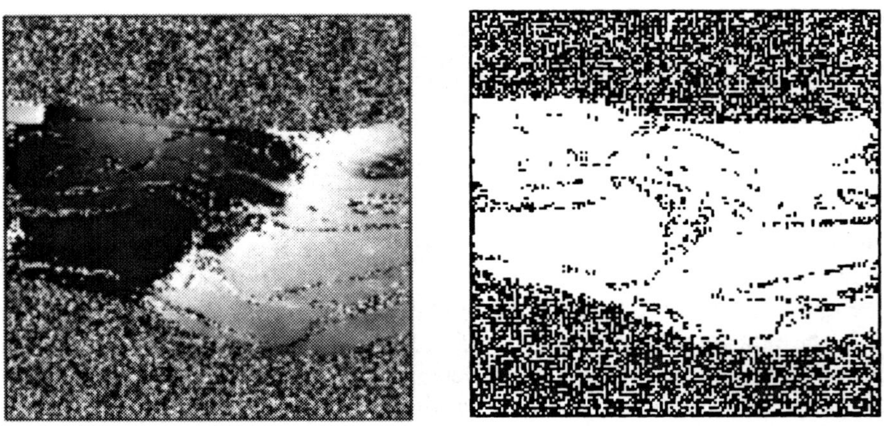

(a) (b)

Figure 3.4 The MRI water/fat separation problem on an experiment involving a knee. (a) The phase data. These data are defined by 2φ in Equation 3.6. (b) The residues. Notice the excessive number of residues induced by the background noise.

When evaluating phase unwrapping algorithms, it is essential to have problems for which the true unwrapped surface is known. This is necessary so that the errors and the problematic regions in the unwrapped phase may be ascertained and so that the accuracy of the algorithm may be quantified. While simulated data sets are not sufficient by themselves to provide an exhaustive evaluation of an algorithm—nothing can take the place of actual data—they have the overwhelming advantage of providing a priori knowledge of the "ground truth."

Table 3.1 Parameter values for the IFSAR simulator (taken from the TOPSAT system).

Parameter	Value
Altitude	600 km
Incidence angle	30 deg
Baseline	1 km
Baseline tilt	30 deg
Wavelength	24 cm
Range resolution	4 m
Azimuth resolution	4.5 m

The two IFSAR phase unwrapping problems in this section were supplied by Michael W. Roth of the Johns Hopkins University Applied Physics Laboratory using his high-fidelity IFSAR simulator [12]. We will describe his simulator and then present the two IFSAR examples.

Roth's simulation follows the work of Rodriguez and Martin [13]. Roth begins with digital terrain elevation models sampled at 30×30-meter intervals, which were obtained from the U.S. Geological Survey. He uses a diffuse scattering model in which the backscatter at each pixel is generated by the coherent addition of many point scatterers with random phases and amplitudes. This is a reasonable model because the scatterers are assumed to be small compared to the image resolution but large compared to the radar wavelength. (See Table 3.1 for the simulation parameters.) The return from each individual scatterer has circular Gaussian statistics. The coherent addition generates the

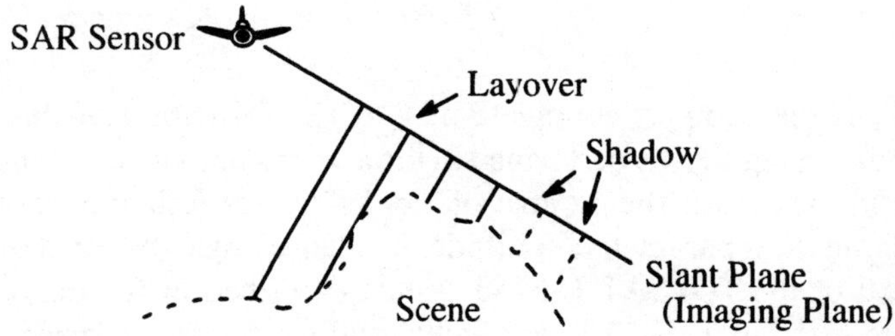

Figure 3.5 The SAR phenomena of radar shadow and layover. Because the radar wavefronts are approximately planar when the sensor is far from the scene, the scene points are projected in parallel lines to the imaging plane. The portions of the scene that are not illuminated by the SAR sensor due to occlusion appear as dark areas called shadows in the SAR image. They are analogous to shadows in optical imagery. Equidistant portions of the scene cause multiple, simultaneous radar returns that create brightly illuminated regions of "layover" in the SAR image.

SAR phenomenon of *speckle* noise, and no additional noise is added. In addition to speckle noise, the simulation accurately models the SAR phenomena of *shadow* and *layover*. Shadows are due to the absence of radar returns from portions of the terrain that are occluded by other portions and thus not illuminated by the transmitted radar signal. Layover is due to the simultaneous return of radar signals from multiple portions of the terrain that are at the same distance from the transmitter. Both of these phenomena are illustrated in Figure 3.5. Examples of radar shadow and layover in a SAR image can be seen in Figure 1.7.

Roth's simulator generates two complex-valued SAR images. The images are formed as follows. The radar return at the image pixel (m, n) is defined to be the coherent sum

$$u_{m,n} = \sum_p A_p \exp(jkr_p) \operatorname{sinc}\left[\frac{r_p - r_m}{R}\right] \operatorname{sinc}\left[\frac{y_p - y_n}{Y}\right], \tag{3.9}$$

where A_p is the complex random amplitude of the scatterer p, and k is the wavenumber $2\pi/\text{wavelength}$. The term r_p is the range from the SAR sensor to the scatterer p, r_m is the offset range of the pixel (m, n) in the SAR image, and R is the range resolution. The terms y_p, y_n, and Y are the analogous quantities in the azimuth, rather than the range, direction. The sinc function is defined by $sinc(x) = \sin \pi x/\pi x$.

The two SAR images $u_{m,n}$ and $v_{m,n}$ are formed at a given baseline distance and "tilt" of the baseline vector (see Table 3.1) and then co-registered into spatial alignment. The complex-valued IFSAR image $z_{m,n}$ is then defined by the normalized sum

$$z_{m,n} = \frac{\sum u_{i,j}\overline{v_{i,j}}}{\sqrt{\sum |u_{i,j}|^2 \sum |v_{i,j}|^2}}, \tag{3.10}$$

where $\overline{v_{i,j}}$ is the complex conjugate of $v_{i,j}$. The formation of this sum, called "multilook averaging," is performed in the $k \times k$ neighborhood centered at each pixel (m, n), where k is the number of "looks" in azimuth and elevation.

The simulation parameters (altitude, incidence angle, baseline distance, etc.) are typical of the TOPSAT L-band system described in Reference [13]. These values are listed in Table 3.1. The individual scatterers are spaced two meters apart. The output resolution of the images is 3.75×7.5 meters in range and azimuth, respectively. The IFSAR image is generated with $k = 4$ looks in range and four in azimuth.

The wrapped phase values at each pixel (m, n) are extracted from Equation 3.10 by the four-quadrant arc tangent operator (see Equation 1.13). The magnitudes of the complex values $z_{m,n}$ of Equation 3.10 are called the *correlation coefficients* and will be discussed in Section 3.3.1.

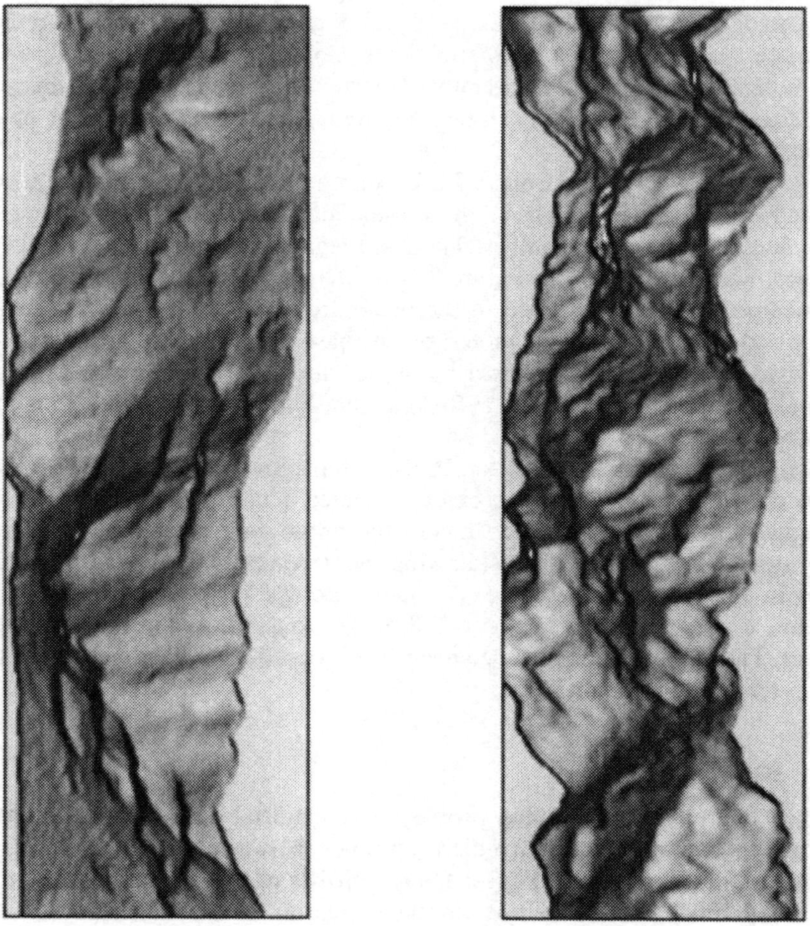
(a) (b)

Figure 3.6 Bird's-eye shaded-surface views of the terrain for the two simulated IFSAR problems. These surfaces were derived from digital terrain elevation models from the U.S. Geological Survey. Both surfaces have been resampled in the SAR slant plane so that they can be compared directly with the surfaces obtained by unwrapping the phase data. The surfaces are flat and undefined near the borders due to the projection of the high terrain relief into the slant plane. (a) Mountainous terrain around Long's Peak, Colorado. (b) Mountainous terrain around Isolation Peak, Colorado.

We will now present the two data sets that were generated by this simulator. Both examples were simulated from high-relief terrain models, and they illustrate the problems posed by SAR layover in mountainous terrain. Figure 3.6(a) shows a bird's-eye (i.e., from a viewpoint directly overhead) shaded-surface rendering of a mountainous area around Long's Peak, Colorado.[3] The terrain elevation data are defined on a 152 × 458 grid. Figure 3.6(b) shows a similar view of a mountainous area around Isolation Peak, Colorado, defined on a 157 × 458 grid. Both surfaces have been resampled in the "slant plane" or SAR imaging plane (see Figure 3.5), so they can be compared directly with the IFSAR phase and correlation data.

The phase data for the Long's Peak example are shown in Figure 3.7(a). Notice the decorrelated regions on the left sides of the mountain peaks where the phase fringes come close together. These are regions of SAR layover. The 776 residues, which are shown in Figure 3.7(b), are concentrated in these regions. The phase data are not defined in the black areas around the borders of the figures and should be masked out before the phase is unwrapped. These areas of undefined phase, which are caused by the terrain relief being projected in the slant plane, are artifacts of the IFSAR simulation. Actual data also may be undefined near the borders.

The phase data for the Isolation Peak example are shown in Figure 3.8(a). These data have many regions of decorrelated phase due to layover. The challenge on this problem is to unwrap the phase data around and between these numerous regions without allowing the decorrelated phase to introduce errors in the unwrapped surface. As in the Long's Peak example, the 1157 residues, which are shown in Figure 3.8(b), are concentrated in the regions of layover. The phase is also undefined near the borders due to the projection of the terrain relief into the slant plane.

3.2.4 Shear

The last two phase unwrapping problems are artificial data sets that contain "shear," or boundaries of sharp discontinuities, between two planar surfaces. The first data set, shown in Figure 3.9(a), consists of two planar surfaces that have been tilted relative to one another, creating a line of shear between them [14]. The line of shear is clearly discernible in the phase data, which are shown in Figure 3.9(b). The twelve residues, which are shown in Figure 3.9(c), are all positive and lie on the line of shear. The phase data corresponding to the two planes must be unwrapped independently without crossing the line of shear.

The last phase unwrapping example is defined by the "spiral shear" surface shown in Figure 3.10(a). As in the previous example, two planar surfaces have been tilted relative to one another, but in this case the boundary between them consists of intertwined spirals. Figure 3.10(b) shows the wrapped phase data. Notice the arms of the spirals twisting around one another, with the boundaries of shear between the arms. The 48 residues, shown in Figure 3.10(c), all lie on

[3] This type of rendering, which essentially shows the surface gradients as gray-scale values, is effective at portraying the local surface structure. Another common method of rendering shows the the surface elevations rather than the gradients as gray-scale values. We will use the latter method in Chapter 5. We have supplied C language software for producing these renderings in Section A.5 of Appendix A.

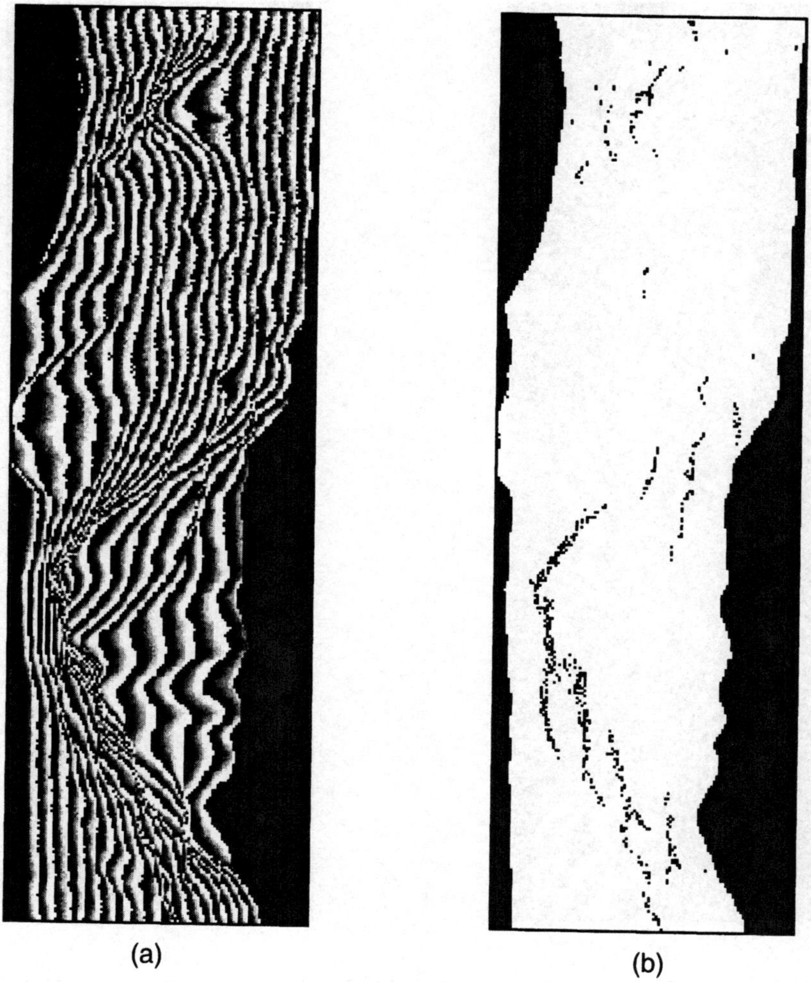

Figure 3.7 (a) The phase of the simulated IFSAR data that were generated from the Long's Peak digital terrain elevation model of Figure 3.6(a). Notice the regions to the left of the mountain peaks where the phase fringes come together and become decorrelated. These regions were caused by SAR layover. The phase data in the black regions near the borders were not defined due to the projection of the terrain relief into the slant plane. (b) The residues are concentrated in the decorrelated regions.

these curved lines of shear. The problem here is to unwrap the arms of the spirals independently without crossing the shear boundaries.

In addition to these seven examples, we will consider variations of them in later chapters. In particular, we will consider the sheared planes example corrupted with Gaussian noise.

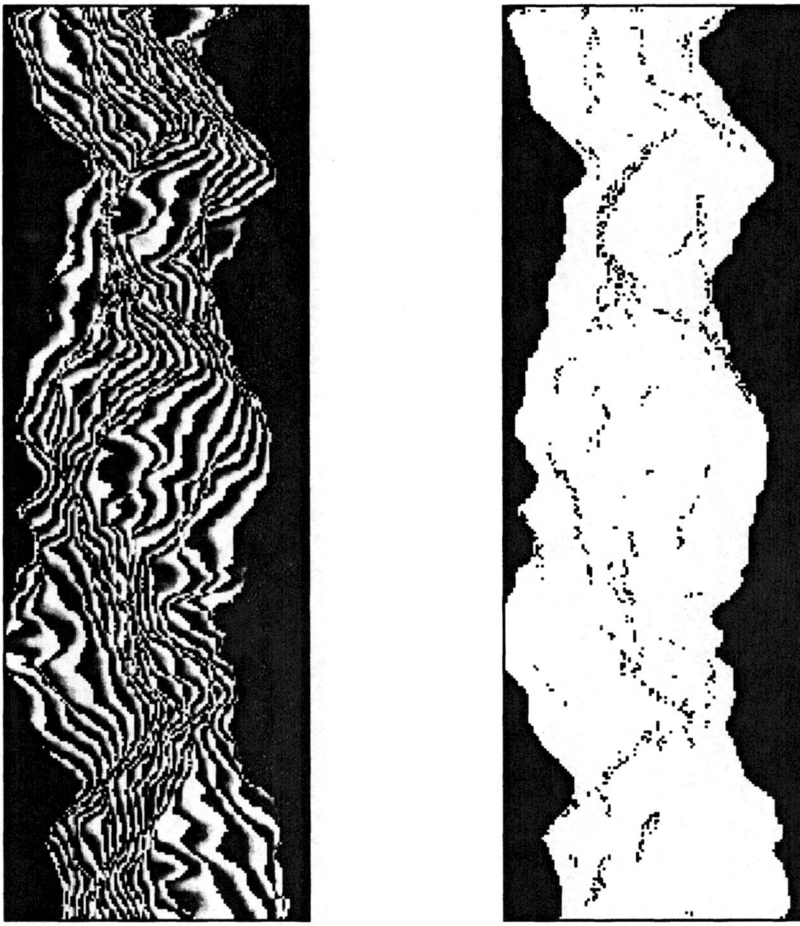

(a) (b)

Figure 3.8 (a) The phase of the Isolation Peak IFSAR problem of Figure 3.6(b). Notice the numerous regions where the phase fringes come together and become decorrelated due to SAR layover. The phase data in the black regions near the borders are not defined due to the projection of the terrain relief into the slant plane. (b) The residues are concentrated in the decorrelated regions.

3.3 QUALITY MAPS

In this section we define several quality maps that are useful in phase unwrapping. *Quality maps* are arrays of values that define the quality or goodness of each pixel of the given phase data. They are necessary for guiding several of the path-following algorithms that will be presented in Chapter 4. They are also necessary for some of the weighted L^p-norm algorithms that will be presented in Chapter 5.

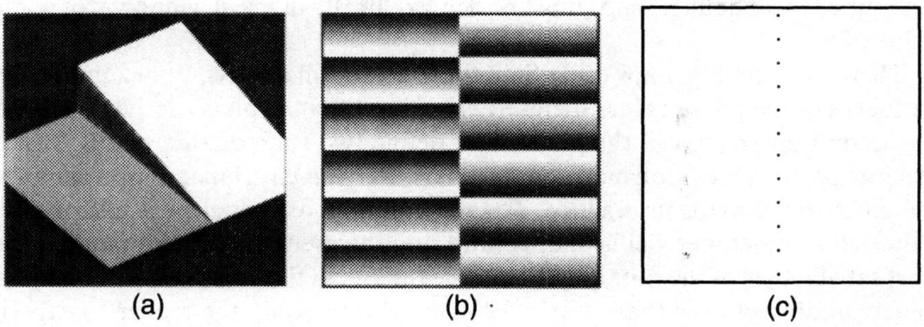

Figure 3.9 (a) A three-dimensional view of the sheared planes. (b) The wrapped phase. (c) The twelve residues are all positive and lie on the line of shear.

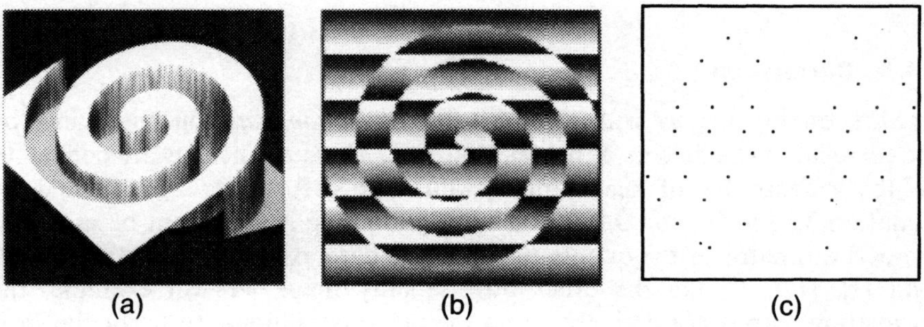

Figure 3.10 (a) A three-dimensional view of the "spiral shear" surface. (b) The wrapped phase. (c) The residues lie on the curved lines of shear.

The first quality map that we will discuss is the *correlation map*, which is defined by the correlation coefficients that are extracted from IFSAR data. This map is only available from IFSAR data and not, for example, from MRI data. Where the correlation map is not available, there are other quality maps that can be derived from the phase data alone. The first such map is the *pseudocorrelation map*, which is identical to the correlation map except that the (unknown) SAR pixel magnitudes are assumed to be unity. The pseudocorrelation map may be derived from any phase data, including MRI data. In practice, it is not a very good estimator of phase quality. A much better estimator is the *phase derivative variance*, which is defined to be the local sample variance of the partial derivatives (wrapped differences) of the phase data. Strictly speaking, the phase derivative variance indicates the "badness" rather than the goodness of the phase data, but we can assume that the variance has been negated so that it represents goodness. A fourth quality map is the *maximum phase gradient*, which is again an indicator of "badness" because the gradients in regions of noisy phase tend to be large. We will assume that the maximum phase gradient, like the phase derivative variance, is always negated so that this measure defines a quality map.

All four of these quality maps must be defined locally in $n \times n$ windows for some value of n.

There exist quality maps other than the four we will discuss. For example, the variances of the phase values themselves (rather than the phase derivatives) and the second differences of the phase data define two such quality maps. In our experience the correlation map and the phase derivative variance map seem to be the most useful maps in practice. The maximum phase gradient is also useful, although it sometimes yields poor results, and the pseudocorrelation map is the least satisfactory of the four quality maps. Our aim in this section is to convey an understanding of how these maps "behave" and to equip the reader to experiment with these and other quality maps on a particular phase unwrapping problem. The C language software for generating these quality maps can be found in Section A.3 of Appendix A.

3.3.1 Correlation

The first quality map we will examine is the *correlation map*, which is defined by the correlation coefficients of the IFSAR data. These coefficients are defined to be the magnitudes of the complex-valued IFSAR pixels $z_{m,n}$ defined by Equation 3.10 in Section 3.2.3. The correlation map is important because it is the best estimator of the quality of the phase data extracted from the IFSAR data [1], [12]. Unlike the other three quality maps we will examine, the correlation map is not derivable from the phase data alone. In particular, it is not available for MRI data.

Notice that the IFSAR pixels $z_{m,n}$ of Equation 3.10 are computed in $k \times k$ windows. The correlation coefficients may be averaged further. Figure 3.11(a) shows the correlation map (defined in 4×4 windows) for the Long's Peak simulated IFSAR data of Figure 3.7. The region marked by the rectangle is enlarged in the small images of Figures 3.11(b)–(d). The phase data are shown in Figure 3.11(b), and the correlation coefficients are shown in Figure 3.11(c). Because the correlation coefficients are "grainy," they were averaged in 3×3 windows to produce the quality map shown in Figure 3.11(d). When used to guide a phase unwrapping algorithm, the averaged coefficients tend to provide better results than the original correlation coefficients.

Comparison of the correlation map of Figure 3.11(a) with the phase data of Figure 3.7(a) shows that the decorrelated (noisy) regions of the phase data correspond to the lowest-quality (darkest) regions of the correlation map. These low-quality regions were caused by SAR layover. The correlation map is a good indicator of these decorrelated regions. The reader also will notice that the upper part of the correlation map is darker than the lower part. These dark pixels are indicative of low-correlation coefficients that were induced by a slight misregistration of the images during the IFSAR processing procedure. We left these errors in the data because they are typical of the errors encountered in actual IFSAR data.

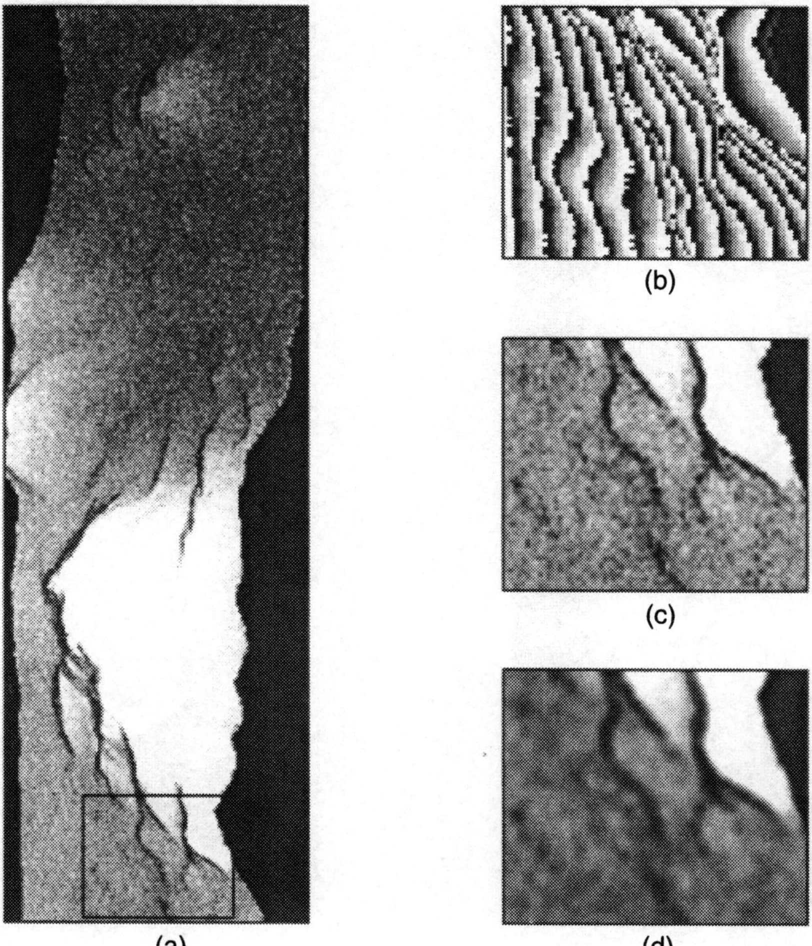

Figure 3.11 (a) The correlation map for the Long's Peak simulated IFSAR problem of Figure 3.7. The low correlation coefficients are shown as dark pixels and the high correlation coefficients as light pixels. The low-correlation regions along the mountain peaks correspond to the regions of decorrelated phase, which were induced by SAR layover. (b) A close-up view of the phase data in the region enclosed by the rectangle. (c) The correlation map within the rectangle. (d) The correlation map averaged in 3×3 windows within the rectangle.

Figure 3.12(a) shows the correlation map for the Isolation Peak example of Figure 3.8. Figures 3.12(b) and (c) show close-up views of the phase data and the correlation map in the region enclosed by the rectangle. Notice once again that the decorrelated regions of the phase data correspond to the low-quality regions of the correlation map. These regions were induced by SAR layover. The 3×3-averaged correlation map appears in Figure 3.12(d).

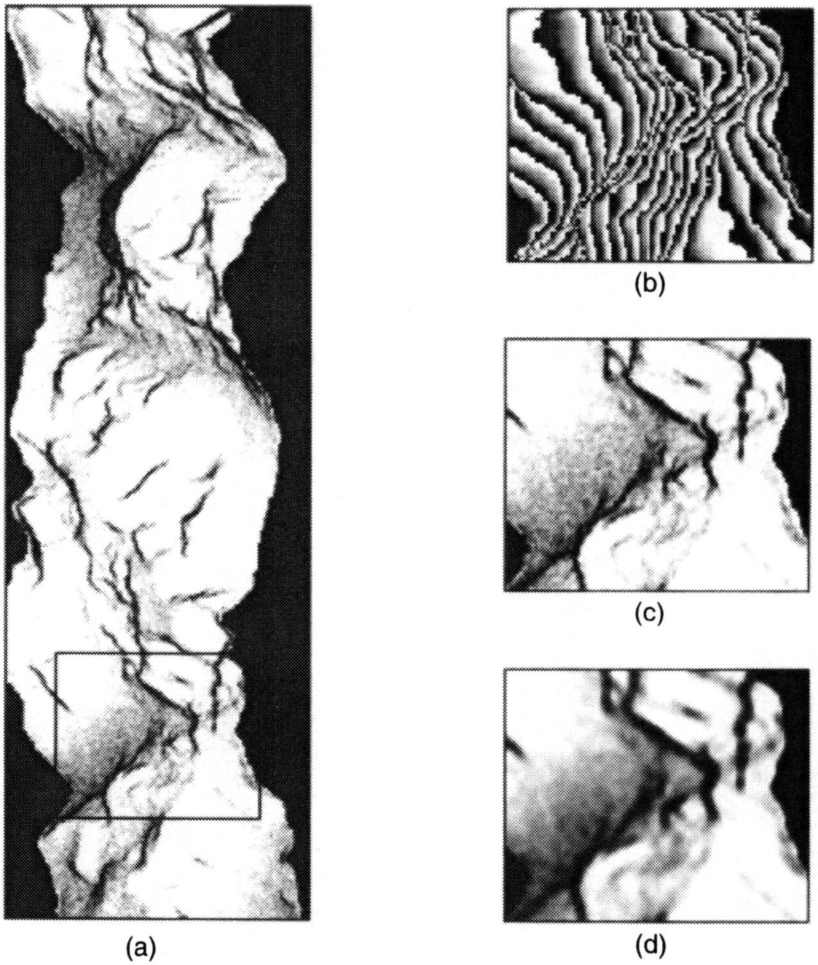

Figure 3.12 (a) The correlation map for the Isolation Peak simulated IFSAR problem of Figures 3.6(b) and 3.8(a). As in Figure 3.11 the low-correlation coefficients are shown as dark pixels and the high-correlation coefficients by light pixels. The low-correlation regions correspond to the regions of decorrelated phase, which are induced by SAR layover. (b) A close-up view of the phase data in the region enclosed by the rectangle. (c) The correlation data within the rectangle. (d) The correlation data averaged in 3 × 3 windows.

3.3.2 Pseudocorrelation

The pseudocorrelation map is designed to mimic the correlation map when the magnitudes of the complex-valued SAR images $u_{m,n}$ and $v_{m,n}$ are unknown. If these magnitudes are assumed to be unity, then the magnitudes of the complex-valued IFSAR image defined by Equation 3.10 become

Figure 3.13 The pseudocorrelation map for the IFSAR phase data of Figure 3.2. The dark, low-correlation regions correspond to the patches of decorrelated (noisy) phase.

$$|z_{m,n}| = \frac{\sqrt{\left(\sum \cos \psi_{i,j}\right)^2 + \left(\sum \sin \psi_{i,j}\right)^2}}{k^2}, \tag{3.11}$$

where the sums are evaluated in the $k \times k$ neighborhood of each pixel (m, n) for some fixed number of looks k. The terms $\psi_{i,j}$ are the phase values extracted by the arc tangent operator from the pixelwise SAR image product $u_{i,j}\overline{v_{i,j}}$. Equation 3.11 defines the values of the pseudocorrelation map.

In all of the examples of this chapter, we compute the pseudocorrelation for 3×3 looks. The pseudocorrelation map of the IFSAR example of Figure 3.2 is shown in Figure 3.13. The low-correlation patches are indicators of the decorrelated patches of phase. The pseudocorrelation maps of the MRI examples of Figures 3.3 and 3.4 are shown in Figures 3.14(a) and (b), respectively. Notice once again that the low-correlation regions in the maps indicate where the noisy regions of phase lie.

The pseudocorrelation maps of the simulated IFSAR examples are shown in Figures 3.15(a) and (b). Compare these maps with the correlation maps of Figures 3.11(a) and 3.12(a), respectively. The maps are quite different. For example, the pseudocorrelation maps mark the steep terrain as low-quality regions, even though the phase data are well-defined and not noisy in these regions. This is due to the fact that the high terrain slopes are regions of high-phase variation, and thus the phase data are not well correlated with one another. This suggests that a superior quality map would result if the phase derivatives rather than the phase values were used to define the quality map. Such a quality map will be examined in the next section.

The pseudocorrelation maps of the sheared planes and spiral shear examples of Figures 3.9 and 3.10 are shown in Figures 3.16(a) and (b). The dashed lines of low-quality values mark the boundaries of shear between the two planes in both

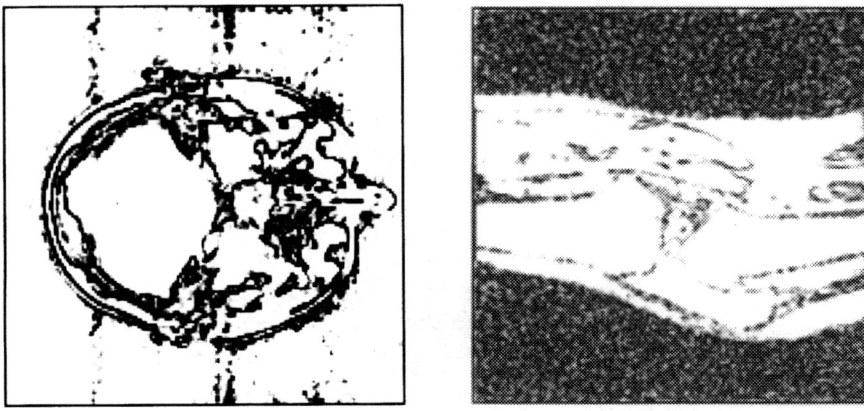

(a) (b)

Figure 3.14 The pseudocorrelation maps for the MRI (a) head and (b) knee examples of Figures 3.3 and 3.4. The dark pixels indicate the regions of decorrelated, highly varying phase. For example, the background noise of the knee data is decorrelated and therefore has low pseudocorrelation values.

examples. Solid rather than dashed lines are to be desired to completely isolate the two planes. However, examination of the phase data shows that they are correlated periodically along the shear line. This periodic correlation accounts for the dashed lines. The pseudocorrelation map fails to isolate the sheared planes (or the arms of the spiral) from one another.

1.3.3 Phase Derivative Variance

As we observed in Section 3.3.2, a disadvantage of the pseudocorrelation map is that it marks highly varying phase (due to steep terrain in IFSAR images, for example) as regions of low quality. This motivates the definition of a new quality map that measures the statistical variance of the phase derivatives [15]. The *phase derivative variance* is defined by the equation

$$z_{m,n} = \frac{\sqrt{\sum\left(\Delta^x{}_{i,j} - \overline{\Delta^x}_{m,n}\right)^2} + \sqrt{\sum\left(\Delta^y{}_{i,j} - \overline{\Delta^y}_{m,n}\right)^2}}{k^2}, \tag{3.12}$$

where for each sum the indexes (i, j) range over the $k \times k$ window centered at the pixel (m, n). The terms $\Delta^x{}_{i,j}$ and $\Delta^y{}_{i,j}$ are the partial derivatives of the phase (i.e., the wrapped phase differences). The terms $\overline{\Delta^x}_{m,n}$ and $\overline{\Delta^y}_{m,n}$ are the averages of these partial derivatives in the $k \times k$ windows. Thus Equation 3.12 is a root-mean-square measure of the variances of the partial derivatives in the x and y directions. For the examples in this section, we computed the phase derivative variances in 3×3 windows.

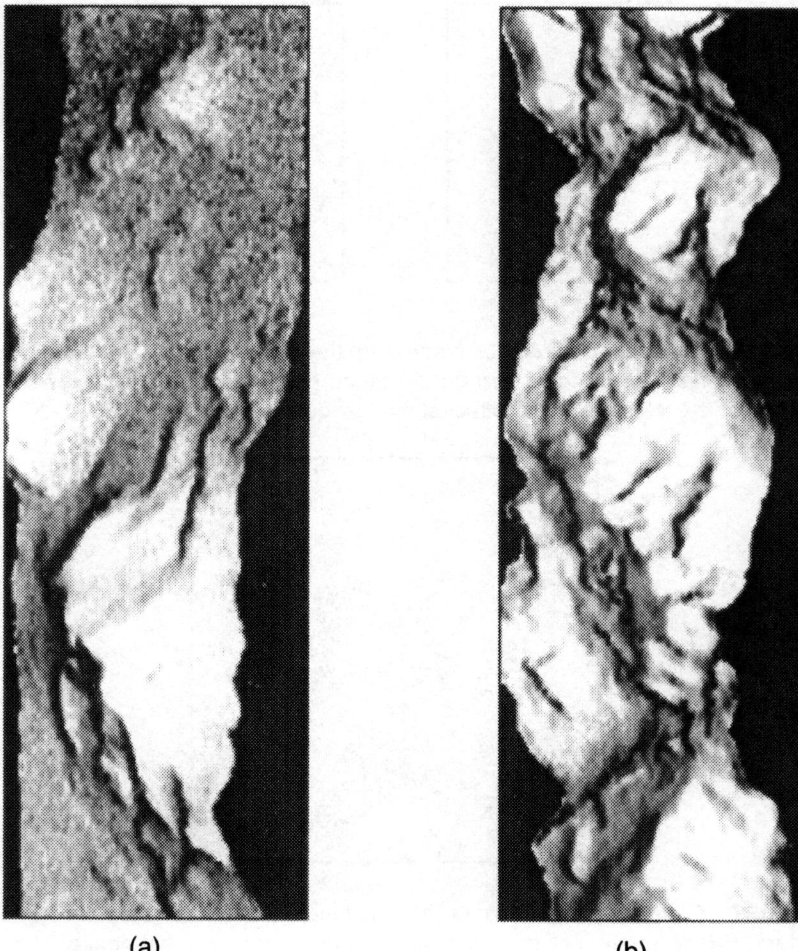

Figure 3.15 The pseudocorrelation maps for the (a) Long's Peak and (b) Isolation Peak simulated IFSAR examples. These maps are quite different from the correlation maps of Figures 3.11(a) and 3.12(a). The dark regions indicate the regions of highly varying phase. This includes the noisy phase regions as well as the regions of high terrain variation.

The phase derivative variance is a quality measure that differs from the correlation and pseudocorrelation measures. For example, on phase that exhibits a constant variation, such as that induced by a tilted plane, the phase derivative variance is zero. The pseudocorrelation, on the other hand, is nonzero. In our experience, the phase derivative variance is generally the most reliable measure of phase quality when the correlation map is not available.

The phase derivative variance map of the IFSAR example is shown in

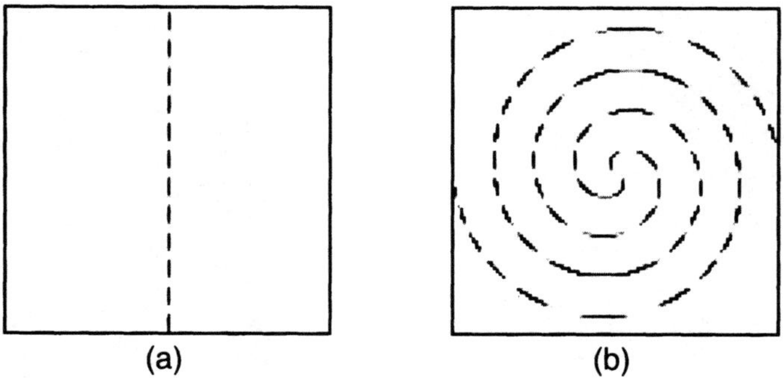

Figure 3.16 The pseudocorrelation maps of (a) the sheared planes and (b) the spiral shear examples. The periodic correlation of the phase data along the shear lines accounts for the "dashed lines" effect of the low-quality values.

Figure 3.17 The phase derivative variances of the IFSAR phase data of Figure 3.2. The dark, low-correlation regions are the regions of decorrelated phase. The map appears very similar to the pseudocorrelation map of Figure 3.13.

Figure 3.17. The phase derivative variances of the MRI examples are shown in Figures 3.18(a) and (b). These maps are quite similar to the pseudocorrelation maps of Figures 3.13 and 3.14. The phase derivative variance maps of the simulated IFSAR examples are shown in Figures 3.19(a) and (b). Like the correlation maps of Figures 3.11(a) and 3.12(a), the phase derivative variance maps highlight the regions of decorrelated phase caused by SAR layover. The phase derivative variance maps are quite different from the pseudocorrelation maps of Figure 3.15, however.

The sheared planes and the spiral shear examples of Figures 3.9 and 3.10 have phase derivative variance maps that appear identical to the pseudocorrelation maps of Figure 3.16. The phase derivative variance, however, really does isolate the planes completely. The quality maps along the lines of shear are shown as

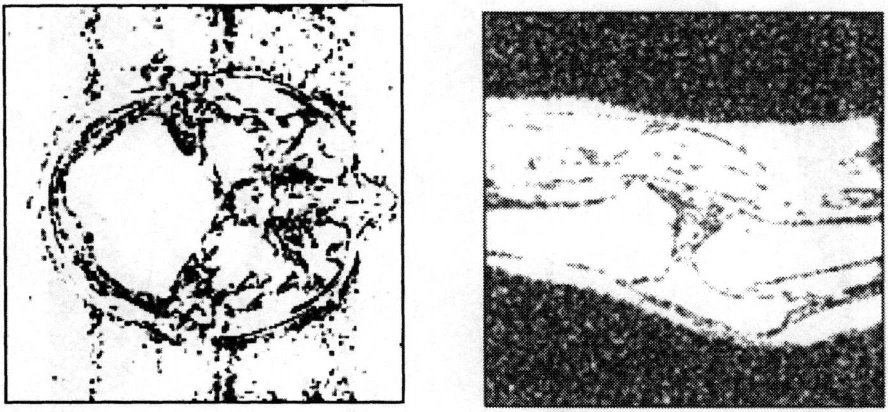

(a) (b)

Figure 3.18 The phase derivative variance maps for the MRI (a) head and (b) knee examples of Figures 3.3 and 3.4. The maps are very similar to the pseudocorrelation maps of Figure 3.14.

three-dimensional surfaces in Figure 3.20. The surfaces have been magnified and inverted for the sake of clarity so that they represent the "badness" rather than the goodness of the phase. In the pseudocorrelation case, shown in Figure 3.20(a), it can be seen that the line dips down periodically to the level of the surrounding quality values. The line for the phase derivative variance shown in Figure 3.20(b), on the other hand, never dips quite down to the level of the surrounding values. Thus the phase derivative variance defines a true barrier between the planes, albeit a fragile one. The addition of noise can destroy the barrier, thereby rendering even the phase derivative variance unsuitable for isolating the sheared planes (or the arms of the spiral) from one another. An example of this will be shown in Chapter 4.

3.3.4 Maximum Phase Gradient

The *maximum phase gradient* measures the magnitude of the largest phase gradient (i.e., partial derivative or wrapped phase difference) in the $k \times k$ neighborhood of each pixel. The motivation for the definition of this quality measure is the observation that in regions of noisy phase, the gradients tend to be large. The maximum phase gradient shares the disadvantage of pseudocorrelation, however, in that steep terrain slopes (i.e., highly varying but noiseless phase) will yield low-quality values.

The maximum phase gradient at a pixel is defined to be the greater of the two values

$$\max\{|\Delta^x{}_{i,j}|\}, \tag{3.13}$$

$$\max\{|\Delta^y{}_{i,j}|\}, \tag{3.14}$$

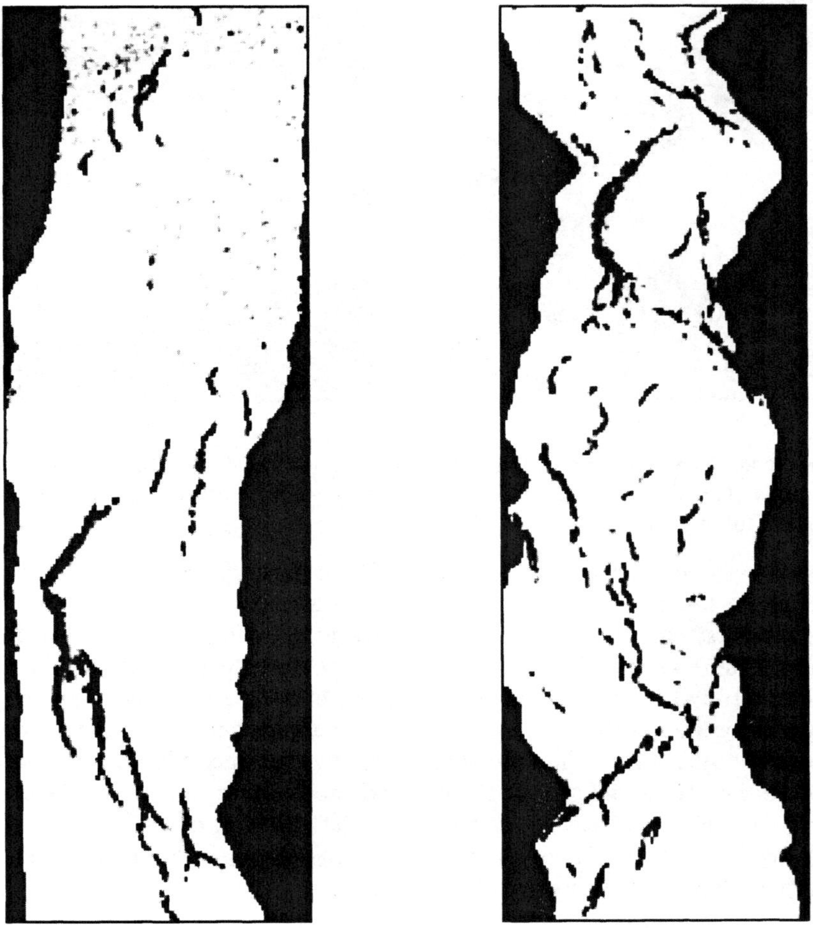

(a) (b)

Figure 3.19 The phase derivative variance maps of the (a) Long's Peak and (b) Isolation Peak simulated IFSAR examples. Compare these maps with the correlation maps of Figures 3.11(a) and 3.12(a). The dark pixels indicate the regions of decorrelated phase due to the SAR layover. Unlike the pseudo-correlation maps of Figure 3.15, the phase derivative variance map does not mark the regions of steep terrain as low quality unless the phase data are decorrelated.

where the maxima are evaluated in $k \times k$ neighborhoods of the given pixel. The terms $\Delta^{x}{}_{i,j}$ and $\Delta^{y}{}_{i,j}$ are the partial derivatives of the phase. Like the phase derivative variance, the maximum phase gradient will be assumed to be negated so that it becomes a measure of goodness.

In the following examples we computed the maximum phase gradients in 3×3 windows. The maximum phase gradient map of the IFSAR example is

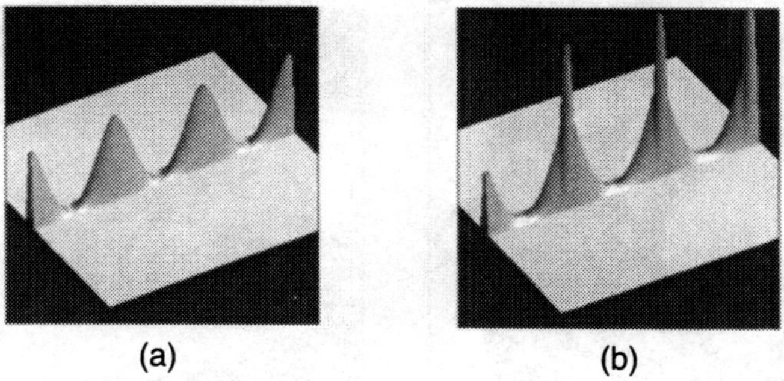

Figure 3.20 Three-dimensional renderings of the quality maps for the (a) pseudocorrelation and the (b) phase derivative variance applied to the sheared planes example. The renderings have been enlarged and inverted for clarity. In both cases the lines of low-quality values that separate the planes vary periodically, but the phase derivative variance never dips as low as the surrounding values. The phase derivative variance thus constitutes a true barrier between the planes.

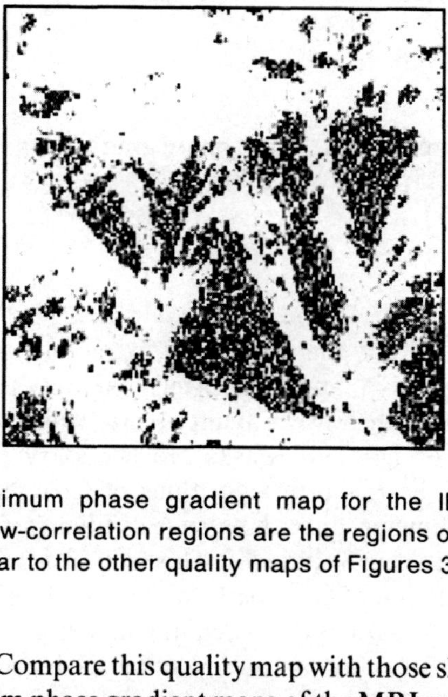

Figure 3.21 The maximum phase gradient map for the IFSAR phase data of Figure 3.2. The dark, low-correlation regions are the regions of decorrelated phase. The map appears similar to the other quality maps of Figures 3.13 and 3.17.

shown in Figure 3.21. Compare this quality map with those shown in Figures 3.13 and 3.17. The maximum phase gradient maps of the MRI examples are shown in Figures 3.22(a) and (b). Compare these with the quality maps of Figures 3.14 and 3.18. The maximum phase gradient maps of the simulated IFSAR examples are shown in Figures 3.23(a) and (b). Compare these maps with the correlation maps of Figures 3.11(a) and 3.12(a), the pseudocorrelation maps of Figure 3.15 and the phase derivative maps of Figure 3.19.

The maximum phase gradient maps for the sheared planes and the spiral shear examples are similar to the pseudocorrelation maps of Figure 3.16. Like the

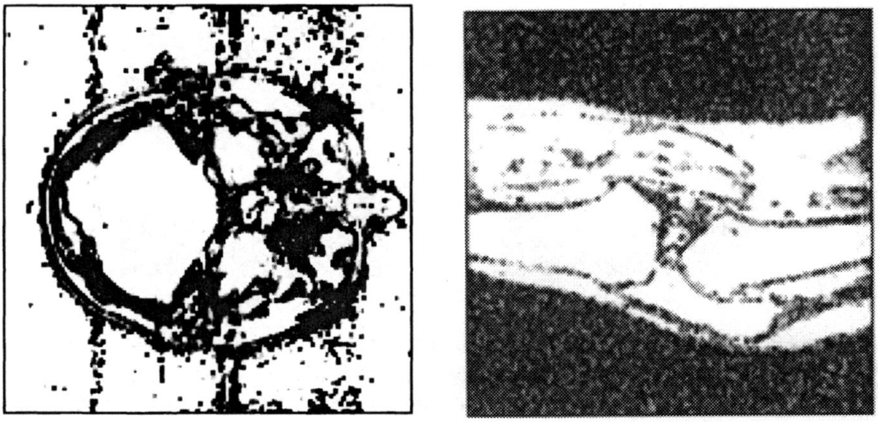

(a) (b)

Figure 3.22 The maximum phase gradient maps for the MRI (a) head and (b) knee examples of Figures 3.3 and 3.4. The maps appear similar to those of Figures 3.14 and 3.18.

pseudocorrelation map, the maximum phase gradients fail to separate the planes (or arms of the spiral) from one another.

3.4 MASKS

In this section we will study the problem of extracting masks from the quality maps. A mask is a quality map whose pixels take only two values: 0 or 1. The zero-valued pixels mark the lowest-quality phase values that are to be masked out, zero-weighted, or ignored. Masks are necessary for defining the zero-weights for some of the L^p-norm algorithms of Chapter 5. They can also be used in some path-following algorithms, such as Flynn's minimum discontinuity algorithm to be described in Section 4.5. This section describes the general issues involved in mask design. Some path-following algorithms impose special requirements on their masks as discussed in Chapter 4.

We will generate a mask by computing and applying a threshold to a quality map. All quality values below the threshold will be defined to be 0 (i.e., masked out), and those values above the threshold will be defined to be 1. The main difficulty with defining a mask is the selection of a threshold. If the threshold is too low, not enough pixels will be masked out. When the mask is then used to define the zero-weights for a path-following algorithm, too many low-quality phase values will be able to corrupt the unwrapped solution. On the other hand, if the threshold is too high, then too many pixels will be masked out, and regions will be isolated from one another.

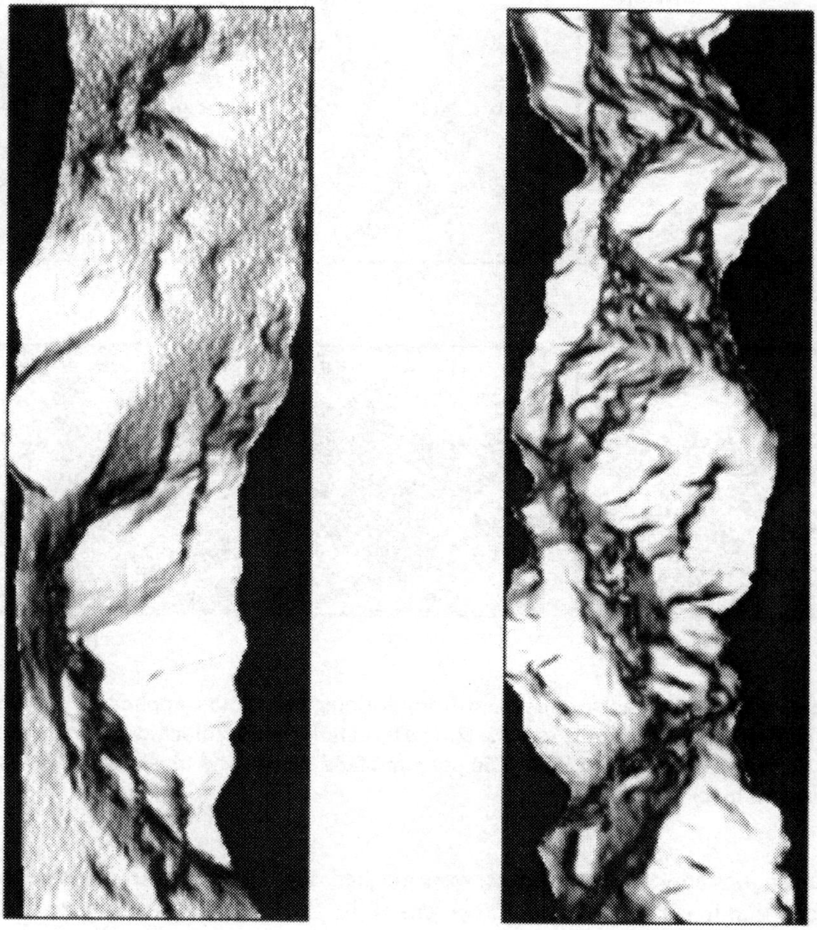

(a) (b)

Figure 3.23 The maximum phase gradient maps for the simulated IFSAR examples of (a) Long's Peak and (b) Isolation Peak. The dark pixels indicate phase data that have steep gradients, which tend to be noisy pixels. Compare these maps with the correlation maps of Figures 3.11(a) and 3.12(a) and the other maps of Figures 3.15 and 3.19.

Figures 3.24(a)–(f) show an example for the IFSAR data. Various thresholds were applied to the phase derivative variance map of Figure 3.17 to mask out 5, 10, 20, 30, 40, and 50 percent, respectively, of the lowest-quality pixels. The 5 percent threshold, whose mask is shown in Figure 3.24(a), is too low, since not enough pixels in the patches of noise were masked out. The 50 percent threshold, on the other hand, is too high. An excessive number of pixels, including high-

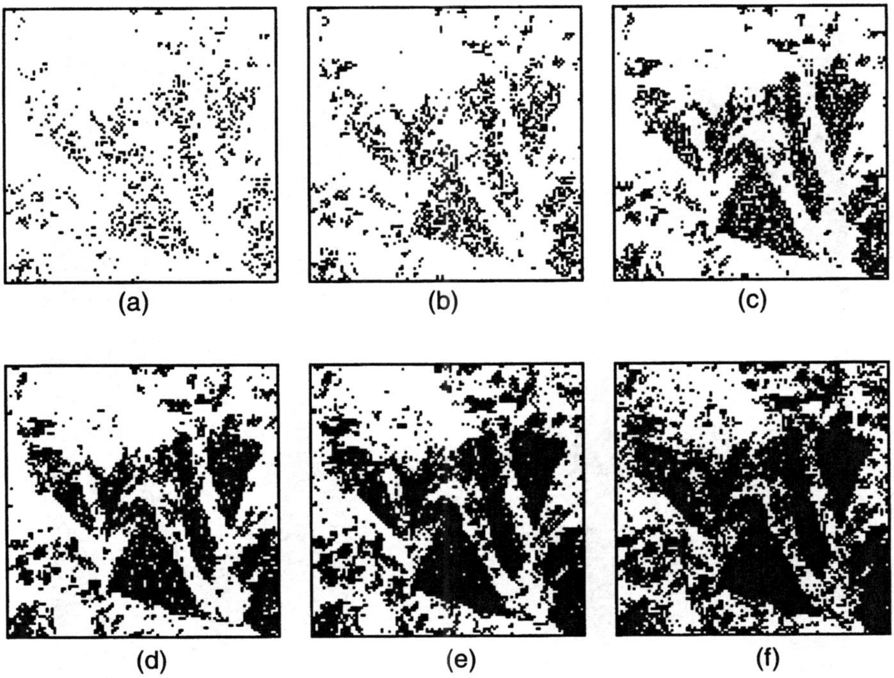

Figure 3.24 The masks generated from various thresholds applied to the phase derivative variance map of Figure 3.17. The thresholds were selected to mask out (a) 5, (b) 10, (c) 20, (d) 30, (e) 40, and (f) 50 percent of the pixels. The masked-out pixels are shown as black pixels.

quality ones between the patches, were masked out. The ideal threshold seems to be one that masked out about 10 percent of the pixels as shown in Figure 3.24(b).

The challenge of defining a good mask is to define the threshold so that the total number of masked pixels is small but the lowest-quality pixels (including the residues) are masked out. The threshold is usually chosen manually by a human operator [12], [16], [17]. We present here an automated method for selecting the threshold that requires no manual intervention [15]. We begin by examining the distribution of the quality map. Figure 3.25(a) shows a histogram of the phase derivative variances for the IFSAR example. These values have been scaled to the range 0–1 and placed in ten histogram "bins." There are very few pixels in the lowest (leftmost) bin, and the distribution increases exponentially to about 80 percent of the pixels in the highest (rightmost) bin. The overwhelming majority of the pixels have high-quality values, and there appears to be no natural place to define the threshold.

The quality values can be "re-mapped," however, to provide a natural point at which to define the threshold. Figure 3.25(b) shows the results of this re-mapping, which produces a U-shaped distribution. The natural choice for the

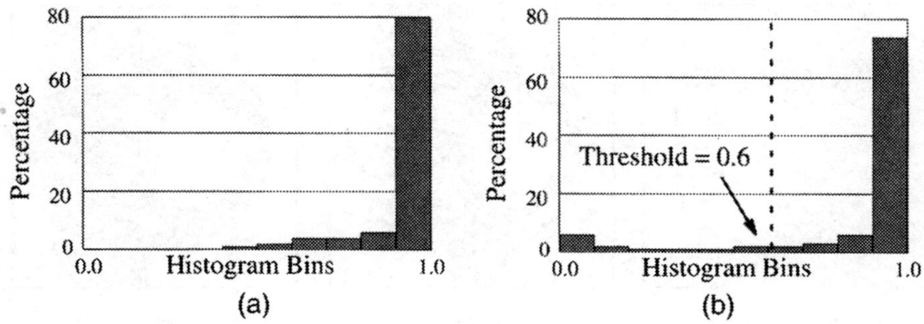

Figure 3.25 (a) The distribution of the phase derivative variances for the IFSAR example. The quality values have been scaled to the range 0–1 and collected in ten histogram "bins". The leftmost bin represents the number of lowest-quality (0.0–0.1) pixels, while the rightmost bin represents the number of highest-quality (0.9–1.0) pixels. (b) The re-mapped distribution and the automatically selected mask threshold.

threshold is the location of the minimum. In practice, it is best to modify this approach slightly by choosing the location at which the distribution begins to increase substantially. This produces a threshold of 0.6 on this example, which results in 13 percent of the pixels being masked out. The resulting mask is similar to that of Figure 3.24(b).

We will describe the procedure for selecting the threshold in more detail. (Our C language implementation of this algorithm can be found in Section A.3 of Appendix A.) The quality values are first scaled to the range 0.0–1.0, as described above, and collected into 10,000 histogram bins. The quality values are then adaptively re-mapped so that about 5 percent of the values lie in the range 0.0–0.1 and 5 percent lie in the range 0.9–1.0. This re-mapping procedure is performed as follows: The lowest bins of the 10,000-bin histogram are accumulated until the sum exceeds 5 percent of the pixels. (If the sum suddenly jumps to greater than 50 percent of the pixels after the addition of a bin, then that bin is not accumulated.) The actual sum may be less than or greater than 5 percent. This procedure is repeated on the highest bins of the histogram. This defines a new 10-bin histogram with about 5 percent of the pixels in the first bin and about 5 percent in the last bin. The remaining pixels are scaled to the range 0.1–0.9 and collected in the eight remaining bins.

The resulting 10-bin histogram of re-mapped quality values usually has a minimum. The location of the minimum is determined by examining each bin from left to right and detecting a decrease and then an increase in the number of pixels. The increase must be "substantial" (i.e., at least 0.5 percent of the pixels) to avoid false minima induced by noise. The location of the minimum defines the threshold. If there are two minima, the leftmost minimum is selected.

Occasionally the procedure for selecting the threshold may fail due to the lack of a minimum in the re-mapped histogram. In this case the threshold is chosen to mask out a predetermined percentage of the lowest-quality pixels, or the user

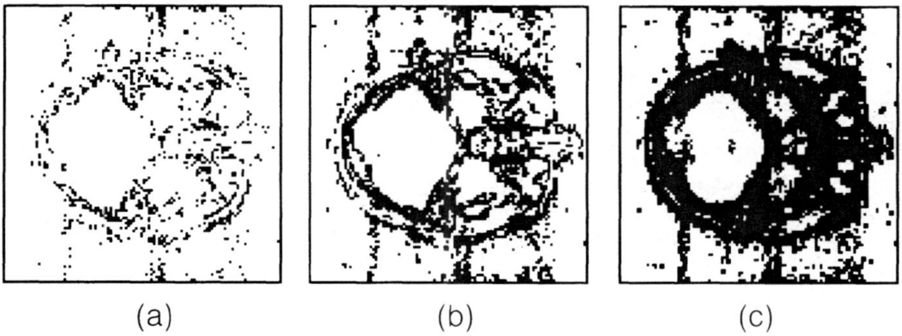

Figure 3.26 The masks generated from the phase derivative variance map of the MRI head data shown in Figure 3.18(a) for various thresholds. The thresholds were selected to mask out (a) 5, (b) 20, and (c) 40 percent of the pixels.

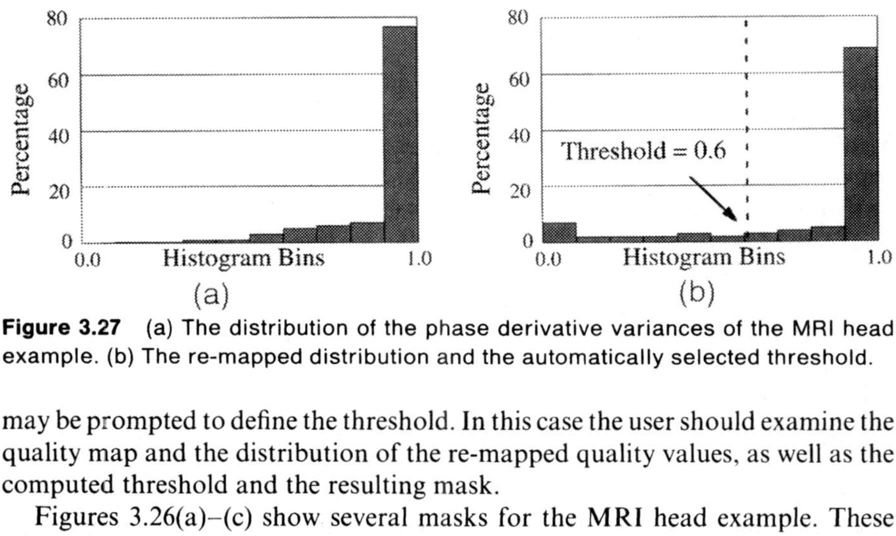

Figure 3.27 (a) The distribution of the phase derivative variances of the MRI head example. (b) The re-mapped distribution and the automatically selected threshold.

may be prompted to define the threshold. In this case the user should examine the quality map and the distribution of the re-mapped quality values, as well as the computed threshold and the resulting mask.

Figures 3.26(a)–(c) show several masks for the MRI head example. These masks were generated from the phase derivative variance map of Figure 3.18(a). The percentage of mask pixels is 5, 20, and 40 percent, respectively. The first mask appears to mask out an insufficient number of pixels, while the last one masks out too many. The 20-percent mask appears to be ideal. Figures 3.27(a) and (b) show the original and re-mapped distributions of the quality values. The algorithm described above selected a threshold of 0.6, which resulted in 18 percent of the pixels being masked out.[4] The resulting mask looks like that of Figure 3.26(b).

Figures 3.28(a)–(c) show several masks for the MRI knee example of Figure 3.18(b). The percentage of mask pixels is 5, 20, and 60 percent, respectively. The 5 percent mask does not mask out enough of the noisy pixels

[4] Although the histogram shown in Figure 3.27(b) increases at the point 0.4 (i.e., between the fourth and fifth bins), this value was not selected for the threshold because the increase was too small.

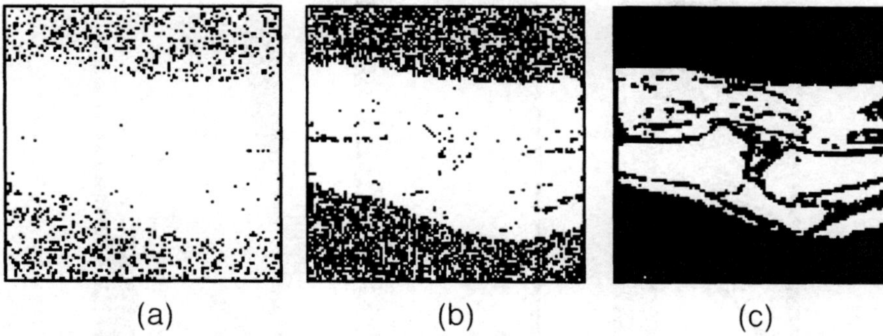

Figure 3.28 The masks generated from the phase derivative variance map of the MRI knee data shown in Figure 3.18(b) for various thresholds. The thresholds were selected to mask out (a) 5, (b) 20, and (c) 60 percent of the pixels.

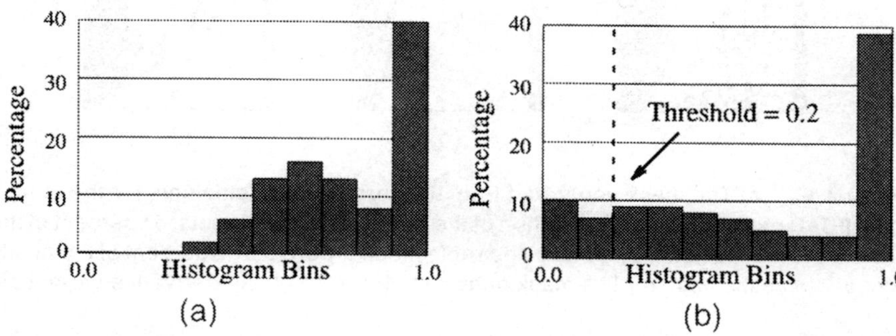

Figure 3.29 (a) The distribution of the phase derivative variances of the MRI knee example. (b) The re-mapped distribution and the automatically selected threshold.

within the knee, while the 60 percent mask isolates the regions within the knee from one another. The 20 percent mask appears to be correct and is in fact approximately the mask that was generated by the threshold selection procedure. The original distribution of quality values is shown in the histogram of Figure 3.29(a), and the re-mapped distribution is shown in Figure 3.29(b). A threshold of 0.2 was selected, which resulted in 17 percent of the pixels being masked out. The resulting mask looks like that of Figure 3.28(b).

Figure 3.30(a) shows the mask that results from the correlation map of the Long's Peak IFSAR example of Figure 3.11(a). Something clearly has gone wrong. The upper half of the image is almost completely masked out, although most of the phase data in this region are valid. The cause of the problem is not the threshold selection procedure but rather the correlation map itself. During the IFSAR simulation procedure, there were misregistration errors in the upper part of the image, which resulted in slight decorrelation and a resulting degradation in the correlation coefficients. The phase values were not corrupted as much as the correlation coefficients. (We noted in Section 3.3.1 that these errors were left in the data because they are typical of the errors to be found in actual IFSAR data.) Figures 3.31(a) and (b) show the distributions of the correlation coefficients

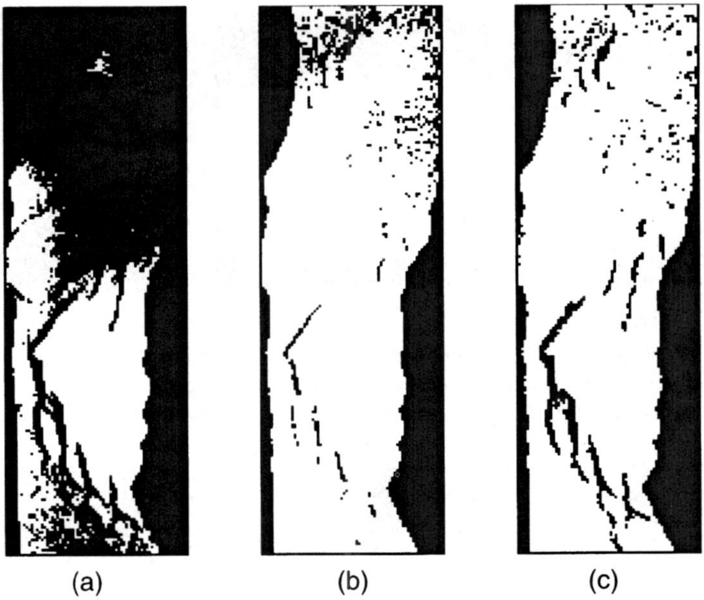

Figure 3.30 (a) The mask generated from the correlation coefficients for the Long's Peak IFSAR data. Due to the corruption of the correlation coefficients, 57 percent of the pixels were masked out. (b) A more desirable mask, where only 5 percent of the pixels have been masked out. (c) The mask generated from the phase derivative variances.

before and after the re-mapping procedure. The percentage of low-correlation pixels is higher than usual, as the figure shows. The degraded correlation map resulted in a threshold that masked out 57 percent of the pixels.

Figure 3.30(b) shows the mask that masks out 5 percent of the pixels. This is much more satisfactory. However, the re-mapped distribution of Figure 3.31(b) shows that there is no minimum that defines a threshold that masks only 5 percent of the pixels. The solution to the problem of automatically generating a satisfactory mask is to employ a different quality map.

Figure 3.30(c) shows the mask that results when the phase derivative variance map is used instead of the correlation map. Notice that the decorrelated regions of layover are cleanly masked out as desired, while there are only scattered mask pixels in the upper part of the image where the phase data are valid. The mask is even better than that of Figure 3.30(b). The distribution of the phase derivative variances is shown in Figure 3.31(c), while the re-mapped distribution is shown in Figure 3.31(d). A threshold of 0.7 was selected, which masked out 7 percent of the pixels.

Figures 3.32(a) and (b) show the masks for the correlation coefficients and the phase derivative variances, respectively, on the Isolation Peak IFSAR example. In both cases the regions of layover are masked out cleanly. The distributions of the correlation coefficients before and after the re-mapping procedure are shown

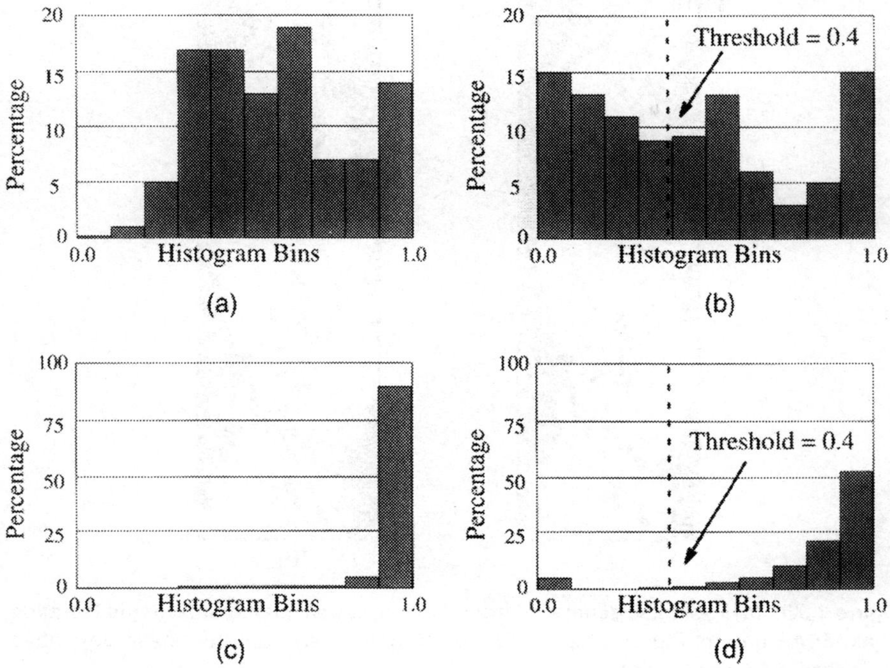

Figure 3.31 The distributions of the quality maps for the Long's Peak IFSAR data. (a) The distribution of the correlation coefficients. (b) The re-mapped distribution and the automatically selected threshold. (c) The distribution of the phase derivative variances. (d) The re-mapped distribution of the phase derivative variances and the automatically selected threshold.

in Figures 3.33(a) and (b). The distributions of the phase derivative variances are shown in Figures 3.33(c) and (d).

The re-mapped distribution of the phase derivative variances of the sheared planes example of Figure 3.9 is shown in Figure 3.34(a). Most of the pixels have the highest-quality value (i.e., a variance of zero), while only about 1 percent of the pixels have a low-quality value. These low-quality pixels all lie on the shear line, as the mask of Figure 3.34(b) shows. Figures 3.35(a) and (b) show the re-mapped distribution and the resulting mask for the spiral shear example of Figure 3.10. If the pseudocorrelation or maximum phase gradient maps were used instead of the phase derivative variance map, the masks would consist of dashed rather than solid lines like those shown in Figures 3.16(a) and (b).

We conclude this section by examining the distributions of the quality values at the residues. We would expect the residues to be concentrated at the low-quality pixels. This expectation is supported by the examples. Figures 3.36(a)–(d) show the distributions of the phase derivative variances of the residue pixels on the IFSAR, MRI head, MRI knee, and sheared planes examples, respectively. The thresholds are also indicated on these distribution plots. In every case except

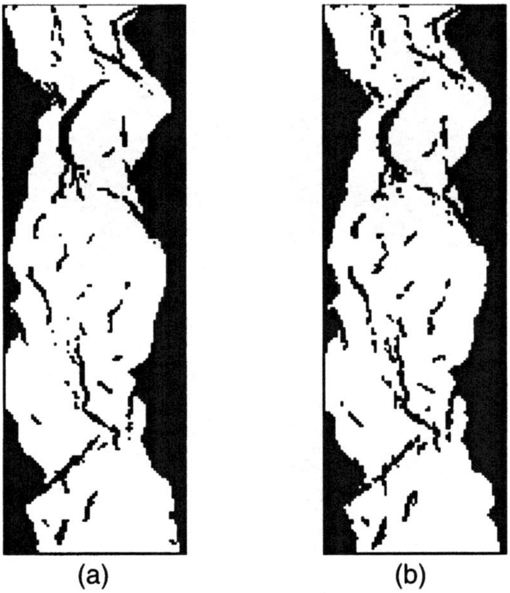

Figure 3.32 (a) The mask generated from the correlation coefficients for the Isolation Peak IFSAR data of Figure 3.12. (b) The mask generated from the phase derivative variances of Figure 3.19(b).

the knee example, the threshold defines a mask that contains all the residues. (The nonmasked residues of the MRI knee example all lie in the background noise.) Figures 3.37(a) and (b) show the distributions of the *correlation coefficients* at the residues on the two simulated IFSAR examples. The thresholds do not mask out all of the residues in these two cases. Figures 3.37(c) and (d) show the distributions of the *phase derivative variances* at the residues on the same two examples. The thresholds now mask out all the residues. These results indicate that the phase derivative variances are superior to the correlation coefficients for masking out the residues, although the deficiency of the correlation coefficients may be due to a defect in the IFSAR simulation. In any case we have seen that the phase derivative variance is an effective measure of the quality of the phase data and the locations of the residues.

We conclude this section by noting that the masks can be refined further. For example, simple morphological operations can be applied to thicken or thin the masks. One of the algorithms of Chapter 5 will require slightly thickened masks to unwrap the examples correctly.

3.5 NOISE FILTERING

In this section we briefly address the problem of filtering the phase data. Other than increasing the signal-to-noise ratio, noise filtering is useful for reducing the

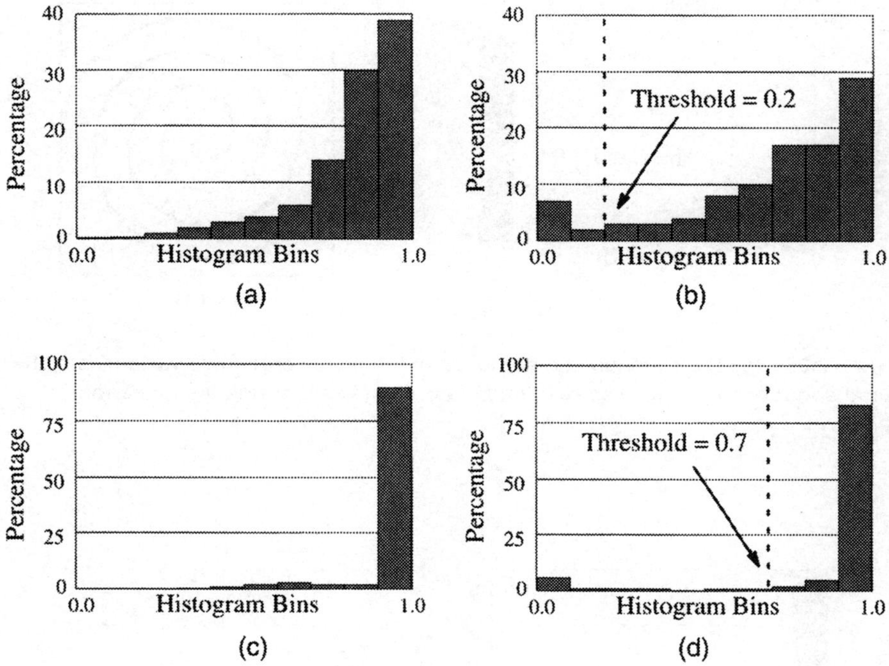

Figure 3.33 The distributions of the quality maps for the Isolation Peak IFSAR data. (a) The distribution of the correlation coefficients. (b) The re-mapped distribution and the automatically selected threshold. (c) The distribution of the phase derivative variances. (d) The re-mapped distribution of the phase derivative variances and the automatically selected threshold.

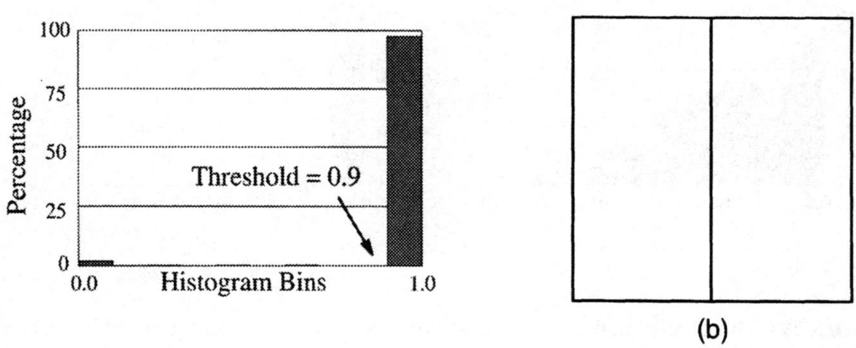

Figure 3.34 (a) The re-mapped distribution of the phase derivative variances of the sheared planes example. (b) The resulting mask, which consists of a line down the middle, cleanly separates the two planes from one another.

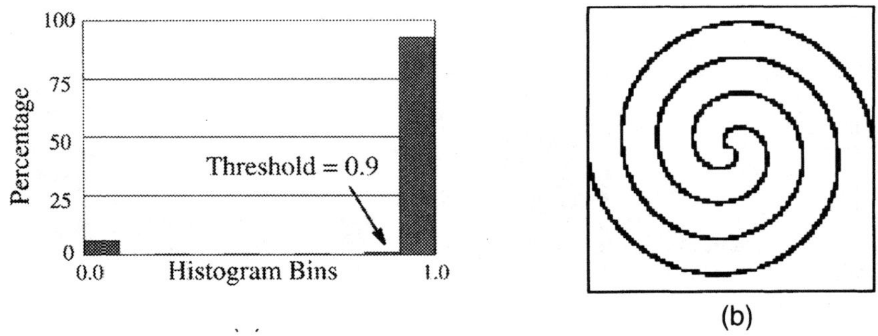

Figure 3.35 (a) The re-mapped distribution of the phase derivative variances of the spiral shear example. (b) The resulting mask cleanly separates the two arms of the spiral.

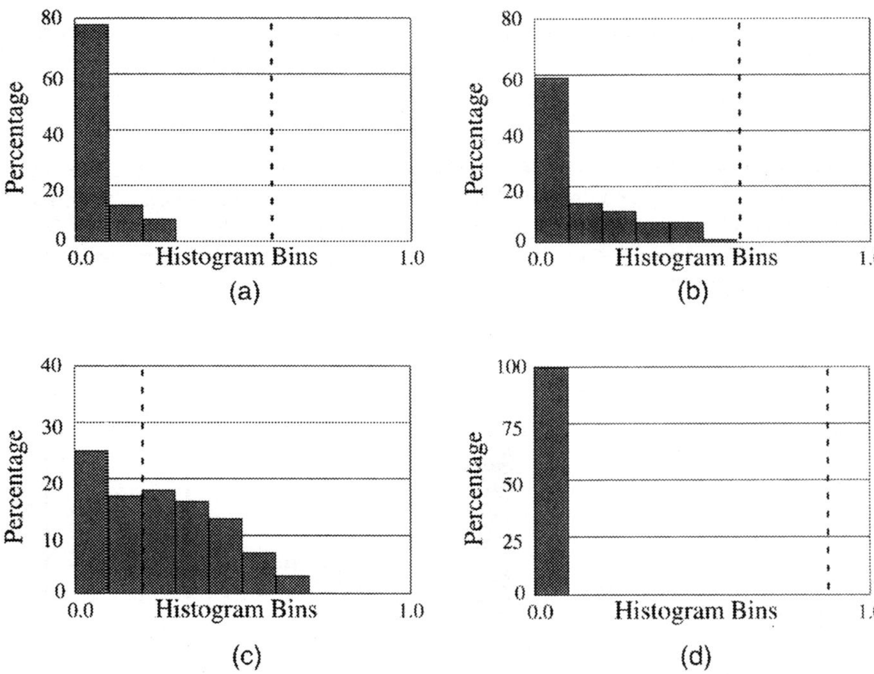

Figure 3.36 The distributions of the phase derivative variances *at the residues* of the (a) IFSAR, (b) MRI head, (c) MRI knee, and (d) sheared planes examples. These plots show that the residues are concentrated at the low-quality pixels. The threshold on each example is indicated to show that all of the residues are masked out except on the excessively noisy MRI knee example.

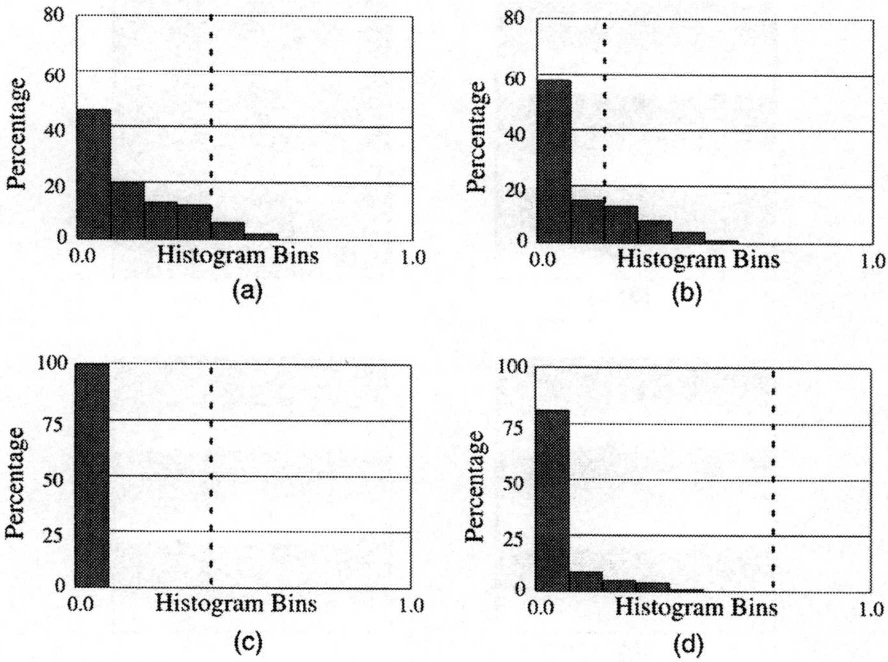

Figure 3.37 The distributions of the quality maps *at the residues* of the two simulated IFSAR examples. The distributions of the correlation coefficients are shown for (a) Long's Peak and (b) Isolation Peak. The distributions of the phase derivative variances are shown for (c) Long's Peak and (d) Isolation Peak. The thresholds are indicated to show that the mask derived from the phase derivative variances masks out all the residues.

number of residues, which can make the phase unwrapping process considerably easier. We must qualify this remark with important observation, however. As we noted in Chapter 1, phase itself is *not* a signal. It is rather a *property* of a signal. Therefore it is the signal itself that should be filtered and not the phase. For example, complex-valued IFSAR data should be filtered in $k \times k$ windows before the phase is extracted.

Having said that, we acknowledge that there do arise situations in which the observations are no longer available and the phase itself must be filtered. We must be careful when doing so, however. The average of the two phase values $\pi/4$ and $7\pi/4$ is not π, but 0.

The simplest filter is the averaging filter. The average phase $\tilde{\psi}_{m,n}$ cannot be defined by the conventional formula

$$\tilde{\psi}_{m,n} = \frac{\sum \psi_{i,j}}{k^2}, \tag{3.15}$$

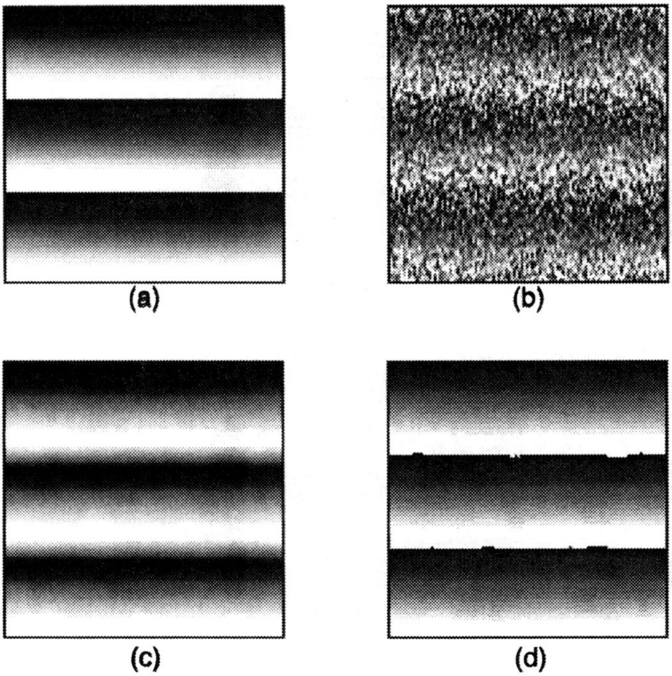

Figure 3.38 (a) The wrapped phase of a slanted plane. (b) The phase corrupted with Gaussian noise. (c) The result of smoothing the phase with the incorrectly defined filter of Equation 3.15. (d) The result of smoothing with the correct filter defined by Equation 3.16.

where the sum is computed in $k \times k$ windows centered at the pixel (m, n). The phase data first must be mapped to vectors $\exp(j\psi_{i,j})$ in the complex plane.[5] These vectors can then be averaged and the resulting phase extracted by means of the arc tangent operator. Thus the correct formula for the average phase is

$$\tilde{\psi}_{m,n} = \arctan\left(\sum \sin \psi_{i,j}, \sum \cos \psi_{i,j}\right). \qquad (3.16)$$

Figure 3.38(a) shows a simple example consisting of the wrapped phase of a tilted plane, with the phase corrupted by Gaussian noise in Figure 3.38(b). If one uses the incorrect Equation 3.15 to filter the phase, the result is shown in Figure 3.38(c). The fringes, or boundaries between the 0 (black) and 1 (white) pixels, have been blurred. This is guaranteed to produce a disastrously incorrect result when the phase is unwrapped. The correct averaging operation defined by Equation 3.16 produces the result of Figure 3.38(d). The operation correctly smoothes the noise without blurring the fringes. (Incidentally, both filters smooth the noise and remove all the residues.)

[5] We are using the symbol j for both the array index and the imaginary term $j = \sqrt{-1}$.

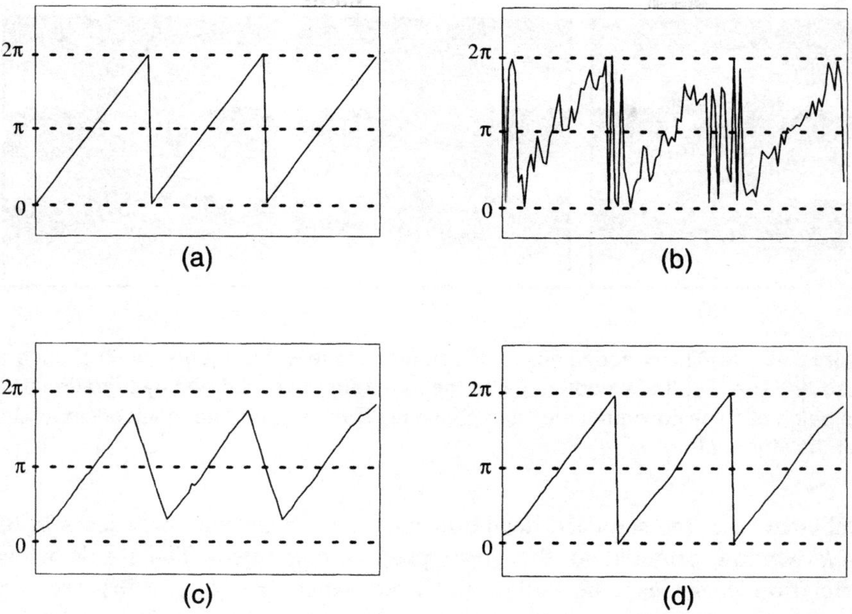

Figure 3.39 These plots show a cross section through the wrapped phase of Figures 3.38(a)–(d). (a) The wrapped phase of a line with positive slope. (b) The phase corrupted with Gaussian noise. (c) The result of smoothing the phase with the incorrectly defined filter of Equation 3.15. (d) The result of smoothing with the correct filter defined by Equation 3.16.

Figures 3.39(a)–(d) show plots of a crosssection through the four phase images of Figures 3.38(a)–(d). Figure 3.39(a) shows the sawtooth plot of a "slice" through the wrapped phase of Figure 3.38(a), and Figure 3.39(b) shows the Gaussian noise-corrupted result. When the incorrect averaging filter is applied, the result is Figure 3.39(c). This plot is no longer the wrapped phase of a line but rather the wrapped phase of a skewed sawtooth function. It cannot be unwrapped because it already is. Figure 3.39(d) shows the result of the application of the correct filter. As in Figure 3.39(a) it is the wrapped phase of a line.

The correctly defined averaging filter can be extended to more complicated types of filters by convolving the complex numbers $\exp(j\psi_{i,j})$ with other filter coefficients. For example, in the case of IFSAR data, the correlation coefficients could be used. This would be equivalent to averaging the original complex-valued IFSAR pixels.

A second type of filter is the adaptive filter that averages only selected phase values within $k \times k$ neighborhoods. One such filter is the sigma filter [18]. This filter is defined only for IFSAR data because it requires correlation coefficients to determine which phase values to average. Given a pixel with a correlation

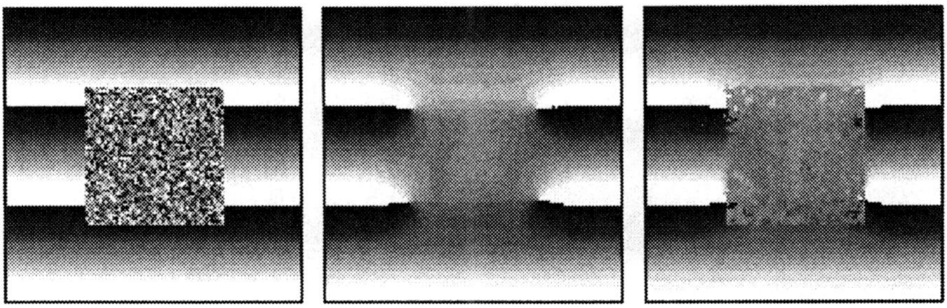

Figure 3.40 (a) The wrapped phase of a slanted plane with a square patch of uniform noise. (b) The result of smoothing with the averaging filter defined by Equation 3.16. The patch of noise corrupted the valid phase in its vicinity. (c) The result of smoothing with the sigma filter.

coefficient of z, the standard deviation σ of the correlation coefficients in the $k \times k$ window centered at the given pixel is computed. The pixels whose correlation coefficients lie within the fixed range $(z - 2\sigma, z + 2\sigma)$ are then selected, and their phases are averaged. The sigma filter can be modified to work with any phase data by employing some other quality map in place of the correlation coefficients.

Figure 3.40(a) shows the wrapped phase of a slanted plane, where the phase has been corrupted with a patch of uniformly distributed noise.[6] If one applies the correct averaging filter of Equation 3.16, the valid phase values near the patch of noise will be corrupted as shown in Figure 3.40(b). If one applies the sigma filter instead, with the phase derivative variances in place of the correlation coefficients, then the phase can be filtered without allowing the patch of noise to corrupt the neighboring phase appreciably, as shown in Figure 3.40(c).

Not all filters can be modified to work with phase data. For example, if the phase data are corrupted with "salt and pepper" noise, the median filter is more effective than the averaging filter for attenuating the noise. Because the complex numbers do not have a natural ordering, however, there is no natural way to define the median filter. An approximation can be defined by the equation

$$\tilde{\psi}_{m,n} = \arctan\left(\text{median}\{\sin \psi_{i,j}\}, \text{median}\{\cos \psi_{i,j}\}\right), \tag{3.17}$$

where the medians are evaluated in the $k \times k$ neighborhood of the pixel (m, n).

We have shown how to define a convolution (averaging) filter, an adaptive filter, and an approximation to a median filter for phase data. These three examples should suffice for introducing the reader to the filtering of phase. We must once again stress our philosophy that one ordinarily should not filter phase

[6] We did not add Gaussian noise to this example because this would have rendered the phase derivative variance useless.

data. Instead, one should filter the observations from which the phase data are extracted. We therefore will not pursue this topic any further in this book.

3.6 SUMMARY AND CONCLUSION

In this chapter we have presented the following seven phase unwrapping problems:

1. IFSAR
2. MRI head
3. MRI knee
4. Long's Peak IFSAR (simulated)
5. Isolation Peak IFSAR (simulated)
6. Sheared planes
7. Spiral shear

These problems encompass a broad range of the typical problems that are encountered in phase unwrapping. For example, the MRI knee example contains a great deal of background noise, while the sheared planes example, although noise-free, contains a line of sharp discontinuity. The IFSAR example contains large patches of noise that the phase unwrapping process must unwrap around, while the Long's Peak data suffer from slight decorrelation induced by image registration errors. These seven problems (and variations of them) will be used to test and evaluate the phase unwrapping algorithms of Chapter 4 and 5.

We also presented four techniques for measuring the quality of the phase data:

1. Correlation coefficients
2. Phase derivative variance
3. Pseudo-correlation coefficients
4. Maximum phase gradient

The correlation coefficients are available only for IFSAR data, while the other three measures can be calculated from any type of phase data. These quality measures will be needed to guide many of the phase-unwrapping algorithms we will study.

Some of the phase unwrapping algorithms of Chapters 4 and 5 require arrays of zero-weights to indicate which corrupted phase values should be ignored during the unwrapping process. These zero-weights, or masks, can be generated by applying a threshold to a quality map. We presented an automated technique for selecting this threshold. The resulting masks are effective at masking out the low-quality phase data and the residues, especially for the phase derivative variance map.

Last, we took a brief look at noise filters for phase data. We stressed that filtering should be performed not on the phase itself but on the observations from which the phase is extracted. If the observations are not available, however, then care must be exercised in defining the filter. Because the phase data are wrapped, they must be treated as vectors $\exp(j\psi)$ in the complex plane. This allows one to define averaging and convolution filters and even some types of adaptive filters. It creates problems for other types of filters, such as the median filter, because there is no natural ordering for the vectors.

This chapter contains many fundamental concepts and examples that will be necessary to understand the phase unwrapping algorithms of Chapters 4 and 5. The reader should be prepared to refer to this chapter repeatedly while proceeding through these algorithms.

REFERENCES

[1] C. V. Jakowatz, Jr., D. E. Wahl, P. H. Eichel, D. C. Ghiglia, and P. A. Thompson, *Spotlight-Mode Synthetic Aperture Radar: A Signal Processing Approach*, Kluwer, Boston, 1996.

[2] D. L. Bickel and W. H. Hensley, "Design, theory and applications of interferometric synthetic aperture radar for topographic mapping," Sandia National Laboratories Technical Report, SAND96–1092, May 1996.

[3] M. D. Pritt, "Automated subpixel image registration of remotely sensed imagery," *IBM Journal of Research and Development*, Vol. 38, No. 2, pp. 157–166, 1994.

[4] J. Way and E. A. Smith, "The evaluation of synthetic aperture radar systems," *IEEE Transactions on Geoscience and Remote Sensing*, Vol. 29, No. 6, pp. 962–985, November 1991.

[5] H. A. Zebker, K. B. Wheeler, T. Miller, Y. Lou, G. Alberti, S. Vetrella, and A. Cucci, "The TOPSAR interferometric radar topographic mapping instrument," *IEEE Transactions on Geoscience and Remote Sensing*, Vol. 30, No. 5, pp. 933–940, September 1992.

[6] E. R. Stofan, D. L. Evans, C. Schmullius, B. Holt, J. J. Plaut, J. van Zyl, S. D. Wall, and J. B. Way, "Overview of results of spaceborne imaging radar–C, X-band synthetic aperture radar (SIR-C/X-SAR)," *IEEE Transactions on Geoscience and Remote Sensing*, Vol. 33, pp. 817–828, July 1995.

[7] D. C. Ghiglia and L. A. Romero, "Minimum L^p-norm two-dimensional phase unwrapping," *Journal of the Optical Society of America A*, Vol. 13, No. 10, pp. 1999–2013, 1996.

[8] G. H. Glover and E. Schneider, "Three-point Dixon technique for true water/fat decomposition with inhomogeneity correction," *Magnetic Resonance Medicine*, Vol. 18, pp. 371–383, 1991.

[9] J. Szumowski, W. R. Coshow, F. Li, and S. F. Quinn, "Phase unwrapping in the three-point Dixon method for fat suppression MR imaging," *Radiology*, Vol. 192, No. 2, pp. 555–561, 1994.

[10] S. M. Song, S. Napel, N. J. Pelc, and G. H. Glover, "Phase unwrapping of MR phase

images using Poisson equation," *IEEE Transactions on Image Processing*, Vol. 4, No. 5, pp. 667–676, 1995.

[11] The MRI data sets shown in Figures 3.3(a) and 3.4(a) are courtesy of Dr. Gang Zhu. They come from the ESTEEM clinical MRI scanner manufactured by Elscint MR Inc., Ft. Collins, CO.

[12] M. W. Roth, "Phase unwrapping for interferometric SAR by the least-error path," Johns Hopkins University Applied Physics Laboratory Technical Report, Laurel, Maryland, March 30, 1995.

[13] E. Rodriguez and J. M. Martin, "Theory and design of interferometric synthetic aperture radars," *IEE Proceedings Part F*, Vol. 139, No. 2, pp. 147–159, April 1992.

[14] D. C. Ghiglia and L. A. Romero, "Robust two-dimensional weighted and unweighted phase unwrapping that uses fast transforms and iterative methods," *Journal of the Optical Society of America A*, Vol. 11, pp. 107–117, 1994.

[15] M. D. Pritt, "Phase unwrapping by means of multigrid techniques for interferometric SAR," *IEEE Transactions on Geoscience and Remote Sensing*, Vol. 34, No. 3, pp. 728–738, 1996.

[16] D. J. Bone, "Fourier fringe analysis: the two-dimensional phase unwrapping problem," *Applied Optics*, Vol. 30, No. 25, pp. 3627–3632, 1991.

[17] T. J. Flynn, "Consistent 2-D phase unwrapping guided by a quality map," *Proceedings of the 1996 International Geoscience and Remote Sensing Symposium,* Lincoln, NE., May 27–31, 1996, IEEE, Piscataway, NJ, pp. 2057–2059, 1996.

[18] J. S. Lee, T. L. Ainsworth, M. R. Grunes, and R. M. Goldstein, "Noise filtering of interferometric SAR images," *Proceedings of the SPIE*, Vol. 2315, Society of Photo-Optical Instrumentation Engineers, Bellingham, WA, pp. 735–742, 1994.

4

PATH-FOLLOWING METHODS

4.1 INTRODUCTION

In Chapter 2 we observed that the unwrapped phase computed at one pixel, starting from another pixel, can depend on the integration path. If we were to follow two different paths from one pixel to the other, we could end up with two different answers for the unwrapped phase. We found that these inconsistencies are caused by pointlike structures called *residues*. A residue is located somewhere inside a "loop" of four pixels for which the integral of the phase derivatives (i.e., the sum of the wrapped phase differences) is not zero, but 2π or -2π. By convention this sum is calculated in a counterclockwise direction around the loop, as shown in Figures 2.5 and 2.6. We refer to the sign of this sum as the *polarity* or *charge* of the residue, and we refer to the residue as *positive* or *negative* depending on whether the polarity is positive or negative.

If there are no residues, the phase[1] can be unwrapped along any path, and the result will be independent of the path. If residues are present, however, then branch cuts must be placed between residues of opposite polarity to avoid path-dependent results. The role of the branch cuts is to "balance" the residues so that any closed path always encloses an equal number of positive and negative residues or no residues at all. The residue theorem (Equation 2.29) then ensures that the path integral is zero for every closed path. In other words, it ensures that the third condition for path independence (see Section 2.2) is satisfied.

Once the branch cuts are in place and all the residues are balanced, the phase can be unwrapped along any path that does not cross the branch cuts. The phase unwrapping problem is thus reduced to the problem of choosing a "good" set of branch cuts.

[1] As in Chapter 3 we use the term *phase* interchangeably with the term *wrapped phase* where there is no possibility of confusion.

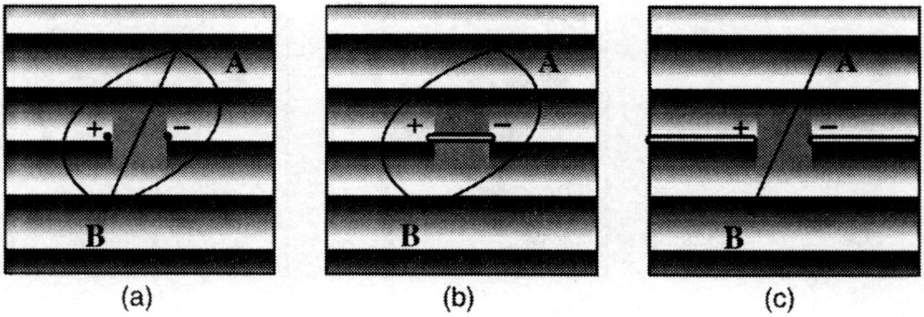

Figure 4.1 Path-dependent phase unwrapping in the presence of residues. (a) Phase data with a positive (+) and a negative (-) residue. The unwrapped value at the point B depends on the path followed from the point A. The straight and the curved paths yield different values. (b) If a branch cut connects the residues, the unwrapped phase is the same for all allowed paths. (c) The branch cuts may also connect the residues to the border. In this case the phase is unwrapped differently.

Figure 4.1(a) provides a simple illustration. As usual, the phase values are shown as gray-level pixels that range from $-\pi$ (black) to π (white). There are two residues of opposite polarity in the image. If the phase is unwrapped from the point marked A to the point marked B along the straight path, a crossover from $-\pi$ to π occurs twice. The unwrapped phase of the point B is therefore approximately 4π plus the unwrapped phase of A. If one of the curved paths is followed, however, then there are *three* crossovers from $-\pi$ to π, and the unwrapped phase of B is approximately 6π rather than 4π. The answer clearly depends on the choice of path. If a branch cut is placed between the two residues as shown in Figure 4.1(b), then the unwrapped value of B will be 6π no matter which path is followed.

The choice of branch cuts is not obvious. For example, we could have placed two cuts that connect the residues to the image border as shown in Figure 4.1(c). In this case the unwrapped value of B would be 4π no matter which path were followed.

Some criteria must be established to guide the placement of branch cuts. It is not enough to require only that the residues be balanced. For example, consider the residues shown in Figure 4.2(a). Positive and negative residues are shown as filled and empty circles, respectively. Figure 4.2(b) shows an arrangement of branch cuts that balances the residues, but these branch cuts are clearly unsatisfactory. Notice that they completely isolate portions of the image from the rest of the image. There is no way to relate the unwrapped phase in these isolated portions to the rest of the image. Furthermore the branch cuts are excessively long. Figure 4.2(c) shows a much more appealing arrangement of branch cuts. No portion of the image is isolated, and the branch cuts are as short as possible.

This example suggests a natural criterion for choosing the branch cuts,

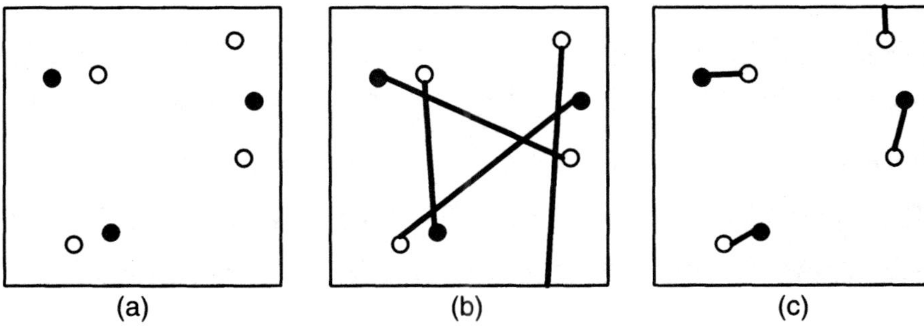

Figure 4.2 Choice of branch cuts. (a) Positive and negative residues marked by filled and empty circles. (b) A poor choice of branch cuts. The cuts balance the residues but isolate portions of the image. (c) A better choice of branch cuts.

namely that they be as short as possible. In some applications it can be shown that this criterion is statistically the best strategy [1]. Examining all possible choices of cuts in order to find the shortest ones is impractical, however. There are $n!$ possible ways to pair up n positive residues with n negative residues. There are even more possibilities if one considers arrangements of cuts in which the residues are connected in "clumps" that contain more than one pair of positive and negative residues. (The phase-unwrapping algorithm in Section 4.2 generates these types of cuts.) In this case, there are about 2^{n^2} ways to connect n residues. Because a typical 500×500-pixel image may contain thousands of residues, faster methods for finding the optimal arrangement of branch cuts must be employed.

In this chapter we present four basic path-following approaches to the phase unwrapping problem. The "classic" path-following approach, in which one identifies the residues and then balances them by connecting them with branch cuts that interdict the integration path, is attributable to Goldstein, Zebker, and Werner [2]. Their algorithm for choosing branch cuts is a fast and effective method. We present this algorithm in Section 4.2 along with a detailed description of their algorithm and its performance on a number of phase unwrapping problems. In Section 4.3 we shift gears and examine a completely different approach that does not generate branch cuts explicitly or even identify residues. Instead, this method relies on a quality map (e.g., the correlation coefficients) to guide the integration path. This method is able to unwrap some types of phase unwrapping problems on which Goldstein's algorithm fails. In Section 4.4 we merge these two methods into a hybrid technique that uses a quality map to guide the placement of the branch cuts. In Section 4.5 we describe a recently introduced algorithm that minimizes the discontinuities in the unwrapped surface. These discontinuities are analogous to branch cuts.

These four approaches constitute a fair representation of the wide variety of algorithms in path following. They are also, in our opinion, the most successful of the path-following methods. In addition to detailed pseudocode descriptions of each algorithm, we have supplied our C language implementations in Sections A.6–A.9 of Appendix A.

4.2 GOLDSTEIN'S BRANCH CUT ALGORITHM

In this section we describe in detail the classic path-following algorithm of Goldstein, Zebker, and Werner [2]. We will refer to this algorithm as *Goldstein's algorithm.*[2] The algorithm is effective at generating optimal (i.e., short) branch cuts, and it is extremely fast. The idea is to connect nearby residues with branch cuts so that the residues are balanced, namely connected in pairs of opposite polarity or "clumps" that contain multiple pairs. Residues can also be balanced by connecting them with branch cuts to the image border. The cuts are generated by a method that attempts to minimize the sum of the cut lengths.

There are other approaches in addition to Goldstein's for generating the branch cuts. Huntley [3] joined pairs of residues of opposite polarity (called *dipoles*) by means of a simple nearest-neighbor algorithm. The goal was to minimize approximately the sum of lengths of the dipole cuts. Cusack, Huntley, and Goldrein [4] investigated several techniques for finding the optimal configuration of dipole cuts, including two nearest-neighbor algorithms, a "stable marriages" approach, and a simulated annealing algorithm. Buckland, Huntley, and Turner [5] used a graph-theory approach called the Hungarian algorithm to find the configuration of dipole cuts that minimizes the sum of the lengths of the cuts. Unlike the other approaches, the latter algorithm finds a true (rather than approximate) solution to the minimization problem.

Before proceeding with our description and evaluation of Goldstein's algorithm, we remark that the algorithms cited above all differ from Goldstein's algorithm in an important respect. These algorithms are restricted to dipole cuts. In contrast to this, Goldstein's algorithm generates more general types of branch cuts that can join the residues in "clumps" rather than pairs. In some cases, such as the one illustrated in Figures 4.3(a)–(c), the latter kinds of branch cuts are preferable to dipole cuts. For this reason we have chosen to concentrate on Goldstein's algorithm as representative of the class of "branch cut" algorithms.

Later in this section we will show how Huntley's nearest-neighbor technique can be applied as a preprocessing step to improve the robustness of Goldstein's algorithm on some types of noisy phase data.

4.2.1 Algorithm

Goldstein's algorithm operates as follows: The phase image is scanned pixel by pixel until a residue is detected. A 3×3-pixel "box" or square is then placed

[2] We apologize if this terminology slights the other two authors of the algorithm, but the name *Goldstein-Zebker-Werner algorithm* is awkward, and our distaste for acronyms precludes the name *GZW algorithm.*

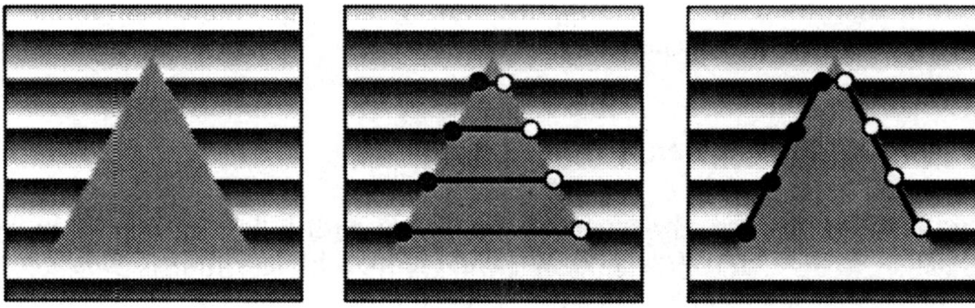

Figure 4.3 Comparison of dipole cuts and the branch cuts generated by Goldstein's algorithm. (a) These wrapped phase data have a set of residues in an inverted V shape. The residues on the left arm of the V are all positive, while those on the right arm are all negative. (b) The optimal set of dipole cuts that balances these residues. (c) The optimal set of Goldstein's branch cuts that balances the residues. These branch cuts have a shorter overall length than the dipole cuts.

around the residue and searched for another residue. If found, a branch cut is placed between the residues, even if the residues have the same polarity. If the residues have opposite polarities, they are designated "balanced" by the branch cut. If the residues have the same polarity, the search of the 3×3 box continues. Whenever a new residue is encountered, it is connected by a branch cut to the residue at the center of the box. This connection is performed even if the residue has already been connected to some other residue. If the residue has not yet been connected to another residue, its polarity (1 or -1) is added to the sum of the polarities of the other residues. If the residue has already been connected to other residues, however, its polarity is *not* added, since it has already been accounted for. When the cumulative "charge" reaches zero, the residues are designated balanced. The residue theorem (Equation 2.29) ensures that these balanced residues will render the phase unwrapping independent of the choice of integration path.

If the cumulative charge is not zero when the search of the 3×3 box ends, the box is centered at each of the other residues (in the box) in turn and searched. If the cumulative charge is still nonzero, then the box is enlarged to 5×5 pixels and centered at each of the residues in turn. This process of enlarging and repositioning the box continues until either the cumulative charge reaches zero or the box touches the image border. In the latter case a branch cut is placed to the border. The residues are then designated balanced because this branch cut will prevent any path integral from encircling the residues.

The original description of Goldstein's algorithm as published in [2] was so brief that it omitted key steps. This omission has resulted in many faulty implementations and premature conclusions that Goldstein's algorithm does not work. For example, Judge and Bryanston-Cross [6] mistakenly concluded that Goldstein's algorithm is suitable only when the phase noise is localized "spike" noise. Roth's implementation [7] resulted in an excessive number of poorly placed branch cuts, most likely due to missing steps. The implementation of Just et al. [8] drastically modified Goldstein's algorithm to the point that its performance was quite poor and probably inferior to that of Goldstein's original algorithm. Indeed our own implementation did not work correctly until a telephone call to Richard M. Goldstein [9] supplied the information necessary to correct the key steps.

It is instructive to examine the original description of Goldstein's algorithm [2] and note carefully the omitted steps. We quote the description in full:

> We have implemented the following algorithm to connect the residues with branch cuts. The interferogram is scanned until a residue is found. A box of size 3 is placed around the residue and is searched for another residue. If found, a cut is placed between them. If the residue is of opposite sign, the cut is designated "uncharged" and the scan continues for another residue. If, however, the sign of the residue is the same as the original, then the box is moved to the new residue and the search continues until either an opposite residue is located and the resulting total cut is uncharged or no new residues can be found within the boxes. In the latter case, the size of the box is increased by 2 and the algorithm repeats from the current starting residue.

This description omits two crucial steps that must be performed, according to Richard M. Goldstein [9]:

1. Whenever a residue is encountered during the search of a box, it is always connected by a branch cut to the residue at the center of the box, *even if the residue has already been joined to some other residue by a branch cut.*
2. If a border pixel is encountered during the search of a box, a branch cut is placed to the border. This serves to balance or discharge the residues because no integration path will be able to encircle the residues.

Before proceeding with our examination of Goldstein's algorithm, we must explain how the residues and the branch cuts are represented in the arrays. Because a residue is defined by four pixels, we adopt the convention that the *upper left* pixel of the four is marked as the residue. We also adhere to the usual convention in image processing that a pixel to the right of another pixel has the greater i index and a pixel below another pixel has the greater j index. Thus, when we say there is a residue at pixel (i, j), we mean the four pixels (i, j), $(i+1, j)$, $(i+1, j+1)$, and $(i, j+1)$ form a residue.

Because of these conventions, a branch cut that connects the two residues shown in Figure 4.4(a) can be defined as in Figure 4.4(b). It need not include the pixel for the residue at the upper left because it only needs to join the "center" of the four pixels where the residue is actually located, as indicated by Figure 4.4(c).

We now give an example to illustrate how Goldstein's algorithm operates. Suppose that a phase image contains the two residues marked in Figure 4.5(a). The image is scanned from top to bottom until the positive residue is detected. A 3×3 box is then placed around the residue as shown in Figure 4.5(b) and

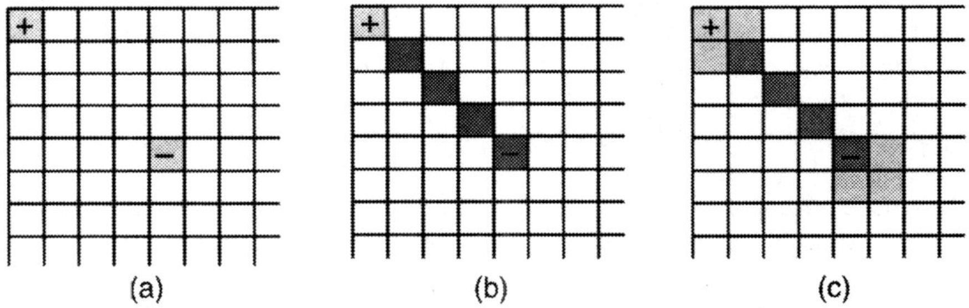

Figure 4.4 Branch cut connecting two residues. (a) A positive and a negative residue. (b) The branch cut that joins these residues is composed of the dark pixels. (c) The branch cut need not include the uppermost (or leftmost) pixel because it needs only join the "center" of the four pixels that define each residue.

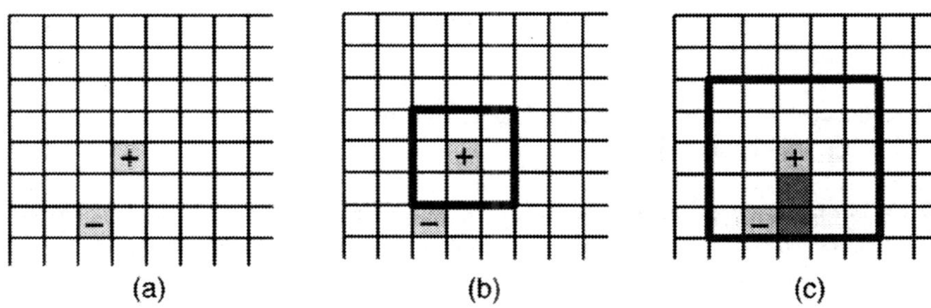

Figure 4.5 The operation of Goldstein's algorithm. (a) A positive and a negative residue. (b) The 3 × 3 box to be searched. (c) The 5 × 5 box to be searched and the resulting branch cut (dark pixels) that the algorithm places between the residues.

searched for additional residues. Because none are found, the box is enlarged to 5 × 5 pixels as shown in Figure 4.5(c) and searched again. The positive residue is detected, and the branch cut shown in Figure 4.5(c) is placed between the residues, thereby balancing them. The scan for other unbalanced residues then continues.

Suppose that there is a third residue as shown in Figure 4.6(a). The algorithm detects the residue and then searches the 3 × 3 box centered there. The box is enlarged to 5 × 5 and then 7 × 7 pixels before another residue is found. In this case the residue marked with the arrow is detected, and the branch cut shown in Figure 4.6(b) is placed between the residues. Because there are no other residues in the 7 × 7 box, the box is centered at the residue marked with the arrow and searched. Because border pixels lie in this box, the branch cut shown in Figure 4.6(c) is placed between the residue and the border. This balances all the residues.

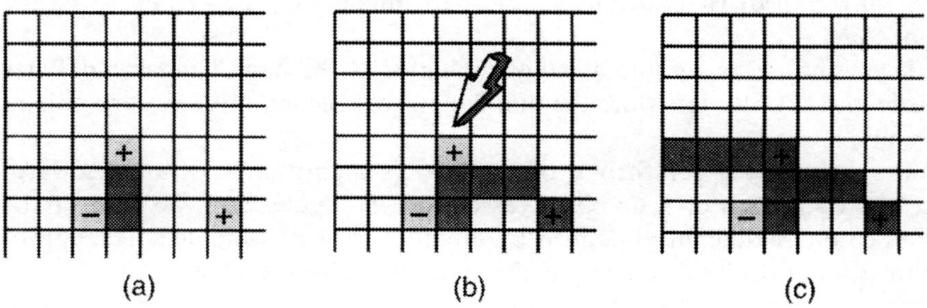

Figure 4.6 The operation of Goldstein's algorithm in the presence of a third residue. (a) The new residue lies on the right. (b) The second branch cut (dark pixels) joins the new residue to the residue marked with the arrow. (c) The third branch cut joins the residue marked with the arrow in (b) to the image border.

4.2.2 Pseudocode

We are now in a position to present the pseudocode for Goldstein's algorithm. At the highest level the algorithm consists of three steps:

Step 1. Identify the residues
Step 2. Generate the branch cuts
Step 3. Path-integrate around the branch cuts

We employ the following means of "bookkeeping" to keep track of the residues, branch cuts, and associated information. Each pixel has seven flags associated with it:

1. Positive residue
2. Negative residue
3. Branch cut
4. Image border
5. Balanced residue
6. Active residue
7. Unwrapped pixel

Most of the flags are self-explanatory. The "image border" flag denotes the undefined pixels near the border as in the simulated IFSAR examples of Figures 3.7(a) and 3.8(a). When there are no such border pixels, as on the other examples, this flag is unused. The "balanced residue" flag keeps track of which residues have been balanced with branch cuts. The "active residue" flag indicates whether a residue is active, which means it is connected to the current set of branch cuts. Any residue that is encountered during the search of the boxes is marked active. The "unwrapped pixel" flag, which marks the pixels that have been unwrapped, is used only in the path integration step (Step 3) of the algorithm.

These seven flags are stored and manipulated as "bit flags," an array of bytes whose lowest-order bits store the first flag, whose second bits store the second flag, and so on.

The pseudocode for Step 1 of Goldstein's algorithm is straightforward. Around each loop of four pixels (as shown in Figure 2.5), the sum of the wrapped phase differences defined by Equation 2.41 is computed. The appropriate pixel of the bit flags array is then labeled if it is a residue.

Step 1. Identification of Residues

```
for (each pixel (m, n) in the image)
    Sum the wrapped phase differences (see Equation 2.41):
      W{ψ(m, n + 1) − ψ(m, n)} + W{ψ(m + 1, n + 1) − ψ(m, n + 1)}
      +W{ψ(m + 1, n) − ψ(m + 1, n + 1)} + W{ψ(m, n) − ψ(m + 1, n)}
    if (the sum is 2π) then mark pixel (m, n) as a positive residue
    if (the sum is −2π) then mark pixel (m, n) as a negative residue
  end
```

The substance of Goldstein's algorithm is the generation of the branch cuts (Step 2). The phase data are not needed for this step. Only the residues, which are stored in the bit flags array, are needed. As the branch cuts are generated, they also are stored in the bit flags array.

Step 2. Generation of Branch Cuts

```
for (each unbalanced residue)
  Mark the residue active
  Let charge = 1 if the residue is positive, −1 if negative
  for (n = 3 to MaxBoxSize} step 2)
    for (each active pixel)
      Center the n × n box at the active pixel
      for (each ''box pixel'', i.e., each pixel in the n × n box)
        if (box pixel is a border pixel) then
          Let charge = 0
          Place branch cut between the active pixel and the border
        else if (the box pixel is an inactive residue) then
        if (the box pixel is unbalanced) then
            Add its polarity (1 or −1) to charge
            Mark the box pixel balanced
          end
          Mark the box pixel active
          Place branch cut between the active pixel and the box pixel
```

```
          end
          if (charge = 0) then jump to the statement marked * below
        end
      end
    end
*   if (charge is nonzero) place a branch cut to the border
    Mark all the active pixels balanced and inactive
  end
```

Our implementation of Step 3 of Goldstein's algorithm, the path integration step, uses a "flood-fill" algorithm. The basic form of the algorithm will be used (with suitable modifications) in Sections 4.3 and 4.4. The algorithm maintains a list of pixel indexes for keeping track of which pixels adjoin the unwrapped pixels. We call this list the "adjoin list." The algorithm stores the unwrapped values in an array called the "solution array."

The flood-fill algorithm is straightforward. A starting pixel is selected, and the phase of this pixel is stored as the unwrapped value in the solution array. The four neighbors of this pixel are then unwrapped, the unwrapped values being placed in the solution array, and the pixels are inserted in the adjoin list. The algorithm proceeds in an iterative fashion by selecting (and removing) a pixel from the adjoin list and unwrapping and inserting its four neighbors in the list, avoiding branch cut pixels and pixels that have already been unwrapped. Eventually the adjoin list becomes empty, which means either that all the pixels have been unwrapped or there is a region of the image that is isolated by a branch cut. Any remaining isolated regions are unwrapped by selecting a starting pixel within the region and repeating the process.

After all the non–branch cut pixels have been unwrapped, the branch cut pixels that adjoin unwrapped pixels are unwrapped. This step must be performed because, strictly speaking, the branch cuts lie *between* the pixels, not on top of them. Unwrapping only the branch cut pixels that adjoin unwrapped pixels avoids the problem of unwrapping *across* the branch cuts.

Step 3. Path Integration around Branch Cuts

```
repeat
  Select a starting pixel
  Store its phase value in the solution array
  Mark the pixel unwrapped
  Update the adjoin list (see below)
  while (the adjoin list is not empty)
    Fetch any pixel from the adjoin list
    Update the adjoin list (see below)
  end
until (all nonborder, non-branchcut pixels have been unwrapped)
```

```
for (each branch cut pixel)
  if (the pixel adjoins an unwrapped pixel)
    Unwrap the branch cut pixel
  end
end
```

The "Update the adjoin list" step of the pseudocode accepts a pixel as an input parameter and performs the following operation:

Update the Adjoin List

```
for (each of the four adjoining pixels)
  if (the pixel is not on the adjoin list and not unwrapped)
    if (the pixel is not a branch cut and not a border pixel)
      Unwrap the pixel and store the value in the solution array
      Insert the pixel in the adjoin list
      Mark the pixel unwrapped
    end
  end
end
```

4.2.3 Implementation

Our C language implementation of Goldstein's algorithm can be found in Section A.6 of Appendix A. The software for most of the basic functions (residue identification, placement of branch cuts, and list management) can be found in Section A.2. The memory requirements for Goldstein's algorithm are as follows:

1. Phase data array
2. Bit flags array
3. Solution array
4. Adjoin list

All arrays are the same size and data type as the input phase data, except for the bit flags array, which is an array of bytes rather than real numbers. (The memory requirements for the adjoin list are negligible if one implements the simple list-trimming procedure described in Section 4.3.1.)

Goldstein's algorithm is extremely fast. Table 4.1 lists the average execution times of our implementation for various array sizes. (These times are given for purposes of comparison with the other phase unwrapping algorithms in this book.) The actual times depend not only on the array size but also on the number and placement of the residues. The times shown in the table were averaged over several different phase unwrapping problems.

Table 4.1 Average execution times of Goldstein's algorithm for various array sizes on an IBM RS/6000 workstation

Array Size (pixels)	Execution Time (min:s)
256 × 256	0:02
512 × 512	0:06
1024 × 1024	0:25
2048 × 2048	3:00

4.2.4 Examples

We will now present the results of Goldstein's algorithm on the phase unwrapping examples of Chapter 3.

The first example is the IFSAR data of Figure 3.2. Goldstein's algorithm seems to unwrap this example correctly, as the branch cuts in Figure 4.7(a) show. Compare these branch cuts with the residues shown in Figure 3.2(b). The branch cuts are confined to the patches of corrupted (noisy) phase, and they do not isolate any regions of valid phase. This enables the path integration step to unwrap around all of the patches. A bird's-eye shaded rendering of the unwrapped surface is shown in Figure 4.7(b).

Figures 4.7(c) and (d) show close-up views of the branch cuts and the unwrapped surface for the portion of the image marked by the box. Compare these figures with the close-up views of the phase and residue data in Figures 3.2(c) and (d). The patches of noise appear as noisy, "dimpled" patches in the unwrapped surface as shown in Figure 4.7(d). Because the branch cuts prevent the path integral from encircling unbalanced residues, one can be assured that any unwrapping errors are confined to these patches without affecting the unwrapped surface outside the patches.

Goldstein's algorithm also appears to unwrap the MRI images correctly. On the MRI head data of Figure 3.3(a), the algorithm produces the branch cuts shown in Figure 4.8(a). As in the IFSAR example the branch cuts are confined to the regions of corrupted phase. Some regions inside the head are isolated by branch cuts, but these are unavoidable due to the corrupted phase that completely surrounds these regions. Compare these branch cuts with the residues shown in Figure 3.3(b). A shaded bird's-eye view of the unwrapped surface produced by the algorithm is shown in Figure 4.8(b).

While the unwrapping appears qualitatively correct, it turns out that some isolated regions inside the head have the wrong 2π multiple, leading to an incorrect water/fat separation. The correct solution is simply not achievable unambiguously from the input data due to the isolation of these regions by corrupt phase. On this example we will therefore judge each phase unwrapping algorithm according to whether its result appears qualitatively correct.

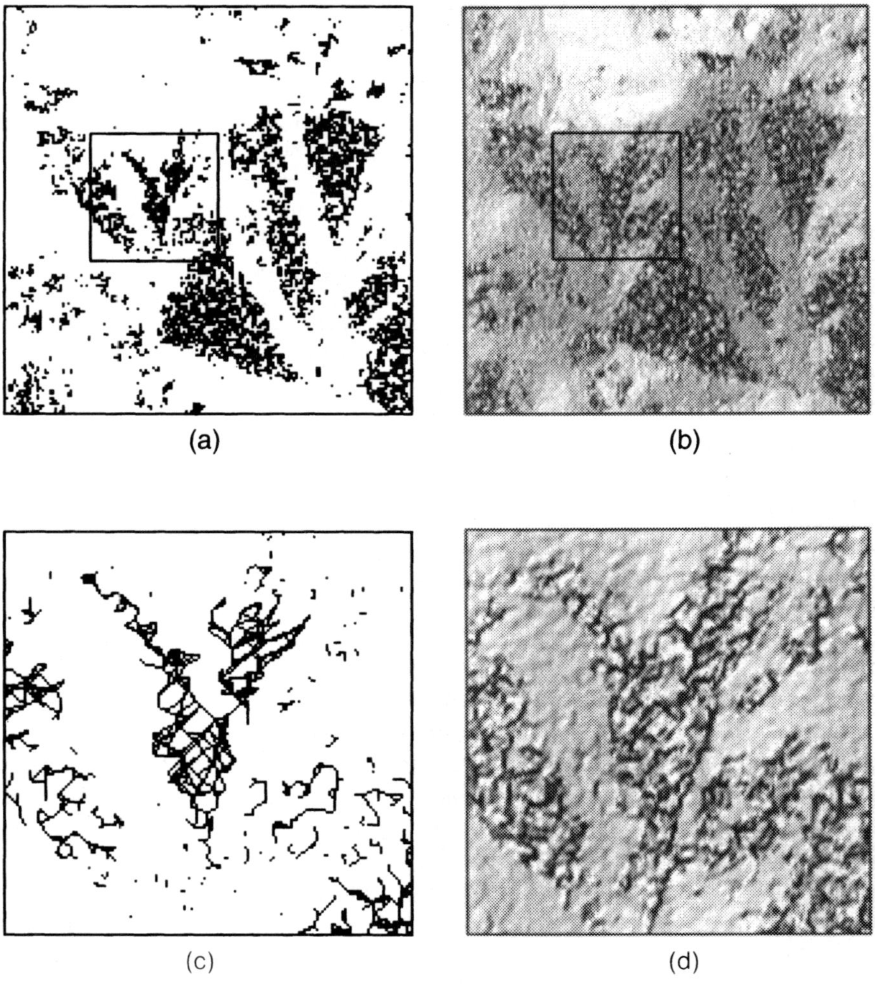

Figure 4.7 The results of Goldstein's algorithm on the IFSAR example of Figure 3.2. (a) The branch cuts are confined to the regions of corrupted phase. (b) A shaded bird's-eye view of the unwrapped surface. (c) A close-up view of the branch cuts in the region marked by the box. (d) A close-up view of the shaded surface in the region marked by the box.

On the MRI knee data of Figure 3.4(a), the algorithm produces the branch cuts of Figure 4.9(a) and the unwrapped surface of Figure 4.9(b). Once again, the branch cuts are confined to the noisy regions, and the extensive regions of corrupted phase that surround the knee do not corrupt the unwrapping of the phase within the knee. Analysis of the unwrapped phase by clinical diagnosticians indicated that the correct water/fat separation was achieved. Compare the branch cuts with the residues shown in Figure 3.4(b).

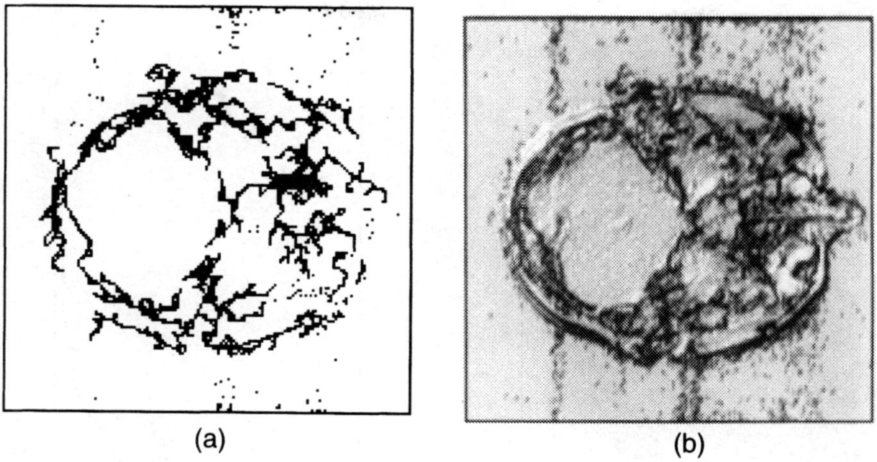

Figure 4.8 Results of Goldstein's algorithm on the MRI head example of Figure 3.3. (a) The branch cuts are generally confined to the regions of corrupted phase. Some regions inside the head are isolated by cuts due to the corrupted phase that surrounds these regions. (b) A shaded bird's-eye view of the unwrapped surface.

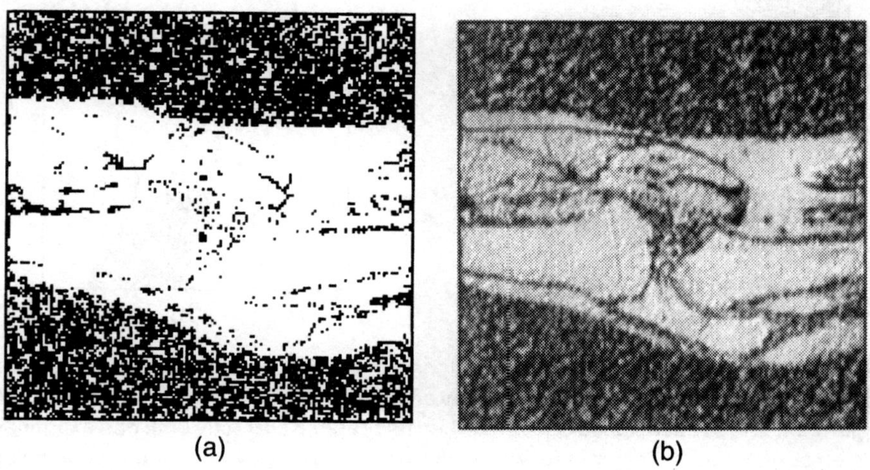

Figure 4.9 The results of Goldstein's algorithm on the MRI knee example of Figure 3.4. (a) The branch cuts are generally confined to the regions of corrupted phase. (b) A shaded bird's-eye view of the unwrapped surface.

We now turn our attention to the simulated IFSAR examples. These challenging phase unwrapping problems will enable us to quantify the accuracy of the surfaces produced by Goldstein's algorithm.

The first such example is the Long's Peak data of Chapter 3. A shaded view of the terrain elevation surface is shown in Figure 3.6(a), and the phase and residue

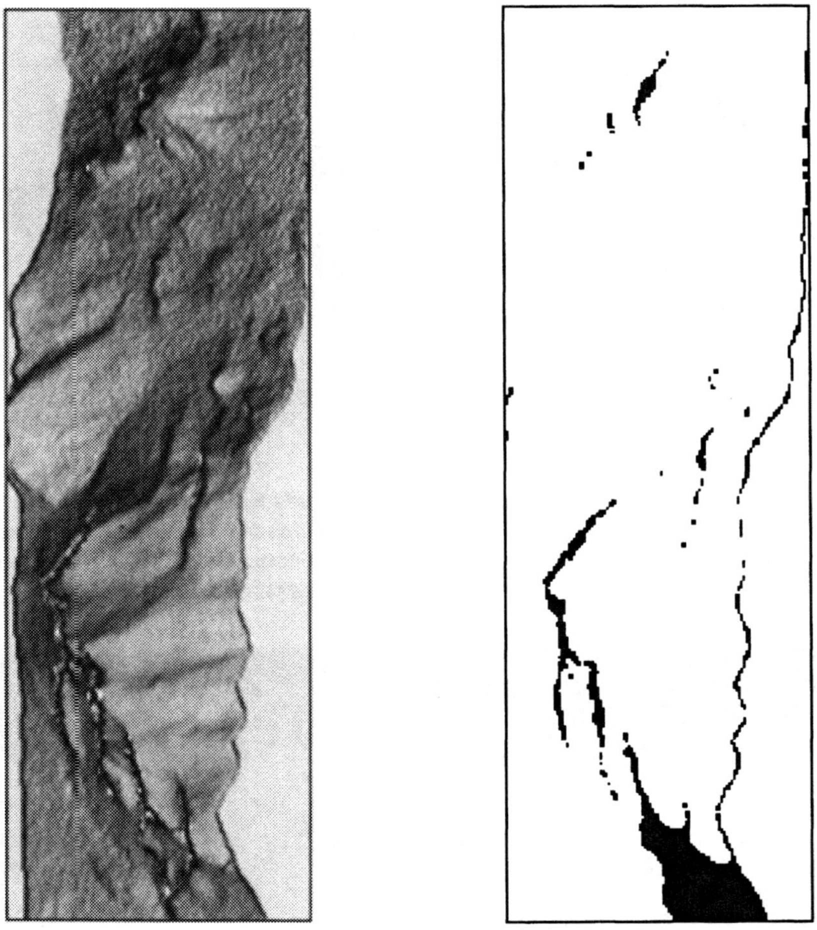

(a) (b)

Figure 4.10 The results of Goldstein's algorithm on the Long's Peak IFSAR example of Figure 3.7. (a) A shaded bird's-eye view of the unwrapped surface produced by the algorithm. The surface appears to have been unwrapped correctly until one examines the error map. (b) The error map, which shows the incorrectly unwrapped regions as black pixels. There are minor errors in the regions of layover as one would expect, but there is also a large region of error in the lower portion of the image.

data are shown in Figures 3.7(a) and (b). Goldstein's algorithm produces the unwrapped surface shown in Figure 4.10(a). This surface appears correct until one examines the "error map," which is shown in Figure 4.10(b). The error map compares the unwrapped surface with the actual surface that was used to generate the phase data. Any regions that differ are marked as black pixels.[3]

[3] We have supplied software for producing these error maps in Section A.5 of Appendix A.

As expected, there are errors in the regions of radar layover where the phase is corrupted. More significant, however, is the large region of error in the lower portion of the image.

To discover what caused this region of error, we first must examine the branch cuts produced by Goldstein's algorithm. Figure 4.11(a) shows these branch cuts. The problematic region is marked with a box and enlarged in Figure 4.11(b). As one can see, part of the region is completely isolated by branch cuts, which certainly would be expected to result in an incorrect unwrapping.

The residues, which are shown in Figure 4.11(c), lie in several clusters. These clusters correspond to the regions of radar layover in the SAR images. Goldstein's algorithm placed branch cuts between these layover regions, thereby isolating part of the image as shown in Figure 4.11(b). Examination of the correlation image in Figure 4.11(d) shows that some of the branch cuts were forced to cross regions of high-quality phase. (In the correlation image the dark pixels represent regions of low-quality phase, while the light pixels represent high-quality phase.)

The isolated portion of the image is only part of the problem. The rest of the error region lies to the lower right, and the problem here seems to have been caused by the isolated residue marked by the arrow in Figure 4.11(c). This residue is joined to the image border by the branch cut marked by the arrow in Figure 4.11(b), leaving a gap through which the path integration step of Goldstein's algorithm must travel to unwrap the phase in the lower-right portion of the image. This gap consists of low-quality phase, however, as the correlation image of Figure 4.11(d) shows. Thus an unwrapping error occurs as this gap is crossed. The branch cuts appear to have been placed in a poor location because they cross high-quality phase. It would have been better if the residue marked by the arrow were joined to the other residues on the left. This would have left a gap of high-quality phase, as indicated by the arrow of Figure 4.11(d), between this residue and the border for the path integration step to cross. In order to place the branch cuts in this manner, however, Goldstein's algorithm would have needed to exploit the information about the quality of the phase data (e.g., the correlation coefficients) to guide the placement of the branch cuts. We will study the idea of exploiting the quality information in Sections 4.3 and 4.4.

On the Isolation Peak example of Figures 3.6(b) and 3.8, the story is similar. Figure 4.12(a) shows the shaded surface produced by Goldstein's algorithm, and Figure 4.12(b) shows the error map. In addition to the unavoidable errors in the regions of layover, there are four small regions of error in the upper part of the image and a very large region of error in the lower part. We will examine the large region of error.

Figure 4.13(a) shows the branch cuts produced by Goldstein's algorithm. The large region of error is almost isolated by branch cuts from the rest of the image, except for two small gaps. The branch cuts are shown in an enlarged view in Figure 4.13(b). The residues that these branch cuts connect are shown in Figure 4.13(c), and the correlation data are shown in Figure 4.13(d). The reader will notice the two small gaps in Figure 4.13(b) through which the path

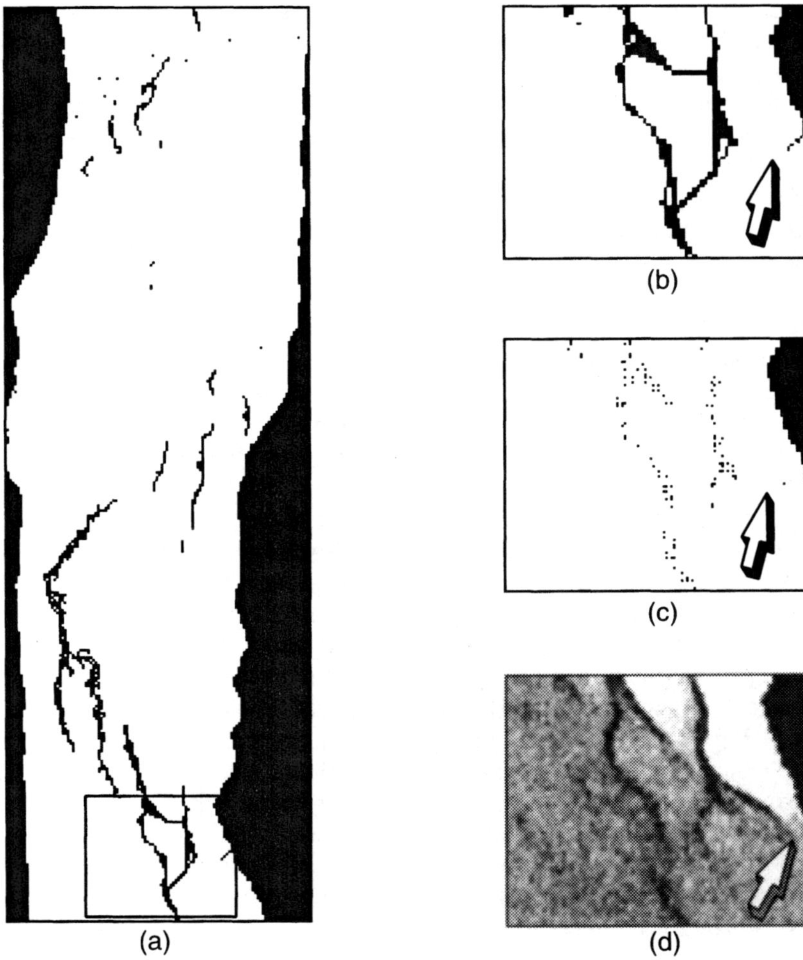

Figure 4.11 (a) The branch cuts produced by Goldstein's algorithm on the Long's Peak example. (b) The branch cuts in the region enclosed by the box in part (a). The arrow marks the misplaced branch cut. (c) The residues in the box. The arrow marks the problematic residue. (d) The correlation data for the region enclosed by the box. Notice that the misplaced branch cut in (b) crosses high-correlation data (marked by the arrow), leaving a gap of low-quality pixels through which the path-integration step must pass.

integration step must pass. These gaps are composed of low-quality phase data, as the correlation map of Figure 4.13(d) shows. As these gaps are crossed, an error is introduced that corrupts all of the phase in the lower portion of the image. It would have been better if the branch cuts marked by the arrow of Figure 4.13(b) had been placed elsewhere, since they interdict the path of

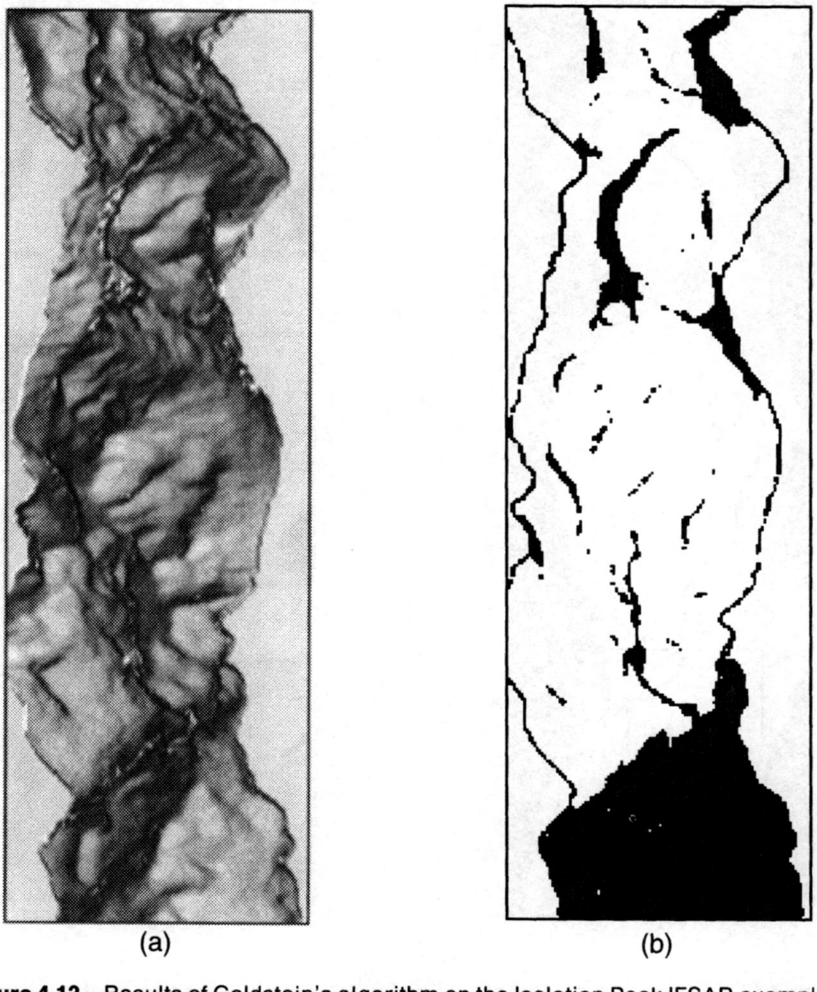

Figure 4.12 Results of Goldstein's algorithm on the Isolation Peak IFSAR example of Figure 3.8(a). (a) A shaded bird's-eye view of the unwrapped surface produced by the algorithm. The surface appears correct until one examines the error map. (b) The error map. The pixels that were unwrapped incorrectly are marked in black. There are minor errors in the regions of layover, as one would expect. However, there are also four small regions of error in the upper part of the image and a very large region of error in the lower part.

highest-quality phase from the upper to the lower portions of the image. This high-quality region is marked by the arrow of Figure 4.13(d).

It should be emphasized that the phase data were unwrapped correctly *within* these regions of error. It is only the multiple of 2π that relates them to the rest of the unwrapped surface that was determined incorrectly.

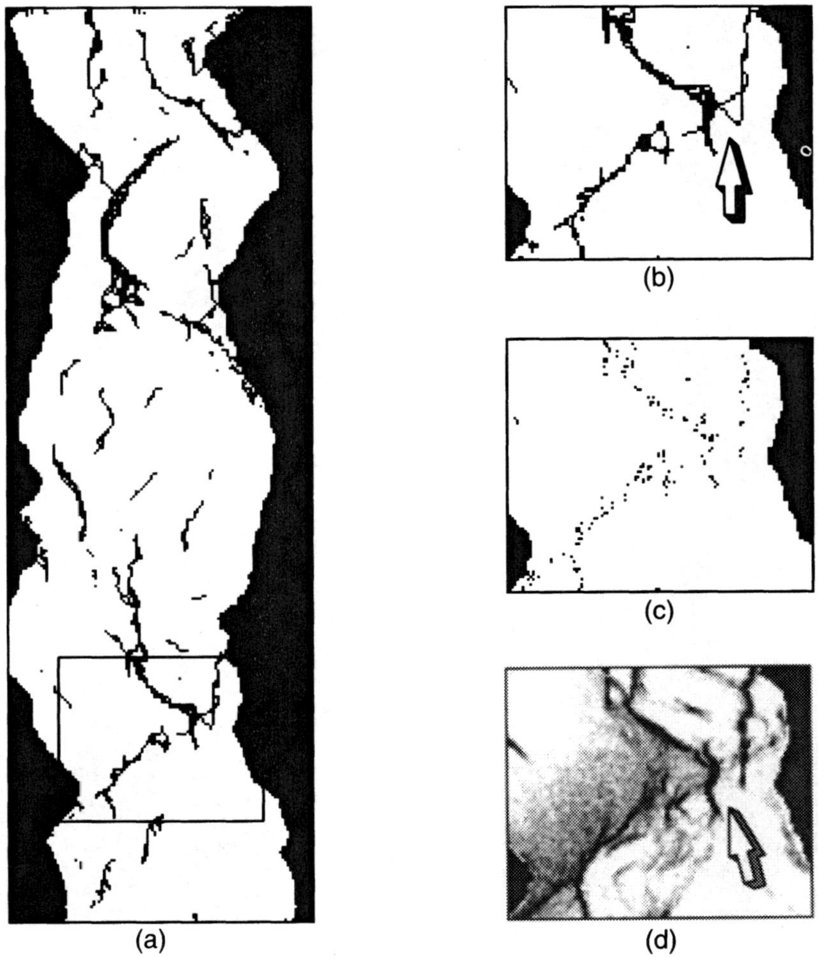

Figure 4.13 (a) The branch cuts produced by Goldstein's algorithm on the Isolation Peak example. (b) The branch cuts in the region enclosed by the box in part (a). There are only two small gaps through which the path integration step can pass, and these gaps are in regions of low-quality pixels. The arrow shows where branch cuts were misplaced, thereby cutting off the path of high-quality phase. (c) The residues. (d) The correlation data. The arrow shows the high-quality phase that the branch cuts of (b) interdict.

We now turn to the shear examples. On the sheared planes example of Figure 3.9, Goldstein's algorithm unwraps the phase flawlessly. It connects the residues along the line of shear with branch cuts, thereby enabling the path integration step to unwrap each plane independently without crossing the shear line.

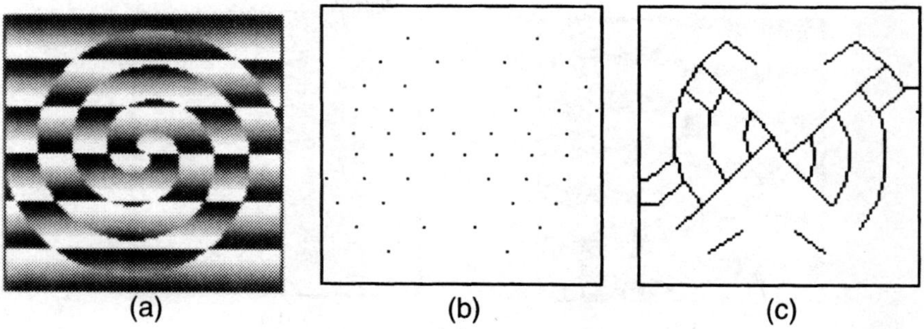

Figure 4.14 "Spiral shear" example. (a) The wrapped phase. (b) The residues all lie on the curved lines of shear. (c) The branch cuts produced by Goldstein's algorithm are completely useless.

On the spiral shear example, however, the algorithm fails completely. The phase data for this example are shown in Figure 4.14(a), and the residues are shown in Figure 4.14(b). Goldstein's algorithm produces the branch cuts shown in Figure 4.14(c). Rather than restricting the branch cuts to the curved lines of shear, the algorithm places the cuts *across* the arms of the spiral. This results in an unwrapped surface that is almost useless. What went wrong is clear: Goldstein's algorithm does not "know" about the shear lines. This is essentially the same problem the algorithm exhibited on the simulated IFSAR examples, where the algorithm misplaced branch cuts because it did not "know" where the low-quality regions were. Goldstein's algorithm is designed to minimize the lengths of the branch cuts by connecting the residues to their nearest neighbors regardless of the quality of the phase values. Its nearest-neighbor strategy is *not* always the best approach. In Sections 4.3 and 4.4 we will consider alternative strategies that are guided by the quality information.

4.2.5 Enhancement

In this section we will show one method for enhancing Goldstein's algorithm. We begin with an example to motivate the need for such an enhancement. Figure 4.15(a) shows the wrapped phase of the sheared planes of Figure 3.9 corrupted with Gaussian noise. The noise has introduced additional residues as shown in Figure 4.15(b). Most of these residues are closely spaced "dipoles," or pairs of residues of opposite polarity. If we apply Goldstein's algorithm to this example, we obtain the branch cuts shown in Figure 4.16(a), which result in the error map of Figure 4.16(b). Most of the dipoles have been connected with branch cuts correctly, but the dipoles near the shear line have caused the algorithm to "wander" in the search for residues to balance the cuts. This observation suggests that the algorithm would perform better if the dipoles were connected with cuts and then *removed* before applying Goldstein's algorithm.

We can perform such a dipole-removing procedure as a preprocessing step.

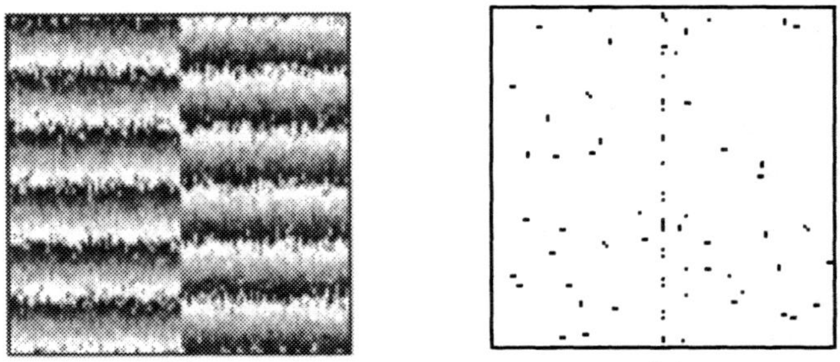

(a) (b)

Figure 4.15 Sheared planes example, corrupted with Gaussian noise. (a) The wrapped phase. (b) The residues.

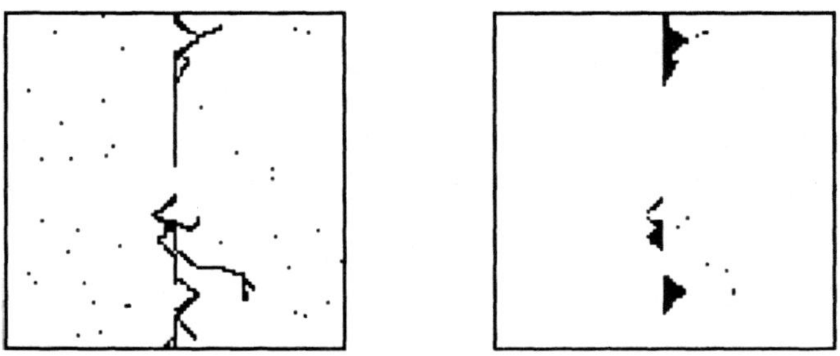

(a) (b)

Figure 4.16 Results of Goldstein's algorithm on the noise-corrupted sheared planes example. (a) The branch cuts produced by the algorithm. (b) The error map.

We use the nearest-neighbor procedure of [3] to connect the dipoles, but we consider only *adjoining* residues. Such a procedure by itself is insufficient to connect all the residues in a satisfactory manner, but it proves to be an effective preprocessing step for Goldstein's algorithm.

Dipole Remover (Pre-processor)

```
for (each residue)
  if (there is an adjoining residue of opposite polarity)
    Place a branch cut between the two residues
    Remove the two residues
  end
end
```

The C language software for this preprocessing step can be found in Section A.2 of Appendix A.

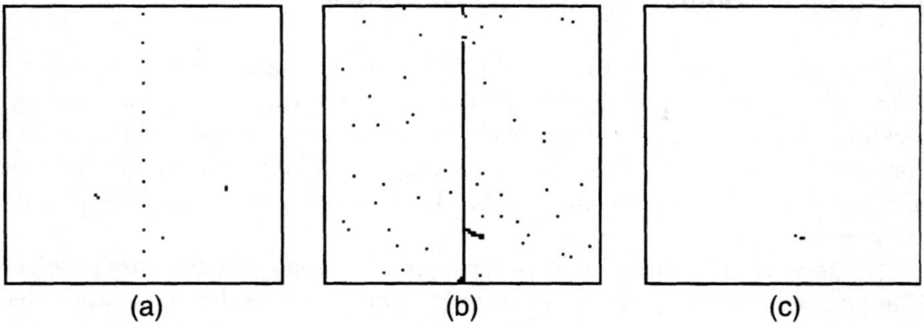

Figure 4.17 Results of Goldstein's algorithm with the dipole-removing preprocessing step applied to the noise-corrupted sheared planes example. (a) The residues that remain after the dipole-removing step. (b) The resulting branch cuts, including the dipole cuts. (c) The error map.

After the dipoles are connected, the residues are removed from the bit flags array because they do not need to be considered by Goldstein's algorithm. (Indeed, if we did not remove them, Goldstein's algorithm would produce exactly the same configuration of branch cuts as it did without the preprocessing step.) The branch cuts remain in the bit flags array to prevent the path integration step from encircling the unbalanced residues.

With this preprocessing step, Goldstein's algorithm produces a much better unwrapping of the noise-corrupted example. Figure 4.17(a) shows the residues that remain after the preprocessing step, and Figure 4.17(b) shows the branch cuts produced by the preprocessing step and Goldstein's algorithm. Compare these with the branch cuts of Figure 4.16(a). The resulting surface is almost flawless, as the error map of Figure 4.17(c) shows.

4.2.6 Conclusion

Goldstein's algorithm is the classic path-following algorithm. It is extremely fast and generally satisfactory. By examining the branch cuts and determining if any appear to be placed poorly or if any isolate a region, the user can usually determine if the algorithm failed. However, the branch cuts do not always indicate when there are problems. An entire region may have the wrong 2π multiple due to a poorly placed branch cut, and the user may not be warned. On some problems, such as the spiral shear example, Goldstein's algorithm fails due to the nearest-neighbor strategy, which places the branch cuts in undesired locations. To avoid these problems, we need algorithms that somehow exploit additional information from the phase data to determine where the branch cuts should lie. We will investigate such algorithms in the following sections of the chapter.

4.3 QUALITY-GUIDED PATH FOLLOWING

In the preceding section we examined examples where the nearest residue approach of Goldstein's algorithm places the branch cuts in the "wrong" locations. In this section we approach the phase unwrapping problem from a completely different point of view by asking the following question: Is there information (besides the residues) in the phase data that can be used to guide the integration path?

A glance at the quality maps of Chapter 3 suggests that the answer is yes. Comparison of these quality maps with the phase and residue data shows that the corrupted phase (and residues) tend to have low-quality values. This suggests an approach to phase unwrapping in which the integration path follows the high-quality pixels and avoids the low-quality pixels. This approach does not identify the residues nor generate branch cuts. Instead, it depends on the assumption that a good quality map will guide the integration path without encircling any unbalanced residues. The assumption is, perhaps, risky but it is surprisingly reliable in practice.

The first researcher to adopt a quality measure to guide the unwrapping seems to have been Bone [10]. He calculated the second differences (i.e., second-order partial derivatives) of the phase data and then applied a threshold to produce a "quality mask" of the pixels to be unwrapped. Only those pixels whose second differences did not exceed the threshold were unwrapped, and they were unwrapped in any order. Xu and Ai [11], working with optical interferometry data, used a quality measure that was based on the modulation of the pattern of phase fringes. Like Bone, they applied a threshold to define the mask of pixels to be unwrapped and then applied morphological operations to enlarge the mask. Quiroga, González-Cano, and Bernabeu [12] enhanced Bone's technique by applying an adaptive threshold. Lim, Xu, and Huang [13] extended these techniques to a true quality-guided path follower by allowing the threshold to increase as the unwrapping progressed. This had the effect of unwrapping the highest-quality pixels first, followed by increasingly lower-quality pixels until all the pixels were unwrapped. Their quality measure was defined to be an L^1 measure of the variance of the phase derivatives. Later Xu and Cumming [14] used the correlation coefficients of IFSAR data as the quality measure. Roth [7] brought these approaches to maturity by letting the quality measures define the order in which the pixels were unwrapped without the need for thresholds. Roth also used IFSAR correlation coefficients as his quality measure and noted that the pseudocorrelation could also be used.

In this section we present our implementation of a quality-guided path follower. It allows the use of any quality map to guide the unwrapping path.

4.3.1 Algorithm

The quality-guided path-following algorithm is essentially the flood-fill algorithm of Section 4.2.1 in which the order of fill is determined by the quality map.

The quality map consists of an array of values that define the quality or goodness of each phase value. The map can be defined from the phase data, such as pseudocorrelation, variance of phase derivatives, or maximum phase gradients, or it can be derived from some other source, such as an IFSAR correlation image. Since we discussed quality maps in detail in Chapter 3, we will assume in this chapter that the quality map is already given.

The algorithm operates as follows: A starting pixel with a high-quality value is selected and its four neighbors are examined. These neighbors are unwrapped and stored in a list called the "adjoin list." The algorithm then proceeds iteratively as follows: The list pixel with the highest-quality value is removed from the list, and its four neighbors are unwrapped and placed in the list. (The pixels are sorted and stored in the list in the order of their quality values.) If a neighbor has already been unwrapped, it is not placed in the list. This iterative process of removing the highest-quality pixel from the list, unwrapping its four neighbors and inserting them in the list continues until all the pixels have been unwrapped.

Figure 4.18 illustrates the adjoin list. If the unwrapped pixels consist of the lightly shaded pixels of Figure 4.18(a), then the adjoin list will consist of the dark pixels shown in Figure 4.18(b). If the pixel in the third row and the second column is selected and unwrapped, then the adjoin list will be updated as shown in Figure 4.18(c).

It is natural to view this algorithm as a region-growing approach. In areas of high-quality pixels, the region of unwrapped pixels grows until only low-quality pixels are left. The algorithm then follows the path of highest-quality pixels through the low-quality regions until the next high-quality region is encountered. In this way, all the high-quality pixels are unwrapped first, and the low-quality pixels are unwrapped last.

Some algorithms (e.g., Roth's [7] and Xu and Cumming's [14]) grow the highest-quality regions independently of one another and then join them in a

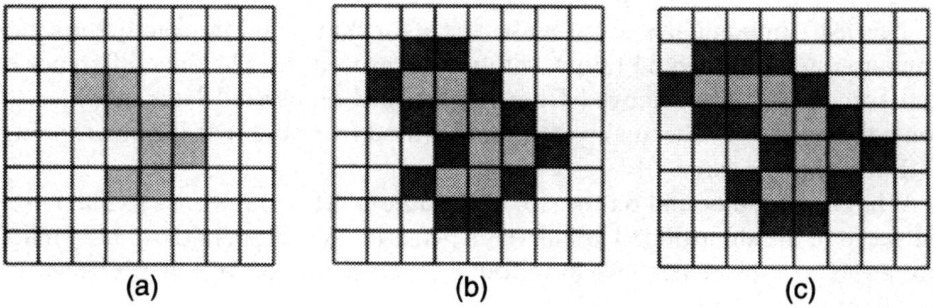

Figure 4.18 The adjoin list. (a) The lightly shaded pixels represent those that have been unwrapped. (b) The adjoin list consists of the dark pixels. (c) If the pixel in the third row and the second column is selected from the list and unwrapped, then the adjoin list will be updated as shown.

postprocessing step. It can be argued that this multiple-region growing approach, if done properly, finds the best places to join the regions. However, this approach entails much "bookkeeping" that may be unnecessary, in our opinion. It also requires a technique for joining the regions together. Our approach is much simpler and seems to achieve comparable results.

The algorithm is very similar to the flood-fill procedure of Goldstein's algorithm (Section 4.2.1). It requires arrays for the phase, the bit flags, and the solution. It also requires an array to store the quality map. The main differences between the algorithms are the ways in which they manage the adjoin list. The flood-fill procedure removes pixels from the list in any order, but the quality-guided algorithm must maintain the list in sorted order based on the quality values. This adds to the execution time significantly and necessitates the second major difference between the algorithms: the quality-guided algorithm must employ a "list-trimming" procedure to keep the size of the list small. Without this procedure, the execution time would be an order of a magnitude longer. It should be noted that the flood-fill procedure of Step 3 of Goldstein's algorithm can also employ this list-trimming procedure, not to speed up Goldstein's algorithm but to reduce its memory requirements.

The list-trimming procedure works as follows: When the size of the list exceeds a predetermined bound, half of the pixels with the lowest-quality values are removed from the list and marked with a special flag. We will refer to these pixels as "postponed" pixels because they will be placed back on the list later. Their unwrapped values are not changed. The "minimum-quality threshold" is then set to the lowest-quality value of the pixels that remain on the list. No pixel whose quality value falls below this threshold will be inserted in the list. Such a pixel, when encountered, will instead be unwrapped and then marked "postponed."

The algorithm proceeds until the adjoin list becomes empty. At this point all of the remaining pixels that adjoin the unwrapped pixels have quality values that are too low. The minimum quality threshold must then be reduced, at which point the postponed pixels whose quality values exceed this new threshold are placed on the list, and the processing continues.

This list-trimming procedure causes the algorithm to follow exactly the same unwrapping path it would follow without the procedure. The only difference is that some pixels are removed from the list and re-inserted later in order to control the list size. The quality threshold is used to ensure that the unwrapping path remains the same.

What is a good bound on the size of the adjoin list if one wishes to maximize the speed of the algorithm? The answer depends on the nature of the quality map. For an $M \times N$-pixel array we have found that a bound of $M + N$ pixels works well in practice.

4.3.2 Pseudocode

The quality-guided path follower does not identify the residues or generate branch cuts. It uses the bit flags array only to manage the following flags:

1. Border pixel
2. Unwrapped pixel
3. Postponed pixel

The "border pixel" flag marks the pixels at which the phase is undefined and should not be unwrapped, as in the simulated IFSAR examples of Chapter 3. The "unwrapped pixel" flag keeps track of which pixels have been unwrapped, and the "postponed pixel" flag keeps track of which pixels have been removed temporarily from the adjoin list, as explained in Section 4.3.1.

Quality-Guided Path Follower

```
Let MinQualityThresh be a small value
Let MaxListSize be a suitable value
Select a starting pixel
Store its phase value in the solution array
Mark the pixel unwrapped
Update the adjoin list (see below)
while (the adjoin list is not empty)
  Fetch the highest quality pixel from the adjoin list
  Update the adjoin list (see below)
  if (the adjoin list is empty)
    for (every postponed pixel)
      Insert the pixel in the adjoin list (see below)
    end
  end
end
```

The "Update the adjoin list" step is defined as follows. It requires a pixel as an input parameter.

Update the Adjoin List

```
for (each of the four adjoining pixels)
  if (the pixel is not unwrapped and not postponed)
    if (the pixel is not a border pixel)
      Unwrap the pixel and store the value in the solution array
      if (the pixel's quality exceeds MinQualityThresh) then
        Insert the pixel in the adjoin list (see below)
        Mark the pixel unwrapped
      end
    end
  end
end
```

The "Insert pixel in adjoin list" step is defined below. It requires a pixel as an input parameter.

Insert Pixel in Adjoin List

```
Place the pixel on the adjoin list, preserving the quality order
if (the number of pixels in the adjoin list exceeds MaxListSize) then
  for (each of the MaxListSize/2 lowest-quality pixels on the list)
    Mark the pixel postponed
    Remove the pixel from the list
  end
  Let MinQualityThresh = lowest-quality value of remaining pixels
end
```

4.3.3 Implementation

Our C language implementation of the quality-guided path-following algorithm can be found in Section A.7 of Appendix A. It gives the user the following choice of quality maps:

1. User-supplied correlation coefficients (or other quality map)
2. Pseudocorrelation
3. Variance of phase derivatives
4. Maximum phase gradient

The correlation coefficients can be averaged locally, and the other three quality maps can be computed in windows of any size. The software can be modified easily to compute other quality maps, such as second differences.

The memory requirements are the same as those of Goldstein's algorithm (see Section 4.2.3) except that an extra array is required to store the quality map. All arrays are the same size and data type as the input phase data, except that the bit flags array is an array of bytes rather than real numbers. The memory required by the adjoin list is insignificant.

1. Phase data array
2. Quality map array
3. Bit flags array
4. Solution array
5. Adjoin list

Table 4.2 lists the average execution times of our implementation for various array sizes. The actual times depend not only on the array sizes but also on the quality maps. The table lists the approximate times that one can expect in practice. Comparison with Table 4.1 shows that the algorithm is slower than Goldstein's by a factor of approximately 8.

Table 4.2 Average execution times of the quality-guided path-following algorithm for various array sizes on an IBM RS/6000 workstation.

Array Size (pixels)	Execution Time (min:s)
256 × 256	00:07
512 × 512	00:36
1024 × 1024	03:00
2048 × 2048	26:00

4.3.4 Examples

We will now examine the results of the quality-guided path-following algorithm on the phase unwrapping examples of Chapter 3. On all of the examples we will use the quality map that is defined by the variance of the phase derivatives in 3 × 3-pixel windows, unless noted otherwise.

Like Goldstein's algorithm, the quality-guided algorithm appears to unwrap the IFSAR data of Figure 3.2 correctly. Using the quality map of Figure 3.17 the algorithm produces the surface shown in Figure 4.19(a). The differences between this surface and that produced by Goldstein's algorithm are shown by the error map of Figure 4.19(b). The dark regions are those in which the two surfaces

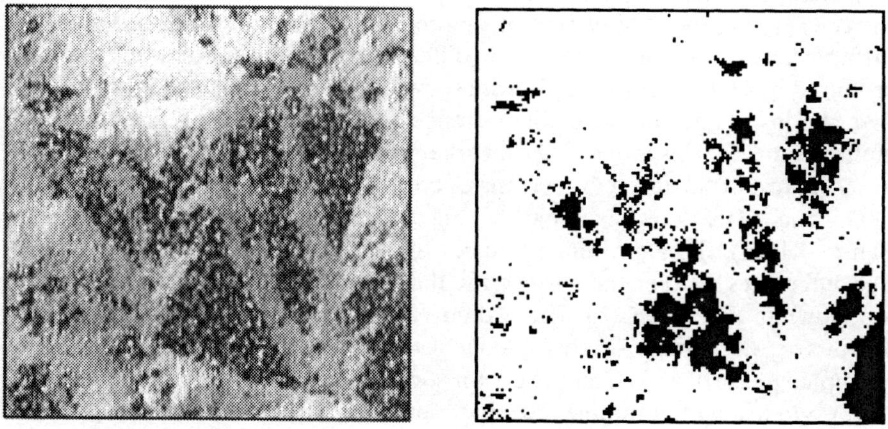

(a) (b)

Figure 4.19 The results of the quality-guided algorithm on the IFSAR example of Figure 3.2. (a) A shaded bird's-eye view of the unwrapped surface. (b) The differences between the surfaces unwrapped by the algorithm and Goldstein's algorithm are marked in black. The unwrapped surfaces are identical outside the regions of corrupted phase.

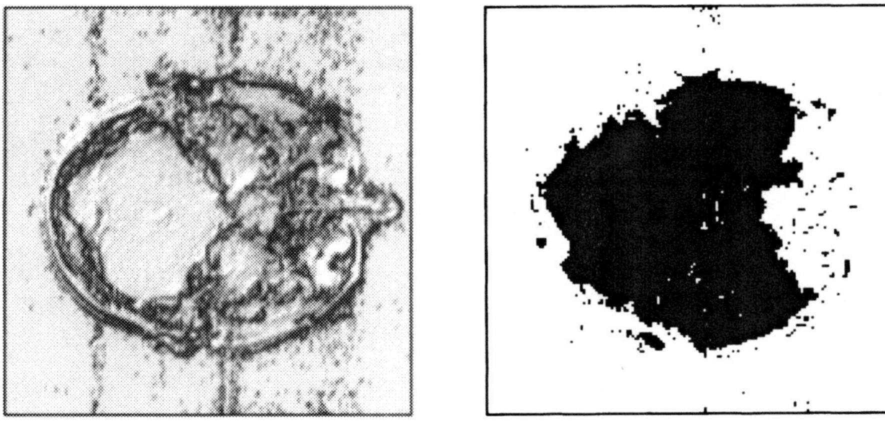

(a) (b)

Figure 4.20 Results of the quality-guided algorithm on the MRI head example of Figure 3.3. (a) A shaded bird's-eye view of the unwrapped surface. (b) The differences between this surface and the surface produced by Goldstein's algorithm. The differences inside the head regions are 2π-multiple differences due to the corrupted phase that completely surrounds these regions.

differ. Notice that the differences are confined to the patches of corrupted phase. Thus, exclusive of these corrupted regions, the two algorithms produce the same unwrapped surface.

The algorithm also appears to unwrap the MRI head example of Figure 3.3 correctly. Using the quality map of Figure 3.18(a), the algorithm produces the surface shown in Figure 4.20(a). The differences between this surface and that produced by Goldstein's algorithm are shown in the error map of Figure 4.20(b). The regions inside the head differ from those of Goldstein's algorithm (by multiples of 2π). This is due to the fact that these regions are completely isolated from the rest of the image by regions of corrupted phase.

On the MRI knee example of Figure 3.4, using the quality map of Figure 3.18(b), the algorithm produces the surface shown in Figure 4.21(a). The differences between this surface and that produced by Goldstein's algorithm are shown in Figure 4.21(b). The two unwrapped surfaces are exactly the same, except for the regions of corrupted phase. Notice that the large regions of corrupted phase outside the knee do not corrupt the unwrapping inside the knee.

It is interesting to examine the unwrapping paths followed by the algorithm. Figures 4.22(a)–(f) show the unwrapped pixels at various stages of the unwrapping process for the IFSAR example of Figure 4.19. The algorithm begins at the upper leftmost pixel of the image and confines itself to the regions of high-quality phase in between the patches of corrupted phase. The low-quality phase in the patches are unwrapped last. Compare these images with the quality map of Figure 3.17.

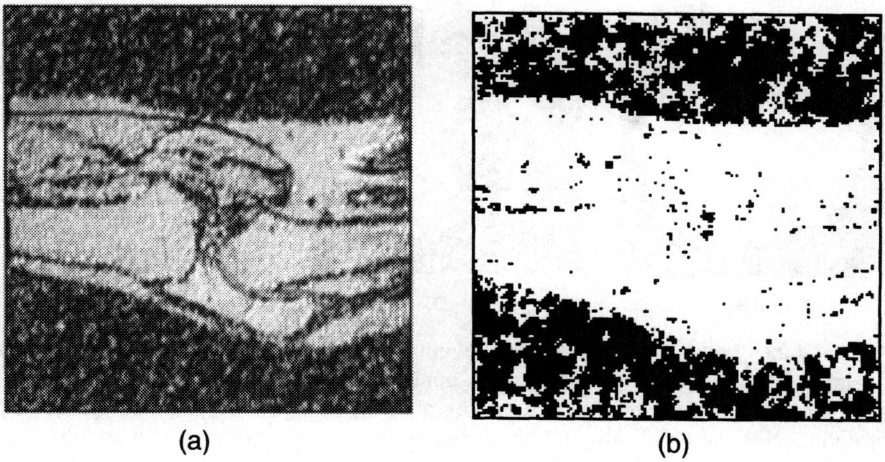

Figure 4.21 Results of the quality-guided algorithm on the MRI knee example of Figure 3.4. (a) A shaded bird's-eye view of the unwrapped surface. (b) The differences between this surface and that produced by Goldstein's algorithm. The two algorithms produce the same unwrapping (except for the corrupted regions.)

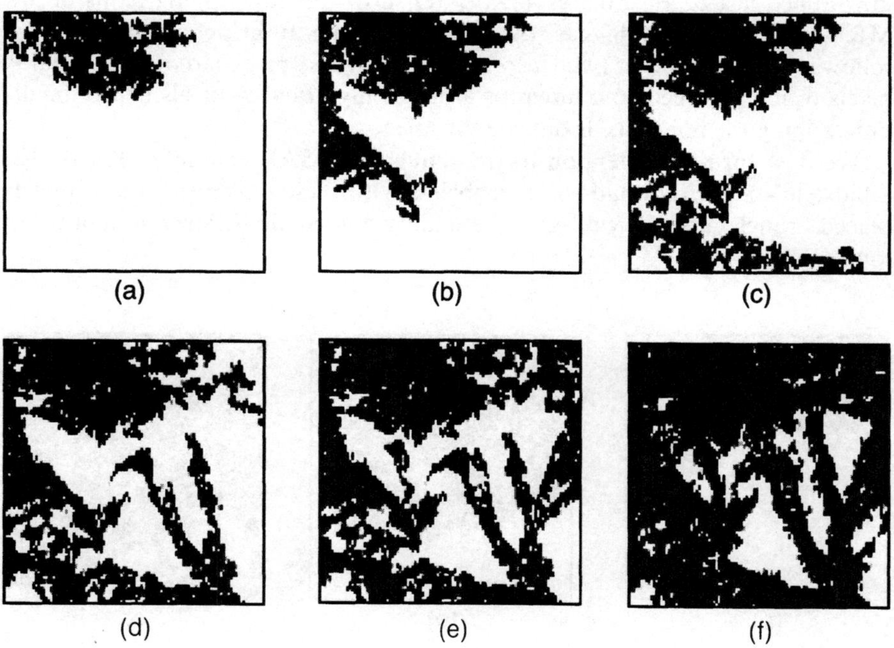

Figure 4.22 Paths followed by the quality-guided algorithm on the IFSAR example. The unwrapping begins at the upper leftmost pixel and then progressively unwraps around and between the patches of corrupted phase as shown in (a)–(f). It unwraps the corrupted patches last of all from the outside inward.

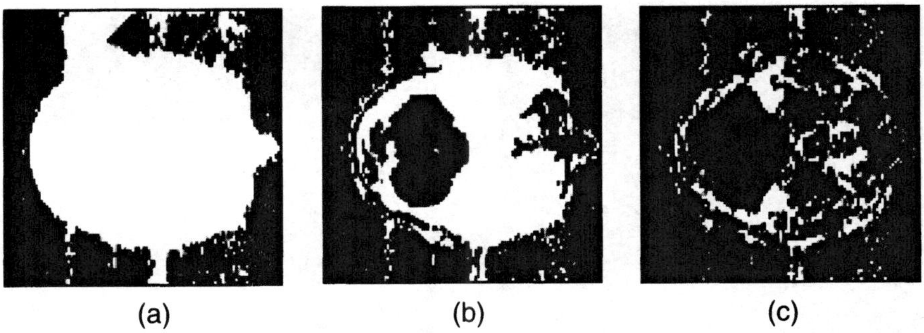

Figure 4.23 Paths followed by the quality-guided algorithm on the MRI head example. The unwrapping begins at the upper leftmost pixel and unwraps the high-quality regions outside the head (a) before entering and unwrapping the high-quality regions inside the head (b)–(c). The corrupted pixels are unwrapped last of all.

The unwrapping paths followed on the MRI head example, which are shown in Figures 4.23(a)–(c), begin at the upper leftmost pixel, and unwrap all the high-quality pixels around the head before reaching the high-quality pixels within. The corrupted, low-quality pixels that divide the various regions of the head are unwrapped last of all. Figures 4.24(a)–(c) show the unwrapping paths on the MRI knee example. In this case the algorithm begins at the upper leftmost pixel, follows the image border (which consists of less-noisy pixels) to the high-quality pixels inside the knee, and unwraps all these high-quality pixels before finally unwrapping the noisy pixels outside the knee.

We now turn our attention to the simulated IFSAR examples. Recall that Goldstein's algorithm had some problems with these examples due to poorly placed branch cuts. We will see if the quality-guided algorithm can do a better job.

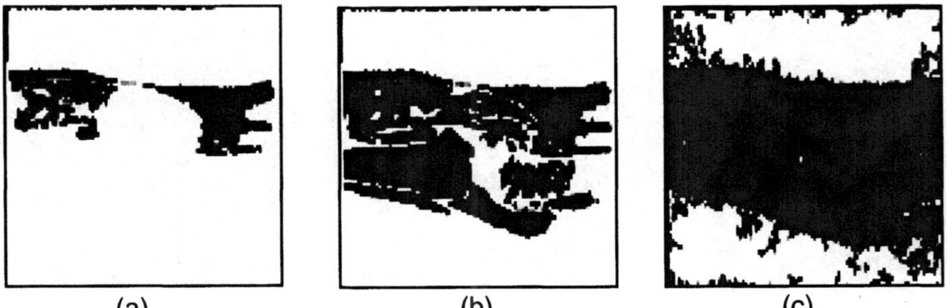

Figure 4.24 Paths followed by the quality-guided algorithm on the MRI knee example. The unwrapping begins at the upper leftmost pixel and makes its way around the high-quality pixels along the image border to the high-quality pixels within the knee (a)–(b). After unwrapping these high-quality regions, it begins to unwrap the corrupted phase outside the knee (c).

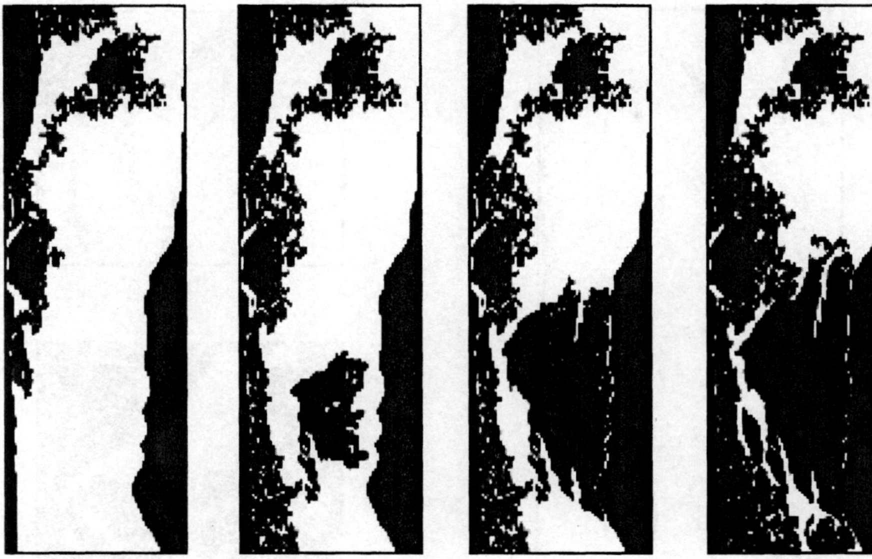

(a) (b) (c) (d)

Figure 4.25 Paths followed by the quality-guided algorithm on the Long's Peak example of Figure 3.7. The algorithm begins at the upper leftmost pixel and unwraps the phase around and between the low-quality pixels as shown in (a)–(d).

The first such example is the Long's Peak example of Figure 3.7(a). The unwrapping paths followed by the algorithm are shown in Figures 4.25(a)–(d). The algorithm begins at the upper leftmost pixel and unwraps the highest-quality pixels around and between the low-quality regions of layover before finally unwrapping these low-quality regions themselves. Compare the unwrapping paths with the quality map of Figure 3.19(a).

The error map is shown in Figure 4.26(a). The unwrapping is flawless except for the unavoidable errors in the regions of layover and the insignificant errors along the edges of the image. The results, when the correlation coefficients (averaged in 3×3-pixel windows) are used in place of the phase derivative variance, are essentially the same.

The region marked by the rectangle in Figure 4.26(a) is the region that was problematic for Goldstein's algorithm. The reader will recall that Goldstein's algorithm produced the branch cuts shown in Figure 4.26(b) with the branch cuts marked by the arrow cutting off the best path from the upper to the lower regions. Compare this with the correlation map shown in Figure 4.26(d), where the arrow points to the ideal path. One can see that the quality-guided algorithm follows this ideal path as shown in Figure 4.26(c), where the arrow marks the crossing of the path to the lower region.

On the Isolation Peak example of Figure 3.8 the results are similar. Recall that Goldstein's algorithm incorrectly unwrapped several regions as shown in

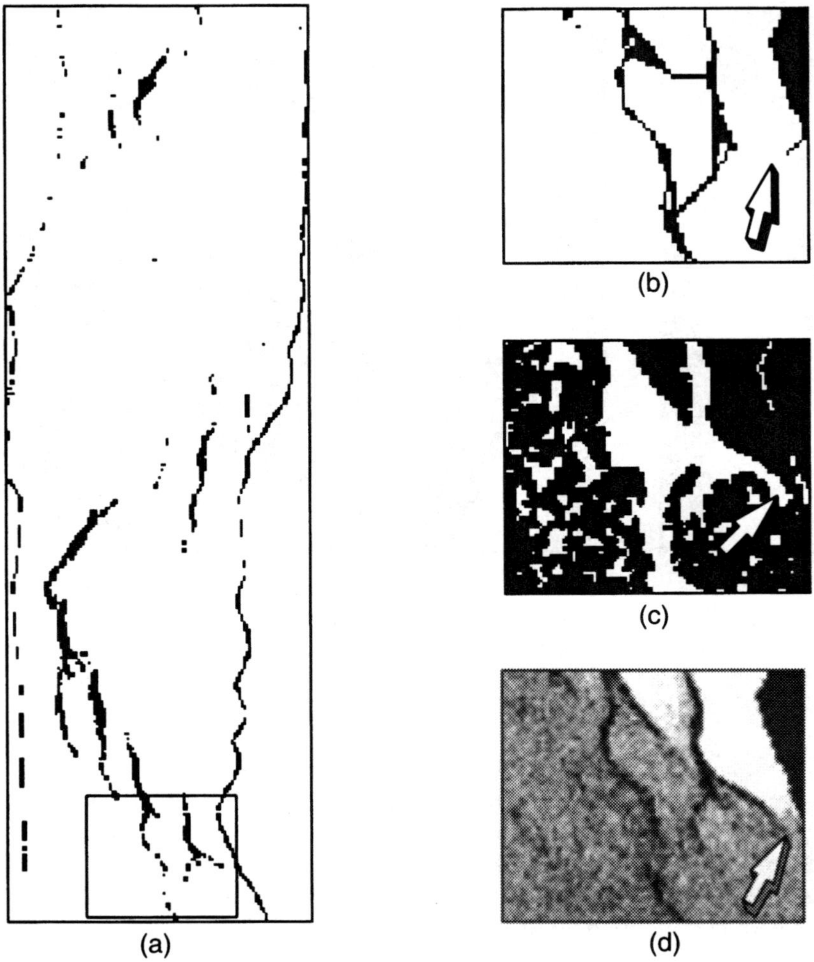

Figure 4.26 The results of the quality-guided algorithm on the Long's Peak example of Figure 3.7. (a) The error map. (b) The branch cuts produced by Goldstein's algorithm in the region marked by the rectangle are shown for comparison. (c) The unwrapping path followed by the quality-guided algorithm to the lower region. (The black pixels denote the pixels that have been unwrapped.) The algorithm crosses to the lower region at the point marked by the arrow. (d) The correlation map, which shows that the point marked by the arrow is actually the best place at which to cross.

Figure 4.12(b). By contrast, the quality-guided algorithm correctly unwraps these troublesome regions. The unwrapping paths followed by the algorithm are shown in Figures 4.27(a)–(d). The algorithm begins at the upper leftmost pixel and unwraps the highest-quality pixels around and between the low-quality regions of layover before finally unwrapping these regions themselves. Compare the unwrapping paths with the quality map of Figure 3.19(b).

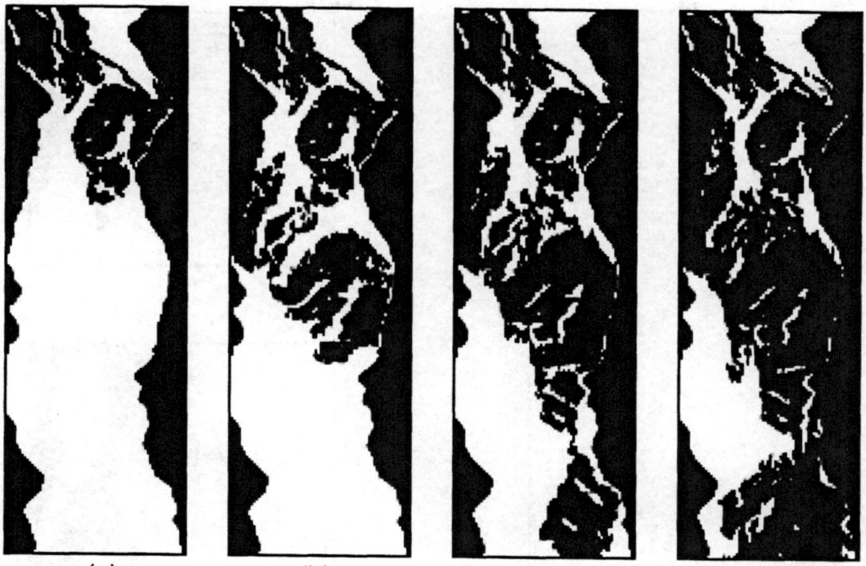

Figure 4.27 Paths followed by the quality-guided algorithm on the Isolation Peak example of Figure 3.8. The algorithm begins at the upper leftmost pixel and progresses as shown in (a)–(d).

The error map is shown in Figure 4.28(a). The unwrapping is correct except for the unavoidable errors in the regions of layover and the insignificant errors at the edges. The results with the quality map defined by the correlation coefficients rather than the phase derivative variances are essentially the same.

The region marked by the rectangle in Figure 4.28(a) is the region that was problematic for Goldstein's algorithm. Goldstein's algorithm produced the branch cuts shown in Figure 4.28(b), which almost completely isolate the lower region from the rest of the image except for two small gaps. These gaps forced the unwrapping process to cross low-quality pixels as shown by the correlation map of Figure 4.28(d), thereby producing the unwrapping error. The quality-guided algorithm, on the other hand, follows the ideal path to the lower region, as the arrows in Figures 4.28(c) and (d) indicate.

The results with the quality map defined by the correlation coefficients of Figure 3.12(a) instead of the phase derivative variances are essentially the same. The error map is shown in Figure 4.29(a). This is not to say that any quality map will work. If one uses the maximum phase gradients of Figure 3.23(b), a few regions are unwrapped incorrectly as the error map of Figure 4.29(b) shows.

We now turn to the shear examples. Like Goldstein's algorithm, the quality-guided algorithm unwraps the sheared planes of Figure 3.9 flawlessly. This is due to the fact that the phase derivative variance map defines a clearly defined line of low-quality pixels along the shear line.

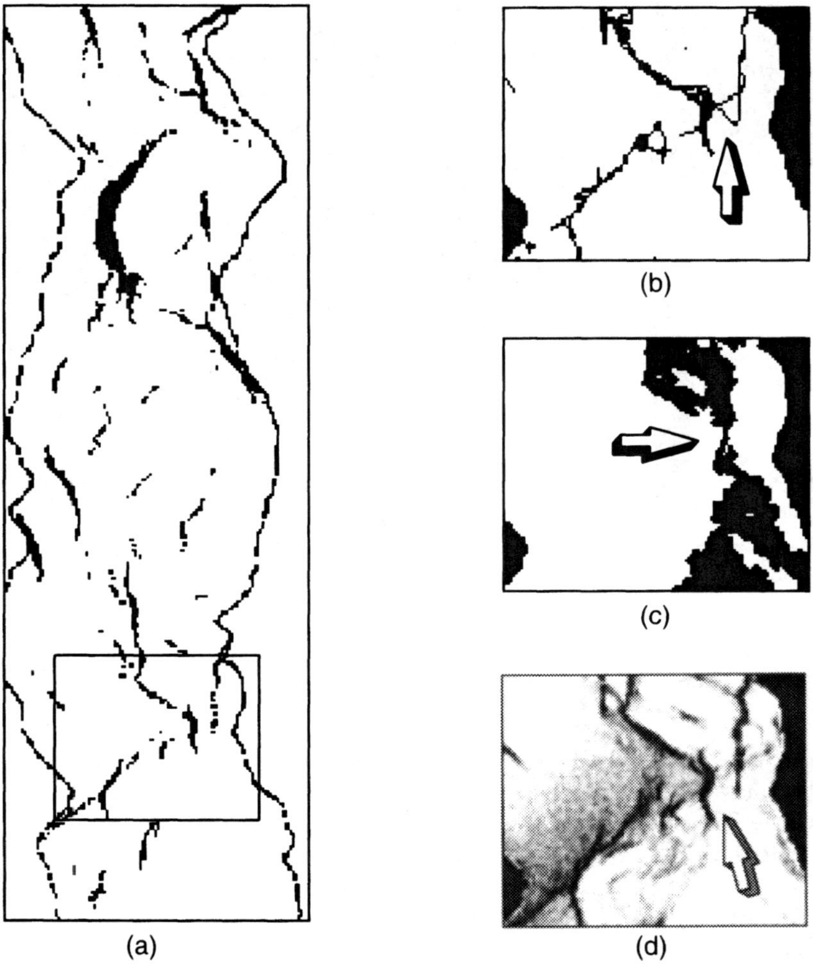

Figure 4.28 Results of the quality-guided algorithm on the Isolation Peak example of Figure 3.8. (a) The error map. The errors are confined to the regions of layover. (b) The branch cuts produced by Goldstein's algorithm in the region marked by the rectangle are shown for comparison. The arrow marks the misplaced branch cuts. (c) The unwrapping path that the algorithm follows from the upper to the lower regions is marked with the arrow. (d) The correlation map, which shows that the point marked by the arrow is actually the best place at which to cross to the lower region.

The algorithm also unwraps the spiral shear example of Figure 3.10 flawlessly—unlike Goldstein's algorithm. This is due to the fact that the quality map defines a boundary of low-quality pixels between the two arms of the spiral. This enables the algorithm to unwrap each arm independently without crossing the boundary.

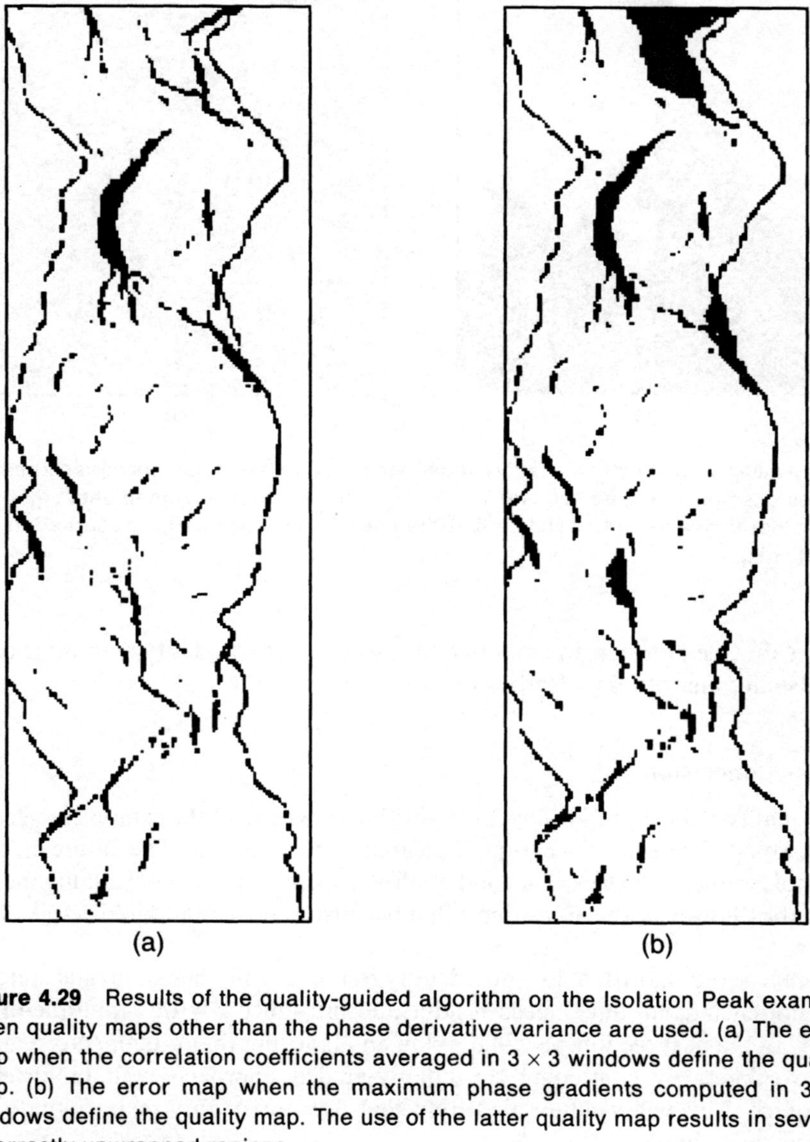

(a) (b)

Figure 4.29 Results of the quality-guided algorithm on the Isolation Peak example when quality maps other than the phase derivative variance are used. (a) The error map when the correlation coefficients averaged in 3×3 windows define the quality map. (b) The error map when the maximum phase gradients computed in 3×3 windows define the quality map. The use of the latter quality map results in several incorrectly unwrapped regions.

The quality-guided algorithm is useless, however, in cases where there is not a good quality map. On the sheared planes example of Figure 4.15 that was corrupted with Gaussian noise, the algorithm fails miserably. The error map of Figure 4.30(a) shows that only half of the image was unwrapped correctly. The cause of the problem is simple to ascertain. The noise corrupts the quality map as shown in Figure 4.30(b), which causes the algorithm to wander back and forth

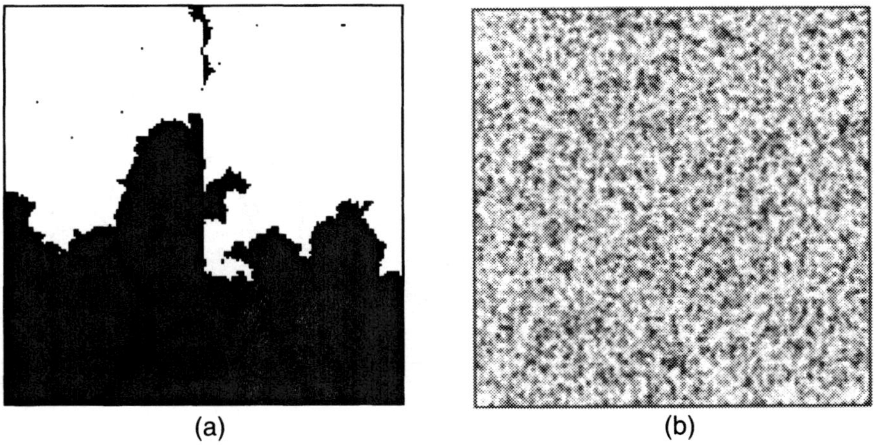

Figure 4.30 Results of the quality-guided algorithm on the noise-corrupted sheared planes example of Figure 4.15. (a) The error map. (b) The quality map defined by the phase derivative variance. Notice that the line of shear cannot be discerned in the quality map.

across the line of shear. In cases like this, where a good quality map is lacking, Goldstein's algorithm is clearly superior.

4.3.5 Conclusion

The quality-guided path-following algorithm unwraps all the examples successfully, except for the noise-corrupted sheared planes example. The failure on that example is due to the lack of a good quality map. When a reliable quality map is available, however, the algorithm often performs better than Goldstein's algorithm.

Because the algorithm does not identify residues at all, there is no guarantee it will not encircle an unbalanced residue and introduce a wrong-multiple-of-2π error. In the next section we will develop an algorithm that attempts to remedy this weakness by combining the quality-guided approach with Goldstein's approach. This will yield an algorithm that uses the quality map to guide the placement of branch cuts.

4.4 MASK CUT ALGORITHM

In the preceding section we learned that the quality-guided path-following algorithm can solve some phase unwrapping problems on which Goldstein's algorithm fails due to the misplacement of branch cuts. This is due to the fact that the quality-guided approach employs more information (the quality map) to guide the unwrapping path. In this section we combine this idea with Goldstein's algorithm to produce a "hybrid" algorithm that uses the quality map to guide the placement of the branch cuts. The resulting algorithm combines the advantages of both algorithms: additional information (besides the residue data) is exploited to guide the unwrapping process, and branch cuts are employed to ensure that no unwrapping errors are introduced by encircling unbalanced residues.

The use of a quality map to guide the placement of branch cuts was suggested by Prati, Giani, and Leuratti [15]. These researchers connected pairs of residues of opposite polarity in IFSAR data with what they called "ghost lines" (actually branch cuts) of minimum cost. They defined the cost of these branch cuts to increase with the length of the cut and the variability of the SAR intensity (or possibly the correlation coefficients—their description is not clear.) However, these researchers did not actually disclose their method for finding the branch cuts of minimal cost.

Derauw [16], working with IFSAR data, devised a modification of Goldstein's algorithm that connected two residues only if there was a path of low correlation connecting them. The branch cuts were then combined with a mask that was generated by applying a preset threshold to the correlation coefficients. Key details of the algorithm, such as how the low-correlation paths between residues were traced, were not provided. However, the author demonstrated the advantage of using the correlation coefficients to guide the placement of the branch cuts.

Flynn [17] appears to have been the first researcher to publish a detailed description of an algorithm for choosing branch cuts guided by a quality map. His method grows pixel masks that play the role of branch cuts that connect the residues. We will refer to these pixel masks as "mask cuts," and we will refer to the algorithm as a "mask cut algorithm." Flynn's mask cut algorithm does not employ any of the branch cut techniques of Goldstein's algorithm. Once the residues have been identified, the algorithm relies completely on the quality map to guide the placement of the mask cuts. In our implementation of the algorithm, we will assume as we did in Section 4.3 that the quality map is given.

4.4.1 Algorithm

The structure of the mask cut algorithm is similar to that of the quality-guided path-following algorithm of Section 4.3. In some ways it can be regarded as the reverse of this algorithm. Rather than unwrapping regions of high-quality pixels, the algorithm starts at a residue and "grows" pixel masks through the low-quality regions. The growth of a mask terminates when the total charge of the residues it connects is zero (i.e., when there are equal numbers of positive and negative residues) or when the mask grows into the image border. In the latter case, no path will be able to encircle the residues. In the former case, the residue theorem (Equation 2.29) ensures that the unwrapped phase will be independent of the choice of integration path.

Because the mask cuts are generated by what is essentially a region-growing approach, they tend to be thick. They must then be thinned by a simple morphological operation (see [18]). After this thinning operation, the flood-fill procedure described in Section 4.2.2 is employed to unwrap the phase around the mask cuts.

The mask-thinning step makes repeated passes through the image, each time removing the pixels from the mask cut that adjoin nonmask pixels. These pixels are removed, however, only if their removal does not change the connectivity of the mask. The connectivity is tested by examining the 3×3 neighborhood of the pixel to be removed. The repeated passes through the image terminate when a pass results in no pixels being removed from the mask. The test for connectivity is described in more detail in Section 4.4.2 below.

4.4.2 Pseudocode

The mask cut algorithm consists of four steps:

Step 1. Identify the residues
Step 2. Generate the mask cuts
Step 3. Thin the mask cuts
Step 4. Path-integrate around the mask cuts

Steps 1 and 4 were described in detail in Section 4.2.2 in our presentation of Goldstein's algorithm. Step 2, which is the main step of the algorithm, is essentially a combination of the quality-guided path follower (Section 4.3) and Goldstein's algorithm (Section 4.2). Like both algorithms it employs the "bit flags array" to do the bookkeeping, the "solution array" to store the unwrapped result, and the "adjoin list" to manage the mask-growing. The bit flags array manages the following flags:

1. Positive residue
2. Negative residue
3. Mask cut
4. Image border
5. Balanced residue
6. Postponed pixel
7. Unwrapped pixel

The first two flags are self-explanatory. The "mask cut" flag simply keeps track of which pixels are masked out by the mask cuts. This flag is completely analogous to the branch cut flag of Section 4.2.2. The "image border" flag defines any pixels whose phase values are undefined, such as the undefined border pixels of the simulated IFSAR examples of Figures 3.7(a) and 3.8(a). The "balanced residue" flag keeps track of which residues have been balanced with mask cuts. The "postponed pixel" flag keeps track of which pixels have been removed temporarily from the adjoin list by the list-trimming procedure described in Section 4.3.1. The "unwrapped pixel" flag keeps track of which pixels have been unwrapped and is used only in Step 4 of the algorithm during path integration.

For the sake of simplicity, the pseudocode below omits the "list trimming" procedure performed on the adjoin list to speed up the algorithm. This list-trimming procedure, which is essential in this algorithm, was defined in Section 4.3.1. Notice that the lowest- rather than the highest-quality pixel is retrieved from the adjoin list on each iteration. Notice also that the inner loop terminates when the total charge of the residues connected by the mask cut becomes zero. At this point the residues are balanced, and there is no further need to grow the current mask cut.

Step 2. Generation of Mask Cuts

```
for (each unbalanced residue)
  Mark the residue balanced
  Let charge = 1 if the residue is positive, −1 if negative
  Mark the pixel a ''mask cut'' pixel
  Update the adjoin list (see below)
  while (the adjoin list is not empty and charge ≠ 0 )
    Fetch the lowest quality pixel from the adjoin list
    Mark the pixel a ''mask cut'' pixel
    if (the pixel is an unbalanced residue) then
      Add its polarity (1 or −1) to charge
       Mark the residue balanced
    end
    if (the pixel is a border pixel) let charge = 0
    Update the adjoin list (see below)
  end
end
```

The "Update the adjoin list" step is defined as follows. It requires a pixel as an input parameter.

Update the Adjoin List

```
for (each of the four adjoining pixels)
  if (the pixel is not in the adjoin list)
    Insert the pixel in the adjoin list
  end
end
```

The pseudocode for the mask-thinning step is defined as follows:

Step 3. Thin the Mask Cuts

```
repeat
  for (each mask pixel that adjoins a nonmask pixel)
    if (the pixel does not adjoin a residue)
      if (the pixel can be removed without changing mask connectivity)
        Remove the pixel from the mask
      end
    end
  end
until (no pixels are removed from the mask)
```

We now describe the criteria for detecting a change in mask connectivity. Such a change is determined by "walking" around the border of the 3 × 3 neighborhood of the pixel to be removed. The number of transitions between mask and non-mask pixels during this walk must be no greater than two. If the number of transitions is greater, the center pixel cannot be removed without disconnecting the mask.

The number of transitions between mask and nonmask pixels is determined as follows. Let A, B, and C be the three pixels of the 3 × 3 neighborhood shown in Figure 4.31(a). Initialize a counter to zero and perform the following steps:

1. If A is a mask pixel but C is not, then increment the counter.
2. If A is not a mask pixel but B is, then increment the counter.
3. If A and B are not mask pixels but C is, then increment the counter.
4. If A and C are not mask pixels but B is, then increment the counter twice.

Figures 4.32(a)–(e) illustrate the situations under which the counter must be incremented. In Figure 4.32(c) the counter must be incremented twice because

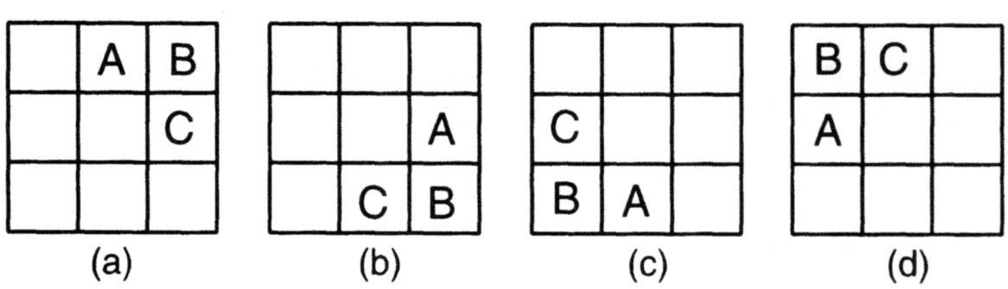

Figure 4.31 The 3 × 3-pixel neighborhood of the mask pixel to be tested for removal. The number of mask-nonmask transitions is determined for the three pixels labeled A, B, and C in (a). The process is repeated in turn for the three labeled pixels in (b), (c), and (d). If the total number of transitions exceeds two, the center pixel cannot be removed.

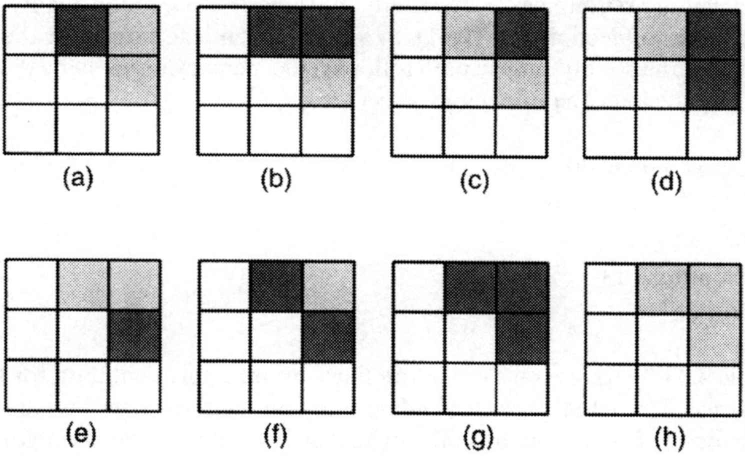

Figure 4.32 Calculation of the number of mask-nonmask transitions on the three pixels labeled A, B, and C in Figure 4.31(a). These eight cases represent all possible configurations of mask and nonmask pixels. The mask pixels are darkly shaded, while the nonmask pixels are lightly shaded. In cases (a)–(e) there is a mask-nonmask transition; in case (c) there are two such transitions, while in cases (f)–(h) there are no transitions.

there are two mask-nonmask transitions. Figures 4.32(f), (g), and (h) illustrate the situations under which the counter must not be incremented. In Figure 4.32(f) there is no mask-nonmask transition because the mask pixels are connected on the diagonal.

After this counting procedure is performed for the pixels labeled A, B, and C in Figure 4.31(a), it is repeated in turn for the pixels labeled in Figures 4.31(b), (c), and (d) without resetting the counter. The final value of the counter determines the number of transitions between mask and nonmask pixels.

4.4.3 Implementation

Our C language implementation of the mask cut algorithm can be found in Section A.8 of Appendix A. It gives the user the following choice of quality maps:

1. User-supplied correlation coefficients (or other quality map)
2. Pseudocorrelation
3. Variance of phase derivatives
4. Maximum phase gradient

These quality maps can be computed in windows of any size. The software can be modified easily to compute other quality maps, such as second differences or other measures of phase variance.

The memory requirements are exactly the same as those of the quality-guided path follower of Section 4.3. The array sizes are identical, except for the adjoin list. As with the quality-guided path follower, the memory required by the adjoin list is insignificant. The following arrays are required:

1. Phase data array
2. Quality map array
3. Bit flags array
4. Solution array
5. Adjoin list

Table 4.3 lists the average execution times of our implementation for various array sizes. (The actual times depend not only on the array sizes but also on the quality maps.) Comparing this table with Table 4.2 shows that the algorithm is about half as fast as the quality-guided path follower.

4.4.4 Examples

We will now examine the results of the mask cut algorithm on the phase-unwrapping examples of Chapter 3. As in Section 4.3 we will use the quality map that is defined by the variance of the phase derivatives in 3×3-pixel windows, unless noted otherwise.

The algorithm unwraps the IFSAR data of Figure 3.2, producing essentially the same result as Goldstein's algorithm and the quality-guided algorithm. The mask cuts are shown in Figure 4.33(a), and the differences between the unwrapped surface and the surface produced by Goldstein's algorithm are shown in Figure 4.33(b). Outside of the corrupted regions, the algorithms produce the same result.

A magnified view of the mask cuts within the rectangle marked in Figure 4.33(a) is shown in Figures 4.33(c) and (d) before and after the thinning operation, respectively. Comparison of Figures 4.33(d) and 4.7(c) shows that the mask cuts tend to be thicker than Goldstein's branch cuts.

Table 4.3 Average execution times of the mask cut algorithm for various array sizes on an IBM RS/6000 workstation.

Array Size (pixels)	Execution Time (min:s)
256×256	00:08
512×512	00:50
1024×1024	04:15
2048×2048	45:00

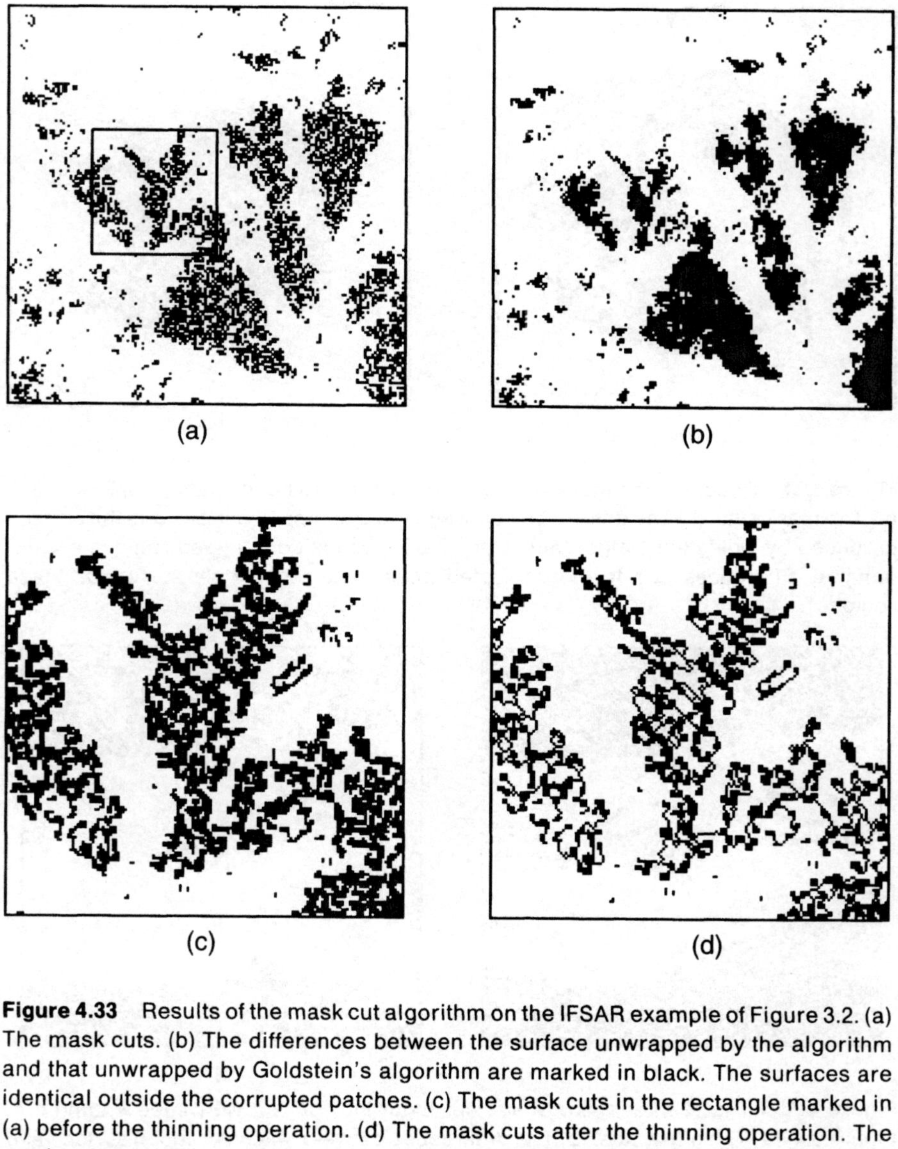

(a) (b) (c) (d)

Figure 4.33 Results of the mask cut algorithm on the IFSAR example of Figure 3.2. (a) The mask cuts. (b) The differences between the surface unwrapped by the algorithm and that unwrapped by Goldstein's algorithm are marked in black. The surfaces are identical outside the corrupted patches. (c) The mask cuts in the rectangle marked in (a) before the thinning operation. (d) The mask cuts after the thinning operation. The mask cuts tend to be thicker than the branch cuts produced by Goldstein's algorithm.

Figures 4.34(a)–(b) show the results on the MRI head example of Figure 3.3(a). The mask cuts are shown in Figure 4.34(a), and the differences between the unwrapped surface and that produced by Goldstein's algorithm are shown in the error map of Figure 4.34(b). The regions inside the head differ from those of Goldstein's algorithm, but only by multiples of 2π. This is due to the fact that these regions are completely isolated from the rest of the image by regions of corrupted phase.

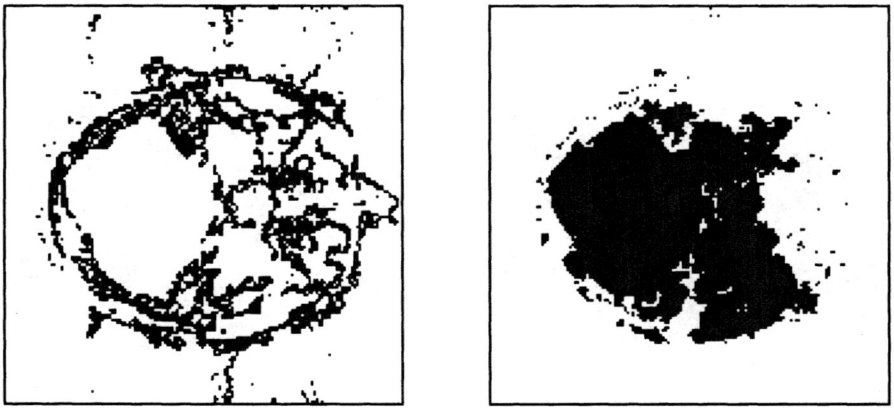

(a) (b)

Figure 4.34 Results of the mask cut algorithm on the MRI head example of Figure 3.3. (a) The mask cuts. (b) The differences between the unwrapped surface and the surface produced by Goldstein's algorithm. The differences inside the head regions are 2π-multiple differences due to the corrupted phase that completely surrounds these regions.

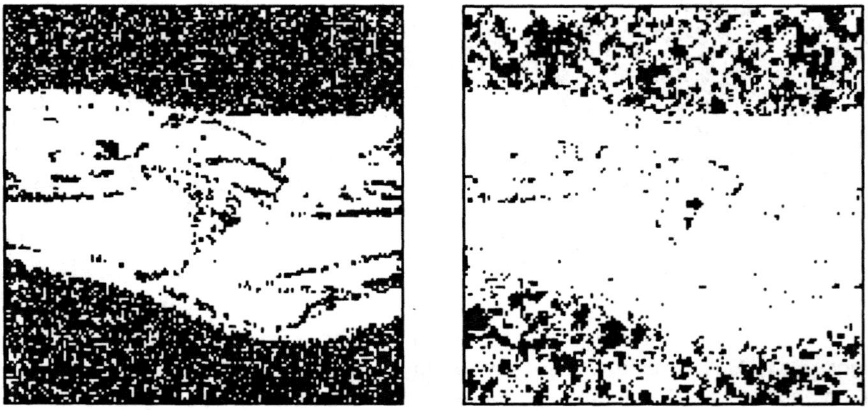

(a) (b)

Figure 4.35 The results of the mask cut algorithm on the MRI knee example of Figure 3.4. (a) The mask cuts. (b) The differences between the unwrapped surface and that produced by Goldstein's algorithm. The two algorithms produce the same unwrapping (except for the corrupted regions.)

On the MRI knee example of Figure 3.4, the algorithm produces the mask cuts shown in Figure 4.35(a). The differences between the unwrapped surface and that produced by Goldstein's algorithm are shown in Figure 4.35(b). The two unwrapped surfaces are the same, except for the regions of corrupted phase. Notice that the large regions of corrupted phase outside the knee do not corrupt the unwrapping inside the knee.

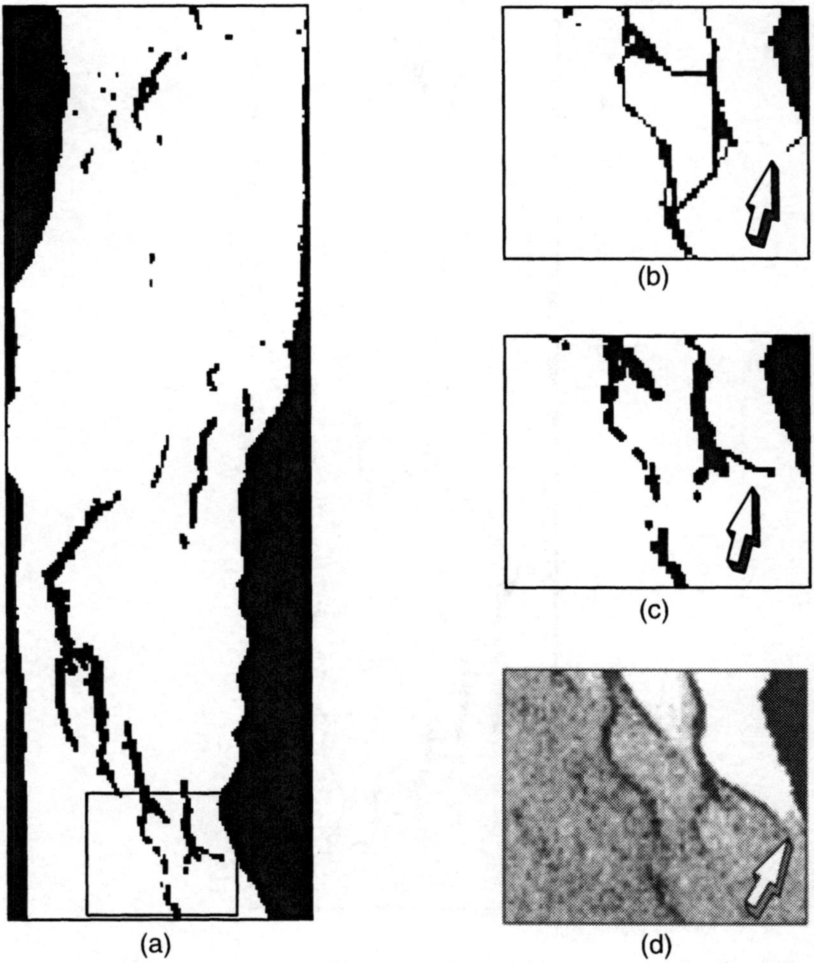

Figure 4.36 Results of the mask cut algorithm on the Long's Peak example of Figure 3.7. The region that caused problems for Goldstein's algorithm is marked by the rectangle and enlarged. (a) The mask cuts. (b) The branch cuts produced by Goldstein's algorithm in the region marked by the rectangle are shown for comparison. (c) The mask cuts produced by the algorithm. (d) The correlation map, which shows that the mask cuts allow the path-integration step to follow the best path of high-quality pixels to the lower region.

We now turn our attention to the simulated IFSAR examples. We recall that Goldstein's algorithm had problems with these examples due to poorly placed branch cuts. We will see if the mask cut algorithm places its branch cuts more successfully.

The first such example is the Long's Peak example of Figure 3.7. The mask cuts are shown in Figure 4.36(a), where the rectangle marks the region that

Figure 4.37 Error map for the mask cut algorithm on the Long's Peak example.

created problems for Goldstein's algorithm. Goldstein's branch cuts in this region are shown in Figure 4.36(b), and the mask cuts are shown in Figure 4.36(c). Notice that the branch cuts marked by the arrows are placed very differently. Examination of the correlation image of Figure 4.36(d) shows that the mask cuts are superior to Goldstein's branch cuts. This is due to the fact that the mask cuts do not block the path-integration step from following the highest-quality pixels marked by the arrow in Figure 4.36(d).

The error map is shown in Figure 4.37. The unwrapping is correct except for the unavoidable errors in the regions of layover. The results for the quality map defined by the correlation coefficients (averaged in 3×3-pixel windows) rather than the phase derivative variance are essentially the same.

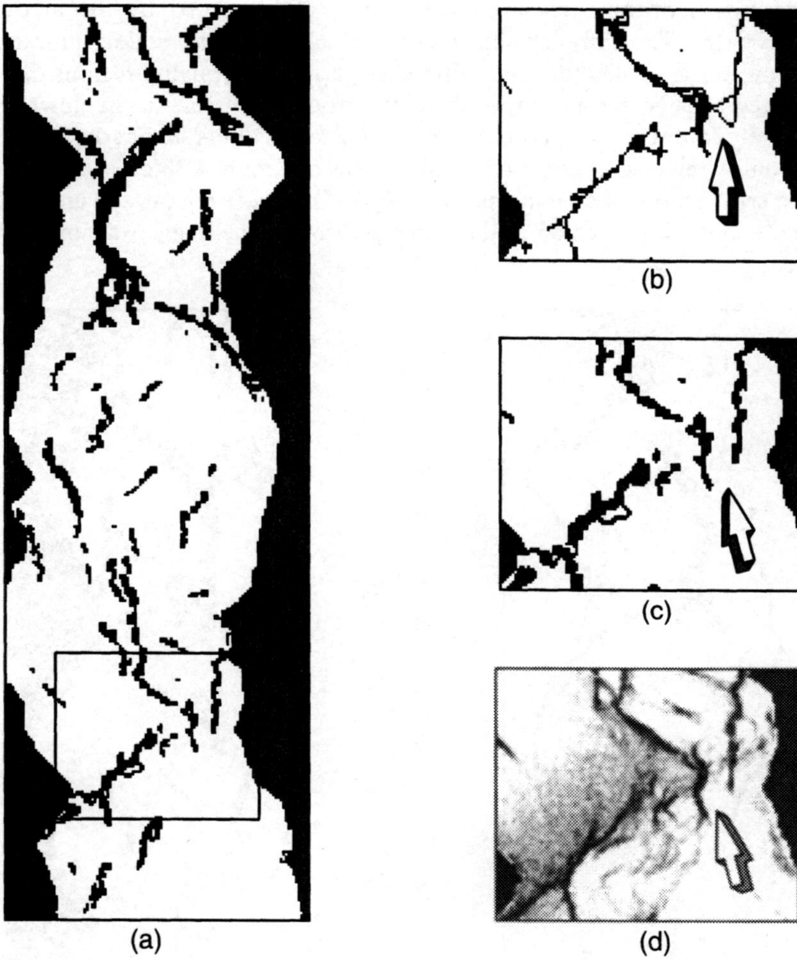

Figure 4.38 Results of the mask cut algorithm on the Isolation Peak example of Figure 3.8. (a) The mask cuts. (b) The branch cuts produced by Goldstein's algorithm in the region marked by the rectangle are shown for comparison. The misplaced branch cuts are marked by the arrow. (c) The mask cuts produced by the mask cut algorithm leave a gap of high-quality phase. (d) The correlation map, with the high-quality gap marked by the arrow.

On the Isolation Peak example of Figure 3.8, we recall that Goldstein's algorithm unwrapped several regions incorrectly as shown in Figure 4.12(b). The mask cuts produced by the mask cut algorithm are shown in Figure 4.38(a), and the region that was problematic for Goldstein's algorithm is marked with a rectangle. In this region Goldstein's algorithm produced the branch cuts shown in Figure 4.38(b), while the mask cut algorithm produced the mask cuts shown in Figure 4.38(c). The correlation map is shown in Figure 4.38(d). We recall that

Goldstein's algorithm misplaced the branch cuts marked by the arrow in Figure 4.38(b). These branch cuts interrupt the high-quality regions marked by the arrow in Figure 4.38(d). They also leave gaps in low-quality regions that the path-integration step must cross, thereby introducing errors in the unwrapped surface. By contrast, the mask cut algorithm places the cuts so that the gaps lie in high-quality regions, as indicated by the arrow of Figure 4.38(c).

The error map is shown in Figure 4.39(a). The lower region was unwrapped correctly, but a large region in the upper portion of the image was unwrapped

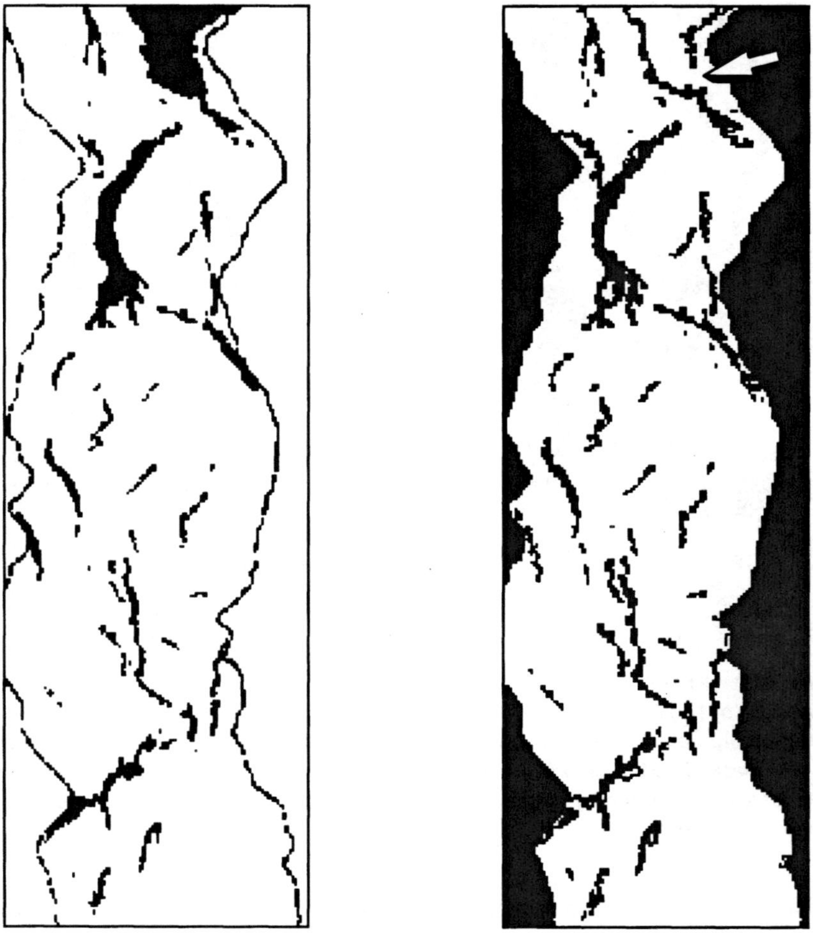

(a) (b)

Figure 4.39 Results of the mask cut algorithm on the Isolation Peak example. The quality map is defined by the phase derivative variance. (a) The error map. Notice the region of incorrectly unwrapped phase in the upper portion of the image. (b) The mask cuts, with the cause of the error marked by the arrow. The path-integration step was forced to cross this narrow gap of low-correlation phase.

incorrectly. An examination of the mask cuts, shown in Figure 4.39(b), reveals that they left only a small gap (marked with an arrow in the figure) in a low-correlation region, as shown in the correlation image of Figure 3.12(a). The path-integration step introduced an error as it crossed this gap.

If the correlation coefficients, rather than the phase derivative variances, define the quality map, then the phase in this region is unwrapped correctly as the error map of Figure 4.40(a) shows. The mask cuts are shown in Figure 4.40(b). The new mask cuts were placed differently, locating the gap marked by the arrow in a region of higher-quality phase.

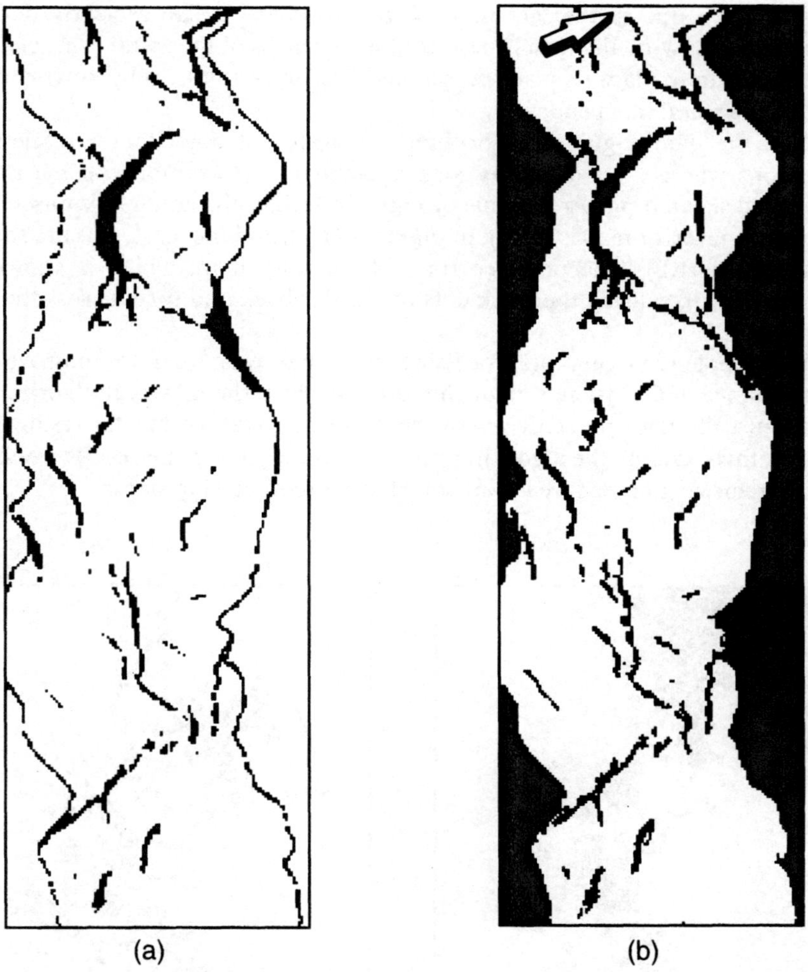

(a) (b)

Figure 4.40 Results of the mask cut algorithm on the Isolation Peak example with the quality map defined by the correlation coefficients averaged in 3 × 3 windows. (a) The error map. Notice that the upper region was unwrapped correctly, although a small region on the right was unwrapped incorrectly. (b) The mask cuts. The gap marked by the arrow differs from the gap marked in Figure 4.39(b).

The reader will notice that a small region of phase on the right was unwrapped incorrectly. The mask cut algorithm does not unwrap this example quite as well as the quality-guided algorithm of Section 4.3.

We now turn to the shear examples. Like the quality-guided algorithm, the mask cut algorithm unwraps the sheared planes example of Figure 3.9 flawlessly because the quality maps define a line of low-quality pixels along the shear line. This in turn enables the algorithm to place the mask cuts along this line, thereby separating the two planes from one another so that they can be unwrapped independently.

The algorithm also unwraps the spiral shear example of Figure 3.10 flawlessly. This is again due to the fact that the quality maps define a clearly defined boundary of low-quality pixels between the two arms of the spiral. This enables the algorithm to place its mask cuts along these lines and thereby unwrap each arm of the spiral independently.

Like the quality-guided algorithm, the mask cut algorithm is useless in situations where a good quality map is lacking. For example, on the noise-corrupted sheared-planes example of Figure 4.15, the algorithm fails miserably. The resulting error map is shown in Figure 4.41(a), and the mask cuts are shown in Figure 4.41(b). The noise corrupts the quality map, which is shown in Figure 4.30(b), causing the mask cuts to wander back and forth across the line of shear.

The dipole-removing preprocessing step that was used to improve the performance of Goldstein's algorithm does not help the mask cut algorithm at all because the algorithm is driven by the quality map rather than the residues. In fact on this example the algorithm produces exactly the same results when the dipole-removing procedure is performed as a preprocessing step.

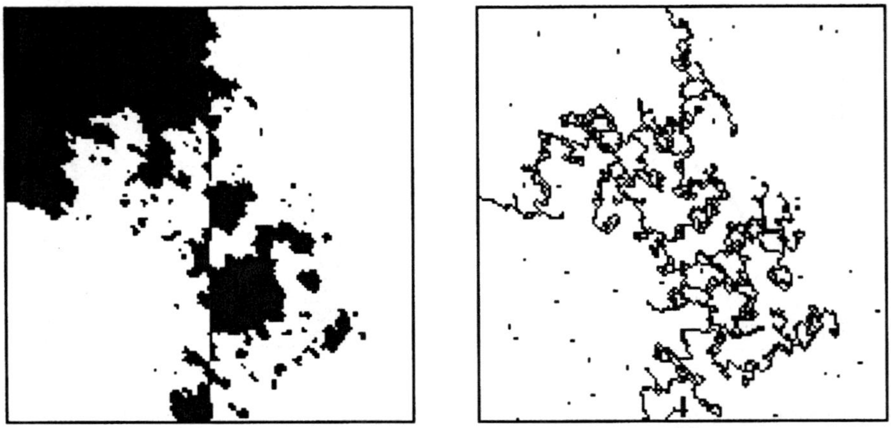

(a) (b)

Figure 4.41 Results of the mask cut algorithm on the noise-corrupted sheared-planes example of Figure 4.15. (a) The error map. (b) The mask cuts.

4.4.5 Conclusion

The mask cut algorithm is essentially a cross between Goldstein's algorithm and the quality-guided path follower of Section 4.3. Like Goldstein's algorithm it identifies the residues and connects them with branch cuts. Unlike Goldstein's algorithm, however, it has more information at its disposal (e.g., the quality map) to guide the placement of the mask cuts. On any phase unwrapping problem, the mask cut algorithm is worth a try if a good quality map is available or can be derived from the phase data.

In general, the mask cut algorithm seems to give the same results as the quality-guided path follower of Section 4.3. The success of both algorithms depends on the accuracy of the quality map. In particular, they can fail if the residues are not confined to low-quality regions. In cases like this, where a good quality map is lacking, Goldstein's algorithm is superior.

4.5 FLYNN'S MINIMUM DISCONTINUITY APPROACH

In our discussion of path-following algorithms thus far, we have passed over a very natural approach, which we will now describe. If one examines the pattern of fringes in the wrapped phase data of Figure 4.42(a), one notices that the "fringe lines," or boundaries between the lightest and darkest pixels that mark the transitions from 0 to 2π, divide the image into the regions shown in Figure 4.42(b). If one plots the wrapped phase data as the surface shown in Figure 4.42(c), one notices "discontinuities" along the fringe lines. The magnitudes of the partial derivatives (slopes) of the surface along the fringe lines exceed π.

Phase unwrapping can be viewed as the process of assigning an integer multiple of 2π to each region to minimize the discontinuities across the fringe lines. The addition of an integer multiple to each region "lifts" that region into place, thereby removing the discontinuities, as shown in Figures 4.42(d)–(f).[4]

This approach to phase unwrapping is simple and appealing. It consists of identifying the fringe lines and then adding multiples of 2π to the regions delineated by the fringe lines. The multiple added to a pixel depends on how many fringe lines lie between the pixel and a reference pixel. This approach was proposed by Lin, Vesecky, and Zebker [19], who used edge detection techniques to find the locations of the fringe lines. However, this approach is intractable for two reasons. First the presence of noise results in fringe lines that are poorly defined and disconnected, as shown in Figures 4.43(a) and (b). Lin, Vesecky, and Zebker used a least-squares curve-fitting technique to fit sinusoidal functions and linear splines to the noisy fringe lines [20], but such curve-fitting techniques fail where the lines are closely spaced. The second reason that this approach is intractable is that the presence of an isolated residue will result in the start (or

[4] This is reminiscent of a flat "camping cup" composed of multiple sections that telescope out to form the cup.

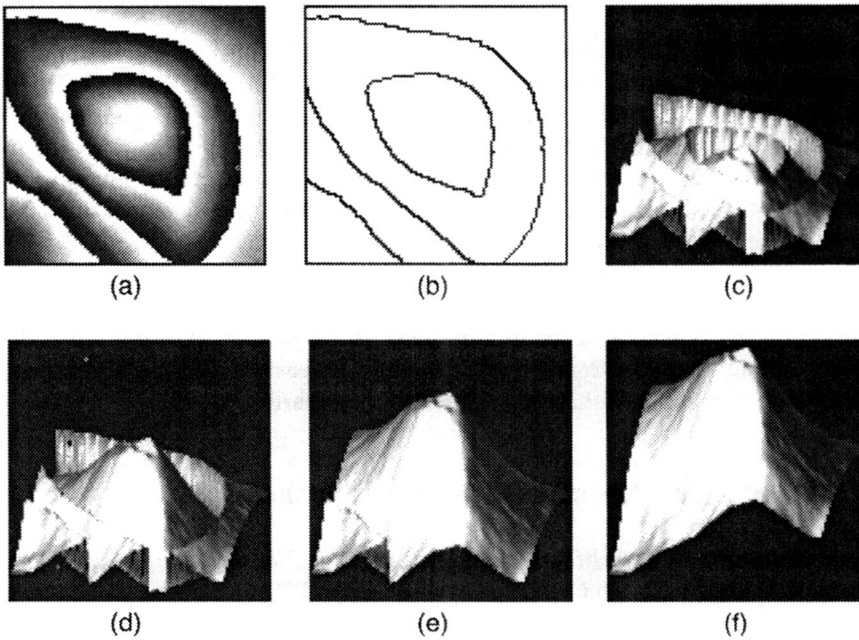

Figure 4.42 Illustration of the "fringe line" approach to phase unwrapping. The fringe lines in the wrapped phase (a) are identified to produce lines (b) that separate the phase data into regions. The initial surface (c) has discontinuities along these lines. Multiples of 2π are added to the regions to remove these discontinuities. A 2π-multiple added to the central region removes the discontinuities along the innermost fringe and results in the surface (d). As these 2π-multiples are added to the other regions, the surface becomes smoother (e) until the final unwrapped surface (f) has no discontinuities.

end) of a fringe line, as shown in Figures 4.44(a) and (b). In the neighborhood of such a residue, two regions that should have different 2π-multiples will be connected with no fringe line to separate them.

The approach of separating the phase image into regions by means of the fringe lines and then adding a multiple of 2π to each region to minimize the discontinuities is not equipped to deal with the problem of isolated residues. Indeed, this isolated residue problem is essentially the central problem of phase unwrapping. Thus the fringe-line approach is unsatisfactory.

It is therefore quite surprising that the "minimum discontinuity" algorithm of Flynn [21] adopts this very approach and makes it successful. There is a crucial difference between Flynn's algorithm and the fringe line approach, however. Rather than attempting to identify the fringe lines, Flynn's algorithm employs a clever scheme for detecting lines of discontinuities that do not necessarily lie on the fringe lines. The algorithm identifies the lines of discontinuities that are

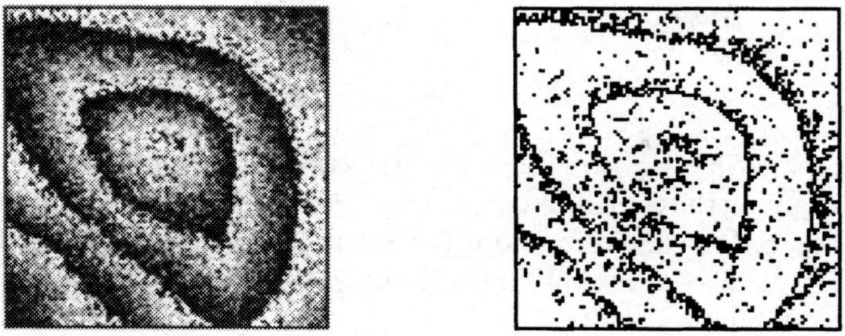

(a) (b)

Figure 4.43 (a) The phase data of Figure 4.42(a) corrupted with Gaussian noise. (b) The fringe lines are indistinct and disconnected.

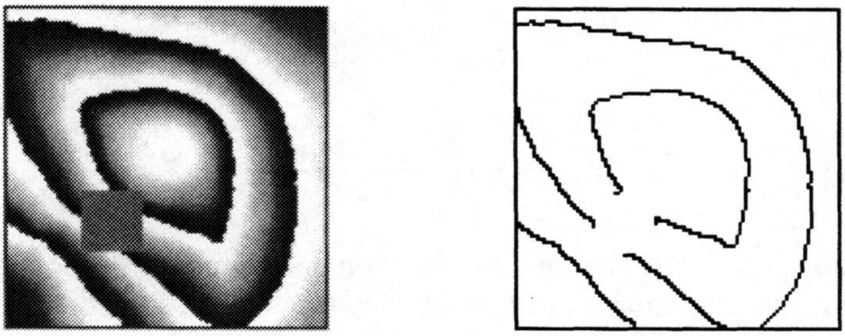

(a) (b)

Figure 4.44 (a) The phase data of Figure 4.42(a) with a patch of constant phase that produces four isolated residues where the fringe lines terminate. (b) Due to these residues the fringe lines fail to separate the regions that should have different 2π-multiples.

joined into loops, and then it essentially adds the appropriate multiple of 2π to the pixels enclosed by the loop if this action will remove more discontinuities than it adds. This process is performed iteratively, reducing the discontinuities at each stage, until no more loops can be identified. At this point the phase data have been unwrapped, and the resulting solution is a "minimum discontinuity" solution, which means that the remaining discontinuities are minimal in the sense of a particular measure.

4.5.1 Algorithm

Before describing Flynn's minimum discontinuity algorithm, we need to develop a number of new concepts. We begin with the observation that phase unwrapping is essentially the process of assigning an integer multiple $c_{m,n}$ of 2π to each pixel of the wrapped phase data $\psi_{m,n}$. The unwrapped solution $\phi_{m,n}$ satisfies the relation

$$\phi_{m,n} = \psi_{m,n} + 2\pi c_{m,n}. \tag{4.1}$$

The integer multiples $c_{m,n}$ are called the *wrap counts*. A restriction or constraint must be placed on the choice of wrap counts. Otherwise, the wrap counts $c_{m,n} = 0$ can be chosen. We therefore adopt the requirement that the wrap counts be chosen to minimize the discontinuities in the unwrapped solution $\phi_{m,n}$. But what exactly is a discontinuity?

We define a *discontinuity* to be a pair of adjoining pixels whose difference exceeds π in magnitude. The adjoining pixels can be oriented vertically or horizontally. We define the *jump count* of the pixels to be the nearest integer multiple of 2π of this difference. Thus the *vertical* and *horizontal jump counts* are defined by the equations

$$v_{m,n} = \text{Int}\left(\frac{\phi_{m,n} - \phi_{m+1,n}}{2\pi}\right), \tag{4.2}$$

$$z_{m,n} = \text{Int}\left(\frac{\phi_{m,n} - \phi_{m,n+1}}{2\pi}\right), \tag{4.3}$$

respectively, where $Int(\cdot)$ denotes the function that rounds to the nearest integer. The sum of the magnitudes of the jump counts,

$$E = \sum |v_{m,n}| + \sum |z_{m,n}|, \tag{4.4}$$

is a measure of the overall severity of the discontinuities, and it is this measure that the algorithm is designed to minimize.

There is a very close relationship between the wrap counts and the jump counts. The wrap counts can be derived from the jump counts, and vice versa. Furthermore an increment in the wrap count $c_{m,n}$ causes increments in the jump counts $v_{m,n}$ and $z_{m,n}$ and decrements in $v_{m-1,n}$ and $z_{m,n-1}$. If the wrap counts of a set of pixels are all incremented by 1, then the jump counts surrounding the set are changed by 1 or -1. Conversely, a change in the jump counts induces a change in the wrap counts.

Figures 4.45(a) and (b) illustrate this simple yet important relationship between the wrap counts and the jump counts. Figure 4.45(a) shows the wrapped phase values (scaled to the range 0–1) of a pyramid-shaped surface. The dots represent the pixels, the labels indicate the phase values, and the dashed lines show the locations of the fringe lines, which form a closed loop in the shape of a square. Since the phase data have not yet been unwrapped, the wrap counts are all zero. Figure 4.45(b) shows the nonzero jump counts with arrows whose labels indicate the jump count values. All the other jump counts are zero. Notice that the nonzero jump counts all lie on the fringe lines.

If the wrap counts of all the pixels enclosed by the fringe lines of Figure 4.45(a)

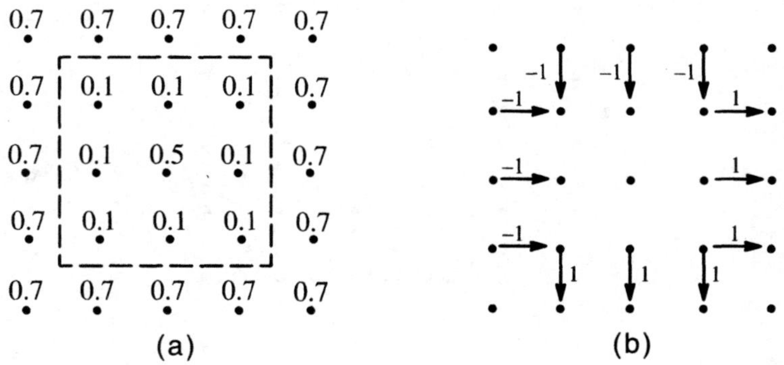

Figure 4.45 Close relationship between the wrap counts and the jump counts. (a) The wrapped phase data of a pyramid. The dots indicate the pixels, the numbers indicate the phase values (scaled to the range 0–1), and the dashed lines indicate the fringe lines. This phase can be unwrapped by incrementing the wrap counts of the region enclosed by the dashed lines. (b) The jump counts are nonzero at the fringe lines, and the numbers indicate the jump count values. Incrementing the wrap counts as explained in (a) is equivalent to incrementing and decrementing the jump counts in the manner described in the text so that the jump counts all become zero, which indicates that the phase has been unwrapped.

are incremented, then the phase values 0.1 and 0.5 will become 1.1 and 1.5, respectively. The -1 jump counts of Figure 4.45(b) will be incremented, and the 1 jump counts will be decremented, resulting in all the jump counts becoming zero. The phase will then be unwrapped, and there will be no discontinuities remaining in the unwrapped surface.

We will represent changes in wrap counts by changes in jump counts. We will need a notation for these jump count changes. We imagine there is an array of grid points that is offset by $\frac{1}{2}$-pixel horizontally and vertically from the phase data array. This array, whose points are called *nodes*, is illustrated by the squares of Figure 4.46. The circles represent the pixels as in Figure 4.45. An *edge* is defined to be an arrow that connects a node to one of its four neighbors. An edge defines a change in the jump count of the pair of pixels it separates. By convention, a leftward (resp. rightward) edge represents an increment (resp. decrement) of a vertical jump count. Similarly a downward (resp. upward) edge represents an increment (resp. decrement) of a horizontal jump count. The *value* of an edge from the node (m, n) to the node (m', n'), which we will denote $\delta V(m, n, m', n')$, is defined to be 1 if the edge induces a decrease in the magnitude of the jump count and -1 if it induces an increase.

An understanding of the concept of nodes and edges is crucial for understanding Flynn's algorithm. Figure 4.46 shows the edges that describe the operation of changing the jump counts of Figure 4.45(b). The edges also describe the operation of incrementing the wrap counts of the pixels enclosed by the fringe lines of Figure 4.45(a). The edges are indicated by arrows whose orientations

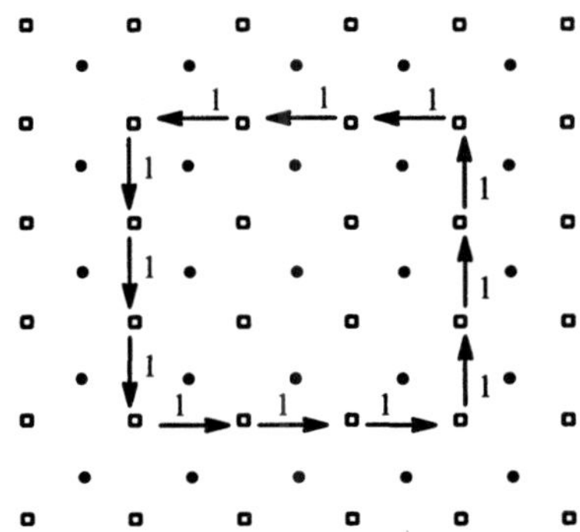

Figure 4.46 Nodes and edges that correspond to the phase unwrapping example of Figure 4.45. The squares represent the nodes, which are offset by $\frac{1}{2}$ pixel horizontally and vertically from the pixels (represented by the dots). The arrows represent the edges, and their labels indicate the edge values. These edges represent the jump count changes that unwrap the phase, as explained in Figure 4.45. Notice that the edges define a loop that coincides with the dashed fringe lines of Figure 4.45(a).

indicate increments or decrements in the jump counts according to our convention above, and the arrows are labeled with the values δV of the edges. Notice that the operation is represented by a loop of edges, and the value of every edge is positive. We call an operation that has more positive edges than negative ones an *improving operation* or an *improving loop*. The fundamental task of Flynn's algorithm is to find and apply these operations. (Flynn calls them IEOs, or improving elementary operations [21].)

Flynn's algorithm does not deal with the wrap counts directly. Instead, the algorithm changes the jump counts, and at the conclusion of the algorithm, the wrap counts are extracted from the jump counts, multiplied by 2π and added to the input phase data to produce the unwrapped solution.

Flynn's algorithm finds improving loops by creating trees of nodes with the edges as the links. If an edge leads from one node to another, the first node is called the parent and the second node is called the child. Every node is isolated, is the root of a tree (i.e., edges lead from it but not to it), or can be reached from the root of its tree by a unique path. Associated with each node (m, n) is its value, which is defined to be the sum of the values of the edges that lead to this node from its root. If the sequence of nodes $(m_0, n_0), (m_1, n_1), \ldots, (m_L, n_L)$ defines the path from the root (m_0, n_0) to the node $(m_L, n_L) = (m, n)$, then the value of the node is defined by

$$\text{value}(m, n) = \sum_{k=0}^{L-1} \delta V(m_k, n_k, m_{k+1}, n_{k+1}), \tag{4.5}$$

where $\delta V(m_k, n_k, m_{k+1}, n_{k+1})$ is the value of the edge that leads from node (m_k, n_k) to (m_{k+1}, n_{k+1}). Roots and isolated nodes have the value 0.

Because each node has only four possibilities for its parent and children, the trees are easily represented and managed by means of an array of bit flags. The trees are created as follows: Each pair of adjoining nodes is tested for the addition of an edge, unless an edge between the two nodes already exists. The addition of an edge from the node (m, n) to the node (m', n') results in $value(m', n')$ being increased by

$$dval = \text{value}(m, n) + \delta V(m, n, m', n') - \text{value}(m', n'). \tag{4.6}$$

If *dval* is negative, then the addition of the edge would result in the value of the node becoming negative or less than its old value. In this case the edge is not added. If *dval* is positive, however, then the edge is added and $value(m', n')$ is increased by *dval*. If the node (m', n') has a subtree (i.e., if there are edges leading from it to other nodes), then *dval* is also added to all the nodes in the subtree. Any existing edge to (m', n') is removed.

After the addition of the new edge from the node (m, n) to (m', n'), the trees are searched for a loop. If there is already a path from (m', n') to (m, n), then the new edge completes a loop. Furthermore the loop is an improving loop. The jump counts of the edges along the loop are then incremented or decremented according to the orientations of the edges, and the edges that comprise the loop are removed. Each node in the loop then becomes isolated or the root of a subtree. The value of each of these nodes is set to zero, and the values of its subtree nodes are reduced by its former value. If the values of any of the nodes become negative as a result of this operation, then all edges leading to the negative node are removed and its value is set to zero.

Figure 4.46 shows that the node array is padded on all sides. This ensures that a loop can enclose a region of pixels even if it adjoins the boundary of the image. Edges between the boundary nodes have the value zero because they do not induce changes in the jump counts.

Figure 4.47 shows an example of a set of trees. The large boldface numbers represent the nodes and their values, the arrows represent the edges, and the small numbers specify the values of the edges. The value of a node is simply the sum of the values of the edges that link the node to the root of its tree. The large, shaded arrows indicate two new edges that can be added.

Figure 4.48 shows the effect of adding these two edges. The new edge at the lower right has modified an existing tree. The edge that led to the destination node of the new edge has been removed, and the node values affected by the change have been updated. The new edge at the upper left of the figure formed a loop, which has been removed. All but one of the subtrees leading from the loop nodes are still present with updated values. The edge that led downward from the 4 node has been removed because the value of its destination node was revised to -1.

The operation of Flynn's algorithm can now be stated concisely. The phase data are read, and the horizontal and vertical jump counts are defined. (At this point the phase data are no longer needed until the very end, at which point they

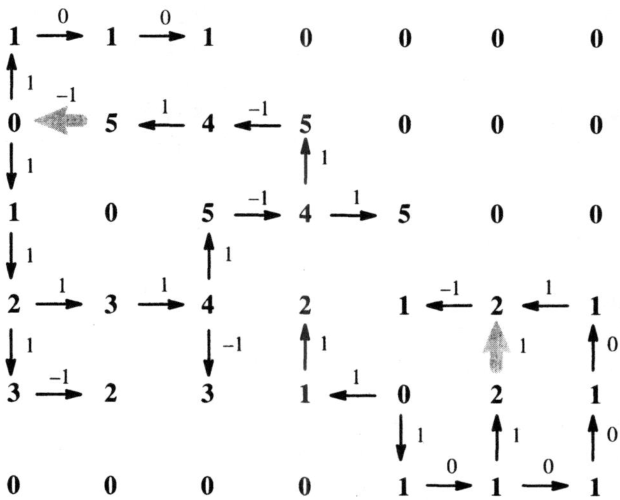

Figure 4.47 An example of a set of trees. The large boldface numbers are the nodes, and the arrows are the edges. The numbers indicate the values of the nodes and the edges. The two large, shaded arrows are new edges that can be added.

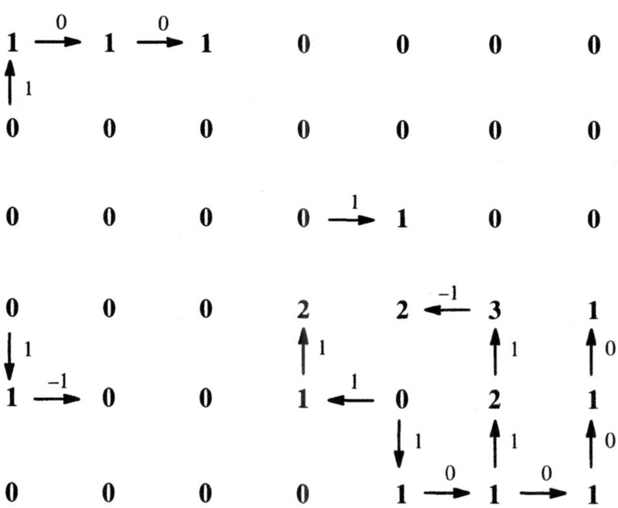

Figure 4.48 Trees that result from the addition of the two new edges indicated by the large shaded arrows of Figure 4.47.

may be read again.) The array of node values is initialized to all zeroes. The nodes are repeatedly scanned, and an edge is added if its value *dval* defined by Equation 4.6 is positive. Whenever a loop is found, its jump counts are all updated, its edges are removed, and the values of its nodes and subtrees are updated accordingly. The algorithm terminates when a complete scan adds no new edges and removes no loops. The wrap counts are then defined by the jump counts and added to the input phase data to yield the unwrapped solution.

Flynn shows that the algorithm is guaranteed to terminate after a finite number of operations [21]. He also shows that the solution is a (globally) minimal solution for the error measure defined by Equation 4.4. In other words, there is no danger that the algorithm will fall prematurely into a "local minimum".

Before we proceed to the pseudocode for this algorithm, we will demonstrate that the algorithm extends naturally to the weighted case. Given a quality map, one can define arrays of weights

$$w^v_{m,n} = \min(q_{m,n}, q_{m+1,n}), \tag{4.7}$$

$$w^z_{m,n} = \min(q_{m,n}, q_{m,n+1}), \tag{4.8}$$

where $q_{m,n}$ are the values of the quality map. The weighted version of the error measure of Equation 4.4 is then defined as follows:

$$E_w = \sum w^v_{m,n}|v_{m,n}| + \sum w^z_{m,n}|z_{m,n}|. \tag{4.9}$$

An unwrapped solution that achieves minimum E_w is called a *minimum weighted discontinuity* solution. Flynn's algorithm can be extended to find a minimum weighted discontinuity solution simply by multiplying each edge value δV by the appropriate weight. The edge values and the term *dval* defined by Equations 4.5 and 4.6, respectively, will be affected by this weighting of δV.

The minimum weighted discontinuity version of the algorithm is useful when one wishes to restrict the discontinuities to the low-quality pixels, such as the regions of layover in the simulated IFSAR examples of Chapter 3.

4.5.2 Pseudocode

Flynn's minimum discontinuity algorithm consists of three main steps:

Step 1. Compute the jump counts from the input phase data

Step 2. Scan the nodes repeatedly, adding edges and removing loops, until no new edges are added and no loops are removed

Step 3. Compute the unwrapped phase from the jump counts

The first and last steps are simple. In the first step, the input phase data are regarded as unwrapped, and the horizontal and vertical jump counts are computed from them by means of Equations 4.2 and 4.3. At this point the input phase data are no longer needed until the last step. In the last step, the wrap counts are computed from the horizontal and vertical jump counts and added to the input phase data to produce the unwrapped solution. We will now concentrate our attention on Step 2, which is the substance of Flynn's algorithm.

The horizontal and vertical jump counts are stored in two arrays. Like the other arrays, except the input phase data and the unwrapped output, the two jump count arrays are indexed on the nodes rather than the pixels. Because the nodes are padded around the border, the array dimensions are $(M+1) \times (N+1)$ for an $M \times N$ input phase array. Associated with each node **node** is its value, which we will denote **value(node)** in the pseudocode. These values are stored in a third array. Associated with each pair of adjoining nodes (**from, to**) is a flag denoted **edge(from, to)** that specifies whether or not there is an edge from node **from** to node **to**. The edge flags are stored in the bit flags array. Four bits of the bit flags array specify whether there is an edge to the lower, upper, rightward, or leftward node, respectively. The trees are stored and managed by means of these flags.

The value of the edge from node **from** to node **to** is denoted δV(**from, to**). It is not stored in an array but is calculated as needed.

If there are any pixels that belong to a border mask as in the simulated IFSAR examples, their neighboring nodes are marked as "border" nodes in the bit flags array. Flynn's algorithm ignores these nodes by assigning zero-weights to them so that the value of any edge that joins a border node is always zero. This ensures that any discontinuities in the border mask will not contribute to the error measure (Equation 4.4 or 4.9) being minimized. The bit flags also keep track of the nodes that change on each iteration. This helps reduce the computational cost because these nodes do not need to be examined on the following iteration.

Thus the bit flags array manages the following flags:

1. Edge to lower node
2. Edge to upper node
3. Edge to right node
4. Edge to left node
5. Border node
6. Value changed

We can now present the pseudocode for Step 2 (the main step) of Flynn's algorithm. The pseudocode is substantially the same as that of [21]. There are three subroutines that are called to handle the details of managing the trees and the jump counts. Subroutine **ChangeValue** changes the values in a tree and tests for loops, **RemoveLoop** removes a loop, and **ChangeJump** changes a jump count.

For each node (denoted **from**), the edge from that node to each of its four neighbors (denoted **to**) is tested in turn. If the edge has a positive value, then **ChangeValue** is called starting at node **to**. This subroutine recursively traverses the subtree of **to**, increasing the values of the nodes by **dval** and checking whether the subtree contains the node **from**. If so, the new edge completes a loop, which is removed by means of subroutine **RemoveLoop**. Another call is then made to **ChangeValue** to undo any inappropriate changes. If a loop is not found, the new edge is added, and any edges that lead to node **to** are removed. Notice that the nodes (**from, to**) must be neighbors.

Step 2 of Flynn's Minimum Discontinuity Algorithm

```
Let value = 0 at all nodes
Let edge(from, to) = false for all pairs of nodes (from, to)
repeat
  Let edge_found = false
  for (all node pairs (from, to))
    if (not edge(from, to)) then
      let dval = value(from) + δV (from, to) - value(to)
      if (dval > 0) then
        let edge_found = true
        let loop_found = ChangeValue(to, dval, from)
        if (loop_found) then
          RemoveLoop(to}, from)
          ChangeValue(to, -dval, NULL)
        else
          for (each neighbor nbr of to)
            let edge(nbr, to) = false
          end
          let edge (from, to) = true
        end
      end
    end
  end
until (not edge_found)
```

Subroutine **ChangeValue** recursively traverses the subtree of node **base**. It changes the value of **base**, adjusting **dval** if necessary, and then invokes itself at all the children of **base** (i.e., all the nodes that are connected to **base** by edges). If **base** is identical to the node **loop**, a loop has been detected, and the **loop_found** flag is set and returned. If **ChangeValue** is called with **loop** = NULL, the loop detection is disabled.

ChangeValue(base, dval, loop)

```
let loop_found = true if base = loop, false otherwise
if (value(base) + dval < 0) then
  let dval = - value(base)
  for (each neighbor nbr of base)
    let edge(nbr, base) = false
  end
end
let value(base) = value(base) + dval
for (each neighbor nbr of base)
  if (edge(base, nbr)) then
```

```
      let loop_found = loop_found OR ChangeValue(nbr, dval, loop)
    end
  end
  return loop_found
```

Subroutine **RemoveLoop** traces the loop backwards, removing each edge and calling subroutine **ChangeJump** to change the jump counts until the base of the loop is reached. Subroutine **ChangeValue** is also called at each step to adjust the values of the nodes in any "orphaned" subtrees.

RemoveLoop(base, last)

```
ChangeJump(last, base)
let tip = last
repeat
  ChangeValue(tip, -value(tip), NULL)
  let prev = the neighbor of tip for which edge(prev, tip) is true
  let edge(prev, tip) = false
  ChangeJump(prev, tip)
  let tip = prev
until (tip = base)
```

Subroutine **ChangeJump** is called by subroutine **RemoveLoop**. It simply increments or decrements the jump count at the edge that joins the nodes **from** and **to**.

ChangeJump(from, to)

```
if (from and to are adjoining nodes but not both border nodes) then
  if (to is above from) then
    decrement the vertical jump count
  else if (to is below from) then
    increment the vertical jump count
  else if (to is on the right of from) then
    decrement the horizontal jump count
  else if (to is on the left of from) then
    increment the horizontal jump count
  end
end
```

4.5.3 Implementation

The C language implementation of Flynn's minimum discontinuity algorithm can be found in Section A.9 of Appendix A. This software was supplied by the author of the algorithm, Thomas J. Flynn, and then modified to match the coding style of this book. Our modification gives the user the following choice of quality maps for defining the weights:

1. User-supplied correlation coefficients (or other quality map)
2. Pseudocorrelation
3. Variance of phase derivatives
4. Maximum phase gradient
5. No quality map

These quality maps can be computed in windows of any size. The software can be easily modified to compute other quality maps, such as second differences or other measures of phase variance. If no quality map is specified, then the unweighted minimum discontinuity solution defined by the error measure of Equation 4.4 is computed.

The memory requirements are slightly higher than those of the quality-guided algorithm of Section 4.3 and the mask cut algorithm of Section 4.4. Because the input phase data, the quality map, and the unwrapped solution are not needed at the same time, these three arrays can all share the same memory. A total of five arrays is therefore required by the algorithm:

1. Phase data, quality map, and solution array
2. Node value array
3. Horizontal jump-count array
4. Vertical jump-count array
5. Bit flags array

Table 4.4 lists the average execution times of the algorithm for various array sizes. The times are considerably greater than those of the three path-following algorithms we have already studied. The times actually depend on the particular data sets to be unwrapped (specifically, the complexity of the discontinuities) more than on the sizes of the data sets. For example, we have recorded execution times of as low as 8 seconds and as high as 55 seconds for a 256 × 256-pixel data set. The execution times can be decreased by inputting a partially unwrapped solution for which most of the discontinuities caused by the phase fringes have

Table 4.4 Average execution times of Flynn's minimum discontinuity algorithm for various array sizes on an IBM RS/6000 workstation.

Array Size (pixels)	Execution Time (hr:min:s)
256 × 256	00:00:20
512 × 512	00:02:30
1024 × 1024	00:20:00
2048 × 2048	03:00:00

been removed. Using the algorithm without partial unwrapping may be comparable to driving a tack with a sledgehammer. Flynn used a simple minimum spanning tree algorithm for the partial unwrapping step [21]. Any unwrapping algorithm can be used for this step as long as it produces an output that is congruent modulo 2π to the input phase data.

4.5.4 Examples

We will now examine the results of Flynn's minimum discontinuity algorithm on the phase-unwrapping examples of Chapter 3. On the first three examples (IFSAR, MRI head, and MRI knee) we will show the results of the unweighted algorithm. On the two simulated IFSAR examples we will show the results of both the unweighted and the weighted algorithms. There are a number of ways in which the weights can be defined. We will use the masks defined by the phase derivative variance computed in 3×3 windows with the threshold computed by the automatic method of Section 3.4. These masks are illustrated in the figures of Section 3.4.

Flynn's minimum discontinuity algorithm unwraps the IFSAR example of Figure 3.2 correctly. The differences between this surface and that produced by Goldstein's algorithm are shown by the error map of Figure 4.49(a). The dark regions are those in which the two surfaces differ. Notice that the differences are confined to the patches of corrupted phase. Exclusive of these corrupted regions, all the path-following algorithms in this chapter produce the same unwrapped surface.

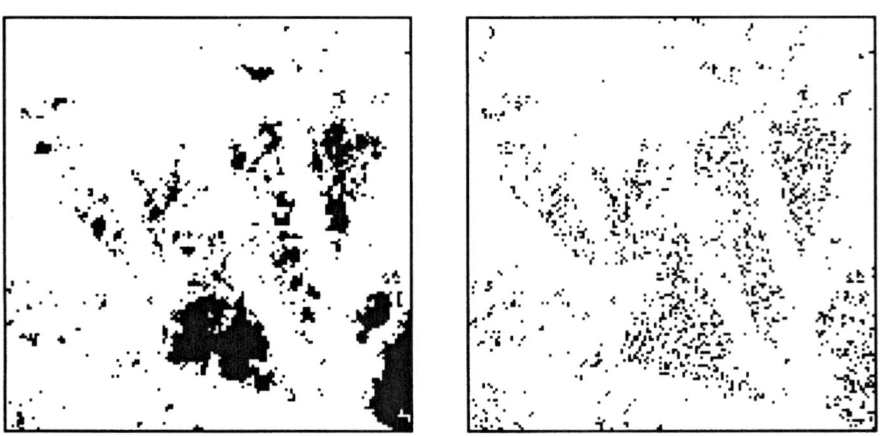

(a) (b)

Figure 4.49 Results of Flynn's minimum (unweighted) discontinuity algorithm on the IFSAR example of Figure 3.2. (a) The differences between the surfaces unwrapped by the algorithm and Goldstein's algorithm are marked in black. The unwrapped surfaces are identical outside the patches of noise. (b) The black pixels mark the discontinuities in the unwrapped surface where the jump counts are nonzero.

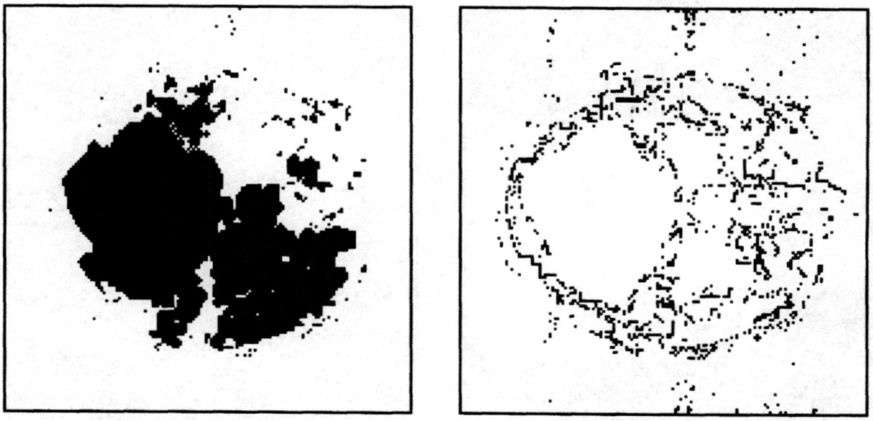

Figure 4.50 Results of Flynn's minimum (unweighted) discontinuity algorithm on the MRI head example of Figure 3.3. (a) The differences between the unwrapped surface and the surface produced by Goldstein's algorithm. The differences inside the head regions are 2π-multiple differences (i.e., wrap count differences) due to the corrupted phase that completely surrounds these regions. (b) The nonzero jump counts, which mark the discontinuities in the unwrapped surface.

It is interesting to examine the pixels where the jump counts are nonzero.[5] These discontinuities are marked as dark pixels in Figure 4.49(b). Notice that they are confined to the patches of noise.

The algorithm also appears to unwrap the MRI head example of Figures 3.3. The differences between the unwrapped surface and that of Goldstein's algorithm are shown in Figure 4.50(a). The regions inside the head differ from those of Goldstein's algorithm. (They have different wrap counts, or multiples of 2π.) This is due to the fact that these regions are completely isolated from the rest of the image by regions of corrupted phase. The nonzero jump counts, which are shown in Figure 4.50(b), generally are confined to the regions of corrupted phase.

On the MRI knee example of Figure 3.4, Flynn's algorithm produces the correct unwrapping. The differences between the unwrapped surface and that produced by Goldstein's algorithm are shown in Figure 4.51(a). The two unwrapped surfaces are exactly the same, except for the regions of corrupted phase and some very small patches within the knee. The large regions of corrupted phase outside the knee do not corrupt the unwrapping inside the knee. The nonzero jump counts are shown in Figure 4.51(b).

We now turn our attention to the simulated IFSAR examples. The unweighted version of Flynn's algorithm was sufficient for unwrapping the

[5] We have supplied software for producing images of the nonzero jump counts in Section A.5 of Appendix A.

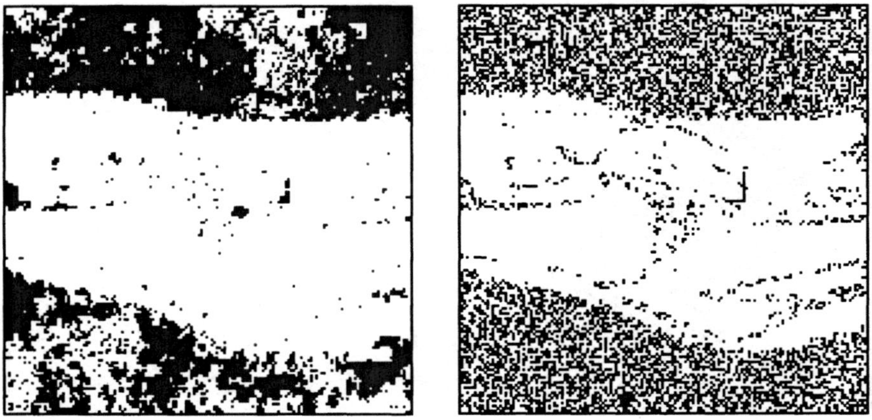

Figure 4.51 Results of Flynn's minimum (unweighted) discontinuity algorithm on the MRI knee example of Figure 3.4. (a) The differences between the unwrapped surface and the surface produced by Goldstein's algorithm. The two algorithms produce the same unwrapping except in the corrupted regions. (b) The nonzero jump counts.

three preceding examples. On the simulated IFSAR examples, however, we will need the weighted version.

On the Long's Peak simulated IFSAR example of Figure 3.7, Flynn's unweighted algorithm produces incorrect results. The error map is shown in Figure 4.52(a). The entire lower half of the image has been unwrapped incorrectly. The locations of the nonzero jump counts, which are shown in Figure 4.52(b), illustrate the cause of the errors. These discontinuities are analogous to the branch cuts of Goldstein's algorithm, which are shown in Figure 4.11(a). To minimize the jump counts, the algorithm was forced to place the discontinuities across the image, thereby isolating the lower half of the image from the upper half. Moreover this minimum-discontinuity solution is a global minimum, as opposed to a local minimum. If the correct surface is input to Flynn's algorithm, it produces the same result.

Like Goldstein's algorithm, Flynn's algorithm may place its "branch cuts" (i.e., discontinuities or jumps) in undesirable locations that completely isolate portions of the image, thereby resulting in incorrect wrap counts (multiples of 2π) in these portions. Unlike Goldstein's algorithm, however, Flynn's algorithm can accept an array of weights that indicate which phase values are more reliable than others. These weights constrain the placement of the discontinuities to the low-quality regions.

There are a number of ways in which the weights can be defined. For example, one could use a quality map derived from the phase derivative variance or the maximum phase gradient, as shown in Figures 3.19(a) or 3.23(a), respectively. We elected, however, to use the mask of Figure 3.30(c) instead, and this choice

(a) (b)

Figure 4.52 Results of Flynn's minimum *unweighted* discontinuity algorithm on the Long's Peak example of Figure 3.7. (a) The error map. (b) The discontinuities produced by the algorithm. This example illustrates the need for weights to guide the placement of the discontinuities.

gave excellent results. This mask, which was generated by applying the thresholding technique of Section 3.4 to the phase derivative variance, defines the locations of zero-weights. It causes Flynn's algorithm to confine the placement of the discontinuities to the zero-weighted (i.e., masked out) regions.[6] The error map of the resulting surface is shown in Figure 4.53(a), and the discontinuities are shown in Figure 4.53(b). The improvement is striking. The discontinuities are

[6] Although the weights in Equation 4.9 can take any value, here we are restricting our attention to binary weights.

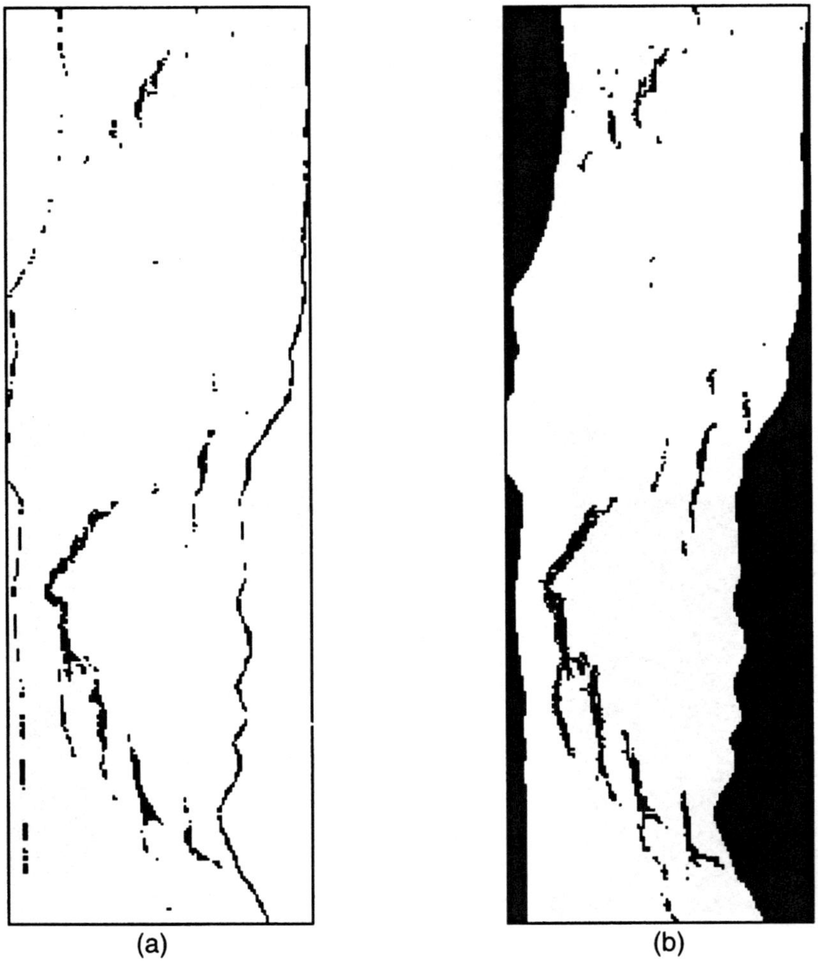

Figure 4.53 The results of Flynn's minimum *weighted* discontinuity algorithm on the Long's Peak example of Figure 3.7. Figure 3.30(c) shows the mask that was used to define the zero-weights. (a) The error map. (b) The discontinuities in the unwrapped surface. The mask restricted the discontinuities to the zero-weighted regions of layover.

confined to the zero-weighted regions of layover, and the unwrapping is flawless outside these regions. Compare the discontinuities with Goldstein's branch cuts of Figure 4.11(a) and the mask cuts of Figure 4.36(a).

On the Isolation Peak example of Figure 3.8, the story is similar. Flynn's *unweighted* algorithm produced an incorrect result whose error map is shown in Figure 4.54(a). The locations of the discontinuities are shown in Figure 4.54(b). Many regions of the image have been isolated by the horizontal lines of

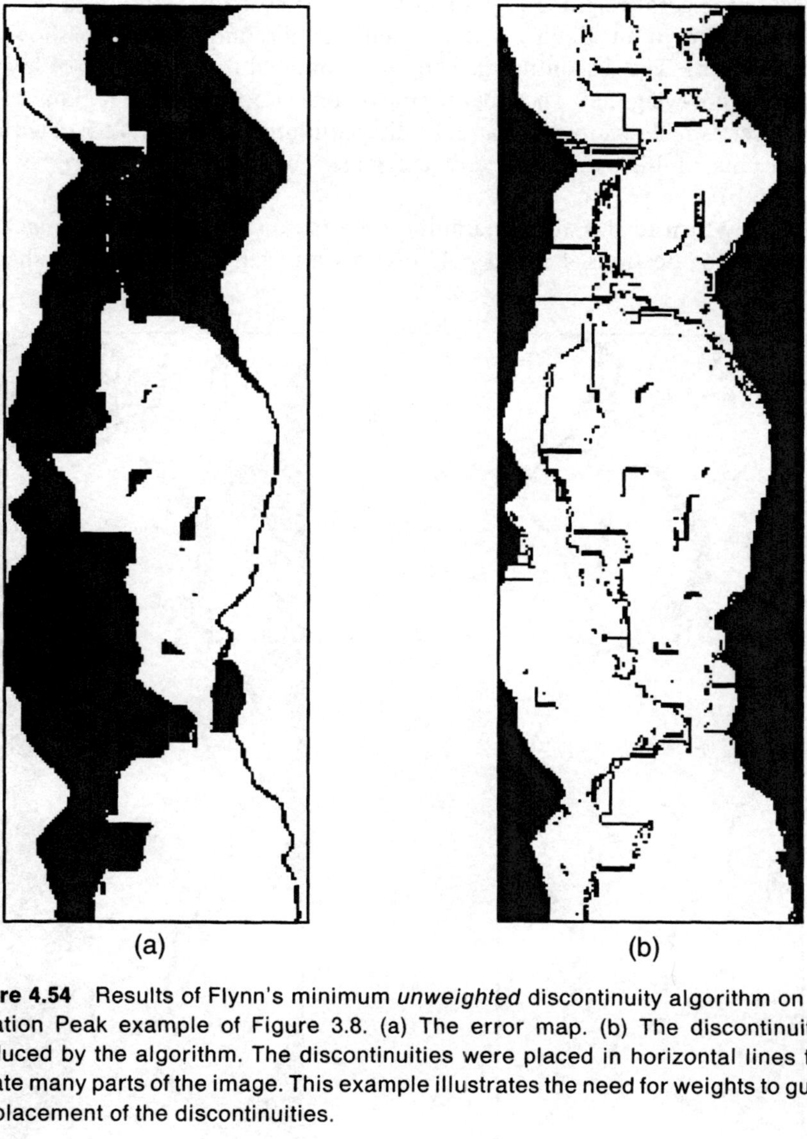

(a) (b)

Figure 4.54 Results of Flynn's minimum *unweighted* discontinuity algorithm on the Isolation Peak example of Figure 3.8. (a) The error map. (b) The discontinuities produced by the algorithm. The discontinuities were placed in horizontal lines that isolate many parts of the image. This example illustrates the need for weights to guide the placement of the discontinuities.

discontinuities, thereby resulting in incorrect wrap counts. Compare the discontinuities of Figure 4.54(b) with Goldstein's branch cuts of Figure 4.13(a).

The results are much more satisfactory if one generates the minimum *weighted* discontinuity solution with the weights defined by the mask of Figure 3.32(b). Like the mask we used for the Long's Peak example, this mask is generated by applying the thresholding technique of Section 3.4 to the phase derivative variance. It causes Flynn's algorithm to confine the placement of the discontinuities to the zero-weighted regions of layover. The error map of the resulting surface is shown in Figure 4.55(a), and the discontinuities are shown in Figure 4.55(b). The discontinuities are now confined to the regions of layover that were zero-weighted. The unwrapping is correct outside these regions except for one very small region. Compare the discontinuities of Figure 4.55(b) with the branch cuts of Figure 4.13(a) and the mask cuts shown in Figures 4.39(b) and 4.40(b).

We now turn to the shear examples. Like the quality-guided and mask cut algorithms of Sections 4.3 and 4.4, Flynn's algorithm unwraps the sheared

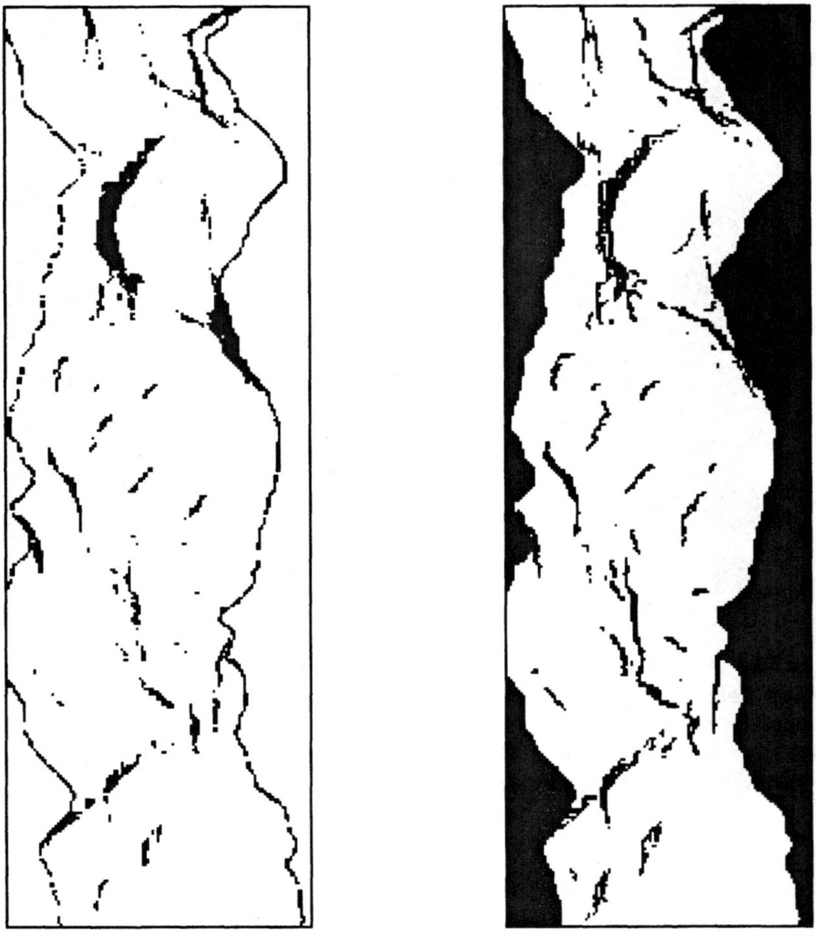

(a) (b)

Figure 4.55 Results of Flynn's minimum *weighted* discontinuity algorithm on the Isolation Peak example of Figure 3.8. Figure 3.32(b) shows the mask that defines the zero-weights. (a) The error map. (b) The discontinuities in the unwrapped surface. The mask restricts the discontinuities to the regions of layover as desired.

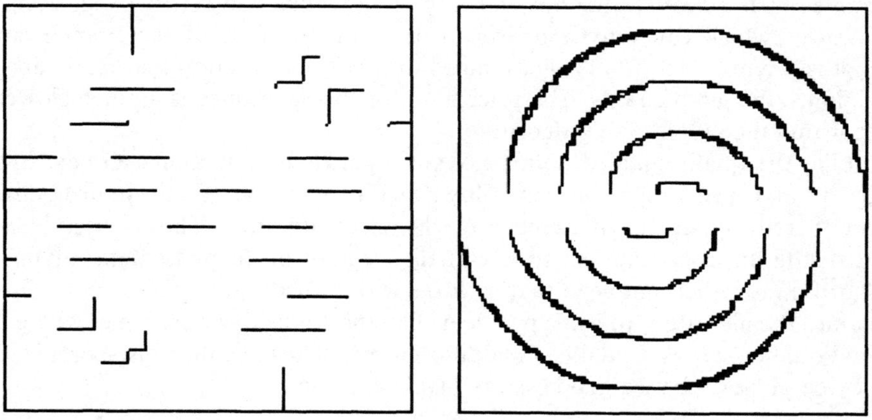

Figure 4.56 (a) The discontinuities produced by Flynn's unweighted algorithm on the spiral shear example of Figure 3.10. The discontinuities cut across the arms of the spiral, and the resulting unwrapped solution is incorrect. (b) The discontinuities produced by Flynn's weighted algorithm with the zero-weights defined by the mask of Figure 3.35(b). The discontinuities are confined to the spiral lines of shear, and the unwrapped solution is correct.

planes data of Figure 3.9 flawlessly. In this case there is no need to use weights. Unlike the quality-guided and mask cut algorithms, Flynn's algorithm also correctly unwraps the noise-corrupted shear example of Figure 4.15.

Unlike Goldstein's algorithm, Flynn's algorithm also unwraps the spiral shear example of Figure 3.10. In this case weights are required to restrict the discontinuities to the spiral lines of shear. If these weights are not supplied, Flynn's algorithm produces the discontinuities of Figure 4.56(a). Like the branch cuts of Goldstein's algorithm shown in Figure 4.14(c), these lines of discontinuities cut across the arms of the spiral. With the zero-weights defined by the mask of Figure 3.35(b), however, the algorithm produces the discontinuities of Figure 4.56(b). These discontinuities lie on the spiral lines of shear and effectively separate the two arms of the spiral so that the phase can be unwrapped correctly. (There are no discontinuities halfway between the top and bottom of the image because this is where the two unwrapped planes "cross" one another.)

Flynn's algorithm is the only "path follower" we know of that successfully unwraps both the noise-corrupted sheared planes and the spiral shear examples.

4.5.5 Conclusion

Flynn's minimum discontinuity algorithm solves the phase unwrapping problem in a systematic, well-defined manner. It is similar to Goldstein's algorithm in that it minimizes a measure of discontinuity. Where Goldstein's algorithm attempts to minimize the branch cut lengths, Flynn's algorithm minimizes the "jump counts" defined by Equations 4.4 and 4.9. Unlike Goldstein's algorithm, however, which can only *approximate* the minimization of the branch cut lengths, Flynn's algorithm is guaranteed to produce a solution that really does minimize the jump counts. The price paid for this guarantee is a much slower algorithm than Goldstein's algorithm.

Like the quality-guided and mask cut algorithms, but unlike Goldstein's algorithm, Flynn's algorithm can utilize a quality map to weight the jump counts and thereby guide the placement of the discontinuities. This is useful for restricting the discontinuities to low-quality regions of the phase data. Flynn's algorithm is rather unique in that it works with or without a quality map. This enables the algorithm to solve problems like the spiral shear example, where a good quality map is available, in addition to problems like the noise-corrupted sheared planes, where a good quality map is lacking.

4.6 SUMMARY AND CONCLUSION

In this chapter we have studied four path-following algorithms. In this concluding section we will highlight the distinctive features of each algorithm and summarize the results on the phase unwrapping examples. We will also make a few suggestions for choosing an algorithm to solve a particular problem and indicate how to evaluate the results of an algorithm when the actual unwrapped surface or "ground truth" is unknown.

The first algorithm, Goldstein's algorithm, is a very fast method that identifies the residues and connects them with branch cuts to balance them. This algorithm approximately minimizes the sum of the lengths of the branch cuts by means of a clever heuristic. This approximately minimizes the discontinuities in the unwrapped phase. Goldstein's algorithm is very fast, requires little memory, and is generally successful at unwrapping the phase correctly. Its weakness is a lack of means to guide the placement of branch cuts with a quality map.

The second algorithm we studied, the quality-guided path-following algorithm, does not identify residues. Instead, it relies completely on the quality map to guide the integration path. It is essentially a region-growing approach that unwraps the highest-quality pixels first and avoids the lowest-quality pixels until the end of the path integration procedure. There is no guarantee the algorithm will not encircle an unbalanced residue and introduce a spurious discontinuity in the unwrapped solution. Instead of relying on branch cuts, the algorithm depends on the quality map to guide its integration path.

The algorithm yields different results depending on which quality map is used. The quality map can be defined by any one of a number of measures. We have examined the phase derivative variances, the maximum phase gradients, and the pseudocorrelation coefficients. The quality map can also be defined by measures that are unique to the observations from which the phase data are extracted. For example, if the phase is extracted from IFSAR data, the quality map can be defined by the correlation coefficients. We obtained the best results with the correlation coefficients (for the IFSAR examples) and the phase derivative variances. If a good quality map is not available, as is the case on the noise-corrupted sheared planes example, the quality-guided path-following algorithm is useless.

The third algorithm we studied was the mask cut algorithm. This algorithm can be regarded as a cross between the first two algorithms. Like Goldstein's algorithm, it identifies the residues, and like the quality-guided algorithm, it uses a quality map to guide it. The quality map guides the generation of pixel masks called mask cuts, which join the residues and balance them. The algorithm does not attempt to minimize the lengths of the mask cuts. Instead, it relies on the quality map to ensure that the masks are sized and placed appropriately. It also uses a morphological thinning operation to reduce the sizes of the masks further. We observed that the mask cuts tend to be thicker than the branch cuts generated by Goldstein's algorithm. However, on some examples the algorithm was able to place the mask cuts more successfully than Goldstein's algorithm due to the quality map. Like the quality-guided algorithm, the mask cut algorithm cannot be used where a good quality map is not available.

The preceding three algorithms approximately minimize the discontinuities in the unwrapped surface. They use various means to achieve this goal. Goldstein's algorithm employs branch cuts, the quality-guided algorithm uses a quality map, and the mask cut algorithm grows mask cuts guided by a quality map. None of these algorithms is guaranteed to find a truly minimal solution.

By contrast, the fourth algorithm we studied, Flynn's minimum discontinuity algorithm, finds a solution that actually (rather than approximately) minimizes the discontinuities. It achieves this goal by attacking the problem directly. The algorithm utilizes a tree-growing approach that traces paths of discontinuities in the phase, detects the paths that form loops, and adds multiples of 2π to the phase values enclosed by the loops to minimize the discontinuities. It performs this process iteratively until no more loops are detected. The process is guaranteed to converge to a "minimum discontinuity" solution.

A unique feature of Flynn's algorithm is that it works with or without a quality map. If a quality map is supplied, then the discontinuities are weighted by the quality values so that the discontinuities in the low-quality regions do not contribute appreciably to the overall measure of discontinuity. As a result the discontinuities in the result tend to be confined to the lowest-quality regions. Although the memory requirements and computational burden of Flynn's algorithm are much greater than those of the other algorithms, it is the only algorithm that successfully unwrapped all of the examples, including the noise-corrupted sheared planes and the spiral shear data sets.

Table 4.5 summarizes several characteristics of the four algorithms. For each algorithm the table lists the memory requirements and the average execution times for a 1024×1024-pixel phase array. The memory requirements are expressed in terms of the number of 1024×1024 arrays for the phase data, the unwrapped surface, the quality map, the bit flags, and other data. The listed execution times are estimates because they really depend on the complexity of the

Table 4.5 Characteristics of the path-following algorithms.

Algorithm	Memory Requirements (Arrays)	Execution Time	Identifies Residues?	Quality Map Required?
Goldstein	3	30 sec	y	n
Quality	4	3 min	n	y
Mask cut	4	4 min	y	y
Flynn	5	20 min	n	y/n

Note: The memory requirements are the number of arrays required. The execution times are given in minutes for a 1024 × 1024-pixel phase array on an IBM RS/6000 workstation. The column labeled "Identifies Residues?" indicates whether or not the algorithm explicitly identifies residues. The last column indicates whether or not the algorithm requires a quality map. Flynn's minimum discontinuity algorithm can be used with or without a quality map.

Table 4.6 Results of the path-following algorithms on the phase unwrapping examples.

	Examples			
Algorithm	IFSAR	MRI Head	MRI Knee	Long's Peak
Goldstein	*y*	*y*	*y*	*n*
Quality	*y*	*y*	*y*	*y*
Mask cut	*y*	*y*	*y*	*y*
Flynn	*y*	*y*	*y*	*y*
Algorithm	Isolation Peak	Sheared Planes	Noisy Shear	Spiral Shear
Goldstein	*n*	*y*	*y*	*n*
Quality	*y*	*y*	*n*	*y*
Mask Cut	(*y*)	*y*	*n*	*y*
Flynn	(*y*)	*y*	*y*	*y*

Note: An entry of *y* indicates the algorithm unwrapped the example correctly, *n* indicates the algorithm did not unwrap the example correctly, and (*y*) indicates the algorithm had minor errors in its unwrapped solution.

phase data—the number of residues, the number of discontinuities, and so on. The table also indicates whether or not the algorithm identifies residues and whether or not a quality map is required by the algorithm.

Table 4.6 summarizes the successes and failures of the algorithms on the seven phase unwrapping examples of Chapter 3 and the noise-corrupted sheared planes example of Figure 4.15. An entry of *y* indicates the algorithm was successful, *n* indicates the algorithm failed, and *(y)* indicates the algorithm unwrapped the example successfully except for minor errors.[7] Flynn's algorithm was the only algorithm to unwrap all of the examples successfully. The two simulated IFSAR examples (Long's Peak and Isolation Peak) presented special difficulties for Goldstein's algorithm. This was due to the fact that the borders of the phase data were irregular and lay very close to the regions of layover.[8] Although Goldstein's algorithm failed on these examples, it actually performs quite well on the typical IFSAR data sets encountered in practice.

Before leaving the topic of the performance of the algorithms on the phase unwrapping examples, we remark that it is straightforward to devise an example on which all the algorithms fail. The example can be constructed by corrupting the spiral shear example with Gaussian noise. Because a good quality map cannot be derived from the noisy phase data, the quality-guided and mask cut algorithms are useless. Goldstein's algorithm places branch cuts across the arms of the spiral, and Flynn's algorithm leaves discontinuities that cut across the arms of the spiral. In both cases the phase is unwrapped incorrectly.

Which algorithm should be used in practice? That depends on the particular phase unwrapping problem to be unwrapped. We are of the opinion that Goldstein's algorithm should always be tried first. This algorithm has simple input requirements (e.g., a quality map is not needed), and it is very fast. When it fails, the user is usually warned by an inspection of the branch cuts to determine if any regions were isolated from the rest of the image. If Goldstein's algorithm fails and a good quality map is available, then the quality-guided or mask cut algorithms should be tried. If the results are not satisfactory, then the user can "tinker" with the quality map to try to obtain a successful unwrapping. We have seen that different quality maps will yield different results on a given problem. The user can choose not only the measure used to compute the quality map but also the type and amount of filtering to be applied to the quality map, the threshold to be applied if a quality *mask* is desired, and whether any morphological operations such as thinning or thickening are to be applied to the mask. Given the numerous possibilities available to generate a quality map, these two algorithms (the quality-guided and the mask cut algorithms) multiply into a myriad of different algorithms.

If these three algorithms fail, then we recommend the use of Flynn's minimum discontinuity algorithm. Although this algorithm takes longer to execute than the others, it can sometimes "crack" tough phase unwrapping problems on which the other algorithms fail.

Because we have supplied `C` language implementations for the algorithms in this book, we cannot resist making the suggestion that *all* the algorithms be run

[7] Because the correct unwrapped solution was unknown for the first three examples, an algorithm was judged a success or failure based on comparison of its result with the input phase and the results of the other phase unwrapping algorithms, as well as the authors' estimation of what constituted success on each example. The results for the other five examples were judged by comparison with the known solutions.

[8] Borders near layover regions cause similar problems for other algorithms, including Flynn's minimum discontinuity algorithm. See Section 5.5.4 for examples of this problem on the unweighted version of the minimum L^p-norm algorithm.

and the results compared. Comparison of the results from several algorithms usually provides insight into whether the problem was correctly unwrapped. For example, the fact that the four algorithms of this chapter produced exactly the same result on the IFSAR example (exclusive of the patches of noise) suggests quite strongly that this result is correct.

After an algorithm has been selected and executed on a particular phase-unwrapping problem, how should the result be evaluated? In addition to comparing it with the results of other algorithms, we suggest that the data produced by the algorithm be employed as part of the evaluation process. Goldstein's algorithm produces a set of branch cuts that should be inspected and compared with the input phase data and quality map (if available). The mask cut algorithm produces a set of mask cuts that should be evaluated in the same way. Similarly Flynn's algorithm produces a set of jump counts (discontinuities). The quality-guided algorithm does not produce branch cuts, but a map of the jump counts can be generated just as one would for Flynn's algorithm. These jump counts indicate where the discontinuities lie in the unwrapped solution. We will return to the topic of evaluation of results in Chapter 6.

We conclude this chapter by answering the following question, which may have occurred to the reader: Why do we classify the algorithms of this chapter as "path-following" algorithms? Goldstein's algorithm is actually a "branch cut placement" algorithm, and the quality-guided algorithm is really a "region-growing" approach. The mask cut algorithm is a "mask cut generation" algorithm, and Flynn's algorithm is a "wrap-count manipulation" technique. While these algorithms employ diverse approaches to solve the phase unwrapping problem, what they all have in common is that they generate paths in the image by means of localized, pixel-by-pixel operations. In contrast to this, the phase unwrapping algorithms of the following chapter employ mathematical techniques to find an unwrapped solution that fits the given phase data on a global, rather than a pixel-by-pixel, basis.

REFERENCES

[1] J. M. Huntley and J. R. Buckland, "Characterization of sources of 2π phase discontinuity in speckle interferograms," *Journal of the Optical Society of America A*, Vol. 12, No. 9, pp. 1990–1996, 1995.

[2] R. M. Goldstein, H. A. Zebker, and C. L. Werner, "Satellite radar interferometry: two-dimensional phase unwrapping," *Radio Science*, Vol. 23, No. 4, pp. 713–720, 1988.

[3] J. M. Huntley, "Noise-immune phase unwrapping algorithm," *Applied Optics*, Vol. 28, No. 15, pp. 3268–3270, 1989.

[4] R. Cusack, J. M. Huntley, and H. T. Goldrein, "Improved noise-immune phase-unwrapping algorithm," *Applied Optics*, Vol. 34, No. 5, pp. 781–789, 1995.

[5] J. R. Buckland, J. M. Huntley, and S. R. E. Turner, "Unwrapping noisy phase maps by use of a minimum cost matching algorithm," *Applied Optics*, Vol. 34, No. 23, pp. 5100–5108, 1995.

[6] T. R. Judge and P. J. Bryanston-Cross, "A review of phase unwrapping techniques in fringe analysis," *Optics and Laser Engineering*, Vol. 21, pp. 199–239, 1994.

[7] M. W. Roth, "Phase unwrapping for interferometric SAR by the least-error path," Johns Hopkins University Applied Physics Lab Technical Report, Laurel, MD, March 30, 1995.

[8] D. Just, N. Adam, M. Schwäbisch, and R. Bamler, "Comparison of phase unwrapping algorithms for SAR interferograms," *Proceedings of the 1995 International Geoscience and Remote Sensing Symposium*, Firenze, Italy, July 10–14, 1995, IEEE, Piscataway, NJ, pp. 767–769, 1995.

[9] R. M. Goldstein, Jet Propulsion Laboratory, Pasadena, CA, personal communication, September 9, 1996.

[10] D. J. Bone, "Fourier fringe analysis: the Two-dimensional phase unwrapping problem," *Applied Optics*, Vol. 30, No. 25, pp. 3627–3632, 1991.

[11] Y. Xu and C. Ai, "Simple and effective phase unwrapping technique," *Interferometry IV: Techniques and Analysis, Proceedings of the SPIE,* Vol. 2003, Society of Photo-Optical Instrumentation Engineers, Bellingham, WA, pp. 254–263, 1993.

[12] J. A. Quiroga, A. González-Cano, and E. Bernabeu, "Phase-unwrapping algorithm based on an adaptive criterion," *Applied Optics*, Vol. 34, No. 14, pp. 2560–2563, 1995.

[13] H. Lim, W. Xu, and X. Huang, "Two new practical methods for phase unwrapping," *Proceedings of the 1995 International Geoscience and Remote Sensing Symposium,* Tokyo, Japan, August 18–21, 1995, IEEE, Piscataway, NJ, pp. 196–198, 1995.

[14] W. Xu and I. Cumming, "A region growing algorithm for InSAR phase unwrapping," *Proceedings of the 1996 International Geoscience and Remote Sensing Symposium,* Lincoln, NE, May 27–31, 1996, IEEE, Piscataway, NJ, pp. 2044–2046, 1996.

[15] C. Prati, M. Giani, and N. Leuratti, "SAR interferometry: A 2-D phase unwrapping technique based on phase and absolute values information," *Proceedings of the 1990 International Geoscience and Remote Sensing Symposium*, IEEE, Piscataway, NJ, pp. 2043–2046, 1990.

[16] D. Derauw, "Phase unwrapping using coherence measurements," *Synthetic Aperture Radar and Passive Microwave Sensing, Proceedings of the SPIE*, Vol. 2584, Society of Photo-Optical Instrumentation Engineers, Bellingham, WA, pp. 319–324, 1995.

[17] T. J. Flynn, "Consistent 2-D phase unwrapping guided by a quality map," *Proceedings of the 1996 International Geoscience and Remote Sensing Symposium*, Lincoln, NE, May 27–31, 1996, IEEE, Piscataway, NJ, pp. 2057–2059, 1996.

[18] A. K. Jain, *Fundamentals of Digital Image Processing*, Prentice-Hall, Englewood Cliffs, NJ, 1989.

[19] Q. Lin, J. F. Vesecky, and H. Zebker, "New approaches in interferometric SAR data processing," *IEEE Transactions on Geoscience and Remote Sensing*, Vol. 30, No. 3, pp. 560–567, May 1992.

[20] Q. Lin, J. F. Vesecky, and H. Zebker, "Phase unwrapping through fringe-line detection in synthetic aperture radar interferometry," *Applied Optics*, Vol. 33, No. 2, pp. 201–208, 1994.

[21] T. J. Flynn, "Two-dimensional phase unwrapping with minimum weighted discontinuity," *Journal of the Optical Society of America A*, Vol. 14, No. 10, pp. 2692–2701, October 1997.

5

MINIMUM-NORM METHODS

5.1 INTRODUCTION

The reader will recall from Chapter 4 that a common path-following methodology for phase unwrapping is to begin at a grid point and integrate the wrapped phase differences over a path that covers the entire array. As we explained in Chapter 2, it is not possible to unwrap the data unambiguously and meet the local gradient[1] assumption everywhere if the wrapped phase data contain residues. Some form of intervention is required to make the unwrapping unique. Path-following schemes explicitly or implicitly generate branch cuts or lines of discontinuity. While schemes such as Goldstein's algorithm (Section 4.2) and the mask cut algorithm (Section 4.4) utilize branch cuts to connect the residues, other schemes, such as the quality-guided algorithm (Section 4.3) and Flynn's minimum discontinuity algorithm (Section 4.5), generate lines of discontinuity without detecting residues. These branch cuts or lines act as barriers that prevent the unwrapping path from crossing the lines. The phase differences can then be larger than π radians in magnitude across the lines. Path-following algorithms are powerful methods for phase unwrapping and should be part of a well-stocked arsenal of phase unwrapping techniques.

In this chapter we take what appears to be a completely different approach to phase unwrapping. This approach imposes constraints on the solution in a mathematically formal manner.[2] Specifically we formulate the phase unwrapping problem in a generalized minimum-norm sense. We seek the unwrapped

[1] We sometimes use the terms gradient and derivative interchangeably, even though at times we should be more precise. Specifically the x and y phase (partial) derivatives should be referred to as gradient *components*. We trust that the context adequately explains the situation and hope no confusion will result.

[2] Some of the material in this chapter is drawn from [1]–[3]. The different mathematical notation used in these references has been made consistent throughout the chapter.

phase whose local derivatives match the measured derivatives "as closely as possible." The term "as closely as possible" will be made precise when we present the generalized error norm.

We also show how phase unwrapping can be formulated as an unweighted or weighted least-squares problem. The least-squares error notion is known formally as a minimum norm in the L^2 sense. That is, the sum (integral) of the squared differences between the gradients of the solution and those of the measurements is minimized. Least-squares error is a specific case of the general L^p-norm minimization problem in which we are free to select p to suit our needs.

The algorithms of this chapter, insofar as they "fit" the solution surface to the wrapped phase data, are similar in approach to other "surface-fitting" phase-unwrapping algorithms. For example, Tarayre, Massonnet, and Sirat [4] divide the input phase array into regions by overlaying a mesh or grid. They iteratively construct a surface composed of facets whose gradients match the average phase gradients within each region. This algorithm has the disadvantage that it cannot deal with discontinuities (e.g., due to shear or SAR layover) in the phase data. The algorithm of Labrousse, Dupont, and Berthod [5] fits a surface to the phase data by minimizing an energy function that is the sum of two functions. The first function measures the discrepancy between the surface and the wrapped phase data, while the second is a "membrane model" that measures the departure from a continuous surface. The minimization problem is solved by means of simulated annealing. This approach is still immature, however, and extremely costly from a computational standpoint.

The algorithms of this chapter are distinguished from these surface-fitting algorithms in that they minimize the differences between the *gradients* of the wrapped phase and the solution surface. We define the gradients of the wrapped phase to be simply the wrapped phase differences, but other definitions are possible [6]. The history of these gradient-fitting algorithms is quite old, at least relative to the short history of two-dimensional phase unwrapping. The unweighted least-squares approach, which minimizes the sums of the squares of the gradients between the wrapped phase and the solution surface, has its origins in the wavefront reconstruction work of Fried [7] and Hudgin [8] in 1977. The subsequent development of the least-squares approach will be presented in Section 5.3. Recent work has focused on new formulations of the least-squares approach, such as the Green's formulation of Fornaro, Franceschetti, and Lanari [9], and methods of injecting weights or masks to improve the least-squares approach, such as the multiple-mask approach of Song et al. [1], the PCG approach of Ghiglia and Romero [1], and the multigrid algorithm of Pritt [11]. The latter two algorithms will be presented in Sections 5.4.3 and 5.4.4. The L^p-norm formulation to be presented in Section 5.5 brings the weighted least-squares formulation to maturity by placing it in a generalized mathematical framework.

In order to set the stage for the general L^p-norm phase unwrapping approach, it is helpful to examine several minimum-norm solutions of a simple data-fitting problem. We wish to find the "best" straight-line fit to the five data points of

Figure 5.1. We define the "best" fit to be the line that minimizes the error defined by the following equation:

$$\epsilon^p = \sum_{i=0}^{4} |y_i - (ax_i + b)|^p, \tag{5.1}$$

where (x_i, y_i) represent the given data points. We must find the slope a and the offset b that minimizes this sum. The common least-square-error ($p = 2$) solution produces the line shown in Figure 5.1(a). Notice that the single outlier point "pulls" the line toward it so as to minimize the *squared* error. In general, L^2-norm solutions do not pass through any of the data points exactly. The minimum *absolute* error ($p = 1$) solution is shown in Figure 5.1(b). The outlier has less influence on the solution and the line more closely follows the remaining points. Because three of the data points given are *collinear*, the minimum L^0-norm solution exactly matches those points as shown in Figure 5.1(c). In other words, the minimum L^0-norm line deviates from the data in as few places as possible. However, since any two points define a straight line, other L^0-norm solutions are possible, but they can only be *local* minima. One (of many) local minimum solutions is shown in Figure 5.1(d). Depending on the problem, there may be many local minima or even many (equal) global minima.

The chapter is structured as follows: In Section 5.2 we present the mathematical formulation of the minimum L^p-norm approach to phase unwrapping. While we develop the equations for general p, we will concentrate on the cases $p = 2$ and $p = 0$ in later sections. In Sections 5.3 and 5.4 we present a number of algorithms for the $p = 2$ case, which we call *least-squares* algorithms. Section 5.3 contains several *unweighted* least-squares algorithms in which the corrupt phase

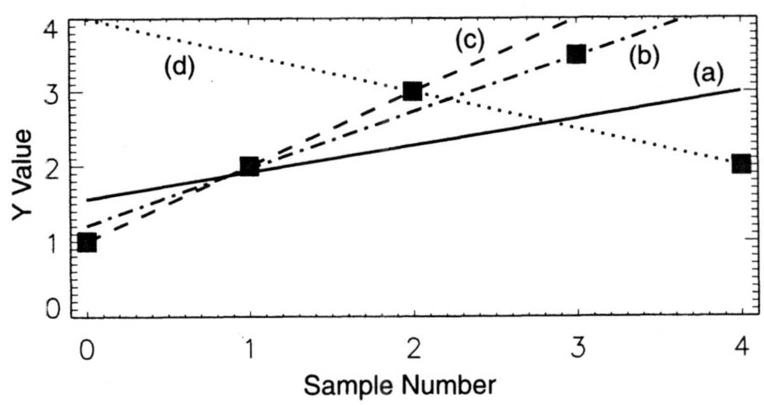

Figure 5.1 Minimum L^p-norm line-fitting solutions to a collection of data points. (a) The least-square-error ($p = 2$) solution. (b) The least-absolute-error ($p = 1$) solution. (c) The minimum L^0-norm solution. In this case, because three data points are collinear, the line deviates from the data in the fewest possible places. (d) One of many other local minimum L^0-norm solutions possible.

values are not ignored. Although these algorithms are generally unsuitable in practice due to the fact that corrupt phase values generally corrupt the unwrapped results on a global scale, they are fundamental to the later algorithms. In Section 5.4 we present several *weighted* least-squares algorithms that are able to zero-weight the corrupt phase values and produce accurate solutions. Like the quality-guided algorithms we studied in Chapter 4, they require quality maps or masks to define the weights.

In Section 5.5 we present an algorithm for the general L^p-norm solution of the phase unwrapping problem. Because this algorithm generates data-dependent weights (if $p \neq 2$), it does not require a quality map to define them. Like Flynn's minimum discontinuity algorithm of Section 4.5, it can be used with or without a quality map. In our evaluation of the algorithm on the phase unwrapping examples, we consider only the $p = 0$ case, which yields the best results in practice. In addition to providing a new phase unwrapping algorithm, the general L^p-norm formulation provides valuable insight into the phase unwrapping problem. It sheds light on the unweighted and weighted least-squares problems and provides a rather surprising link to the path-following methods, which we will explore in Chapter 6.

5.2 MINIMUM L^p-NORM FORMULATION

We will now examine the general case of L^p-norm minimization as applied to two-dimensional phase unwrapping. The reader is encouraged to bear with our discussion, through what may appear to be a rather tedious development. We assure the reader that it will be worth the effort. The general development, when simplified with certain assumptions, will provide valuable insight into minimum-norm methods and phase unwrapping.

Suppose that we are given the wrapped phase values $\psi_{i,j}$, and we wish to determine the unwrapped phase values $\phi_{i,j}$, at the same grid locations. We assume that the phase is defined on a rectangular grid

$$
\begin{aligned}
i &= 0, 1, 2, \ldots, \; M-1, \\
j &= 0, 1, 2, \ldots, \; N-1,
\end{aligned}
\tag{5.2}
$$

and we require that the phase differences of the $\phi_{i,j}$ agree with those of the $\psi_{i,j}$ in the minimum L^p-norm sense. That is, the minimum L^p-norm solution $\phi_{i,j}$ must minimize

$$
\begin{aligned}
J = \epsilon^p &= \sum_{i=0}^{M-2} \sum_{j=0}^{N-1} |\phi_{i+1,j} - \phi_{i,j} - \Delta^x_{i,j}|^p \\
&+ \sum_{i=0}^{M-1} \sum_{j=0}^{N-2} |\phi_{i,j+1} - \phi_{i,j} - \Delta^y_{i,j}|^p,
\end{aligned}
\tag{5.3}
$$

where $\Delta^x_{i,j}$ and $\Delta^y_{i,j}$ are the wrapped phase differences with respect to the i and j indexes, respectively (see Equations 2.36 and 2.37).[3] Equation 5.3 can be further generalized with the inclusion of user-defined weights. This yields a *weighted* minimum L^p-norm problem. Thus the minimum L^p-norm problem can be formulated with or without weights. We will give examples of the results on various phase unwrapping examples both with and without weights in Section 5.5.4.

5.2.1 Continuous Variational Approach

We will take two slightly different approaches to the development of the solution of the minimization problem expressed by Equation 5.3. Each approach provides a slightly different insight. The first approach rewrites Equation 5.3 in integral form as if we were dealing with continuous functions. Thus we wish to find the function $\phi(x, y)$ that minimizes

$$J = \iint f(\phi_x, \phi_y, x, y)\, dx\, dy, \tag{5.4}$$

where the function f is defined by

$$f = |\phi_x - \psi_x|^p + |\phi_y - \psi_y|^p, \tag{5.5}$$

and the phase differences have been replaced with the corresponding continuous partial derivatives.[4] The calculus of variations is used to find the conditions under which J assumes a stationary value. The function ϕ that minimizes J must therefore be the solution of the partial differential equation (PDE) that results from the appropriate Euler-Lagrange equation [13]–[14],

$$\frac{\partial}{\partial x}\left(\frac{\partial f}{\partial \phi_x}\right) + \frac{\partial}{\partial y}\left(\frac{\partial f}{\partial \phi_y}\right) = 0. \tag{5.6}$$

We use the identity

$$\frac{\partial}{\partial z}(|z|^p) = \frac{\partial}{\partial z}[(z^2)^{p/2}] = pz|z|^{p-2} \tag{5.7}$$

[3] Strictly speaking, Equation 5.3 does not describe an error *norm* for $p < 1$ [12]. However, for practical purposes, any value of p can be used in the minimization problem. We will use the term *L^p-norm* regardless of the specific value of p chosen.

[4] Note that ϕ_x and ϕ_y, etc., refer to partial derivatives and should not be confused with the subscripted indexes $\phi_{i,j}$. The same observation applies to ψ_x and ψ_y.

to compute the partial derivatives of f with respect to ϕ_x and ϕ_y and obtain

$$\frac{\partial f}{\partial \phi_x} = p(\phi_x - \psi_x)|\phi_x - \psi_x|^{p-2},$$

$$\frac{\partial f}{\partial \phi_y} = p(\phi_y - \psi_y)|\phi_y - \psi_y|^{p-2}. \tag{5.8}$$

Substituting Equation 5.8 into Equation 5.6 yields the nonlinear PDE that must be solved to obtain the minimum L^p-norm solution for ϕ:

$$\frac{\partial}{\partial x}[(\phi_x - \psi_x)|\phi_x - \psi_x|^{p-2}] + \frac{\partial}{\partial y}[(\phi_y - \psi_y)|\phi_y - \psi_y|^{p-2}] = 0. \tag{5.9}$$

With the following substitutions

$$U(x, y) = |\phi_x - \psi_x|^{p-2},$$

$$V(x, y) = |\phi_y - \psi_y|^{p-2}, \tag{5.10}$$

Equation 5.9 becomes

$$\frac{\partial}{\partial x}[U(x, y)(\phi_x - \psi_x)] + \frac{\partial}{\partial y}[V(x, y)(\phi_y - \psi_y)] = 0. \tag{5.11}$$

The functions $U(x, y)$ and $V(x, y)$ can be considered as generalized *weights* on the x and y phase derivatives. (Do not confuse these data-dependent weights with the optional user-supplied weights.)

In the special case where $p = 2$ (i.e., least-square error), the weights $U(x, y)$ and $V(x, y)$ are both equal to 1, and we obtain Poisson's equation

$$\phi_{xx} + \phi_{yy} = \psi_{xx} + \psi_{yy} \tag{5.12}$$

or

$$\nabla^2 \phi = \rho, \tag{5.13}$$

where ∇^2 is the Laplacian operator and ρ is the function on the right-hand side of Equation 5.12. Therefore unweighted least-squares phase unwrapping amounts to solving Poisson's equation. We will present direct and iterative methods for solving this equation in Section 5.3.

Now let us assume for the moment that both $U(x, y)$ and $V(x, y)$ are independent of the solution ϕ. With this assumption Equation 5.11 is equivalent to the characteristic equation that describes *weighted* least-squares phase unwrapping. Efficient algorithms for solving this equation will be developed in Section 5.4.

It can be seen from Equation 5.10 that (except for least-*squares* problems) the generalized weights $U(x, y)$ and $V(x, y)$ are functions of the original data ψ and the desired solution ϕ. Data-dependent weights make Equation 5.11 a nonlinear problem. We will show how to solve this equation in Section 5.5.

5.2.2 Discrete Variational Approach

Let us now look at the variational approach starting with the actual discretized minimization problem embodied in Equation 5.3. We will include a fair amount of detail in this approach because it will form the basis for understanding the specifics of the unweighted and weighted least-squares algorithms as well as the more general minimum L^p-norm algorithm to be developed in later sections of this chapter. In addition this detailed derivation will show how the boundary conditions are established for the resulting PDE.

The total variation of J, denoted by δJ, consists of the contributions of the variations depending on ϕ. With reference to Equation 5.3 (and with use of Equation 5.7) we obtain

$$\delta J = p(\delta_1 + \delta_2), \tag{5.14}$$

where δ_1 and δ_2 are defined by the equations

$$\delta_1 = \sum_{i=0}^{M-2}\sum_{j=0}^{N-1}(\phi_{i+1,j} - \phi_{i,j} - \Delta^x_{i,j})|\phi_{i+1,j} - \phi_{i,j} - \Delta^x_{i,j}|^{p-2}(\delta\phi_{i+1,j} - \delta\phi_{i,j}), \tag{5.15}$$

$$\delta_2 = \sum_{i=0}^{M-1}\sum_{j=0}^{N-2}(\phi_{i,j+1} - \phi_{i,j} - \Delta^y_{i,j})|\phi_{i,j+1} - \phi_{i,j} - \Delta^y_{i,j}|^{p-2}(\delta\phi_{i,j+1} - \delta\phi_{i,j}). \tag{5.16}$$

We now introduce the following equation for economy of notation:

$$a_{i,j} = (\phi_{i+1,j} - \phi_{i,j} - \Delta^x_{i,j})|\phi_{i+1,j} - \phi_{i,j} - \Delta^x_{i,j}|^{p-2} \tag{5.17}$$

when $0 \le i \le M-2$ and $0 \le j \le N-1$. For the "boundary conditions" we define

$$a_{i,j} = 0, \quad i = -1, M-1, \quad 0 \le j \le N-1. \tag{5.18}$$

We also define

$$b_{i,j} = (\phi_{i,j+1} - \phi_{i,j} - \Delta^y_{i,j})|\phi_{i,j+1} - \phi_{i,j} - \Delta^y_{i,j}|^{p-2} \tag{5.19}$$

when $0 \le i \le M-1$ and $0 \le j \le N-2$, with the boundary conditions

$$b_{i,j} = 0, \quad j = -1, N-1, \quad 0 \le i \le M-1. \tag{5.20}$$

Substitution of Equations 5.17 and 5.19 into Equations 5.15 and 5.16, respectively, yields

$$\delta_1 = \sum_{i=0}^{M-2} \sum_{j=0}^{N-1} a_{i,j}(\delta\phi_{i+1,j} - \delta\phi_{i,j}), \tag{5.21}$$

$$\delta_2 = \sum_{i=0}^{M-1} \sum_{j=0}^{N-2} b_{i,j}(\delta\phi_{i,j+1} - \delta\phi_{i,j}). \tag{5.22}$$

Now, to sum the contributions above and set $\delta J = 0$, we need to shift the i and j indexes of Equations 5.21 and 5.22, respectively, so that we have a common variation $\delta\phi_{i,j}$ that multiplies the individual contributions. Rearranging the summations yields the following equations. Note the change in the summation limits.

$$\delta_1 = -\sum_{i=0}^{M-1} \sum_{j=0}^{N-1} (a_{i,j} - a_{i-1,j})\delta\phi_{i,j}, \tag{5.23}$$

$$\delta_2 = -\sum_{i=0}^{M-1} \sum_{j=0}^{N-1} (b_{i,j} - b_{i,j-1})\delta\phi_{i,j}. \tag{5.24}$$

The above equations assume that $a_{i,j} = b_{i,j} = 0$ on the boundary (i.e., when $i = -1$ or $M-1$, and $j = -1$ or $N-1$ as defined by Equations 5.18 and 5.20).

Substitution of Equations 5.23 and 5.24 into Equation 5.14 yields the equation

$$\delta J = -p \sum_{i=0}^{M-1} \sum_{j=0}^{N-1} (a_{i,j} - a_{i-1,j} + b_{i,j} - b_{i,j-1})\delta\phi_{i,j}. \tag{5.25}$$

Assuming that the perturbation $\delta\phi_{i,j}$ is arbitrary, we conclude that

$$a_{i,j} - a_{i-1,j} + b_{i,j} - b_{i,j-1} = 0 \tag{5.26}$$

when $0 \le i \le M-1$ and $0 \le j \le N-1$. Equation 5.26 requires that we know the values of the terms $\phi_{-1,j}$, $\phi_{M,j}$, $\phi_{i,-1}$, and $\phi_{i,N}$ as well as the corresponding values of ψ. However, Equations 5.18 and 5.20 can be thought of as boundary conditions that determine these values in terms of the interior values. Thus

$$
\begin{aligned}
\phi_{-1,j} - \phi_{0,j} &= \psi_{-1,j} - \psi_{0,j}, \\
\phi_{M,j} - \phi_{M-1,j} &= \psi_{M,j} - \psi_{M-1,j}, \qquad 0 \leq j \leq N-1, \\
\phi_{i,-1} - \phi_{i,0} &= \psi_{i,-1} - \psi_{i,0}, \\
\phi_{i,N} - \phi_{i,N-1} &= \psi_{i,N} - \psi_{i,N-1}, \qquad 0 \leq i \leq M-1.
\end{aligned} \tag{5.27}
$$

Finally, expanding Equation 5.26 and imposing the boundary conditions of Equation 5.27 yields the following equation, whose solution is the solution of the minimum L^p-norm problem:

$$
\begin{aligned}
&(\phi_{i+1,j} - \phi_{i,j} - \Delta^x_{i,j})\, U(i,j) + (\phi_{i,j+1} - \phi_{i,j} - \Delta^y_{i,j})\, V(i,j) \\
&- (\phi_{i,j} - \phi_{i-1,j} - \Delta^x_{i-1,j})\, U(i-1,j) - (\phi_{i,j} - \phi_{i,j-1} - \Delta^y_{i,j-1})\, V(i,j-1) = 0,
\end{aligned} \tag{5.28}
$$

where

$$
\begin{aligned}
U(i,j) &= |\phi_{i+1,j} - \phi_{i,j} - \Delta^x_{i,j}|^{p-2} \\
&\qquad \text{if } i = 0, \ldots, M-2,\ j = 0, \ldots, N-1, \\
U(i,j) &= 0 \quad \text{otherwise,}
\end{aligned} \tag{5.29}
$$

$$
\begin{aligned}
V(i,j) &= |\phi_{i,j+1} - \phi_{i,j} - \Delta^y_{i,j}|^{p-2} \\
&\qquad \text{if } i = 0, \ldots, M-1,\ j = 0, \ldots, N-2, \\
V(i,j) &= 0 \quad \text{otherwise.}
\end{aligned} \tag{5.30}
$$

The rather formidable-looking results shown in Equations 5.28 through 5.30 are merely the discrete counterpart to Equation 5.11 with the boundary conditions imposed and the indexes properly aligned.

When $p = 2$, Equation 5.28 reduces to the discrete version of Poisson's equation:

$$
(\phi_{i+1,j} - 2\phi_{i,j} + \phi_{i-1,j}) + (\phi_{i,j+1} - 2\phi_{i,j} + \phi_{i,j-1}) = \rho_{i,j}, \tag{5.31}
$$

where

$$
\rho_{i,j} = (\Delta^x_{i,j} - \Delta^x_{i-1,j}) + (\Delta^y_{i,j} - \Delta^y_{i,j-1}). \tag{5.32}
$$

An algorithm for solving the general form of Equation 5.28 will be developed in Section 5.5.

5.2.3 Summary and Discussion

Unweighted Least Squares We have shown that the general minimum L^p-norm phase unwrapping problem reduces to the solution of Poisson's equation, Equation 5.31, for the least-squares ($p = 2$) case. It is very important to note that Equation 5.31 is valid for all indexes on the rectangular grid, $i = 0, \ldots, M - 1$, $j = 0, \ldots, N - 1$, and that the appropriate phase differences used to compute $\rho_{i,j}$ in Equation 5.32 are nonzero only if they come from phases entirely within the rectangular grid. This requirement results directly from the least-squares formulation and forms the discrete equivalent of imposing Neumann boundary conditions on Poisson's equation. In other words, we require the following boundary conditions:

$$\Delta^x_{-1,j} = 0,$$
$$\Delta^x_{M-1,j} = 0, \qquad j = 0, \ldots, N - 1, \tag{5.33}$$

$$\Delta^y_{i,-1} = 0,$$
$$\Delta^y_{i,N-1} = 0, \qquad i = 0, \ldots, M - 1. \tag{5.34}$$

Computationally efficient methods for solving the unweighted least-squares phase-unwrapping problem will be developed in Section 5.3.

Weighted Least Squares The weighted least-squares problem is a natural extension of minimum L^2-norm problems. In our case the data-dependent weights $U(i, j)$ and $V(i, j)$ disappear when $p = 2$ in Equations 5.29 and 5.30, but they can be replaced with an arbitrary set of gradient weights. These weights can be derived from a quality map (Section 3.3) or a mask (Section 3.4). Methods for solving the weighted least-squares phase-unwrapping problem will be developed in Section 5.4.

General Minimum L^p-Norm We will see that the general minimum L^p-norm phase unwrapping problem admits an iterative solution *if* unweighted and weighted least-squares solvers can be built. We will show in Section 5.5 that the computationally efficient methods developed in Sections 5.3 and 5.4 can be used to build an algorithm for solving the L^p-norm problem.

In contrast to the weighted least-squares algorithms, which require explicit weights to be provided, L^p-norm algorithms generate their own weights from the input data (if $p \neq 2$). These weights can be combined with user-supplied weights (or quality map) to yield an algorithm that, like Flynn's minimum discontinuity algorithm, works with or without a quality map. Examples will be provided in Section 5.5.

5.3 UNWEIGHTED LEAST-SQUARES PHASE UNWRAPPING

Least-squares phase unwrapping problems, formulated in both unweighted and weighted forms, amount to solving discretized PDEs with efficient mathematical methods [1], [15]–[19]. Additional constraints, such as smoothness, can be imposed through regularization methods [20], [21]. L^2-norm (least-squares) minimization problems are generally favored because they lead to linear equations where the resulting mathematics are tractable and amenable to efficient methods of solution.

In this section we explain how to solve the Poisson equation, which is the equation that embodies the unweighted least-squares phase unwrapping problem. We present two methods that solve the equation directly by means of the fast Fourier transform (FFT). There are two corresponding methods that use fast cosine transforms, and we then concentrate on one of these. Last we present an iterative algorithm that uses multigrid methods to solve the equation. Although the iterative algorithm is slower than the direct methods, it places no power-of-two restriction on the grid sizes.[5] Because these five algorithms generate essentially identical solutions, the results for the algorithms on several phase unwrapping examples are presented in a single section (Section 5.3.4).

We now examine the least-squares approach to phase unwrapping more closely. This approach obtains an unwrapped solution by minimizing the differences between the discrete partial derivatives of the (wrapped) phase data and the discrete partial derivatives of the unwrapped solution. The solution $\phi_{i,j}$ that minimizes

$$\epsilon^2 = \sum_{i=0}^{M-2} \sum_{j=0}^{N-1} (\phi_{i+1,j} - \phi_{i,j} - \Delta^x_{i,j})^2 + \sum_{i=0}^{M-1} \sum_{j=0}^{N-2} (\phi_{i,j+1} - \phi_{i,j} - \Delta^y_{i,j})^2 \tag{5.35}$$

is the least-squares solution. We have shown in Section 5.2 that the solution that minimizes Equation 5.35 is the solution to the discretized Poisson's equation (Equation 5.31).

Equation 5.31 can also be written in matrix notation as a $K \times K$ linear system

$$\mathbf{P}\vec{\phi} = \vec{\rho}, \tag{5.36}$$

where **P** is sparse matrix that represents the discrete Laplacian operator defined by the left-hand side of Equation 5.31, $\vec{\phi}$ and $\vec{\rho}$ are regarded as one-dimensional vectors, and $K = MN$ is the number of grid points. The elements ϕ_k of the one-dimensional vector $\vec{\phi}$ are related to the two-dimensional array elements $\phi_{i,j}$

[5] Most efficient FFT algorithms require the array dimensions to be powers of two. In general, efficient algorithms can be devised where the array dimensions are highly composite numbers (i.e., a product of small primes). Our FFT implementations, however, are restricted to power-of-two grid sizes.

through the indexing equation

$$k = i + jM, \qquad i = 0, \ldots, M-1, \qquad j = 0, \ldots, N-1. \tag{5.37}$$

Similarly the vector elements ρ_k are related to the array elements $\rho_{i,j}$ through the same indexing scheme.

The classical method for solving the discretized Poisson equation (5.31) is called Gauss-Seidel relaxation [22]. In this method one initializes the solution array $\phi_{i,j}$ to zero and then performs the following updates in an iterative fashion:

$$\phi_{i,j} = \frac{(\phi_{i+1,j} + \phi_{i-1,j} + \phi_{i,j+1} + \phi_{i,j-1}) - \rho_{i,j}}{4}. \tag{5.38}$$

This procedure is iterated until convergence. The relaxation equation (Equation 5.38) was obtained simply by solving for $\phi_{i,j}$ in Equation 5.31. The new value of $\phi_{i,j}$ is used as soon as it is available (e.g., in the computation of $\phi_{i+1,j}$), or one may use the "red-black" Gauss-Seidel method,[6] which generally yields better convergence results in practice [22]. Although Gauss-Seidel relaxation is not practical due to its extremely slow convergence, it turns out to be important in the multigrid methods to be introduced in Sections 5.3.3 and 5.4.4.

Unweighted least-squares phase unwrapping hinges on the computationally efficient solution of the discretized Poisson equation (Equation 5.31). We will first present solution methods based on the FFT. We will show in Section 5.3.1 that FFT methods require extending the input data through some form of "mirror" reflection to form a periodic function. It will also be shown that FFTs on extended periodic functions can be mathematically "collapsed" into certain discrete cosine transforms (DCTs). The exact means of periodic extension defines the specific type of DCT required. Because the DCT methods do not employ mirror reflections, they must impose "boundary conditions." DCT methods yield slightly more efficient algorithms because no mirror reflections are required.

Another efficient computational approach for solving Poisson's equation is *multigrid.* This technique, as it applies to phase unwrapping, is covered extensively in Sections 5.3.3 and 5.4.4.

It should be noted that virtually any fast and efficient algorithm for solving Poisson's equation is a candidate for phase unwrapping. The transform-based and multigrid techniques developed for unweighted least-squares phase unwrapping in Sections 5.3.1 and 5.3.3 are used extensively in the solution of the

[6] Gauss-Seidel makes iterative "sweeps" through an array, updating the values of the array in place. In updating the value of each element $c_{i,j}$, some of the array values with their *new* values are used (i.e., $c_{i-1,j}$ and $c_{i,j-1}$) and some of the array values with their *old* values are used (i.e., $c_{i+1,j}$ and $c_{i,j+1}$). The red-black approach uses only the *old* values of the array. To do this, it turns each sweep of the array into two sweeps, where each sweep updates only alternate elements of the array. The terminology "red-black" is derived from the colors of the squares on a checkerboard. On the first sweep, only the "red" squares are updated, while on the second sweep, only the "black" squares are updated.

weighted least-squares and minimum L^p-norm problems to be discussed in Sections 5.4 and 5.5.

5.3.1 FFT-Based Methods

In this section we present a solution of the unweighted least-squares phase unwrapping problem that is simple to understand and implement. It extends the input data to a periodic function using a mirror reflection, and the resulting equation is solved directly using the Fourier transform. Because there are two ways to perform the mirror reflection (as we will show), there are two corresponding FFT-based algorithms for solving the problem. This approach will appeal to those readers familiar with the FFT. As motivation for this approach, we will first discuss the application of Poisson's equation to *periodic* functions, where no boundary conditions are required.

As we have shown before in Equations 5.12 and 5.31, Poisson's equation is a partial differential equation of the form

$$\frac{\partial^2}{\partial x^2}\phi(x,y) + \frac{\partial^2}{\partial y^2}\phi(x,y) = \rho(x,y) \tag{5.39}$$

which is defined on a closed domain D of the plane. The solution $\phi(x,y)$ of this equation is unique up to an additive constant if the Neumann boundary conditions are given, which specify the values of the normal derivative $\partial\phi/\partial\nu$ on the boundary of D. Under certain conditions, however, the boundary conditions are not necessary. Suppose that one limits the set of possible solutions $\phi(x,y)$ to the periodic functions. A periodic function is a function that satisfies $\phi(x,y) = \phi(x+M,y) = \phi(x,y+N)$ for all points (x,y) of the plane, where M and N are fixed constants. Because such a function attains all of its values on the closed region defined by $0 \le x \le M$ and $0 \le y \le N$, it also attains its maximum value there. Now suppose that ϕ_1 and ϕ_2 are two different periodic functions that solve Poisson's equation (Equation 5.39). Because the difference function $\phi = \phi_1 - \phi_2$ is also periodic, it achieves its maximum value at a point $\mathbf{a}$ of the plane. Furthermore, because ϕ_1 and ϕ_2 both satisfy Equation 5.39, ϕ is a harmonic function, which means that its Laplacian $\partial^2\phi/\partial x^2 + \partial^2\phi/\partial y^2$ is zero. By Gauss's mean value theorem [23], the value $\phi(\mathbf{a})$ must equal the mean value of ϕ along the circumference of any disk centered at $\mathbf{a}$:

$$\phi(\mathbf{a}) = (\tfrac{1}{2\pi})\int_0^{2\pi} \phi(\mathbf{a} + r_0 e^{i\theta})d\theta. \tag{5.40}$$

Because $\phi(\mathbf{a})$ is the maximum value of ϕ, this is possible only if ϕ is a constant function. Thus, when restricted to periodic functions, Poisson's equation has a unique solution (up to an additive constant) and boundary conditions are not necessary.

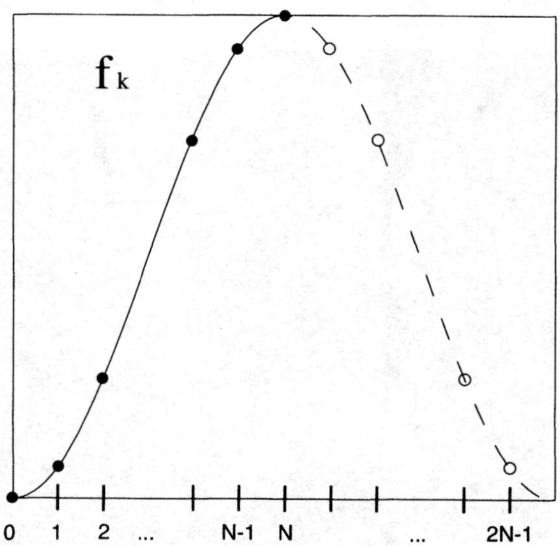

Figure 5.2 Mirror reflection of a function f_k (defined at $k = 0, ..., N$) about the line $k = N$. This reflects the function about the line $x = N$.

Algorithm We will now present an FFT-based algorithm for solving the unweighted least-squares phase unwrapping problem. Suppose that the wrapped phase function $\psi_{i,j}$ of an unknown real-valued function $\phi_{i,j}$ is given on the rectangular grid of points defined by $0 \le i \le M$ and $0 \le j \le N$.[7] The wrapped phase function is defined to be the function whose values lie in the interval $(-\pi, \pi]$ and that satisfies the relation $\exp(j\psi) = \exp(j\phi)$.[8] Extend $\psi_{i,j}$ to a periodic function $\tilde{\psi}_{i,j}$ by performing a mirror reflection (Figure 5.2) about the lines $i = M$ and $j = N$ in the plane (Figure 5.3). The resulting function is then defined by

$$\tilde{\psi}_{i,j} = \begin{cases} \psi_{i,j}, & 0 \le i \le M, \quad 0 \le j \le N, \\ \psi_{2M-i,j}, & M < i < 2M, \quad 0 \le j \le N, \\ \psi_{i,2N-j}, & 0 \le i \le M, \quad N < j < 2N, \\ \psi_{2M-i,2N-j}, & M < i < 2M, \quad N < j < 2N. \end{cases} \tag{5.41}$$

The unknown function $\phi_{i,j}$ can also be regarded as the restriction of a periodic

[7] It is extremely important to note that in this section we depart from the assumption that the wrapped phase values exist on an $M \times N$ grid. Instead, we define them on an $(M + 1) \times (N + 1)$ grid so that the periodic extension $\tilde{\psi}_{i,j}$ defined by Equation 5.41 is defined on a $2M \times 2N$ grid for subsequent Fourier transformation. Later in this section we will define a different periodic extension that utilizes data on the usual $M \times N$ grid.

[8] We trust that no confusion will result with the use of $j = \sqrt{-1}$ in the argument of the complex exponential. The use of i or j as *indexes* should be obvious from the context.

Figure 5.3 Mirror reflection of a function $f_{i,j}$ defined on the grid $0 \leq i \leq M$, $0 \leq j \leq N$. The function is first reflected about the line $i = M$, then about the line $j = N$.

function $\tilde{\phi}_{i,j}$ that is defined by this mirror reflection. Furthermore, because $\tilde{\psi}_{i,j}$ and $\tilde{\phi}_{i,j}$ are periodic functions, they can be defined for all indexes i, j by replacing them with $i \bmod 2M, j \bmod 2N$.

We will now derive a least-squares estimate of $\phi_{i,j}$. Because of the mirror reflection defined by Equation 5.41, the wrapped phase function $\tilde{\psi}_{i,j}$ is defined at all (integer) points of the plane. There is thus no need to formulate boundary conditions. The x and y phase differences $\Delta^x_{i,j}$ and $\Delta^y_{i,j}$ are defined for all indexes i and j by the equations

$$\Delta^x_{i,j} = \mathcal{W}\{\tilde{\psi}_{i+1,j} - \tilde{\psi}_{i,j}\}, \quad \Delta^y_{i,j} = \mathcal{W}\{\tilde{\psi}_{i,j+1} - \tilde{\psi}_{i,j}\}, \tag{5.42}$$

where $\mathcal{W}$ is the wrapping operator. We seek a function $\tilde{\phi}_{i,j}$ that minimizes the sum of squared differences between these phase differences and those of the solution function. Just as we saw in Section 5.3, the least-squares solution is the solution of the discrete form of Poisson's equation (Equation 5.31):

$$(\tilde{\phi}_{i+1,j} - 2\tilde{\phi}_{i,j} + \tilde{\phi}_{i-1,j}) + (\tilde{\phi}_{i,j+1} - 2\tilde{\phi}_{i,j} + \tilde{\phi}_{i,j-1}) = \tilde{\rho}_{i,j}, \tag{5.43}$$

where the extended function $\tilde{\rho}_{i,j}$ is defined by Equation 5.32. Because the function $\tilde{\rho}_{i,j}$ is periodic like $\tilde{\phi}_{i,j}$, the Fourier transform can be used to solve this equation. Applying the two-dimensional Fourier transform on the $2M \times 2N$ grid to both sides of Equation 5.43 yields the equation

$$\Phi_{m,n} = \frac{P_{m,n}}{2\cos(\pi m/M) + 2\cos(\pi n/N) - 4}, \tag{5.44}$$

where $\Phi_{m,n}$ and $P_{m,n}$ are the two-dimensional Fourier transforms of $\tilde{\phi}_{i,j}$ and $\tilde{\rho}_{i,j}$, respectively. (Φ_{00} is undefined, but it can be set to zero because the solution of Equation 5.43 is defined only up to an additive constant.) The solution $\tilde{\phi}_{i,j}$ is then obtained by applying the inverse Fourier transform to Equation 5.44, and the solution $\phi_{i,j}$ is obtained by restricting the result to the grid defined by $0 \leq i \leq M$ and $0 \leq j \leq N$.

Although this formulation does not require explicit boundary conditions, the mirror reflection actually imposes *implicit* boundary conditions. It is easy to see that the partial derivatives (Equation 5.42) are defined by periodicity on the boundary by

$$\begin{aligned} \Delta^x_{-1,j} &= -\Delta^x_{0,j}, \\ \Delta^x_{M,j} &= -\Delta^x_{M-1,j}, \\ \Delta^y_{i,-1} &= -\Delta^y_{i,0}, \\ \Delta^y_{i,N} &= -\Delta^y_{i,N-1}. \end{aligned} \tag{5.45}$$

Note that these boundary conditions differ from those defined by Equations 5.33 and 5.34. There are two choices of boundary conditions depending on how the mirror reflection is defined. We will define a mirror reflection that produces the boundary conditions of Equations 5.33 and 5.34.

The key idea in the preceding derivation is the use of the mirror reflection (Equation 5.41) to extend $\psi_{i,j}$ and $\phi_{i,j}$ to periodic functions. The usual method of periodic extension, which is by translation and repetition of the function without reflection, results in an incorrect solution because it does not preserve boundary derivatives. An example for one-dimensional functions will serve to illustrate this point. Suppose that the wrapped phase function $\psi_k = \mathcal{W}\{2\pi k/3\}$ is defined at the integer points $k = 0, \ldots, 4$. The phase function ψ_k defines a wrapped linear ramp function as shown in Figure 5.4(a). If the periodic extension $\tilde{\psi}_{i,j}$ is defined in the usual way without a mirror reflection (i.e., $\tilde{\psi}_k = \phi_{k \bmod 5}$), then the discrete derivatives (Equation 5.42) will be defined by $\Delta^x_k = 2\pi/3$ for $0 \leq k \leq 3$ and $\Delta^x_4 = -2\pi/3$. The sum to be minimized will then be

$$\epsilon^2 = \sum_{k=0}^{3} \left(\phi_{k+1} - \phi_k - \frac{2\pi}{3}\right)^2 + \left(\phi_0 - \phi_4 + \frac{2\pi}{3}\right)^2, \tag{5.46}$$

and the least-squares solution will be $\phi_k = 4\pi k/15$, as shown in Figure 5.4(b). This is obviously an incorrect unwrapping because the slope is incorrect. If the extension $\tilde{\phi}_k$ is defined instead by the mirror reflection (Equation 5.41), then the

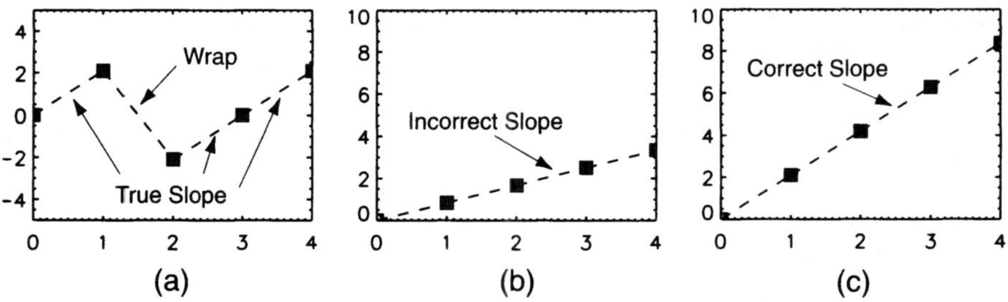

Figure 5.4 FFT methods for least-squares phase unwrapping yield incorrect results if the periodic extension is not performed properly. (a) The wrapped phase of a linear ramp function. (b) An incorrect unwrapping is obtained because of the incorrect periodic extension. See the text. (c) The correct unwrapping. The unwrapped phase slope matches the true slope shown in (a).

sum to be minimized will be

$$\epsilon^2 = 2\sum_{k=0}^{3}\left(\phi_{k+1} - \phi_k - \frac{2\pi}{3}\right)^2 \tag{5.47}$$

and the least-squares solution will be the correct unwrapping $\phi_k = 2\pi k/3$, as shown in Figure 5.4(c).

Pseudocode We now present the pseudocode for an FFT-based algorithm for solving the unweighted least-squares phase unwrapping problem. It requires an array of size $(M+1) \times (N+1)$, where M and N must be powers of two due to the use of the FFT. All operations on the array are performed in place, thus removing the need for any other arrays.

FFT Method 1 for Unweighted Least-Squares Phase Unwrapping

Step 1. Calculate the values $\rho_{i,j}$ defined by Equation 5.32 for $0 \le i \le M$ and $0 \le j \le N$.

Step 2. Compute the two-dimensional FFT of $\rho_{i,j}$ as follows. For each row i of the array $\rho_{i,j}$, define the $2N$-point vector f_j to be the mirror reflection of the row. That is, $f_j = \rho_{i,j}$ for $0 \le j \le N$, and $f_j = \rho_{i,2N-j}$ for $N \le j \le 2N$. Since f_j is a real and even function (i.e., $f_j = f_{2N-j}$), its FFT is also real and even. Compute the FFT F_n, and replace the ith row of the array $\rho_{i,j}$ with the restriction of F_n to $0 \le n \le N$. After all of the rows are processed in this way, repeat the process on the columns.

Step 3. Replace the values of the transformed array $\rho_{i,j}$ with the values $\Phi_{m,n}$ defined by Equation 5.44.

Step 4. Compute the inverse FFT of the array using the procedure outlined in Step 2 (except with row-wise and columnwise inverse FFTs). The result is the desired solution function $\phi_{i,j}$.

Table 5.1 Execution times of the FFT-based algorithm for various array sizes on an IBM RS/6000 workstation.

Array Size (pixels)	Execution Time (min:s)
256 × 256	00:02
512 × 512	00:06
1024 × 1024	00:23
2048 × 2048	01:30

The two-dimensional FFT computed in Step 2 is actually a two-dimensional fast cosine transform. The $2N$-point Fourier transform of the function f_j, when restricted to $0 \leq n \leq N$, is the cosine transform

$$F_n = f_0 + 2\sum_{j=1}^{N-1} f_j \cos\left(\frac{\pi jn}{N}\right) + (-1)^n f_N, \qquad 0 \leq n \leq N. \tag{5.48}$$

The algorithm does not require an explicit formulation of the boundary conditions, nor does it require the use of a cosine transform. Instead, it uses a mirror reflection and an FFT. It is straightforward to implement the algorithm with a cosine transform instead of an FFT. In fact, we have done this very thing in our implementation of the algorithm, which is discussed in the following section.

Implementation Our C language implementation of the FFT-based algorithm for unweighted least-squares phase unwrapping can be found in Section A.10 of Appendix A.[9] Because only one array is required, all the operations being performed in place, the memory requirements are minimal. The input phase data must be defined in an $(M+1) \times (N+1)$ array, where M and N are powers of two. (The second FFT-based algorithm to be discussed in the following section accepts an $M \times N$ array.) Table 5.1 lists the execution times of our implementation for various array sizes. (Unlike those of the path-following algorithms, these times depend only on the array sizes.) These times are comparable to those of Goldstein's path-following algorithm, which was presented in Chapter 4.

Second FFT-Based Algorithm We now present a second FFT-based algorithm for solving the unweighted least-squares phase unwrapping problem. This algorithm results from a change in the definition of the mirror reflection. Figure 5.5 illustrates this second type of mirror reflection. It extends a function defined on N points to one defined on $2N$ points. (The mirror reflection of

[9] Our implementation uses a fast cosine transform rather than a mirror reflection and FFT. Strictly speaking, it is a DCT-based algorithm, but it is functionally equivalent to the FFT-based algorithm.

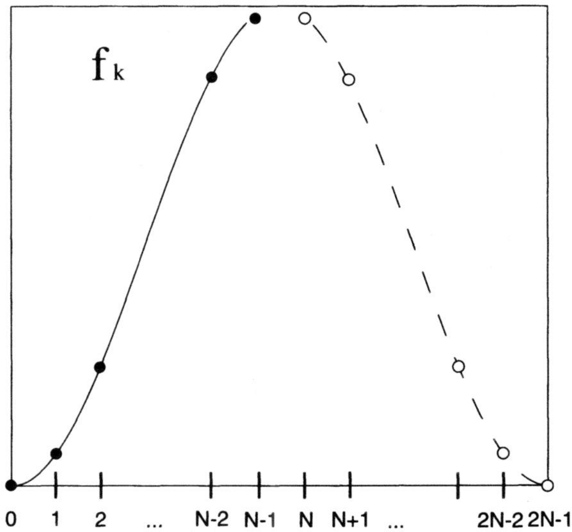

Figure 5.5 Second type of mirror reflection. Compare with Figure 5.2. This reflects the function about the line $x = N - \frac{1}{2}$ rather than the line $x = N$.

Figure 5.2, on the other hand, requires a function defined on $N + 1$ points.) Does this new mirror reflection give rise to a different Fourier transform-based method (and a correspondingly different cosine transform)?

The answer is yes. For simplicity we will restrict ourselves to one-dimensional functions. The results extend easily to two dimensions. Given a function ψ_j defined for $0 \leq j < M$, define its mirror reflection $\tilde{\psi}_j$ by

$$\tilde{\psi}_j = \begin{cases} \psi_j & (0 \leq j < N), \\ \psi_{2N-j-1} & (N \leq j < 2N - 1). \end{cases} \tag{5.49}$$

This is the mirror reflection illustrated in Figure 5.5. The Fourier transform Ψ_k of $\tilde{\psi}_j$ is defined by

$$\tilde{\Psi}_k = \sum_{j=0}^{2N-1} \tilde{\psi}_j W^{jk}, \tag{5.50}$$

where W is the complex number $\exp(j\pi/N)$.[10] Splitting the sum in two parts and using the fact that $W^{2N} = 1$, we have

$$\Psi_k = \sum_{j=0}^{N-1} \psi_j W^{jk} + \sum_{j=0}^{N-1} \psi_j W^{(2N-j-1)k}. \tag{5.51}$$

[10] Again, it should be clear from the context if j is to be interpreted as an index or as $\sqrt{-1}$ in the complex exponential.

Multiplying this expression by the complex number $W^{k/2}$ and denoting the result by $\bar{\Psi}_k$ yields the equation

$$\bar{\Psi}_k = \sum_{j=0}^{N-1} \psi_j [W^{(j+1/2)k} + W^{-(j+1/2)k}]. \tag{5.52}$$

Because $W^x + W^{-x} = 2\cos(\pi x/N)$, this equation is equivalent to

$$\bar{\Psi}_k = 2\sum_{j=0}^{N-1} \psi_j \cos\left[\frac{\pi k(j+\frac{1}{2})}{N}\right]. \tag{5.53}$$

This equation is a second type of cosine transform and should be compared with Equation 5.48. The two equations are the most commonly used cosine transforms [24]. We leave it as an exercise to the reader to prove that $\bar{\Psi}_N = 0$ and $\bar{\Psi}_k = -\bar{\Psi}_{2N-k}$ for $N < k < 2N$.

The process of applying the mirror reflection defined by Equation 5.49, computing the FFT, and then multiplying by the complex numbers $W^{k/2}$ is thus equivalent to computing the cosine transform of Equation 5.53. This process is analogous to the method we gave earlier for computing the cosine transform (Equation 5.48) of a function by means of the first mirror reflection in Equation 5.41. We see here a nice parallel between the two ways to define a mirror reflection, the two ways to define the cosine transform, and the two ways to define the boundary conditions. The boundary conditions that correspond to the second type of mirror reflection (Equation 5.49) are those of Equations 5.33 and 5.34, while the boundary conditions of the first mirror reflection (Figure 5.2 and Equation 5.41) are defined by Equation 5.45.

The inverse of the cosine transform (Equation 5.53) is defined by the equation

$$\hat{\Psi}_k = \frac{1}{N}\sum_{k=0}^{N-1} \psi_k \cos\left[\frac{\pi k(j+\frac{1}{2})}{N}\right], \tag{5.54}$$

where the term for $k = 0$ of the sum is understood to have the coefficient $\frac{1}{2}$ in front of it. This inverse transform can be computed by means of an FFT as follows: First extend the original function ψ_k defined for $0 \leq k < N$ to the function $\bar{\psi}_k$ defined by

$$\bar{\psi}_k = \begin{cases} \psi_k & (0 \leq k < N), \\ 0 & (k = N), \\ -\psi_{2N-k} & (n < k < 2N). \end{cases} \tag{5.55}$$

Then multiply the extended function by the complex numbers $W^{-k/2}$, and compute the inverse FFT of the product. The result is a real-valued mirror-reflected function, and the restriction to $0 \leq j < N$ equals the inverse cosine transform (Equation 5.54). We leave this as an exercise for the reader to prove.

Pseudocode The pseudocode for the second FFT-based algorithm is analogous to the first algorithm. It requires one $M \times N$ array, where M and N are powers of two.

FFT Method 2 for Unweighted Least-Squares Phase Unwrapping

Step 1. Calculate the values $\rho_{i,j}$ defined by Equation 5.32 for $0 \leq j < M$ and $0 \leq k < N$.

Step 2. Compute the two-dimensional cosine transform as follows: For each row i of the array $\rho_{i,j}$, define the $2N$-point vector f_j, for $0 \leq j < 2N$, to be the mirror reflection defined by Equation 5.49.

Step 3. Then compute the cosine transform (Equation 5.53) by multiplying by the complex numbers $W^{j/2}$ and computing the FFT. Replace the ith row of the array with the restriction of the resulting function to $0 \leq j < N$. After all the rows are processed in this way, repeat the process on the columns.

Step 4. Replace the values of the transformed array $\rho_{i,j}$ with the values $\Phi_{m,n}$ defined by Equation 5.44.

Step 5. Compute the two-dimensional inverse cosine transform of the array by means of the procedure outlined in Step 2, substituting the steps for the inverse cosine transform in place of those of the cosine transform.

The second algorithm unwraps data on an $M \times N$ instead of an $(M+1) \times (N+1)$ array, where M and N are powers of two. The differences arise from the different ways the mirror reflection is performed. The restrictions arise because FFT lengths are usually restricted to powers of two. Thus, the first algorithm requires an input array size of $(2^m + 1) \times (2^n + 1)$, while the second algorithm requires the more conventional array size of $2^m \times 2^n$. The array size of the second algorithm is usually more convenient in practice.

Implementation We do not provide a C language implementation of the second FFT-based algorithm. The implementation we have provided (in Section A.10 of Appendix A) for the first FFT-based algorithm can be modified to produce the second algorithm. The input phase data for this algorithm must be defined on an $M \times N$ grid, where M and N are powers of two. Like the first FFT algorithm, the second algorithm requires only one array, and all the operations are performed in place. The execution times are the same as those listed in Table 5.1.

5.3.2 DCT-Based Methods

In this section we provide the details needed to solve the unweighted least-squares phase-unwrapping problem by means of the DCT algorithm rather than the FFT. We will employ a two-dimensional version of the cosine transform defined by Equation 5.53. The algorithm we develop will be equivalent to the FFT Method 2 (Section 5.3.1). The DCT algorithm that is equivalent to the FFT Method 1 can also be derived, but we leave this as an exercise to the interested reader.

Algorithm The two-dimensional form of the cosine transform defined by Equation 5.53 is given by the following equations [25]:

$$C_{m,n} = \begin{cases} \displaystyle\sum_{i=0}^{M-1}\sum_{j=0}^{N-1} 4\phi_{i,j} \cos\left[\frac{\pi}{2M} m(2i+1)\right] \cos\left[\frac{\pi}{2N} n(2j+1)\right], \\ \quad \text{if } 0 \le m \le M-1,\ 0 \le n \le N-1, \\ 0 \quad \text{otherwise.} \end{cases} \tag{5.56}$$

The inverse cosine transform is defined by the equation

$$\phi_{i,j} = \begin{cases} \displaystyle\frac{1}{MN}\sum_{m=0}^{M-1}\sum_{n=0}^{N-1} w(m,n) C_{m,n} \cos\left[\frac{\pi}{2M} m(2i+1)\right] \cos\left[\frac{\pi}{2N} n(2j+1)\right], \\ \quad \text{if } 0 \le i \le M-1, \qquad 0 \le j \le N-1, \\ 0 \quad \text{otherwise,} \end{cases} \tag{5.57}$$

where $w(m, n) = w_1(m)w_2(n)$, and the terms $w_1(m)$ and $w_2(m)$ are defined by

$$\begin{aligned} w_1(m) &= \tfrac{1}{2}, \quad m = 0, \\ w_1(m) &= 1, \quad 1 \le m \le M-1, \\ w_2(m) &= \tfrac{1}{2}, \quad n = 0, \\ w_2(m) &= 1, \quad 1 \le n \le N-1. \end{aligned} \tag{5.58}$$

The cosine transform can be used to solve the Poisson equation (Equation 5.31) as follows. Let us expand the desired solution $\phi_{i,j}$ of Equation 5.31 in terms of its cosine transform coefficients $\hat{\phi}_{m,n}$:

$$\phi_{i,j} = \frac{1}{MN}\sum_{m=0}^{M-1}\sum_{n=0}^{N-1} w(m,n)\hat{\phi}_{m,n} \cos\left[\frac{\pi}{2M} m(2i+1)\right] \cos\left[\frac{\pi}{2N} n(2j+1)\right], \tag{5.59}$$

where $\hat{\phi}_{m,n}$ is defined by Equation 5.56. Substituting Equation 5.59 into Equation 5.31 and performing a similar expansion and substitution for the right-hand side of Equation 5.31 yields the following equation:

$$\hat{\phi}_{i,j} = \frac{\hat{\rho}_{i,j}}{2\cos(\pi i/M) + 2\cos(\pi j/N) - 4}. \tag{5.60}$$

The unwrapped phase $\phi_{i,j}$ is now easily obtained by the inverse DCT of Equation 5.60.

It can be shown by simple substitution that the expansion given by Equation 5.59 automatically imposes the discrete equivalent of the required Neumann boundary conditions:

$$\begin{aligned} \phi_{0,j} - \phi_{-1,j} &= 0, \\ \phi_{M,j} - \phi_{M-1,j} &= 0, \qquad j = 0, ..., N-1, \\ \phi_{i,0} - \phi_{i,-1} &= 0, \\ \phi_{i,N} - \phi_{i,N-1} &= 0, \qquad i = 0, ..., M-1. \end{aligned} \tag{5.61}$$

These boundary conditions correspond to those of Equations 5.33 and 5.34.

Pseudocode The DCT-based algorithm is summarized by the following pseudocode. Like the implementation of the FFT-based algorithms, this implementation requires a single array. The dimensions of the array must be $M \times N$, where M and N are powers of two.

DCT Method for Unweighted Least-Squares Phase Unwrapping

Step 1. Compute the array of values $\rho_{i,j}$ by Equation 5.32 for $0 \leq i \leq M-1$ and $0 \leq j \leq N-1$.

Step 2. Compute the two-dimensional forward DCT (Equation 5.56) of the array of values $\rho_{i,j}$ to yield the transform values $\hat{\rho}_{i,j}$. The DCT can be performed "in place" on the array $\rho_{i,j}$.

Step 3. Replace the values $\hat{\rho}_{i,j}$ with the values on the right-hand side of Equation 5.60 to obtain $\hat{\phi}_{i,j}$.

Step 4. Compute the two-dimensional inverse DCT (Equation 5.57) of $\hat{\phi}_{i,j}$ to obtain the least-squares unwrapped solution $\phi_{i,j}$.

Because the term $\hat{\phi}_{0,0}$ is indeterminate, Poisson's equation cannot be solved for a constant bias component. In practice, we set $\hat{\phi}_{0,0} = 0$.

Implementation We do not provide an implementation of this algorithm. The implementation we have provided in Section A.10 of Appendix A for the first FFT-based algorithm (which is actually a DCT-based algorithm) can be modified to produce the second DCT-based algorithm. The input phase data for this algorithm must be defined on an $M \times N$ grid, where M and N are powers of two. Like the FFT algorithms, the second DCT algorithm requires only one array, in which all the operations are performed in place. The execution times are the same as those listed in Table 5.1.

5.3.3 Unweighted Multigrid Algorithm

In this section we present a fourth algorithm for unweighted least-squares phase unwrapping. Unlike the three previous algorithms, this is an iterative algorithm that forms the basis of the multigrid technique to be presented in Section 5.4 for weighted least-squares phase unwrapping. This iterative algorithm adopts the boundary conditions of the first FFT method, although it could be modified to adopt those of the second FFT method. We will begin the section with a detailed introduction to multigrid methods, including the pseudocode for a typical multigrid algorithm. Because multigrid algorithms are naturally expressed in terms of recursion, the pseudocode will employ recursive subroutines. We will then present a multigrid algorithm for solving the unweighted least-squares phase unwrapping problem.

Before describing multigrid techniques, we make a few remarks to motivate their use. Multigrid methods are a class of techniques for rapidly solving PDEs on large grids. These methods are based on the idea of applying Gauss-Seidel relaxation schemes on coarser, smaller grids. Multigrid methods are typically as fast as the direct Fourier-based methods, solving elliptic PDEs discretized on $N \times N$ grid points in $O(N^2 \log N)$ operations [22], [26]. Unlike the direct methods, however, multigrid methods can solve a much wider class of problems, including nonlinear PDEs. Perhaps just as important from a practical point of view is the fact that multigrid methods are not restricted to power-of-two grid sizes, as is the case for most FFT and DCT methods. The reader is referred to [26] for an excellent introduction to multigrid methods.

We begin by analyzing the convergence of the classical Gauss-Seidel relaxation scheme (Equation 5.38) on the noise-free and unweighted least-squares phase unwrapping problem shown in Figure 5.6. The wrapped phase data of Figure 5.6(a) are defined on a small 200×200-pixel grid. When correctly unwrapped, the phase data produce the surface shown in Figure 5.6(b). The solution obtained by the Gauss-Seidel method after 10 iterations is shown in Figure 5.7(a), while Figure 5.7(b) shows the error surface. As can be seen, the Gauss-Seidel method extracts the high-frequency components of the surface (i.e., the surface detail) but not the low-frequency components (i.e., the smooth, global structure of the surface). Gauss-Seidel relaxation propagates the surface information very slowly because, at each step of the iteration, each grid point receives information only from its immediate neighbors.

Gauss-Seidel relaxation is essentially a local smoothing operator that removes the high-frequency components of the error very quickly but the low-frequency components extremely slowly. Because it converges in $O(N^4)$ operations, it is clearly impractical except for very small grid sizes. Even the faster simultaneous overrelaxation (SOR) method [22], which requires $O(N^3)$ operations, is too slow in practice. The key idea of the multigrid approach, on the other hand, is to

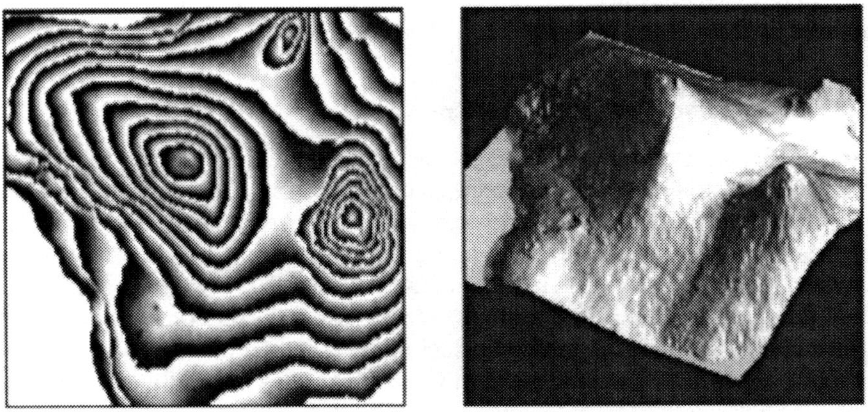

(a) (b)

Figure 5.6 Phase unwrapping example. (a) Wrapped phase data. (b) Three-dimensional view of the unwrapped surface.

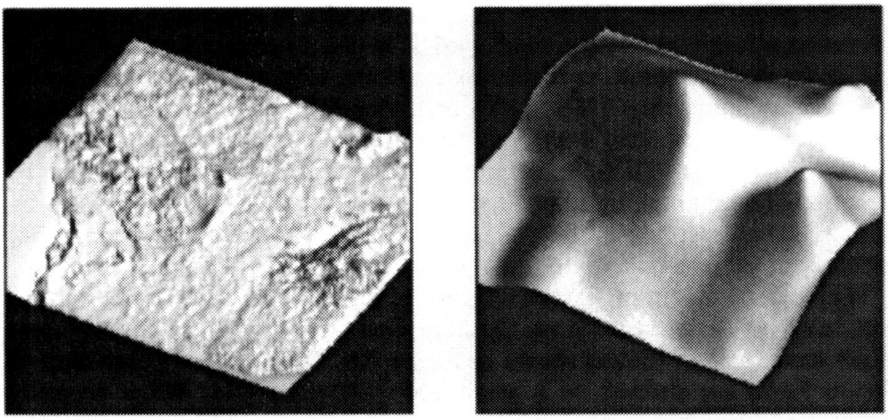

(a) (b)

Figure 5.7 Solution to the phase unwrapping problem of Figure 5.6 obtained by the Gauss-Seidel relaxation method. (a) Gauss-Seidel solution after 10 iterations. (b) Error surface. The error surface looks mountainous and smooth because at this point in the iteration the Gauss-Seidel method has recovered the localized surface detail but not the global surface structure. Compare with Figure 5.6(b).

transform the low-frequency components of the error into high-frequency components, which can be removed quickly by Gauss-Seidel relaxation. This is accomplished by transferring the problem to coarser grids. The lower sampling rates of the coarser grids increase the spatial frequencies of the residual error. If one has a pyramid of grids, with each grid one-half the resolution of its predecessor as shown in Figure 5.8, then the low spatial frequencies on the

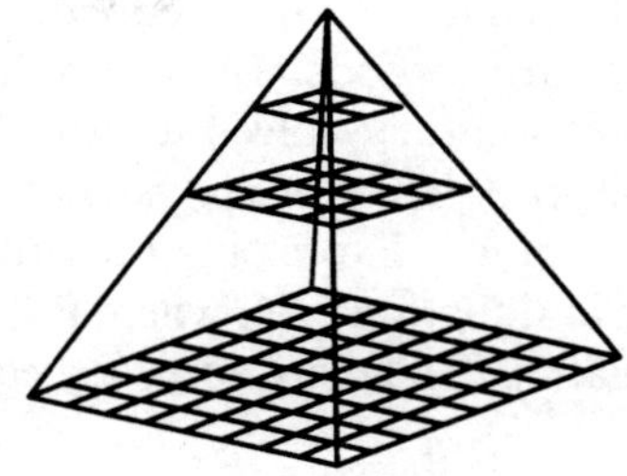

Figure 5.8 Pyramid of grids of decreasing resolution (as one proceeds upward) in a multigrid scheme. Each grid is one-half the resolution of the grid below it.

fine grid will propagate up to increasingly higher frequencies on the coarse grids. (In other words, the "global" information on the fine grid will become "local" information on the coarse grids.)

The operation of transferring the problem to a coarser grid is called *restriction*, while the operation of transferring the solution to a finer grid is called *prolongation*. A natural choice for the restriction operator is the *injection operator*, which resamples every other point. It is defined by $c_{i,j} = f_{2i,2j}$, where $f_{i,j}$ is the $M \times N$ "fine" grid and $c_{i,j}$ is the $M/2 \times N/2$ "coarse" grid. An alternative restriction operator, however, and one that usually gives better results in practice [26], is called *full weighting* and is defined by

$$
\begin{aligned}
c_{i,j} = {} & \tfrac{1}{16}(f_{2i-1,2j-1} + f_{2i+1,2j-1} + f_{2i-1,2j+1} + f_{2i+1,2j+1}) \\
& + \tfrac{1}{8}(f_{2i,2j-1} + f_{2i,2j+1} + f_{2i-1,2j} + f_{2i+1,2j}) + \tfrac{1}{4} f_{2i,2j}.
\end{aligned}
\tag{5.62}
$$

The full weighting operator essentially smoothes the fine grid with the filter template

$$
\begin{bmatrix}
\frac{1}{16} & \frac{1}{8} & \frac{1}{16} \\
\frac{1}{8} & \frac{1}{4} & \frac{1}{8} \\
\frac{1}{16} & \frac{1}{8} & \frac{1}{16}
\end{bmatrix}
\tag{5.63}
$$

and then resamples every other point. A natural choice for the prolongation operator is bilinear interpolation [26]:

$$
\begin{aligned}
f_{2i,2j} &= c_{i,j}, \\
f_{2i+1,2j} &= \frac{c_{i,j} + c_{i+1,j}}{2}, \\
f_{2i,2j+1} &= \frac{c_{i,j} + c_{i,j+1}}{2}, \\
f_{2i+1,2j+1} &= \frac{c_{i,j} + c_{i+1,j} + c_{i,j+1} + c_{i+1,j+1}}{4}.
\end{aligned}
\tag{5.64}
$$

Before proceeding, we will need to explain in more detail how the problem to be solved is managed on the various grids. The restriction operator transfers the "residual error" of the equation to be solved to the coarser grids. Relaxation is performed on this residual error, which is equivalent to relaxing on the equation itself. This can be seen as follows: Expressing the equation to be solved as the linear problem $\mathbf{P}\phi = \rho$ as in Equation 5.36, where $\mathbf{P}$ is a matrix and ϕ and ρ are vectors, the "residual equation" is defined to be the equation

$$\mathbf{P}\epsilon = \rho - \mathbf{P}\hat{\phi}, \tag{5.65}$$

where $\hat{\phi}$ is an approximation to the exact solution ϕ and ϵ is the (unknown) error $\phi - \hat{\phi}$. Relaxation on the equation $\mathbf{P}\phi = \rho$ with the initial guess $\hat{\phi}$ is equivalent to relaxing on the residual equation $\mathbf{P}\epsilon = r$ with the initial guess $\epsilon = 0$, where r is the (known) residual error $r = \rho - \mathbf{P}\hat{\phi}$ [26]. In practice, the approximation $\hat{\phi}$ is the intermediate solution obtained by Gauss-Seidel relaxation on one grid, and the initial guess $\epsilon = 0$ is the starting point for the Gauss-Seidel relaxation on the next coarser grid. The resulting solution on the coarser grid can in turn be regarded as an intermediate solution whose residual error is restricted to the next coarser grid. Alternatively, the solution can be prolonged (interpolated) to the finer grid and added to the approximation $\hat{\phi}$ to yield a better solution on the finer grid.

We can now present a multigrid algorithm for solving the linear problem $\mathbf{P}\phi = \rho$ (i.e., Poisson's equation). For reasons that will soon become apparent, this particular algorithm is called a V-cycle scheme [26]. The algorithm and the ones that follow will be presented in a recursive form similar to the algorithms of [26]. We will need the following notation: Let ϕ_N and ρ_N denote the vectors ϕ and ρ defined on an $N \times N$ grid. For simplicity we will assume the grids are square, although the algorithm works for rectangular grids as well. Let R and P denote the restriction and prolongation operators, respectively. Let $\mathbf{P}$ denote the Laplacian operator (not to be confused with the prolongation operator P), and let $\leftarrow$ denote the assignment operator. Finally, let MV denote the name of the V-cycle routine, since it calls itself recursively. The MV algorithm accepts three parameters: the two vectors ϕ and ρ and the grid dimension N.

Multigrid V-Cycle Algorithm

$\phi_N \leftarrow \mathrm{MV}(\phi_N, \rho_N)$

Step 1. Relax ν_1 times on the equation $\mathbf{P}\phi_N = \rho_N$ with the given initial guess ϕ_N. The vector ϕ_N will then contain new values.

Step 2. If $N \times N$ is the coarsest grid, then go to Step 7.

Step 3. $\rho_{N/2} \leftarrow R(\rho_N - \mathbf{P}\phi_N)$.

Step 4. $\phi_{N/2} \leftarrow 0$.

Step 5. $\phi_{N/2} \leftarrow \mathrm{MV}(\phi_{N/2}, \rho_{N/2})$.

Step 6. $\phi_N \leftarrow \phi_N + P(\phi_{N/2})$.

Step 7. Relax ν_2 times on the equation $\mathbf{P}\phi_N = \rho_N$ with the "new" initial guess ϕ_N, which was computed in Step 6.

In Step 3 the restriction operator transfers the residual error $\rho - \mathbf{P}\phi$ from the $N \times N$ grid to the $N/2 \times N/2$ grid. Step 5 calls the V-cycle routine recursively to solve the residual equation on this coarser grid. In Step 6 the prolongation operator transfers the solution of the residual equation from the $N/2 \times N/2$ grid to the $N \times N$ grid and adds it to the current solution approximation ϕ. The number of Gauss-Seidel relaxation sweeps specified by ν_1 and ν_2 in Steps 1 and 7 is typically two [26].

The operation of the multigrid V-cycle algorithm proceeds as follows: It performs Gauss-Seidel relaxation sweeps on the finest grid and restricts the residual of the intermediate solution to the next coarser grid. It repeats this process with the coarse grid in place of the fine grid until it reaches the coarsest grid, whose size is very small (typically 3×3). It then works its way back up to the finest grid by a repetitive process of prolongation of the intermediate solutions and relaxation on each grid. The schedule for the grids in the order in which they are visited is shown pictorially in Figure 5.9(a). The complete process is called a *V cycle*.

The so-called full multigrid (FMG) algorithm takes the V-cycle algorithm a step further by visiting the grids according to the schedule shown in Figure 5.9(b). A complete cycle through this schedule is called an *FMG cycle*. The FMG algorithm generally has better convergence properties than the V-cycle algorithm [26]. The algorithm for one FMG cycle is given below. For n FMG cycles, the algorithm is called n times. Note that it calls the V-cycle algorithm in Step 6.

Full Multigrid (FMG) Algorithm

$\phi_N \leftarrow \text{FMG}(\phi_N, \rho_N)$

Step 1. If $N \times N$ is the coarsest grid, then go to Step 6.

Step 2. $\rho_{N/2} \leftarrow R(\rho_N - \mathbf{P}\phi_N)$.

Step 3. $\phi_{N/2} \leftarrow 0$.

Step 4. $\phi_{N/2} \leftarrow \text{FMG}(\phi_{N/2}, \rho_{N/2})$.

Step 5. $\phi_N \leftarrow \phi_N + P(\phi_{N/2})$.

Step 6. $\phi_N \leftarrow \text{MV}(\phi_N, \rho_N)$.

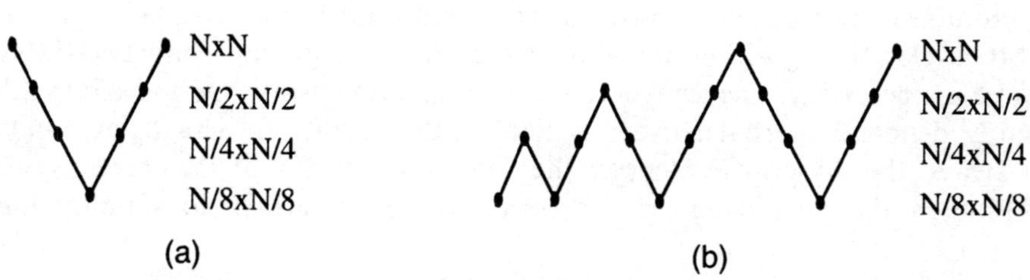

Figure 5.9 (a) The order in which the grids are visited in a multigrid V-cycle scheme. (b) The schedule of grid visits in a full multigrid (FMG) scheme. These schedules should be read from the left to the right. The labels denote the grid sizes, with the smaller grids below the larger.

Multigrid is a remarkable integration of simple ideas and techniques: Gauss-Seidel relaxation, the residual equation, and coarse-to-fine and fine-to-coarse grid transfers. The result is a very powerful algorithm, however, as will be demonstrated.

Algorithm It is now straightforward to define a multigrid algorithm for unweighted least-squares phase unwrapping [11].[11] This algorithm will form the basis of the weighted least-squares algorithm to be presented in Section 5.4.4.

The restriction operator is defined to be the full weighting operator performed on the residuals of the partial derivatives with the following modification: Because each coarse grid is half the resolution of its predecessor, the derivatives must be multiplied by two after each restriction operation to preserve the correct scale. The prolongation operator is defined by bilinear interpolation on the intermediate solutions. The Gauss-Seidel relaxation is performed on Equation 5.38. The algorithm requires three arrays to store (1) the unwrapped solution ϕ, (2) the row derivatives Δ^x, and (3) the column derivatives Δ^y. Each of these three arrays has its own pyramid of decreasing-resolution grids as shown in Figure 5.8.

It is absolutely essential to store the partial derivatives $\Delta^x_{i,j}$ and $\Delta^y_{i,j}$ of the input phase data (*not* the input phase data themselves) in separate arrays and to "correct" these derivatives at the grid boundaries after each restriction operation. The boundary conditions (Equation 5.45) require the partial derivatives Δ^x and Δ^y to satisfy the following relations:

$$\Delta^x_{M,j} = -\Delta^x_{M-1,j}, \quad \Delta^y_{i,N} = -\Delta^y_{i,N-1}. \tag{5.66}$$

These boundary conditions require the first column vector of the array Δ^x to equal the negative of its second column, and the first row of the array Δ^y to equal the negative of its second row. They are necessary to ensure that the Gauss-Seidel relaxation will converge to the correct solution [11]. These corrections must be applied to the residuals of the partial derivatives after each restriction operation so that they satisfy Equation 5.66. (Because the restriction operator is a nonlinear operator, it does not preserve the boundary conditions.)

Pseudocode The pseudocode, which is expressed in the form of a recursive subroutine named UFMG (unweighted full multigrid), is presented below. For n FMG cycles, the algorithm should be called n times. In practice, only two FMG cycles are necessary. The array ϕ denotes the solution array, and the arrays Δ^x and Δ^y denote the partial derivatives. Because these arrays are actually pyramids of arrays, the subscript N denotes the dimensions $N \times N$ of the current grid. (Although the pseudocode is expressed for square arrays to simplify the

[11] One might be tempted to use the "Numerical Recipes" multigrid algorithm described in Reference [24], pp. 871–887. However, this algorithm does not work for phase unwrapping because it assumes homogeneous boundary conditions. This means that the partial derivatives must be zero at the boundaries. This is almost never the case in practice.

notation, it works for rectangular grids as well.) The restriction operator R and the prolongation operator P are defined to be the usual full weighting and bilinear interpolation operators. The partial differentiation operators D^x and D^y are defined by

$$(D^x\phi)_{i,j} = \phi_{i+1,j} - \phi_{i,j}, \quad (D^y\phi)_{i,j} = \phi_{i,j+1} - \phi_{i,j}. \tag{5.67}$$

These operators are needed to compute the residuals in Step 2.

FMG (Multigrid) Algorithm for Unweighted Phase Unwrapping

$\phi_N \leftarrow \mathrm{UFMG}(\phi_N, \Delta_N^x, \Delta_N^y)$

Step 1. If $N \times N$ is the coarsest grid, then go to Step 7.
Step 2. $\Delta_{N/2}^x \leftarrow R(\Delta_N^x - D^x\phi_N)$, $\Delta_{N/2}^y \leftarrow R(\Delta_N^y - D^y\phi_N)$
Step 3. Boundary derivative correction step: Set the first column of the array $\Delta_{N/2}^x$ equal to the negative of its second column, and set the first row of the array $\Delta_{N/2}^y$ equal to the negative of its second row. (See Equation 5.66.)
Step 4. $\phi_{N/2} \leftarrow 0$.
Step 5. $\phi_{N/2} \leftarrow \mathrm{UFMG}(\phi_{N/2}, \Delta_{N/2}^x, \Delta_{N/2}^y)$.
Step 6. $\phi_N \leftarrow \phi_N + P(\phi_{N/2})$.
Step 7. $\phi_N \leftarrow \mathrm{UMV}(\phi_N, \Delta_N^x, \Delta_N^y)$.

The V-cycle subroutine, which is called in Step 7 and is denoted UMV ("Unweighted Multigrid V-cycle"), is defined as follows:

Multigrid V-Cycle Algorithm for Unweighted Phase Unwrapping

$\phi_N \leftarrow \mathrm{UMV}(\phi_N, \Delta_N^x, \Delta_N^y)$

Step 1. Relax 2 times on Equation 5.38 with the given initial guess ϕ_N.
Step 2. If $N \times N$ is the coarsest grid, then go to Step 8.
Step 3. $\Delta_{N/2}^x \leftarrow R(\Delta_N^x - D^x\phi_N)$, $\Delta_{N/2}^y \leftarrow R(\Delta_N^y - D^y\phi_N)$
Step 4. Boundary derivative correction step: Set the first column of the array $\Delta_{N/2}^x$ equal to the negative of its second column, and set the first row of the array $\Delta_{N/2}^y$ equal to the negative of its second row.
Step 5. $\phi_{N/2} \leftarrow 0$.
Step 6. $\phi_{N/2} \leftarrow \mathrm{UMV}(\phi_{N/2}, \Delta_{N/2}^x, \Delta_{N/2}^y)$.
Step 7. $\phi_N \leftarrow \phi_N + P(\phi_{N/2})$.
Step 8. Relax 2 times on Equation 5.38 with the new initial guess ϕ_N.

Implementation Our `C` language implementation of the unweighted multigrid algorithm can be found in Section A.10 of Appendix A. The algorithm requires three arrays, with each array the same size as the input phase array:

1. ϕ-solution array
2. Δ^x-row derivatives
3. Δ^y-column derivatives

Each of these three arrays has its own pyramid of decreasing resolution grids as shown in Figure 5.8. The memory space required by the algorithm is therefore $3 \times (1 + \frac{1}{4} + \frac{1}{16} + \cdots) = 4$ times the memory occupied by the input phase data. Unlike the FFT or DCT algorithms, the multigrid approach has the practical advantage of not requiring power-of-two grid sizes. The input phase data can be defined on any grid size.

Figures 5.10(a)–(d) show the sequence of increasingly detailed surfaces that result from relaxation on the coarse grids for the phase unwrapping problem of Figure 5.6. This sequence shows the intermediate solution surfaces that result from the first few grids of a single FMG cycle. After one complete FMG cycle, the rms error of the result is just 1 per cent, and after two FMG cycles the rms error is 0.001 per cent. It is typical for the algorithm to converge in only one or two FMG cycles regardless of the grid size.

Table 5.2 lists the execution times for various array sizes. The times are for one FMG cycle and two relaxation sweeps per grid, which is generally sufficient for convergence. The algorithm is nearly as fast as the FFT-based or DCT-based algorithms (see Table 5.1).

5.3.4 Examples

In this section we will present the results of the unweighted least-squares algorithms on four phase unwrapping problems. These examples will demonstrate that the unweighted least-squares approach is not accurate enough when there are many residues in the phase data. The importance of unweighted least-squares is that it is the foundation for the weighted least-squares methods to be presented. We therefore will not examine all of the phase unwrapping problems.

We begin with the wrapped phase values in the 512 × 512-pixel array shown in Figure 5.11(a). These phase values increase linearly from the upper left corner to the lower right corner of the array. Unwrapping these values with any one of the unweighted least-squares algorithms yields the slanted phase plane shown in Figure 5.11(b).

Because the dynamic range of the unwrapped result is large, visual display is not always convincing that the unwrapping is correct. Therefore we have chosen to "rewrap" the unwrapped values to allow a direct comparison with the wrapped input.[12] The rewrapping will be visually convincing that the unwrapping is qualitatively correct and will also show how the least-squares formulation

[12] Rewrapping makes sense for least-squares methods but not for path-followers. This is because a least-squares solution is not always congruent to the original data. Rewrapping shows this discrepancy if it exists. A rewrapped path-follower result always looks like the input data.

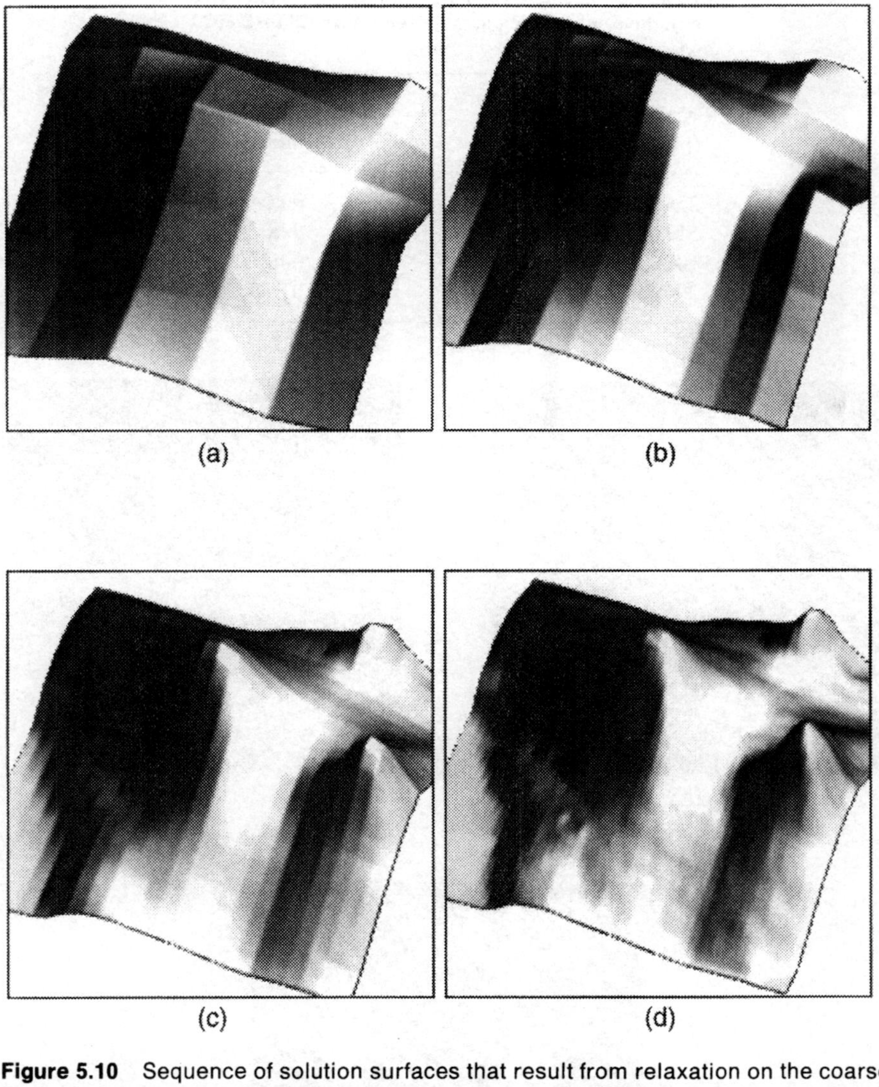

Figure 5.10 Sequence of solution surfaces that result from relaxation on the coarse grids of the phase unwrapping problem of Figure 5.6. The grid sizes are (a) 6 × 6, (b) 12 × 12, (c) 25 × 25, and (d) 50 × 50. Compare with the actual solution shown in Figure 5.6(b).

manifests itself on noisy phase data. The rewrapped values depicted in Figure 5.11(b) are shown in Figure 5.11(c) and are consistent with the original input depicted in Figure 5.11(a). Since the original input was noise-free and completely consistent (i.e., no residues were present), the least-squares unwrapping was perfect.

In the second example we have modified the "perfect" data depicted in

Table 5.2 Execution times of the unweighted multigrid algorithm for various array sizes on an IBM RS/6000 workstation.

Array Size (pixels)	Execution Time (min:s)
256 × 256	00:04
512 × 512	00:07
1024 × 1024	00:27
2048 × 2048	01:55

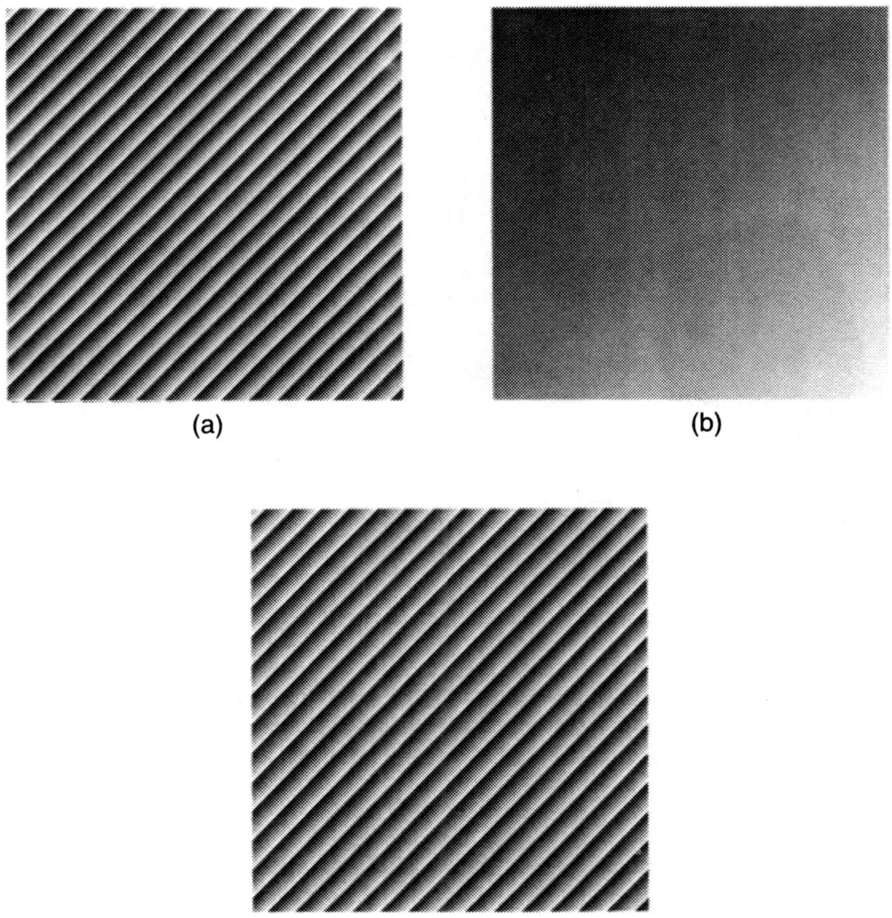

Figure 5.11 Wrapped and unwrapped phase examples. (a) Wrapped phase data. (b) Unwrapped solution obtained by the unweighted least-squares algorithms. (c) "Rewrapped" values of the unwrapped solution. Compare with (a).

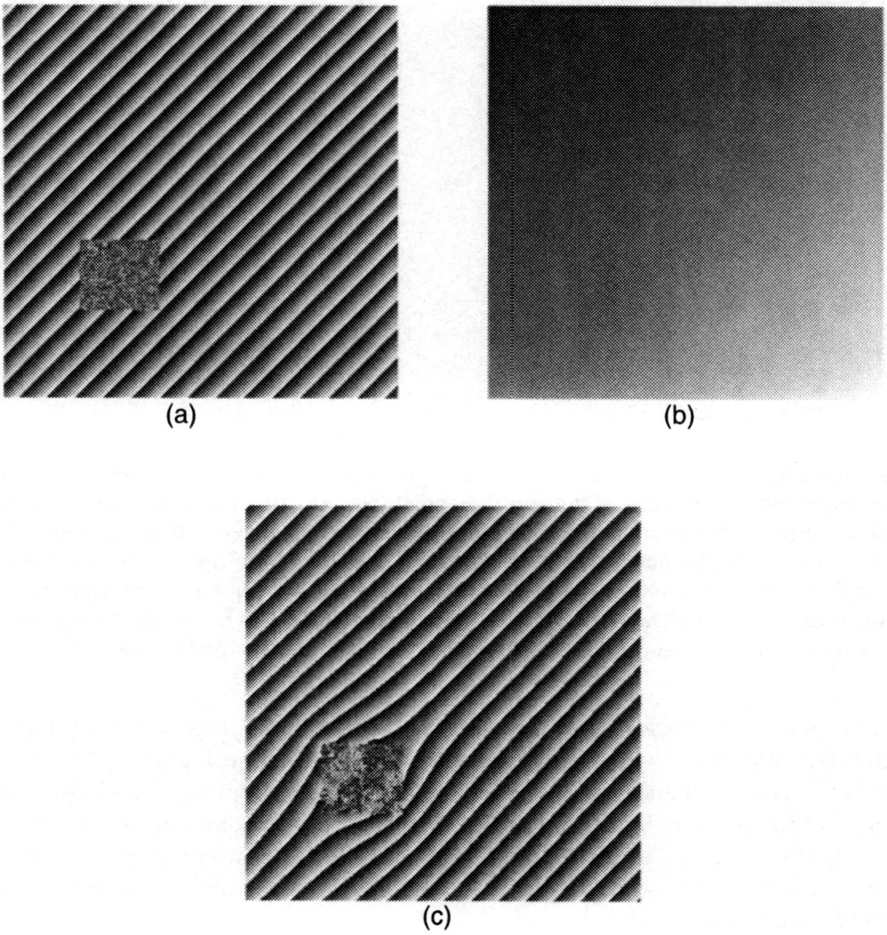

Figure 5.12 Wrapped and unwrapped phase examples with noise. (a) The wrapped phase of a tilted plane with a small rectangular patch containing uniformly distributed noise. The patch contains a high density of residues. (b) The unwrapped solution produced by the least-squares algorithms shows the influence of the noisy patch. (c) The rewrapped solution. The residues in the patch corrupt the solution locally with diminishing effects farther away.

Figure 5.11(a) by replacing the phase values in a rectangular region with uniform noise as shown in Figure 5.12(a).

The phase values contain residues in the rectangular region and around its boundary. These inconsistencies are accommodated by the least-squares formulation and influence the unwrapping regionally. The unwrapped values are shown in Figure 5.12(b) without rewrapping to illustrate the apparent quality of the least-squares product. The two-dimensional planar aspect is qualitatively

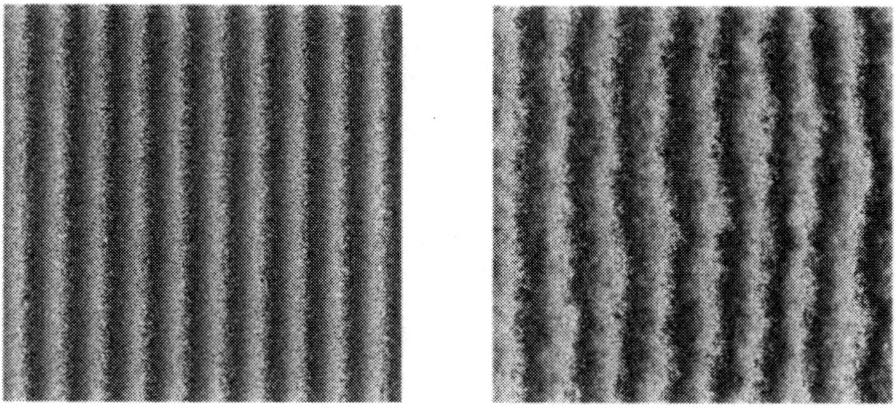

(a) (b)

Figure 5.13 When residues are present, least-squares solutions always underestimate the magnitudes of the local phase slopes. (a) The wrapped phase values of a noise-corrupted, tilted plane. Sufficient noise has been added to generate residues uniformly distributed over about 4.2 percent of the data. Note that there are 10 cycles of phase change in the horizontal direction. (b) The rewrapped least-squares solution indicates that the average phase slope has been underestimated. There now are only about 8 cycles of phase change in the horizontal direction.

captured, and the influence of the noisy region is minimal. Figure 5.12(c) shows the rewrapped phase after unwrapping the data depicted in Figure 5.12(a). It is easy to see how the least-squares approach accommodates noise and how the unwrapped phase is influenced in the vicinity of inconsistent data with diminishing effects further away. Figure 5.12(c) also illustrates why it is important to rewrap the results produced by least-squares algorithms so that subtle deviations can be seen clearly.

The third example, which is shown in Figure 5.13, illustrates that least-squares solutions always *underestimate* the local–average phase slope when residues are present. Figure 5.13(a) shows the wrapped phase of a noise-corrupted, tilted plane. Sufficient noise has been added to generate a moderate number of residues. The rewrapped least-squares solution is shown in Figure 5.13(b). It is evident from the spacing of the fringes that the average phase slope has been underestimated. In order to accommodate the inconsistent phase in a globally optimal manner, the least-squares solution must produce a "smoother" estimate than the measurements dictate. This behavior is analogous to least-squares curve fitting, in which the fitted curve is always smoother than the measured data.

The fourth example shows the behavior of the least-squares algorithm on a sheared planes example. Figure 5.14(a) depicts the wrapped (and noise-free) phase data with the shear line midway between the top and bottom of the image. Above the shear line the phase data represent a planar surface increasing in value from the top left corner toward the right and down. Below the shear line the

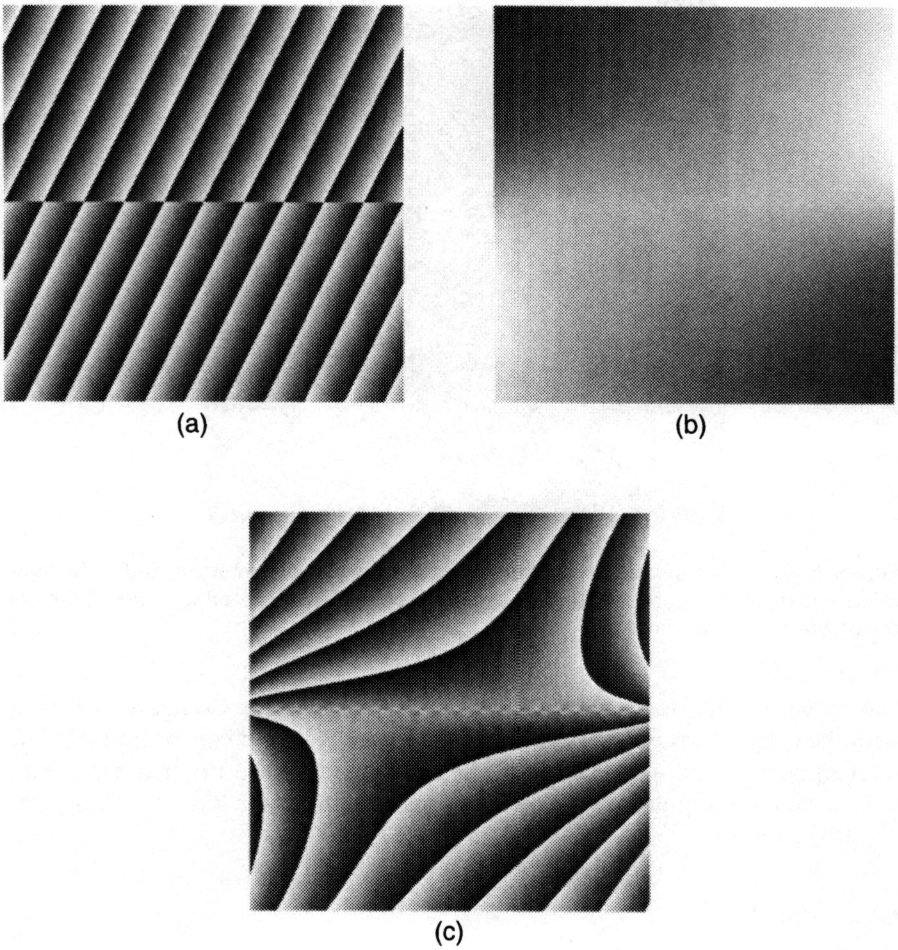

Figure 5.14 Unweighted least-squares phase unwrapping of a sheared planes example. (a) Wrapped, noise-free phase values with a horizontal shear line midway between the top and bottom of the figure. (b) Unweighted least-squares result. (c) The rewrapped values reveal subtleties in the solution and least-squares "seaming" across the horizontal shear line.

phase values are sloping linearly downward to the right and down. It is obvious that unweighted least-squares phase unwrapping cannot reconcile the phase differences everywhere across the shear boundary and yield the "true" solution. The unweighted least-squares solution is shown in Figures 5.14(b) and (c). Without prior knowledge of the shear, this is the best one can hope for (in the least-squares sense). The solution "seams" the phase values together across the horizontal shear line and tends to capture the true phase away from this line.

The final example, which is shown in Figure 5.15, shows the (rewrapped)

Figure 5.15 (Rewrapped) unweighted least-squares solution for the SAR interferometry example of Figure 3.2(a). This solution has underestimated the magnitudes of the phase slopes everywhere.

least-squares solution for the SAR interferometry problem. Compare this result with the original wrapped phase shown in Figure 3.2(a). It can be seen that the least-squares solution has fewer fringe contours than the original data. This means that this solution is a smoother estimate than the raw data can conceivably support.

5.3.5 Conclusion

In this section we presented a mathematical formulation of the unweighted least-squares approach to phase unwrapping and proved that the equation that must be solved is Poisson's equation. We then presented five algorithms for solving this equation. Four of these algorithms find the solution directly by means of FFTs or DCTs. Different ways of performing the mirror reflection yield different algorithms. The fifth algorithm is an iterative approach that is based on multigrid techniques.

Unweighted least-squares algorithms generally are not suitable in practice because of their smoothing behavior.[13] They are, however, fundamental for the weighted least-squares algorithms. We will see that *weighted* least-squares algorithms offer means for mitigating the undesirable smoothing behavior of unweighted algorithms.

[13] Some applications, however, require a smoother estimate than the measurements dictate. Deforming mirrors in adaptive optics applications may fall into this category.

5.4 WEIGHTED LEAST-SQUARES PHASE UNWRAPPING

Least-squares methods, unlike path-followers, do not deal directly with the residue problem because they obtain solutions by integrating through the residues to minimize the gradient differences. Weighted least-squares methods, on the other hand, use predetermined weights (e.g., quality maps) to avoid integrating through the residues. The weights are selected to accommodate the residues in some fashion, to isolate the regions of low signal-to-noise, or to impose other properties or preferences on the desired solution.[14] In some applications the unweighted least-squares solution may be preferred to a path-following result because of the global smoothness constraint. In most cases, however, least-squares solutions that have not adequately accommodated the residue problem provide undesirable results. In Section 5.3.4 we observed that uncompensated residues produce least-squares solutions with underestimated (in magnitude) phase gradients everywhere. The severity of this problem depends on the density of the uncompensated residues and on the specific phase unwrapping problem.

We mentioned at the end of the preceding section that weighted least-squares algorithms offer means for mitigating some of the undesirable behavior of unweighted algorithms. When certain phase values are known to be corrupted by regionally varying noise, measurement errors, aliasing or other degrading influences that generate residues, these values should be zero-weighted. In addition *a priori* information based on the physics of the underlying problem may dictate that certain regions of the unwrapped phase be isolated from one another. For example, phase surfaces that contain regional shearing or isolated regions of no phase information (e.g., the background of the MRI knee example) should be weighted so that the unwrapped phase solution is not required to meet continuity constraints across these boundaries.

In this section we will present the following three solution methods for the weighted least-squares phase unwrapping problem:

1. Picard iteration method
2. Preconditioned conjugate gradient (PCG) method
3. Multigrid method

All of these methods are iterative, in contrast to the direct (i.e., noniterative) FFT and DCT methods discussed previously. Before discussing these approaches, we present a mathematical formulation of the weighted least-squares problem.

5.4.1 Mathematical Formulation

The primary disadvantage of *unweighted* least-squares phase unwrapping is that it unwraps *through* the phase inconsistencies rather than around them. This

[14] Methods for defining the weights were presented in Chapter 3.

disadvantage can be remedied with the introduction of weights. When certain phase values are known to be corrupted due to noise, aliasing, or other degradations, the corrupted phase values should be zero-weighted so that they do not affect the unwrapping. In practice, one is given an array of weights $0 \leq w_{i,j} \leq 1$ that correspond, on a sample-by-sample basis, to the given phase data. (These weights can be defined by the quality maps or masks discussed in Sections 3.3 and 3.4.) The least-squares problem then becomes a *weighted* least-squares problem, and the quantity to be minimized becomes

$$\epsilon^2 = \sum_{i,j} U(i,j)(\phi_{i+1,j} - \phi_{i,j} - \Delta^x_{i,j})^2 + \sum_{i,j} V(i,j)(\phi_{i,j+1} - \phi_{i,j} - \Delta^y_{i,j})^2, \tag{5.68}$$

where the gradient weights, $U(i, j)$ and $V(i, j)$, are defined by

$$U(i,j) = \min(w^2_{i+1,j}, w^2_{i,j}),$$
$$V(i,j) = \min(w^2_{i,j+1}, w^2_{i,j}). \tag{5.69}$$

The least-squares solution $\phi_{i,j}$ to this problem is defined by the equation

$$U(i,j)(\phi_{i+1,j} - \phi_{i,j}) - U(i-1,j)(\phi_{i,j} - \phi_{i-1,j})$$
$$+ V(i,j)(\phi_{i,j+1} - \phi_{i,j}) - V(i,j-1)(\phi_{i,j} - \phi_{i,j-1}) = c_{i,j}, \tag{5.70}$$

where $c_{i,j}$ is the "weighted phase Laplacian" defined by

$$c_{i,j} = U(i,j)\Delta^x_{i,j} - U(i-1,j)\Delta^x_{i-1,j} + V(i,j)\Delta^y_{i,j} - V(i,j-1)\Delta^y_{i,j-1}. \tag{5.71}$$

Equation 5.70 is identical in form to Equation 5.28 except that in this case the gradient weights $U(i, j)$ and $V(i, j)$ are user-defined. Equation 5.70, unlike the discretized Poisson equation (Equation 5.31), cannot be solved directly by means of Fourier or cosine transforms. It must be solved iteratively.

The classical Gauss-Seidel relaxation method solves Equation 5.70 by iterating the equation

$$\phi_{i,j} = \frac{U(i,j)\phi_{i+1,j} + U(i-1,j)\phi_{i-1,j} + V(i,j)\phi_{i,j+1} + V(i,j-1)\phi_{i,j-1} - c_{i,j})}{\nu_{i,j}}, \tag{5.72}$$

where $\nu_{i,j}$ is defined by

$$\nu_{i,j} = U(i,j) + U(i-1,j) + V(i,j) + V(i,j-1). \tag{5.73}$$

As in the unweighted case the Gauss-Seidel relaxation method converges too slowly to be of practical use. It is important for the multigrid method, to be discussed in Section 5.4.4, however.

The weighted least-squares phase unwrapping problem can of course be described by simple matrix operations. It is well known that the *unweighted* least-squares solution to the overdetermined set of linear equations

$$\mathbf{A}\mathbf{x} = \mathbf{b}, \tag{5.74}$$

is the solution to the normal equations

$$\mathbf{A}^T\mathbf{A}\mathbf{x} = \mathbf{A}^T\mathbf{b} \tag{5.75}$$

where, for our problem, $\mathbf{x}$ is a length MN solution vector of phase values, $\mathbf{b}$ is a vector of length $N(M-1) + M(N-1)$ containing the wrapped phase differences, and T refers to matrix transposition. We can rewrite Equation 5.75 in the form

$$\mathbf{P}\phi = \rho, \tag{5.76}$$

where $\mathbf{P} = \mathbf{A}^T\mathbf{A}$ is a matrix that performs the discrete Laplacian operation on the vector ϕ as illustrated in Equation 5.31, and $\rho = \mathbf{A}^T(\mathbf{b})$ is a vector containing the discrete Laplacian operation on the wrapped phase differences as illustrated in Equation 5.32.

The *weighted* form of Equation 5.74 is the equation

$$\mathbf{W}\mathbf{A}\mathbf{x} = \mathbf{W}\mathbf{b}, \tag{5.77}$$

where $\mathbf{W}$ is a matrix of weights. The resulting normal equations (Equation 5.75 are

$$\mathbf{A}^T\mathbf{W}^T\mathbf{A}\mathbf{x} = \mathbf{A}^T\mathbf{W}^T\mathbf{W}\mathbf{b}. \tag{5.78}$$

If we define $\mathbf{Q} = \mathbf{A}^T\mathbf{W}^T\mathbf{W}\mathbf{A}$ and $\bar{\mathbf{b}} = \mathbf{W}^T\mathbf{W}\mathbf{b}$ and substitute these definitions into Equation 5.78, we obtain

$$\mathbf{Q}\mathbf{x} = \mathbf{A}^T\bar{\mathbf{b}}. \tag{5.79}$$

If we then make the substitution

$$\mathbf{c} = \mathbf{A}^T\bar{\mathbf{b}}, \tag{5.80}$$

and associate the desired solution ϕ with $\mathbf{x}$, we obtain

$$\mathbf{Q}\phi = \mathbf{c}. \tag{5.81}$$

It is important to note that $\bar{\mathbf{b}}$ is simply a vector consisting of *weighted* phase differences and $\mathbf{c}$ is nothing more than a vector formed from the discrete phase Laplacian of the *weighted* phase differences. This is completely analogous to what was done to form the right-hand side of Equation 5.76. The left-hand side of Equation 5.81, when expanded in array notation, is identical to the left-hand side of Equation 5.70. Similarly the right-hand side of Equation 5.81, when expanded in array notation, is identical to Equation 5.71. In summary, Equation 5.81 is the matrix description of the weighted least-squares problem.

5.4.2 Picard Method

We present in this section an iterative approach for solving the weighted least-squares phase unwrapping problem. The Picard iterative method, as this approach is called, is easy to implement and requires little memory. However, because it requires many iterations, it is not a practical algorithm. Nevertheless, we present it because it serves as a good motivation for the PCG algorithm to be introduced in Section 5.4.3. After describing the algorithm and the pseudocode, we will examine its performance on several phase unwrapping examples.

Algorithm Incorporation of the weighting matrix makes it impossible to solve Equation 5.81 using the discrete Fourier or cosine transform expansion directly as was done for the unweighted case. However, we can incorporate our ability to solve the unweighted problem into a simple iterative scheme that solves the weighted problem. The technique is based on a simple Picard iteration. Let us split the matrix $\mathbf{Q}$ into the sum of the matrix $\mathbf{P}$ (from Equation 5.76), and a difference matrix, $\mathbf{D}$, to obtain

$$\mathbf{Q} = \mathbf{P} + \mathbf{D}. \tag{5.82}$$

Substitution of Equation 5.82 into Equation 5.81 yields

$$(\mathbf{P} + \mathbf{D})\phi = \mathbf{c} \tag{5.83}$$

or

$$\mathbf{P}\phi = \mathbf{c} - \mathbf{D}\phi. \tag{5.84}$$

Making Equation 5.84 iterative yields

$$\mathbf{P}\phi_{k+1} = \mathbf{c} - \mathbf{D}\phi_k = \rho_k, \tag{5.85}$$

where k refers to the iteration number. Recall that

$$\mathbf{D}\phi_k = (\mathbf{Q} - \mathbf{P})\phi_k, \tag{5.86}$$

where $\mathbf{Q} = \mathbf{A}^T\mathbf{W}^T\mathbf{W}\mathbf{A}$ performs the modified discrete Laplacian operation on the weighted phase differences (Equation 5.71) and $\mathbf{P} = \mathbf{A}^T\mathbf{A}$ performs the ordinary discrete Laplacian operation on the unweighted phase differences (Equation 5.31). Therefore $\mathbf{D}\phi_k$ is nothing more than a matrix difference operator operating on ϕ_k to produce a vector containing the modified Laplacian of the weighted phase values at iteration k minus the ordinary Laplacian of the unweighted phase values at iteration k:

Thus the right-hand side of Equation 5.85 when written out in two-dimensional array notation becomes

$$\begin{aligned}\rho_{i,j} = c_{i,j} - [&U(i,j)(\phi_{i+1,j} - \phi_{i,j}) \\ &- U(i-1,j)(\phi_{i,j} - \phi_{i-1,j}) \\ &+ V(i,j)(\phi_{i,j+1} - \phi_{i,j}) \\ &- V(i,j-1)(\phi_{i,j} - \phi_{i,j-1})],\end{aligned} \tag{5.87}$$

where $c(i, j)$ is defined by Equation 5.71. Equation 5.85 is the basis for a simple iterative scheme for solving the weighted least-squares unwrapping problem by means of one of the fast algorithms of Section 5.3.

Pseudocode In addition to an array for storing the weights, the Picard algorithm requires only one additional array for storing the intermediate solutions. All of the operations can be performed in place in this array. The pseudocode is as follows. The final solution is ϕ_n, where n is the maximum number of iterations.

Picard Algorithm for Weighted Phase Unwrapping

```
Compute the array c of weighted phase Laplacians (Equation 5.71).
Initialize the solution array φ0 to zero or an initial guess.
for (k = 0 to MaxNumberOfIterations - 1)
  Compute the vector ρk = c - Dφk (Equation 5.87).
  Solve the equation Pφk+1 = ρk (Equation 5.76) by means of one of the
    unweighted algorithms of Section 5.3 to obtain the new solution φk+1.
end
```

Implementation The Picard algorithm has the advantages of being easy to implement and requiring little memory because most of the operations can be done in place. Furthermore, if it converges (which is not guaranteed), it converges to the correct solution of the weighted least-squares problem. It is possible to monitor the convergence by computing the norm of the residual $\| \mathbf{D}\phi_k \|$ at each iteration. The disadvantages of the Picard algorithm are the lack of guaranteed convergence and the possibility that an excessive number of iterations may be required for convergence on a given problem.

The primary value of the Picard algorithm is the illumination it sheds on the

PCG algorithm to be presented in Section 5.4.3. We do not recommend that the algorithm be used in practice. Accordingly we have not supplied software for the algorithm, nor have we provided a table of execution times.

Examples The unweighted least-squares example of Figure 5.12 showed that if the phase values are corrupted or destroyed in a region, the unwrapped results are influenced most heavily in the vicinity of the corrupted values. This example also showed that the phase values are influenced globally as well. Assuming we have some additional information that allows us to define a corresponding weighting array, we can eliminate completely the noisy phase values from having any influence on the result simply by assigning a weight of zero to the inconsistent data. Figure 5.16 depicts the corresponding weighting array to accompany the phase values of Figure 5.12(a). The black rectangular region corresponds to zero-valued weights, and the remainder of the 512×512 array is set to unity weight and is shown as white. (The one-pixel-wide dark outside border is superimposed just to locate the proper array size in the figure.)

Figure 5.17 depicts the result of applying the Picard algorithm to the weighted least-squares phase unwrapping problem. In this case the algorithm has relatively fast convergence, requiring about 10 iterations to obtain a perfect solution. Again, the rewrapped results are shown to highlight subtle behavior. It is interesting that the unwrapped phase is continuous across the noisy region even though the weights were zero.

Figure 5.16 Two-dimensional weighting array to accompany the wrapped phase data depicted in Figure 5.12(a). Black corresponds to zero-valued weights, and white corresponds to unity weight. The dark border defines the array boundary.

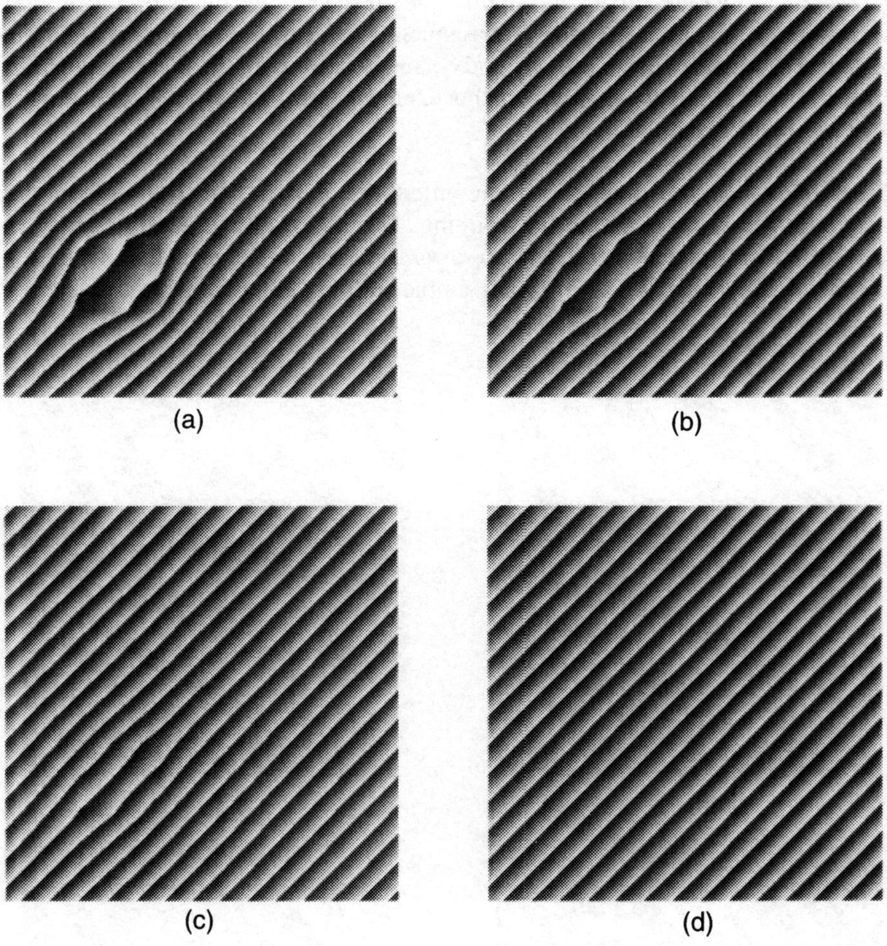

Figure 5.17 Progress of the Picard algorithm on the example of Figure 5.12(a) with the corresponding weighting array of Figure 5.16 after (a) 1, (b) 2, (c) 3, and (d) 10 iterations. Note the correct solution in the regions corresponding to the unity weights, and note also the continuity across the region of zero-weights. The unwrapped results are rewrapped to highlight the subtle features.

The sheared planes example can be used to illustrate the power and robustness of weighted (as opposed to unweighted) phase unwrapping. Recall that Figure 5.14(a) depicts the wrapped (and noise-free) phase values of a sheared surface. The shear line is horizontal midway between the top and bottom of the image. If we know (from the physics of the problem) that a shear line exists, it is a simple matter to define a weighting array containing a line of zero-valued weights along the shear line. Alternatively, we can use the methods of Chapter 3 to define a mask of zero-weights along the shear line. Because the zero-weights extend across the entire phase array, it is impossible for the phase data on opposite sides of the line to influence each other. Consequently the top and bottom halves of the phase array unwrap independently and correctly. However, the Picard algorithm does not converge very quickly on this example (see Figure 5.18), although it does converge eventually.

Conclusion In this section we presented the Picard iterative method for weighted least-squares phase unwrapping. The algorithm generally converges too slowly for practical use. There are several reasons we have included it in our discussion, however. The algorithm is simple and easy to implement. It illustrates

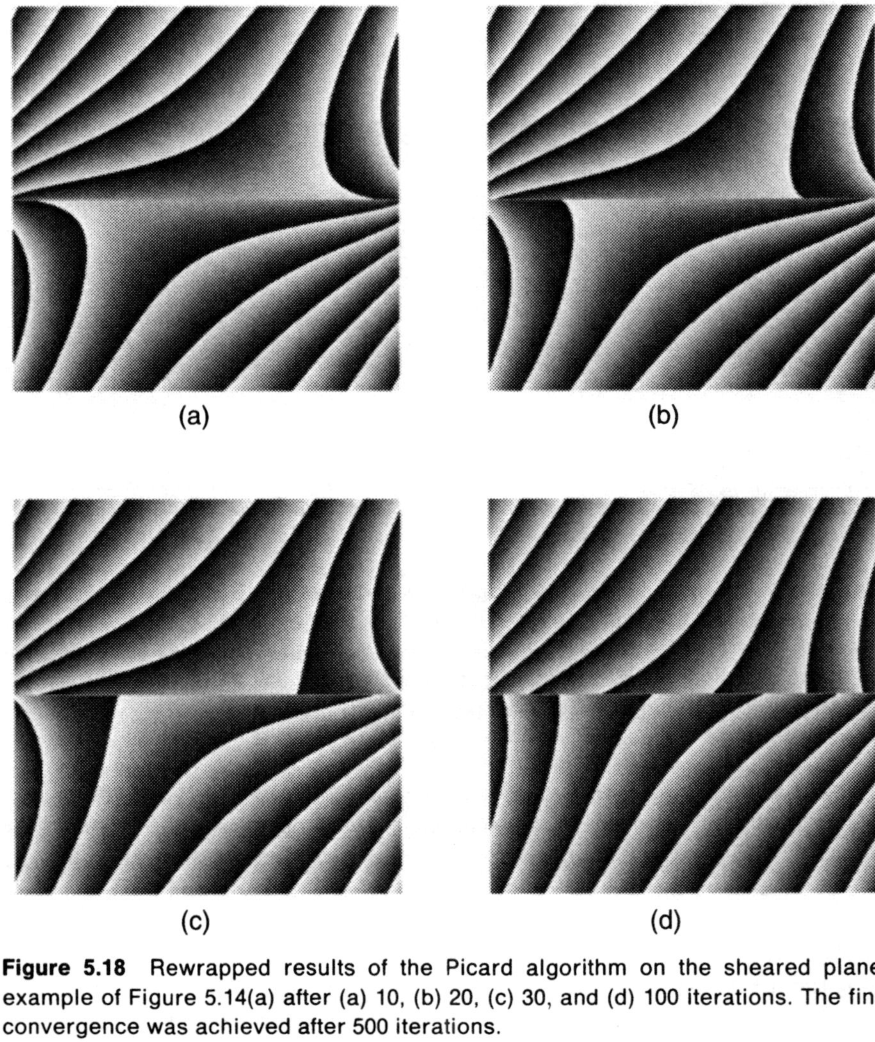

(a) (b) (c) (d)

Figure 5.18 Rewrapped results of the Picard algorithm on the sheared planes example of Figure 5.14(a) after (a) 10, (b) 20, (c) 30, and (d) 100 iterations. The final convergence was achieved after 500 iterations.

the ability of weighted least squares to produce correct results. Furthermore it leads naturally to the PCG algorithm to be described in the following section. Unlike the Picard algorithm, the PCG algorithm converges in a reasonable number of iterations.

5.4.3 PCG Algorithm

In this section we present an iterative algorithm for solving the weighted least-squares phase unwrapping problem. This algorithm converges much faster than the Picard algorithm. After a brief description of conjugate gradient methods, we describe the PCG algorithm followed by the pseudocode and the implementation details. Before examining the performance of the algorithm on a number of phase unwrapping problems, we describe a very important postprocessing operation that must be performed to make an unwrapped result congruent to the wrapped input phase.

Conjugate gradient (CG) methods are a means of iteratively solving a set of sparse linear equations, of which the discretized version of Poisson's equation (Equation 5.31) is one. The conjugate gradient technique poses the solution of the linear equations as a minimization problem and obtains much faster convergence than simpler minimization schemes such as steepest descent [27]. The CG algorithm has robust convergence properties with well-defined termination conditions. Ignoring roundoff error, CG converges in exactly N iterations for an $N \times N$ problem. However, for large N the number of iterations may still be excessive and exact convergence is not guaranteed because of numerical roundoff error.

The actual number of iterations required to obtain an accurate solution depends on the condition number of the underlying matrix. If the underlying matrix is "near" the identity matrix, CG converges very quickly. Preconditioned conjugate gradient (PCG) methods apply a "preconditioning" step that effectively transforms the underlying matrix to one very near the identity matrix; hence very rapid convergence is obtained.

Algorithm The PCG algorithm is similar to the Picard algorithm. Complications arise from the need for the preconditioner. To be useful, the preconditioner must be "close" to the matrix of interest and must represent an easily solvable linear system. In our case the preconditioning is obtained by solving an approximate problem—the *unweighted* least-squares phase unwrapping problem—and then updating the solution at each iteration. The unweighted problem is solved by means of one of the unweighted algorithms of Section 5.3. The solution of this "preconditioning step" is then used to obtain an estimate for the true solution to the exact problem.

The beauty of the algorithm is that it allows us to efficiently incorporate the DCT or FFT methods within the robust structure of conjugate gradients to solve the weighted least-squares problem iteratively. The reader is encouraged to consult [27] for a thorough treatment of conjugate gradient methods.

Pseudocode The pseudocode for the PCG algorithm requires an array for storing the intermediate (and final) solutions ϕ. It also requires arrays for storing the residual $\mathbf{r}$, the preconditioner $\mathbf{z}$, and the conjugate gradient direction vector $\mathbf{p}$. The weights, which must also be stored in an array, are used only in the computations of the weighted phase Laplacian and the weighted Laplacian $\mathbf{Q}$. The final solution is ϕ_{k+1}.

PCG Algorithm for Weighted Phase Unwrapping

```
Compute the array r_0 of weighted phase Laplacians defined by Equation 5.71.
Initialize the solution array φ_0 to zero (or an initial guess).
for (k = 0 to MaxNumberOfIterations − 1)
  Solve the equation Pz_k = r_k (Equation 5.76) by means of one of the
    unweighted algorithms of Section 5.3 to obtain the preconditioner z_k.
  if (k = 0), then
    Let p_1 = z_0.
  else
    Let β_{k+1} = r_k^T z_k / r_{k-1}^T z_{k-1}.
    Let p_{k+1} = z_k + β_{k+1} p_k.
  end
  Let α_{k+1} = r_k^T z_k / p_{k+1}^T Q p_{k+1}.
  Let φ_{k+1} = φ_k + α_{k+1} p_{k+1}.
  Let r_{k+1} = r_k − α_{k+1} Q p_{k+1}.
  if (|| r_{k+1} || < ε || r_0 ||), where ε is a predefined convergence factor,
    then exit the loop.
end
```

Implementation Our C language implementation of the PCG algorithm for weighted least-squares phase unwrapping can be found in Section A.11 of Appendix A. It gives the user the following choice of quality maps, which define the weights:

1. User-supplied correlation coefficients (or other user-supplied quality map)
2. Pseudocorrelation
3. Variance of phase derivatives
4. Maximum phase gradient

The correlation coefficients can be averaged locally, and the other three quality maps can be computed in windows of any size. The user can also specify that a threshold be calculated and applied to the quality map to produce a mask that defines zero weights. The algorithm used to compute the threshold is described in Section 3.4. We recommend that the user always use this option or input a mask because the algorithm seems to converge faster with binary-valued weights.

The memory requirements are as follows. All arrays are the same size and data type as the input phase data. Because the phase data and the residual array can share the same array, a total of five arrays are required:

1. Array for phase data and residual **r**
2. Array for preconditioner **z**
3. Array for conjugate gradient direction vector **p**
4. Array for quality map (weights)
5. Solution array ϕ

The software can accept one or two arrays for the quality map. If two arrays are supplied, then the arrays are assumed to contain two sets of weights $U(i, j)$ and $V(i, j)$ as in Equation 5.68. The L^p-norm algorithm to be described in Section 5.5 invokes the PCG algorithm with two such sets of weights. If only one set of weights $w_{i,j}$ is supplied, however, then the weights are defined by Equation 5.69.

Table 5.3 lists the average execution times of our implementation for various array sizes. The actual times depend on the number of iterations required for convergence. The times are given for 20 iterations, which is sufficient for most phase unwrapping problems. Comparison with Table 4.3 shows that the speed of the algorithm is comparable to that of the mask cut algorithm.

Postprocessing Congruence Operation In this section we present a postprocessing operation that makes the unwrapped surface congruent to the wrapped input phase. We recommend that this operation be applied to the results produced by the PCG and the multigrid algorithms (and indeed any least-squares algorithm) because it greatly reduces the errors in the unwrapped surface.

Unlike path-following algorithms, least-squares algorithms do not produce an unwrapped surface that is congruent to the wrapped phase. That is, if one

Table 5.3 Execution times of the PCG algorithm for weighted least-squares phase unwrapping for various array sizes on an IBM RS/6000 workstation.

Array Size (pixels)	Execution Time (min:s)
256 × 256	00:40
512 × 512	02:40
1024 × 1024	10:00
2048 × 2048	40:00

Note: The times are given for 20 iterations of the algorithm.

re-wraps the output surface, the resulting phase is not necessarily identical to the wrapped input phase. This is because the least-squares formulation minimizes the *squares* of the differences between the gradients. The resulting gradients of the output surface do not match those of the input phase, and as a result the output surface is not congruent to the input phase.

Typically one desires the unwrapped surface to be congruent to the wrapped phase. In this case it is necessary to make the surface congruent by means of a postprocessing operation. This operation approximately minimizes the number of discontinuities and thereby transforms the least-squares solution into approximately a minimum L^0-norm solution. We will refer to such an operation as a *congruence operation*.

The simplest way to define the congruence operation is to subtract the unwrapped surface from the wrapped phase, wrap the result, and then add it to the unwrapped surface. In terms of continuous functions this operation is defined as follows.[15] Given the unwrapped surface $\varphi(x, y)$ and the wrapped phase $\psi(x, y)$, define the *congruence* $\tilde{\varphi}(x, y)$ of $\varphi(x, y)$ by

$$\tilde{\varphi}(x, y) = \varphi(x, y) + \mathcal{W}[\psi(x, y) - \varphi(x, y)], \tag{5.88}$$

where the wrapping operator $\mathcal{W}(\cdot)$ wraps the values of its argument into the range $(-\pi, \pi]$ by adding or subtracting an integral multiple of 2π. It is easy to verify that the surface $\tilde{\varphi}(x, y)$ is congruent to the wrapped phase $\psi(x, y)$.

This simple congruence operation is unsatisfactory in the presence of noise, however, and it must be amended. A simple example will suffice to illustrate this point. Let $\psi(x, y)$ be the wrapped phase of the function $f(x, y) = x$, and consider the surface $\varphi(x, y) = x + \pi + \varepsilon(x, y)$, where $\varepsilon(x, y)$ is a zero-mean, Gaussian noise function (with any nonzero variance). For simplicity, assume that $|\varepsilon(x, y)| < \pi$. Observe that the rewrapping $\mathcal{W}[\psi(x, y) - \varphi(x, y)]$ is equal to $\mathcal{W}[-\pi - \varepsilon(x, y)]$, which is in turn equal to $-\varepsilon(x, y) + \pi$ when $\varepsilon(x, y) > 0$ and $-\varepsilon(x, y) - \pi$ when $\varepsilon(x, y) < 0$. If we apply the simple congruence operation defined by Equation 5.88 to the surface $\varphi(x, y)$, the result will be

$$\tilde{\varphi}(x, y) = \begin{cases} x + 2\pi, & \varepsilon(x, y) > 0, \\ x, & \varepsilon(x, y) < 0. \end{cases} \tag{5.89}$$

Thus the congruent solution $\tilde{\varphi}(x, y)$ is *not* equal to the function $f(x, y) = x$. In fact, it is extremely discontinuous!

The desired solution $\tilde{\varphi}(x, y) = x$ can be obtained by adding the constant $-\pi$ to the surface $\varphi(x, y)$ *before* performing the congruence operation. In other words, the congruence should be defined to be $\tilde{\varphi}_\pi(x, y)$, where the function $\varphi_h(x, y)$ is defined to be $\varphi(x, y) + h$ for any constant h.

This example illustrates a surprising yet important fact: $\tilde{\varphi}_h(x, y)$ is not equal

[15] The results of this section can easily be extended to discrete functions.

to $\tilde{\varphi}(x, y) + h$ for every constant h. In other words, the congruence operation defined in Equation 5.88 is not a linear operator. This example also suggests the correct way to define the congruence operation. The congruence of the function $\varphi(x, y)$ should be defined to be $\tilde{\varphi}_h(x, y)$, where the constant h, $0 < h < 2\pi$, is chosen to minimize the number of discontinuities in the result.

We now shift to discrete functions and present the pseudocode for computing the congruence of the unwrapped surface $\varphi_{i,j}$ that represents the (weighted) least-squares solution of the phase $\psi_{i,j}$. This algorithm approximates the optimal value of the constant h by means of a series of simple tests. The integer N, which controls the accuracy of the approximation, equals 10 in our implementation. The discontinuities that are counted in the loop are defined to be the neighboring grid points whose differences exceed π radians in magnitude. Note that the measure D_k is the L^0 counterpart to the error measure of Equation 4.4 in Chapter 4, which may be regarded as an L^1 measure.

Congruence Operation

```
for (k = 0 to N − 1)
   Let h = 2πk/N.
   Compute the surface function φ_{i,j} + h + W(ψ_{i,j} − φ_{i,j} − h).
   Let D_k = the number of discontinuities in this surface.
end
Define the congruence to be the function computed above for the value k
   that minimizes D_k.
```

We will investigate this operation in more detail in Section 6.2. Our C language implementation can be found in Section A.4 of Appendix A.

Examples We now examine the results of the PCG algorithm on the phase unwrapping examples of Chapter 3. On all of the examples we apply the postprocessing congruence algorithm of Section 5.3, and we utilize binary-valued weights. The zero-weights are defined to be the masks that result from the phase derivative variances (computed in 3×3-pixel windows) with the automatic threshold algorithm of Section 3.4. These masks, which are illustrated in Section 3.4, are the same masks that were used for Flynn's minimum discontinuity algorithm of Section 4.5. The only modification we make is to "fatten" the masks by two pixels (i.e., we change every pixel that borders a mask pixel to a mask pixel and then repeat this operation). This mask thickening operation leads to more accurate results.

The PCG algorithm appears to unwrap the IFSAR data of Figure 3.2 correctly. The differences between this surface and that produced by Goldstein's algorithm (Section 4.2) before the postprocessing congruence operation are shown by the error map of Figure 5.19(a). This figure shows the differences as various levels of gray, where white indicates perfect agreement and black indicates differences of 2π or more. The differences after the congruence

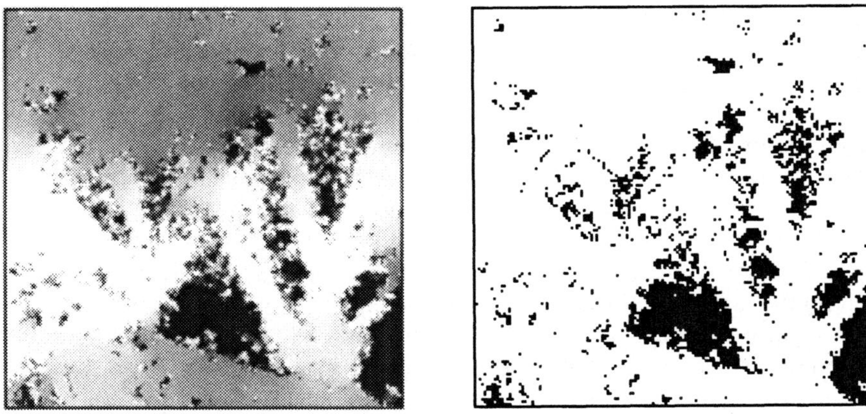

(a) (b)

Figure 5.19 Differences between the surfaces produced by the PCG algorithm after three iterations and Goldstein's algorithm on the IFSAR example (a) *before* and (b) *after* the postprocessing congruence operation. The various gray levels indicate the magnitudes of the differences, with white indicating no difference and black indicating differences of 2π or more.

operation are shown in the binary image of Figure 5.19(b). This image shows all errors in black. As can be seen, the congruence operation reduces the errors dramatically.

Only three iterations of the PCG algorithm were required to produce this result. Notice that the differences are confined to the patches of corrupted phase. Thus, exclusive of these corrupted regions, the PCG algorithm and Goldstein's algorithm produce the same unwrapped surface.

The error maps (i.e., the differences between the surfaces produced by the PCG algorithm and Goldstein's algorithm) after 1, 2, and 3 iterations are shown in Figures 5.20(a), (b), and (c), respectively. The figures, which show the differences before application of the postprocessing congruence operation, illustrate the convergence of the algorithm to the correct solution as the PCG iterations progress.

The PCG algorithm also appears to unwrap the MRI head example of Figure 3.3 correctly. The differences between the surface produced by five iterations of the algorithm and that produced by Goldstein's algorithm before and after the congruence operation are shown in Figures 5.21(a) and (b), respectively. As usual, the regions inside the head differ from those of Goldstein's algorithm by multiples of 2π due to the fact that these regions are completely isolated from the rest of the image by corrupted phase. This example also illustrates the dramatic error reduction achieved by the congruence operation.

On the MRI knee example of Figure 3.4, the PCG algorithm also produces the correct result. The differences between the surface produced by the algorithm

(a) (b) (c)

Figure 5.20 Differences between the surfaces produced by the PCG algorithm and Goldstein's algorithm on the IFSAR example after (a) one, (b) two, and (c) three iterations of the PCG algorithm. These are the differences before application of the congruence operation. The differences are shown as shades of gray with black signifying differences of 2π or more.

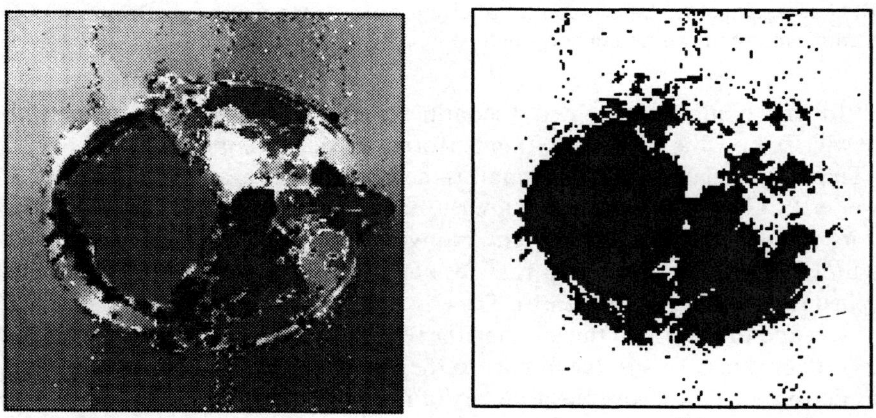

(a) (b)

Figure 5.21 Differences between the PCG algorithm and Goldstein's algorithm on the MRI head example (a) before and (b) after the congruence operation. Five PCG iterations were required for convergence.

and that produced by Goldstein's algorithm are shown in Figures 5.22(a) and (b). As usual, these figures show the errors before and after the congruence operation, and black indicates an error of 2π or more, while white signifies perfect agreement. After the congruence operation the two unwrapped surfaces are exactly the same except for the regions of corrupted phase. Only one PCG iteration was required to produce this result. Notice that the large regions of corrupted phase outside the knee do not corrupt the unwrapping inside the knee.

We now turn our attention to the simulated IFSAR examples. Recall that Goldstein's algorithm had some problems with these examples due to poorly placed branch cuts, while the quality-guided algorithm, the mask cut algorithm

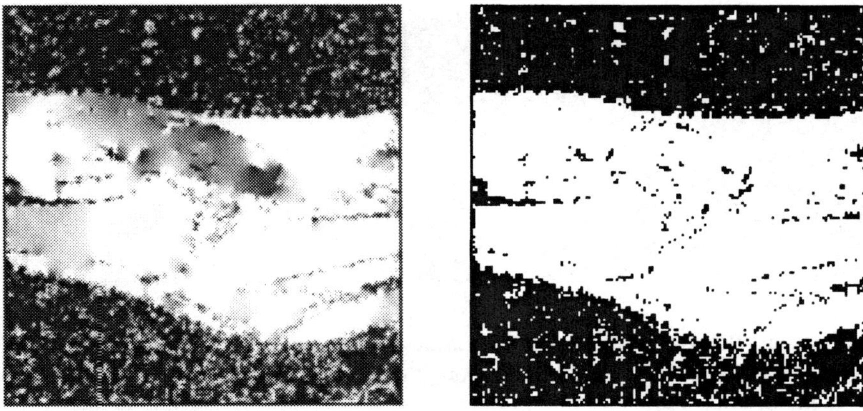

(a) (b)

Figure 5.22 Dfferences between the PCG algorithm and Goldstein's algorithm on the MRI knee example (a) before and (b) after the congruence operation. Only one PCG iteration was required for convergence.

and Flynn's minimum weighted discontinuity algorithm were more successful. We will see how the PCG algorithm performs on this example.

The first simulated IFSAR example is the Long's Peak example of Figure 3.7. Twenty PCG iterations were required to unwrap this example correctly. The errors before application of the congruence operation are shown in Figure 5.23(a). The error map that results after the congruence operation is applied is shown in Figure 5.23(b). The errors are concentrated at the regions of layover, and they are a bit thicker than the regions of error that resulted from the other algorithms. This is due in part to the thickening of the zero-weights that was performed to improve the accuracy of the algorithm. The errors that result after 5, 10, 15, and 20 iterations of the PCG algorithm, and before the congruence operation is applied, are shown in Figures 5.24(a)–(d), respectively. These figures illustrate that the errors diminish as the PCG algorithm iterates closer and closer to the solution.

On the Isolation Peak example of Figure 3.8 the results are similar. Twenty iterations of the PCG algorithm were required. The error maps before and after the congruence operation are shown in Figures 5.25(a) and (b). After the congruence operation, the unwrapping is correct except for two small regions in the upper half of the image. Recall that Flynn's minimum weighted discontinuity algorithm produced an incorrect unwrapping in exactly the same two regions, as did Goldstein's algorithm.

We now turn to the shear examples. The PCG algorithm unwraps the sheared planes example flawlessly. This is due to the fact that the mask derived from the phase derivative variance map defines an unbroken line of zero-weights along the shear line. The results after 10 and 20 iterations are shown in Figure 5.26. A plot of the convergence of the algorithm can be found in Figure 5.38(a). This graph

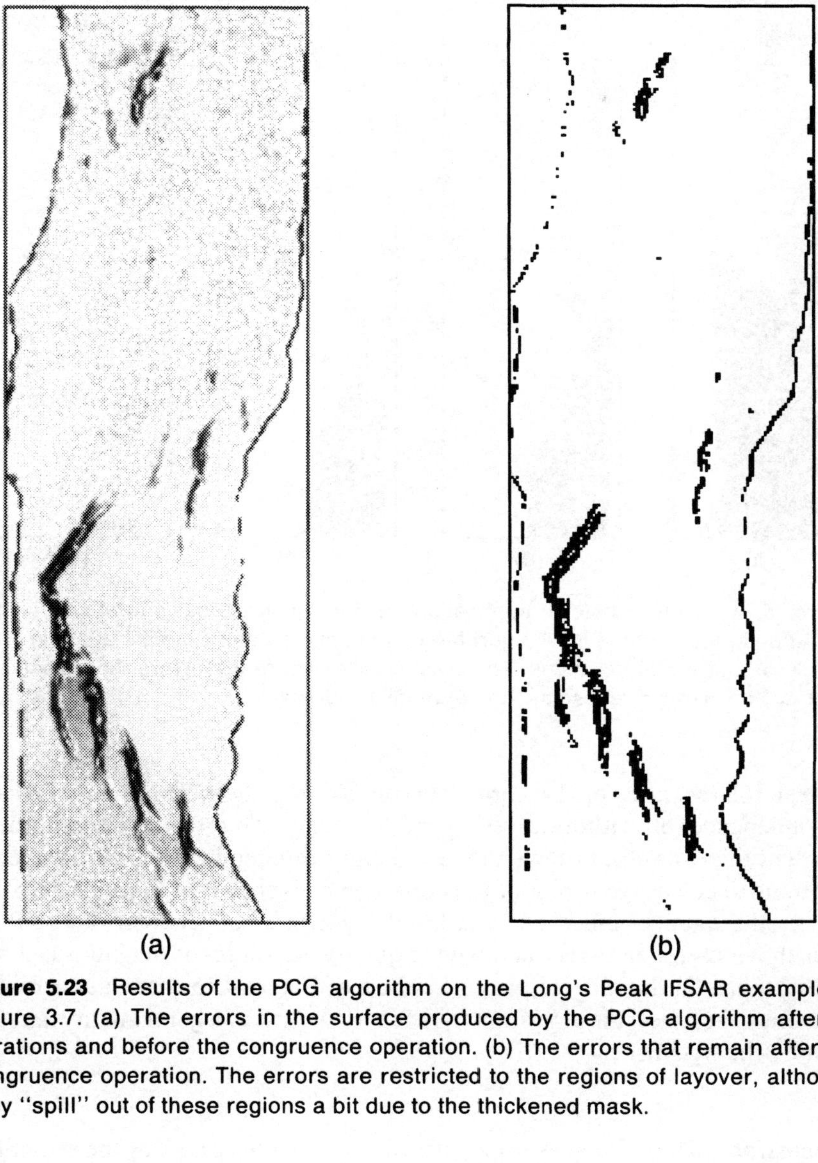

Figure 5.23 Results of the PCG algorithm on the Long's Peak IFSAR example of Figure 3.7. (a) The errors in the surface produced by the PCG algorithm after 20 iterations and before the congruence operation. (b) The errors that remain after the congruence operation. The errors are restricted to the regions of layover, although they "spill" out of these regions a bit due to the thickened mask.

plots the root-mean-square (rms) error of the unwrapped result as a function of the number of iterations.[16]

The algorithm also unwraps the spiral shear example of Figure 3.10 flawlessly, due to the fact that the quality map defines a boundary of low-quality pixels

[16] Because the shear line separates the two planes, this example actually consists of two different surfaces, and the offset between them cannot be derived from the phase data. Therefore we compute the rms error of each plane and average them.

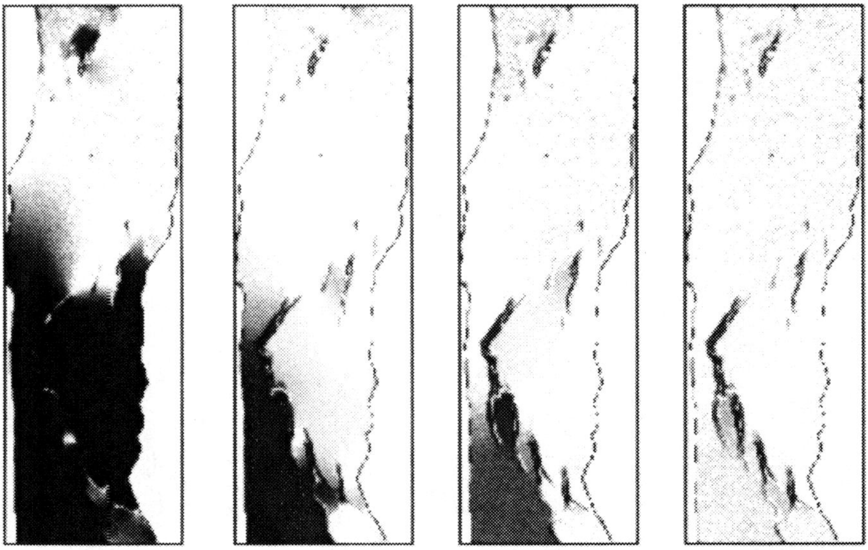

(a) (b) (c) (d)

Figure 5.24 Errors (before application of the congruence operation) in the unwrapped results of the PCG algorithm on the Long's Peak example after (a) 5, (b) 10, (c) 15, and (d) 20 iterations. The black pixels denote the errors that exceed 2π radians. The errors diminish as the iterations progress.

between the two arms of the spiral. This enables the algorithm to unwrap each arm independently without crossing the boundary. Due to the discontinuous nature of this example, however, the PCG algorithm requires a large number of iterations to converge. A plot of the convergence is shown in Figure 5.38(b).

Like the quality-guided and mask cut algorithms of Chapter 4, the PCG algorithm is useless in cases where a good quality map or set of weights is lacking. For example, on the sheared planes example of Figure 4.15 that was corrupted with Gaussian noise, the algorithm fails due to the lack of a good quality map for defining the weights.

Conclusion The PCG algorithm performs successfully on all of the examples, except for the noise-corrupted sheared planes. The algorithm unwraps phase data by means of the weighted least-squares method, which fits the gradients of the unwrapped surface to those of the wrapped phase data. Zero-weights are employed to mask out corrupt values of phase or to allow the unwrapped solution to contain discontinuities along shear lines. The PCG algorithm iterates on unweighted solutions to the least-squares problem. These unweighted solutions define the preconditioners, which enable the PCG algorithm to converge so quickly. A congruence operation is employed as a postprocessing operation to

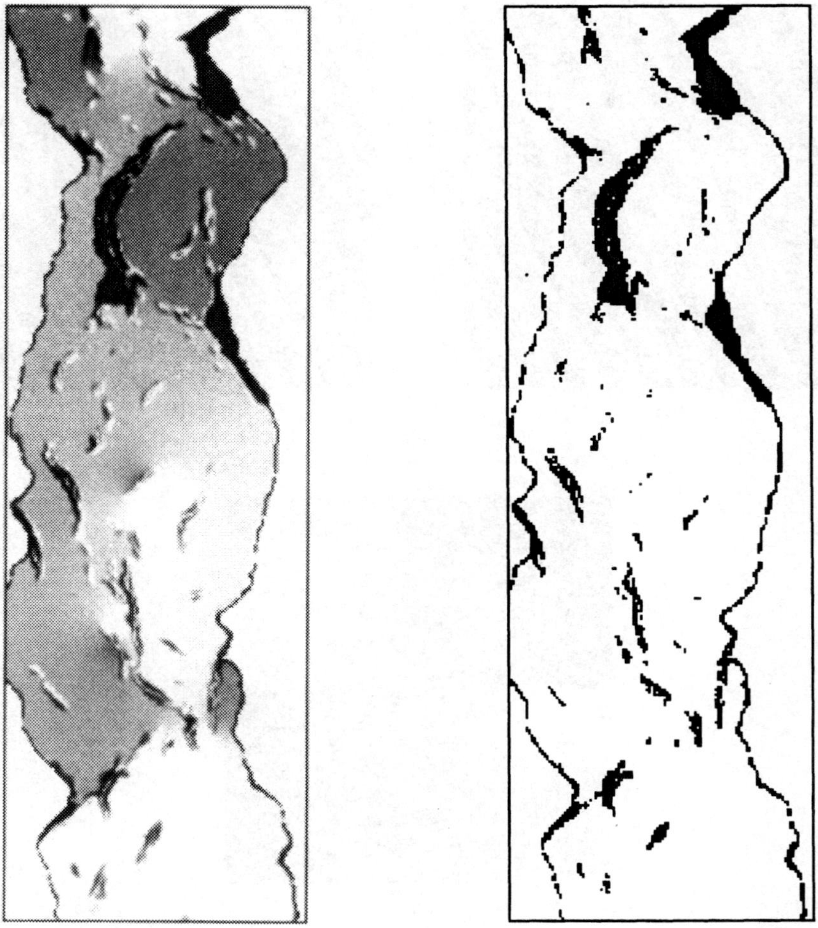

(a) (b)

Figure 5.25 Results of the PCG algorithm on the Isolation Peak example of Figure 3.8. (a) The errors in the surface produced by the PCG algorithm after 20 iterations and before the congruence operation. (b) The errors that remain after application of the congruence operation.

force the unwrapped solution into congruence with the wrapped input phase. This operation also reduces the errors substantially.

The weighted least-squares approach employed by the PCG algorithm differs significantly from the approaches employed by the path-following algorithms of Chapter 4. However, like all of these algorithms, except Goldstein's algorithm and the unweighted version of Flynn's minimum discontinuity algorithm, the PCG algorithm requires a good quality map for defining the weights. The algorithm is roughly comparable to the quality-guided algorithm of Section 4.3 and the mask cut algorithm of Section 4.4 in terms of its data requirements and

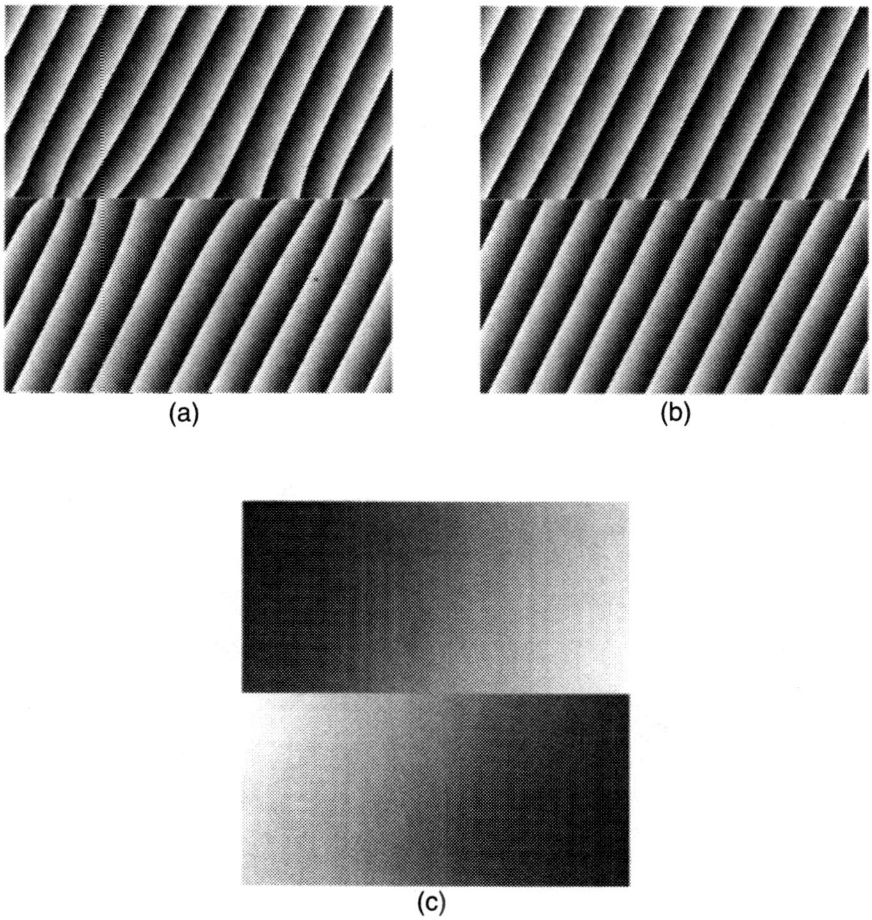

Figure 5.26 Rewrapped results of the PCG algorithm on the sheared planes example of Figure 5.14(a) after (a) 10, (b) 20 iterations, and (c) the unwrapped phase. Note that the top and bottom halves unwrapped independently with no influence across the shear line.

performance. The algorithm works best when the quality map is thresholded to produce zero-weights and when the zero-weights are slightly thickened.

5.4.4 Weighted Multigrid Algorithm

In this section we extend the multigrid algorithm of Section 5.3.3 to the *weighted* least-squares phase unwrapping problem defined by Equation 5.70. Like the unweighted algorithm, the key idea of the weighted algorithm is to store the phase partial derivatives (i.e., the row and column phase differences) in separate arrays and to correct these derivatives at the boundaries of the coarser grids after each fine-to-coarse operation. This procedure preserves the boundary conditions necessary for convergence to the correct solution on the coarser grids. The most challenging problem that must be solved is correctly restricting the weights to the coarser grids. The algorithm converges with a typical root mean square (rms) error of 1 per cent in only one FMG cycle, even on phase data that contain severe discontinuities or shear. The algorithm can be up to 25 times faster than the PCG method. Also, because it does not require the use of fast Fourier or cosine transforms, it places absolutely no restrictions on the grid sizes.

After presenting the algorithm, including our method for the weight restriction operation, we present the pseudocode and the implementation details. We then examine the results of the algorithm on a number of phase unwrapping examples and compare these with those of the PCG algorithm.

Algorithm Like the unweighted algorithm, the weighted multigrid algorithm stores the partial derivatives $\Delta^x_{i,j}$ and $\Delta^y_{i,j}$ of the input phase data in separate arrays and corrects these derivatives at the grid boundaries after each restriction operation. The algorithm requires two additional arrays for storing the weights. One array stores the weights σ to be applied to the row derivatives, and the other stores the weights τ to be applied to the column derivatives.

The algorithm is nearly identical to the unweighted algorithm presented in Section 5.3.3 except for three main differences:

1. The Gauss-Seidel relaxation must be performed on the weighted equation (Equation 5.72) instead of the unweighted equation (Equation 5.38).
2. The restriction of the partial derivatives Δ^x and Δ^y to the coarse grids must be weighted.
3. The weights must be restricted to the coarser grids by means of a carefully defined restriction operator.

We call the weighting arrays σ and τ the "d/dx weights" and the "d/dy weights," respectively. The initial values of these weights are defined to be the input weights. Their values on the coarser grids differ due to the definition of the weight restriction operator, which is denoted $\mathcal{R}$ and defined later in this section. The weighting coefficients in Equation 5.72 are defined as follows:

$$U(i,j) = \min(\sigma^2_{i+1,j}, \sigma^2_{i,j}), \quad V(i,j) = \min(\tau^2_{i,j+1}, \tau^2_{i,j}), \tag{5.90}$$

where σ and τ are the d/dx weights and the d/dy weights, respectively. The restriction operator (denoted R_w) for restricting the residuals of the derivatives to the coarser grids must weight the fine-grid values before applying the full weighting operator (and dividing by an appropriate term to normalize the weighting values) to produce the new coarse-grid values. For restricting the d/dx residual derivatives, R_w resamples every other point after smoothing the fine-grid values by the weighted version of the filter template

$$\begin{bmatrix} \frac{1}{16}\sigma_{i-1,j-1} & \frac{1}{8}\sigma_{i,j-1} & \frac{1}{16}\sigma_{i+1,j-1} \\ \frac{1}{8}\sigma_{i-1,j} & \frac{1}{4}\sigma_{i,j} & \frac{1}{8}\sigma_{i+1,j} \\ \frac{1}{16}\sigma_{i-1,j+1} & \frac{1}{8}\sigma_{i,j+1} & \frac{1}{16}\sigma_{i+1,j+1} \end{bmatrix} \Big/ S_{i,j}, \tag{5.91}$$

where the normalization term $S_{i,j}$ is the sum of the matrix entries. (If the weights are all zero, the unweighted filter template, described by Equation 5.63, is used instead.) The d/dy residual derivatives are restricted to the coarser grids by means of the same filter template but with the d/dy weights τ in place of the d/dx weights σ.

We now describe the weight restriction operator [11], which restricts the weights from the fine to the coarse grids. This operator is denoted by $\mathcal{R}$ and applied in Step 4 of the WFMG algorithm and Step 5 of the WMV algorithm. Because the weights never change after restriction, these restrictions need only be performed once for each grid size. One possible definition of this operator is the full weighting operator defined by Equation 5.62. However, the full weighting operator has the undesirable property of not always restricting zero-weights to zero-weights. For example, due to the smoothing effect of this operator, a single column of zero-weights would be restricted to nonzero values. This would be undesirable if the zero-weights were meant to separate two independent regions of phase (as in the sheared planes of Figure 3.9). Thus one desirable property of the weight restriction operator is that it be "line preserving."

A second desirable property of the weight restriction operator is that it be "area preserving." It should preserve the relative areas of regions of zero-weights on the coarser grids. In particular, this property ensures that scattered zero weights, which may be caused by noise, do not propagate to zero-weights that dominate the coarser grids.

These two properties of line- and area-preservation are satisfied by the following definition of the weight restriction operator. The operator is defined in two parts, one for restricting the d/dx weights σ and the other for restricting the d/dy weights τ. The definition of the d/dx part of the operator is illustrated in Figure 5.27.

1. Weight Restriction Operator: d/dx Weights

Define the weight $\sigma_{i,j}$ of the coarse grid to be 0 if at least one of the following six conditions hold. Otherwise, define $\sigma_{i,j}$ by the full weighting operator.

1. The two fine-grid weights $\sigma_{2i,2j}$ and $\sigma_{2i+1,2j+1}$ are 0.
2. The two fine-grid weights $\sigma_{2i+1,2j}$ and $\sigma_{2i,2j+1}$ are 0.
3. The two fine-grid weights $\sigma_{2i+1,2j}$ and $\sigma_{2i+1,2j+1}$ are 0.
4. The two fine-grid weights $\sigma_{2i,2j}$ and $\sigma_{2i,2j+1}$ are 0.

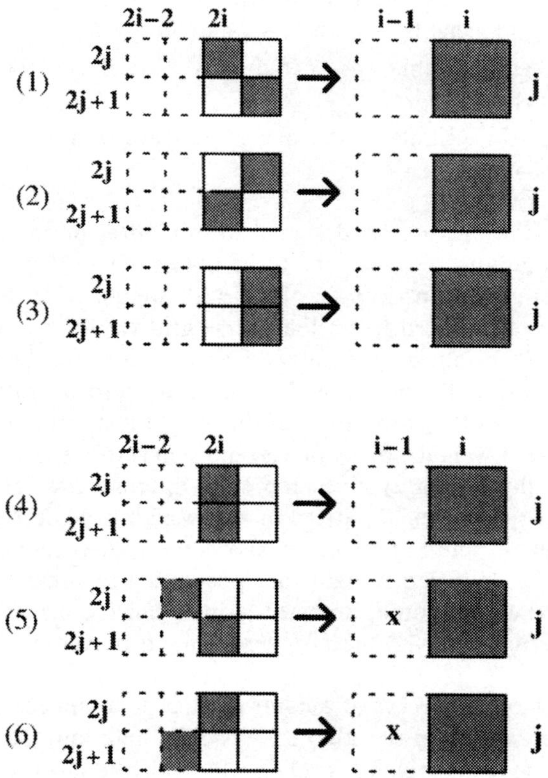

Figure 5.27 Six conditions under which the d/dx weight restriction operator defines a zero-weight. (The d/dy weight restriction operator is defined in an analogous fashion.) The grids on the left of the arrows are the fine grids, while those on the right are the coarse grids. The shaded squares represent the zero-weights, and the two squares marked x must be nonzero-weights.

5. The two fine-grid weights $\sigma_{2i-1,2j}$ and $\sigma_{2i,2j+1}$ are 0 and the coarse-grid weight $\sigma_{i-1,j}$ is nonzero.
6. The two fine-grid weights $\sigma_{2i,2j}$ and $\sigma_{2i-1,2j+1}$ are 0 and the coarse-grid weight $\sigma_{i-1,j}$ is nonzero.

2. Weight Restriction Operator: d/dy Weights

Define the weight $\tau_{i,j}$ of the coarse grid to be 0 if at least one of the following six conditions hold. Otherwise, define $\tau_{i,j}$ by the full weighting operator.

1. The two fine-grid weights $\tau_{2i,2j}$ and $\tau_{2i+1,2j+1}$ are 0.
2. The two fine-grid weights $\tau_{2i,2j+1}$ and $\tau_{2i+1,2j}$ are 0.

3. The two fine-grid weights $\tau_{2i,2j+1}$ and $\tau_{2i+1,2j+1}$ are 0.
4. The two fine-grid weights $\tau_{2i,2j}$ and $\tau_{2i+1,2j}$ are 0.
5. The two fine-grid weights $\tau_{2i,2j-1}$ and $\tau_{2i+1,2j}$ are 0 and the coarse-grid weight $\tau_{i,j-1}$ is nonzero.
6. The two fine-grid weights $\tau_{2i,2j}$ and $\tau_{2i+1,2j-1}$ are 0 and the coarse-grid weight $\tau_{i,j-1}$ is nonzero.

This weight restriction operator is designed to minimize the number of zero-weights while maintaining zero-weights in the places where they are required. For example, under conditions 5 and 6 of the d/dx weight restriction operator defined above, there is no need to set the coarse-grid weight $\sigma_{i,j}$ to zero if the preceding coarse-grid weight $\sigma_{i-1,j}$ is already zero. As a result the unwrapping process is not impeded by the presence of unnecessary zero weights.

Figure 5.28 illustrates the application of the weight restriction operator to a circular pattern of zero-weights. Black pixels represent zero-weights, white pixels represent unity weights, and gray pixels represent intermediate-valued weights. Figure 5.28(a) shows the original pattern of zero-weights, while Figure 5.28(b) shows the d/dx weights, and Figure 5.28(c) shows the d/dy weights on a coarser grid. Notice that the d/dx zero-weights are positioned to prevent the flow of d/dx information (i.e., horizontal derivatives) between the inside and outside regions of the circle, and similarly the d/dy zero-weights prevent the flow of d/dy information.

The maintenance of two arrays of weights may seem an unnecessary complication. In an earlier version of our algorithm, we did maintain the weights in a single array with the same weights applying to both the d/dx and the d/dy derivatives [11], but the result was an inferior algorithm in terms of convergence rate and reliability. It appears that keeping the d/dx and the d/dy weights separate is necessary to achieve a successful algorithm.

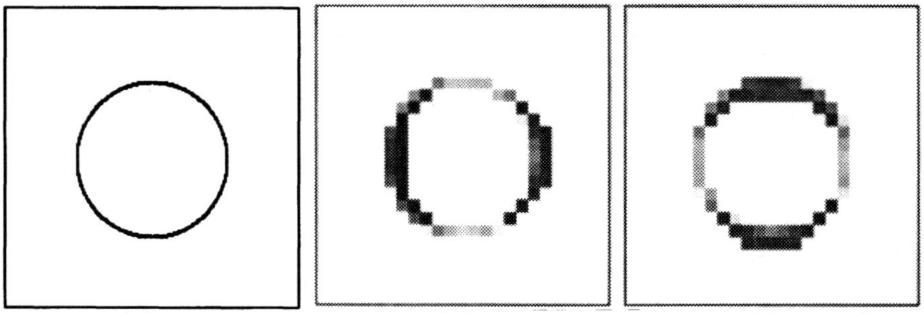

(a) (b) (c)

Figure 5.28 Weight restriction operator applied to (a) a circular pattern of zero-weights. (b) The d/dx, and (c) the d/dy weights as defined by the weight restriction operator on a coarser grid. (Black pixels denote zero-weights, while gray pixels denote weights between 0 and 1.)

Pseudocode The pseudocode for the weighted multigrid algorithm, which is denoted WFMG (weighted full multigrid), is presented below. For n FMG cycles, the algorithm should be called n times. In practice, only two FMG cycles are necessary. The array ϕ denotes the solution array, the arrays Δ^x and Δ^y denote the partial derivatives, and the arrays σ and τ denote the weights. Because these arrays are actually pyramids of arrays, the subscript N denotes the dimensions $N \times N$ of the current grid. (Although the pseudocode is expressed for square arrays to simplify the notation, it works for rectangular grids as well.) The restriction operator is denoted R_w, the weight restriction operator is denoted $\mathcal{R}$, and the prolongation operator (defined to be bilinear interpolation, as in the unweighted multigrid algorithm) is denoted P. The partial differentiation operators D^x and D^y are defined by Equation 5.67. (As in the unweighted multigrid algorithm, the restriction operation R_w multiplies the coarse-grid derivatives by two to preserve the proper scale.)

FMG (Multigrid) Algorithm for Weighted Phase Unwrapping

$\phi_N \leftarrow \text{WFMG}(\phi_N, \Delta_N^x, \Delta_N^y, \sigma_N, \tau_N)$

Step 1. If $N \times N$ is the coarsest grid, then go to Step 8.

Step 2. $\Delta_{N/2}^x \leftarrow R_w(\Delta_N^x - D^x\phi_N)$, $\Delta_{N/2}^y \leftarrow R_w(\Delta_N^y - D^y\phi_N)$.

Step 3. Boundary derivative correction step: set the first column of the array $\Delta_{N/2}^x$ equal to the negative of its second column, and set the first row of the array $\Delta_{N/2}^y$ equal to the negative of its second row. (See Equation 5.66.)

Step 4. $\sigma_{N/2} \leftarrow \mathcal{R}(\sigma_N)$, $\tau_{N/2} \leftarrow \mathcal{R}(\tau_N)$ (weight restriction operations).

Step 5. $\phi_{N/2} \leftarrow 0$.

Step 6. $\sigma_{N/2} \leftarrow \text{WFMG}(\phi_{N/2}, \Delta_{N/2}^x, \Delta_{N/2}^y, \sigma_{N/2}, \tau_{N/2})$.

Step 7. $\phi_N \leftarrow \phi_N + P(\phi_{N/2})$

Step 8. $\phi_N \leftarrow \text{WMV}(\phi_N, \Delta_N^x, \Delta_N^y, \sigma_N, \tau_N)$.

The V-cycle algorithm is called in Step 8. It is denoted WMV (weighted multigrid V-cycle) and is defined as follows:

Multigrid V-Cycle Algorithm for Weighted Phase Unwrapping

$\phi_N \leftarrow \text{WMV}(\phi_N, \Delta_N^x, \Delta_N^y, \sigma_N, \tau_N)$

Step 1. Relax 2 times on Equation 5.72 with the given initial guess ϕ_N.

Step 2. If $N \times N$ is the coarsest grid, then go to Step 9.

Step 3. $\Delta_{N/2}^x \leftarrow R_w(\Delta_N^x - D^x\phi_N)$, $\Delta_{N/2}^y \leftarrow R_w(\Delta_N^y - D^y\phi_N)$.

Step 4. Boundary derivative correction step: Set the first column of the array $\Delta_{N/2}^x$ equal to the negative of its second column, and set the first row of the array $\Delta_{N/2}^y$ equal to the negative of its second row.

Step 5. $\sigma_{N/2} \leftarrow \mathcal{R}(\sigma_N)$, $\tau_{N/2} \leftarrow \mathcal{R}(\tau_N)$ (weight restriction operations)

Step 6. $\phi_{N/2} \leftarrow 0$

Step 7. $\sigma_{N/2} \leftarrow \mathrm{WMV}(\phi_{N/2}, \Delta^x_{N/2}, \Delta^y_{N/2}, \sigma_{N/2}, \tau_{N/2})$

Step 8. $\phi_N \leftarrow \phi_N + P(\phi_{N/2})$

Step 9. Relax 2 times on Equation 5.72 with the new initial guess ϕ_N.

Implementation Our C language implementation of the weighted multigrid algorithm can be found in Section A.12 of Appendix A. It gives the user the following choice of quality maps, which define the weights:

1. User-supplied correlation coefficients (or other quality map)
2. Pseudocorrelation
3. Variance of phase derivatives
4. Maximum phase gradient

The correlation coefficients can be averaged locally, and the other three quality maps can be computed in windows of any size. The user can also specify that a threshold be calculated and applied to the quality map to produce a mask that defines zero-weights. The algorithm used to compute the threshold is described in Section 3.4. We recommend that the user always use the threshold option or input a quality mask because in practice the algorithm (like the PCG algorithm) seems to converge faster with binary-valued weights.

The algorithm requires a total of five arrays, where each array is the same size as the input phase data array:

1. ϕ-solution array
2. Δ^x-row derivatives
3. Δ^y-column derivatives
4. σ-array of weights for the row derivatives
5. τ-array of weights for the column derivatives

Each of these arrays has its own pyramid of decreasing resolution grids as shown in Figure 5.8. The memory space required by the algorithm is $5 \times (1 + \frac{1}{4} + \frac{1}{16} + \cdots) \approx 7$ times the space occupied by the input phase data array. By comparison, the PCG algorithm requires 5 times the space of the input array.

Table 5.4 lists the execution times of our implementation for various array sizes. The times depend not only on the array sizes but also on the number of FMG cycles (iterations) and the number of Gauss-Seidel relaxation sweeps performed on each grid in each FMG cycle. The times are given for two FMG cycles and two relaxation sweeps because these suffice for most phase unwrapping problems. Notice that the times are much faster than those of the PCG algorithm, and that the algorithm is not much slower than the unweighted least-squares algorithms (see Table 5.1).

Table 5.4 Execution times of the weighted multigrid algorithm for various array sizes on an IBM RS/6000 workstation.

Array Size (pixels)	Execution Time (min:s)
256 × 256	00:07
512 × 512	00:25
1024 × 1024	01:35
2048 × 2048	06:20

Note: The times are given for two FMG cycles and two relaxation sweeps per cycle of the algorithm.

Examples We now examine the results of the weighted multigrid algorithm on the phase unwrapping examples of Chapter 3. On all of the examples the number of FMG cycles is two, and the number of Gauss-Seidel relaxation sweeps is also two. As with the PCG algorithm, the weights are binary-valued, and the postprocessing congruence algorithm of Section 5.3 is applied to the resulting surface. The zero-weights are defined by the masks that result from the phase derivative variances (computed in 3 × 3-pixel windows) with the automatic threshold algorithm of Section 3.4. Unlike the PCG algorithm, the multigrid algorithm does not require the zero-weights to be thickened.

The multigrid algorithm unwraps the IFSAR data of Figure 3.2 correctly. The differences between the surface produced by the algorithm and that produced by Goldstein's algorithm (Section 4.2) before the postprocessing congruence operation are shown by the error map of Figure 5.29(a). This figure shows the differences as various levels of gray, where white indicates perfect agreement and black indicates differences of 2π or more. The differences after the congruence operation are shown in the binary image of Figure 5.29(b). This image shows all errors in black. As can be seen, the congruence operation reduces the errors dramatically. Notice that the differences are confined to the patches of corrupted phase. Thus, exclusive of these corrupted regions, the multigrid algorithm and Goldstein's algorithm produce the same unwrapped surface.

Figures 5.30(a), (b), and (c) show the weights on the coarse grids. In these figures the row and column weights have been averaged. The black pixels represent the zero-weights, the white pixels represent the unity weights, and the gray pixels indicate weights with values between 0 and 1. These intermediate-valued weights result from the averaging effect of the weight restriction operator. Notice that the patches of noise are covered with zero weights (i.e., black pixels) that are easily discernible even on the 16 × 16-pixel grid. The larger patches are discernible even on the 8 × 8-pixel grid, although the smaller patches have disappeared.

We now examine the results of the algorithm on the MRI head example of

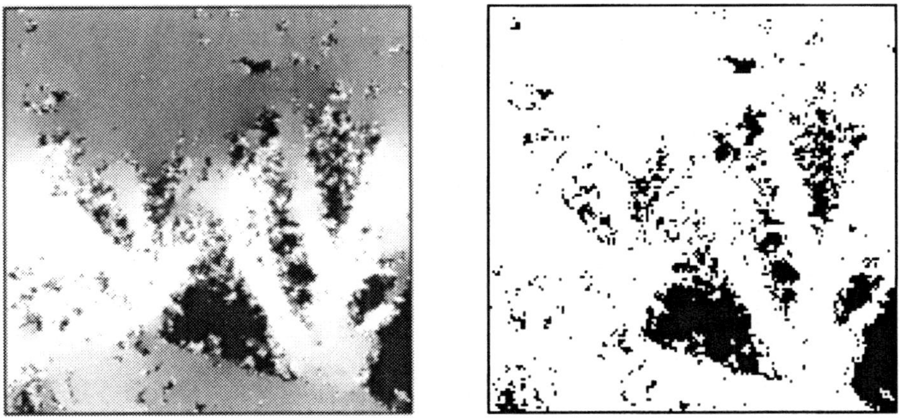

(a) (b)

Figure 5.29 Differences between the surfaces produced by the weighted multigrid algorithm and Goldstein's algorithm on the IFSAR example (a) before and (b) after the postprocessing congruence operation. The various gray levels indicate the magnitudes of the differences with white indicating no difference and black indicating differences of 2π radians or more.

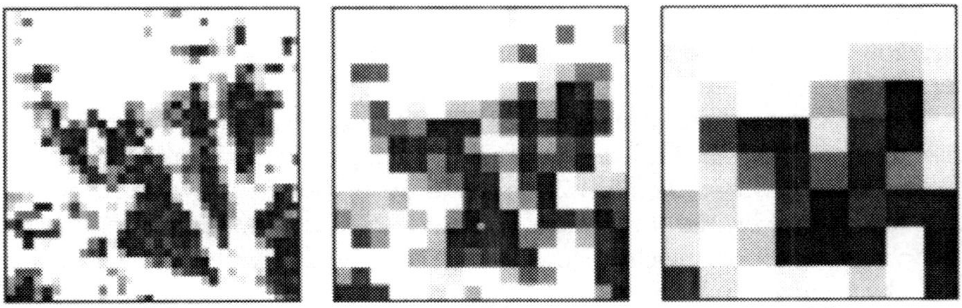

(a) (b) (c)

Figure 5.30 Weights for the IFSAR example on the coarse grids of dimensions (a) 32×32, (b) 16×16, and (c) 8×8 pixels. The black pixels denote the zero-weights. For simplicity the row and column weights have been averaged in these images.

Figure 3.3. The differences between the surfaces produced by the algorithm and Goldstein's algorithm before and after the congruence operation are shown in Figures 5.31(a) and (b), respectively. As usual, some of the regions inside the head differ from those of Goldstein's algorithm by 2π-multiples because these regions are completely isolated from the rest of the image by corrupted phase. Figures 5.32(a), (b), and (c) show the weights on the coarse grids.

We now examine the MRI knee example of Figure 3.4. The differences between the surface produced by the algorithm and that produced by Goldstein's algorithm are shown in Figures 5.33(a) and (b). As usual, these figures show the

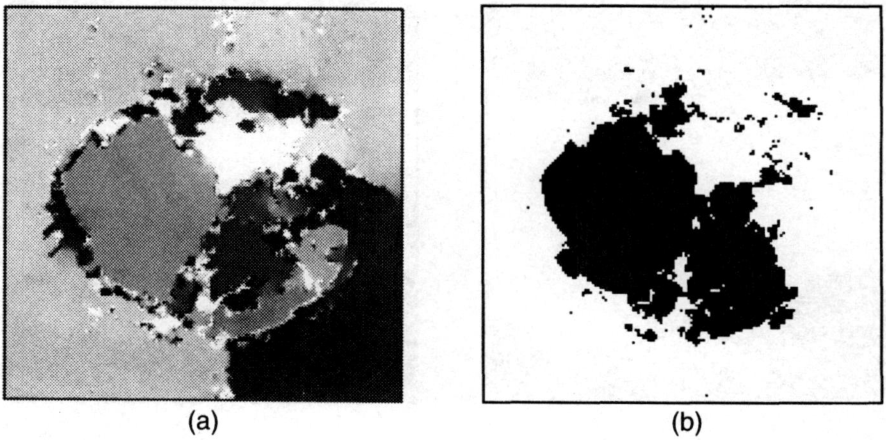

Figure 5.31 Differences between the weighted multigrid algorithm and Goldstein's algorithm on the MRI head example (a) before and (b) after the congruence operation.

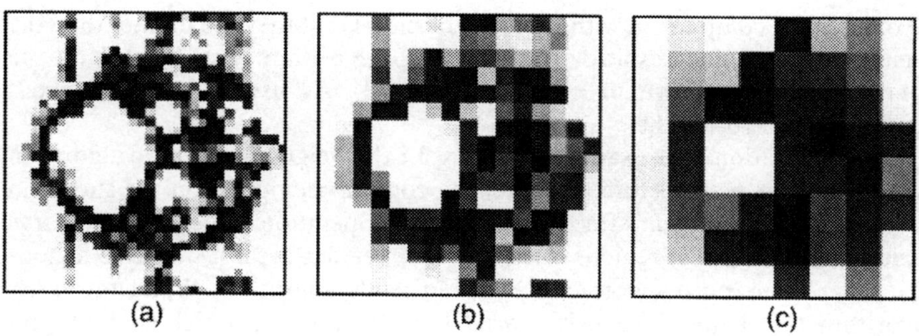

Figure 5.32 Weights for the MRI head example on the coarse grids of dimensions (a) 32×32, (b) 16×16, and (c) 8×8 pixels. Notice that the lower-right quadrant of the image is completely isolated by the zero-weights on the 8×8-pixel grid and nearly isolated on the other grids.

errors before and after the congruence operation, and black indicates an error of 2π or more, while white signifies perfect agreement. After the congruence operation, the two unwrapped surfaces are exactly the same except for the regions of corrupted phase and a few very small regions within the knee. Notice that the zero-weights are effective at preventing the large regions of corrupted phase outside the knee from corrupting the unwrapping inside the knee.

We now turn to the simulated IFSAR examples. The weighted multigrid algorithm unwrapped the Long's Peak example of Figure 3.7 correctly. The errors before application of the congruence operation are shown in Figure 5.34(a). The error map that results after the congruence operation is applied is shown in Figure 5.34(b). The errors are confined to the regions of

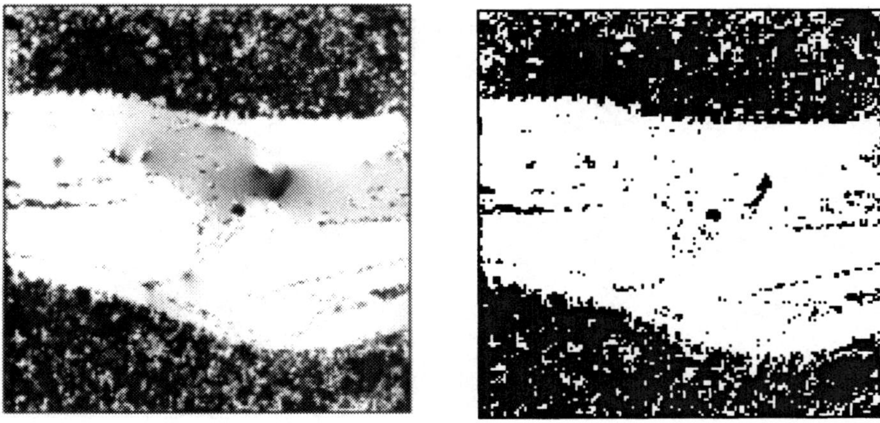

(a) (b)

Figure 5.33 Differences between the weighted multigrid algorithm and Goldstein's algorithm on the MRI knee example (a) before and (b) after the congruence operation.

layover, and a comparison with Figure 5.23 shows that they are thinner than the regions of error that resulted from the PCG algorithm. The weights on the coarser grids are shown in Figures 5.35(a)–(d). As usual, the black pixels represent the zero-weights.

On the Isolation Peak example of Figure 3.8 the weighted multigrid algorithm fails. The error maps before and after the congruence operation are shown in Figures 5.36(a) and (b). After the congruence operation there is a very large region and two smaller regions of error. These regions are not the same regions that were unwrapped incorrectly by Goldstein's algorithm. While the latter algorithm failed due to poorly placed branch cuts, the weighted multigrid algorithm fails because the zero-weights isolate portions of the image on the coarser grids. The weights on the coarser grids are shown in Figure 5.37(a)–(d). It can be seen that the large region of error is almost completely isolated from the rest of the image by the zero-weights on the 19×57-pixel and coarser grids.

We now turn to the shear examples. The weighted multigrid algorithm unwraps the sheared planes example flawlessly and much faster than the PCG algorithm. This is due to the fact that the line of zero-weights along the shear line is maintained on the coarser grids. A plot of the convergence of the algorithm is shown in Figure 5.38(a) and compared with that of the PCG algorithm. This graph plots the rms error of the unwrapped result as a function of the number of iterations.[17] Although three iterations of the PCG algorithm can be performed in the time it takes for one FMG cycle of the multigrid algorithm, the multigrid algorithm converges much faster.

[17] Because the shear line separates the two planes, this example actually consists of two different surfaces, and the offset between them cannot be derived from the phase data. Therefore we compute the rms error of each plane and average them.

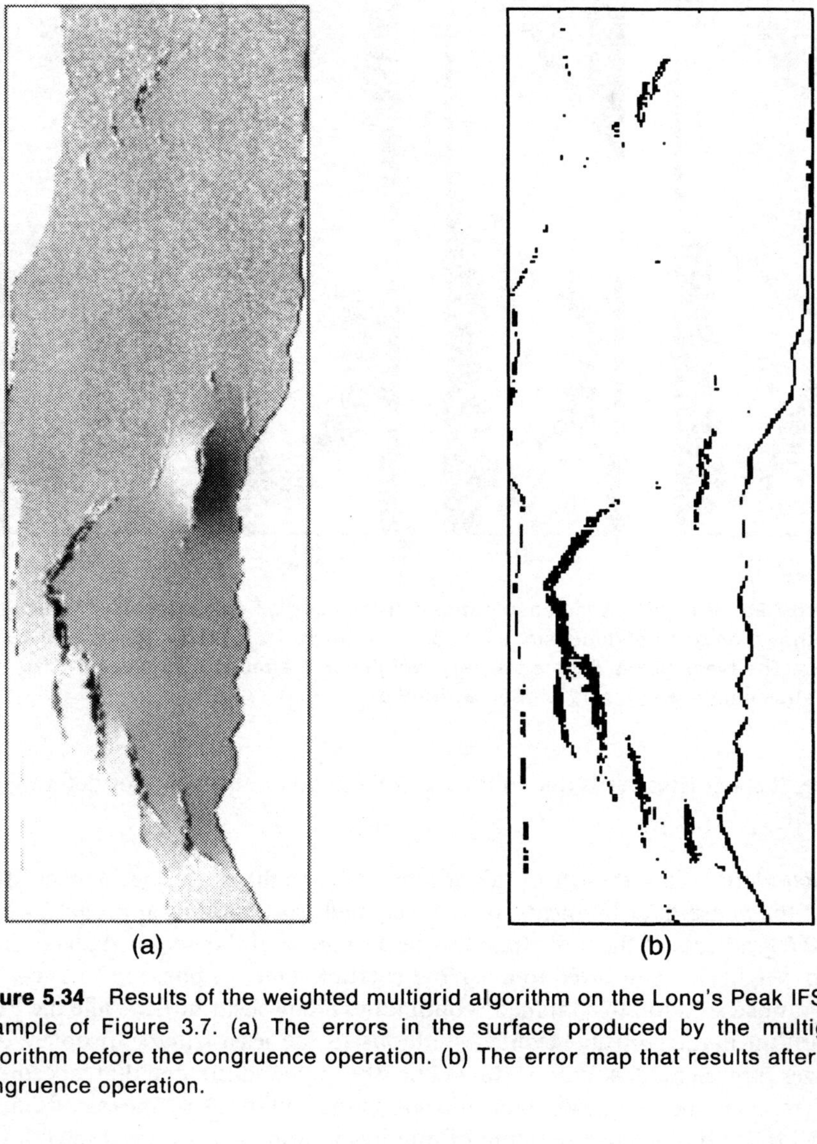

(a) (b)

Figure 5.34 Results of the weighted multigrid algorithm on the Long's Peak IFSAR example of Figure 3.7. (a) The errors in the surface produced by the multigrid algorithm before the congruence operation. (b) The error map that results after the congruence operation.

The weighted multigrid algorithm also unwraps the spiral shear example of Figure 3.10 correctly. However, like the PCG algorithm, the multigrid algorithm requires many iterations to converge. A plot of the convergence is shown in Figure 5.38(b).

Like the quality-guided and mask cut algorithms of Chapter 4 and the PCG algorithm of Section 5.4.3, the weighted multigrid algorithm is useless in cases where a good quality map or set of weights is lacking. For example, on the sheared planes example of Figure 4.15, which was corrupted with Gaussian

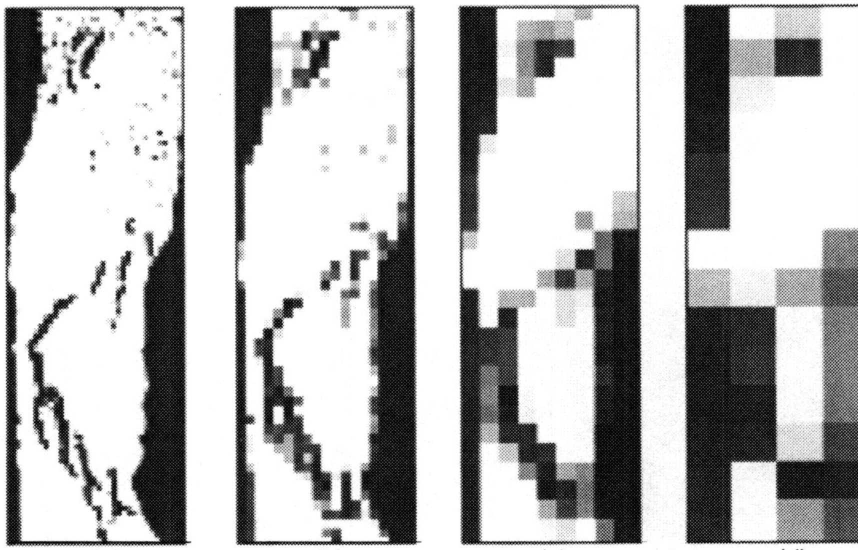

(a) (b) (c) (d)

Figure 5.35 Weights on the coarse grids for the Long's Peak example. The weights are shown on grids of dimensions (a) 38 × 114, (b) 19 × 57, (c) 9 × 28, and (d) 4 × 14 pixels. The black pixels denote the zero-weights. For simplicity in these figures, the row and column weights have been averaged.

noise, the algorithm fails due to the lack of a good quality map for defining the weights.

Conclusion Like the PCG algorithm the weighted multigrid algorithm unwraps phase data by means of the weighted least-squares approach, which fits the gradients of the unwrapped surface to those of the wrapped phase data. Zero-weights are employed to mask out corrupt values of phase or to allow the unwrapped solution to contain discontinuities along shear lines. While the PCG algorithm iterates on unweighted solutions to the least-squares problem and utilizes unweighted solutions to define the preconditioners that speed the convergence, the multigrid algorithm solves the equations by means of Gauss-Seidel relaxation using a pyramid of multiresolution grids to speed the convergence. As in the case of the PCG algorithm, a congruence operation is employed as a postprocessing operation to force the unwrapped solution into congruence with the wrapped input phase.

The multigrid algorithm failed on the Isolation Peak IFSAR example. This failure is not due to the weighted least-squares approach, since the PCG algorithm was able to unwrap the example. The problem was caused by the failure of the multigrid algorithm to solve the weighted least-squares equation properly, since the zero-weights isolated portions of the image from the rest of

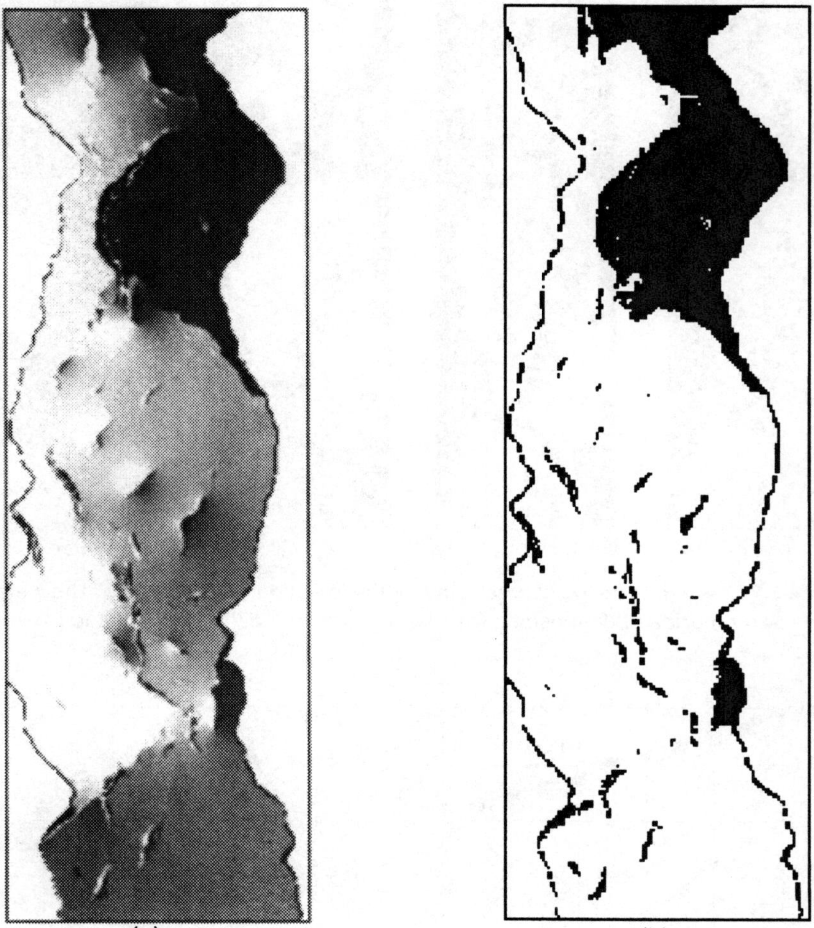

(a) (b)

Figure 5.36 Results of the weighted multigrid algorithm on the Isolation Peak example of Figure 3.8. (a) The errors in the surface produced by the algorithm before the congruence operation. (b) The error map that results after the congruence operation.

the image as the weights were restricted to the coarser grids. This isolating effect, although it does not occur on every data set, appears to be the "Achilles heel" of the algorithm.

Due to this weakness of the weighted multigrid algorithm, we recommend that the PCG algorithm be tried initially when one desires a weighted least-squares solution to the phase unwrapping problem at hand. If the PCG algorithm fails or is slow to converge, then the multigrid algorithm may be tried.

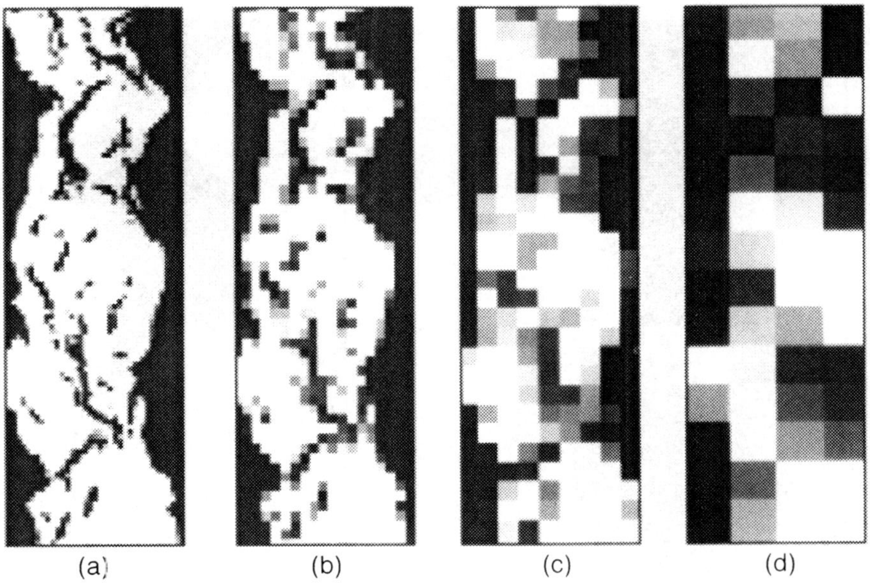

Figure 5.37 Weights on the coarse grids for the Isolation Peak example. The weights are shown on grids of dimensions (a) 39 × 114, (b) 19 × 57, (c) 9 × 28, and (d) 4 × 14 pixels.

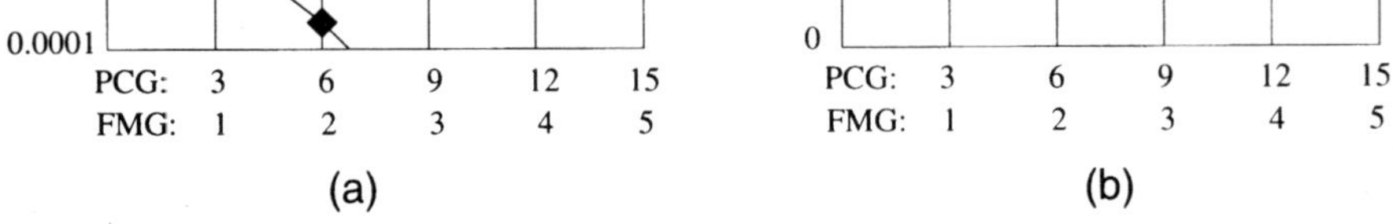

Figure 5.38 Convergence of the weighted multigrid algorithm, which is labeled "FMG," and the PCG algorithm on the (a) sheared planes and (b) spiral shear phase unwrapping problems. Each square represents three iterations of the PCG algorithm, while each triangle represents one FMG cycle of the multigrid algorithm. The rms errors of the unwrapped surfaces are plotted as functions of the number of iterations. The rms errors are normalized so that a flat plane has an rms error of 1.0.

5.5 MINIMUM L^p-NORM PHASE UNWRAPPING

We have shown that the minimum L^p-norm solution entails the solution of a nonlinear PDE. (See Equations 5.11 and 5.28.) The PDE is nonlinear because the weights U and V are functions of the input data and of the solution. We have hinted that this problem can be solved iteratively in the general case. Specifically, $U_0(x, y)$ and $V_0(x, y)$ can be computed from the original data ψ and an initial guess for ϕ_0. These initial weights are held fixed in Equation 5.11, which is then solved by the PCG or weighted multigrid algorithm. The current solution ϕ_1 and the original data ψ are then used to compute a new set of weights, $U_1(x, y)$ and $V_1(x, y)$, from which the next estimate ϕ_2 is obtained. This process is repeated until convergence is achieved. Although the previous discussion has provided the general notion for a workable algorithm (also known as iterative reweighted least squares, or IRLS, in the literature [28]), many subtleties will need to be addressed.

After describing the algorithm, we present the pseudocode and the implementation details. We then present the results on a number of phase unwrapping examples. Departing from our usual order, we examine the performance of the algorithm on several examples of sheared planes before considering the IFSAR and MRI examples. The sheared planes examples provide valuable insight into the behavior of the algorithm. We then conclude by noting the close similarities with Flynn's minimum discontinuity algorithm of Section 4.5.

5.5.1 Algorithm

We first seek to make the structure of the algorithm explicit. We will then address a number of practical issues to complete the definition of the algorithm. Let us begin by rearranging Equation 5.28 in the desired form as follows:

$$\begin{aligned}(\phi_{i+1,j} - \phi_{i,j})\, U(i, j)& \\ + (\phi_{i,j+1} - \phi_{i,j})\, V(i, j)& \\ - (\phi_{i,j} - \phi_{i-1,j})\, U(i-1, j)& \\ - (\phi_{i,j} - \phi_{i,j-1})\, V(i, j-1) &= c(i, j),\end{aligned} \tag{5.92}$$

where $c(i, j)$ is defined to be the "weighted phase Laplacian"

$$\Delta^x_{i,j} U(i, j) - \Delta^x_{i-1,j} U(i-1, j) + \Delta^y_{i,j} V(i, j) - \Delta^y_{i,j-1} V(i, j-1), \tag{5.93}$$

and the weights $U(i, j)$ and $V(i, j)$ are defined by Equations 5.29 and 5.30.

We represent Equation 5.92 in matrix form as

$$\mathbf{Q}\phi = \mathbf{c}, \tag{5.94}$$

which is the same form as the weighted least-squares problem described by Equation 5.81. Notice that the matrix $\mathbf{Q}$ depends on the weights.

The simplified algorithm for solving the minimum L^p-norm two-dimensional phase unwrapping problem is defined as follows: After initializing the solution array $\phi_{i,j}$ to zero, the weights $U(i, j)$ and $V(i, j)$ are defined by Equations 5.29 and 5.30. The weighted phase Laplacian $c(i, j)$ is computed according to Equation 5.93, and the weighted Laplacian of the current solution $\phi_{i,j}$, which is defined by the left-hand side of Equation 5.92, is computed and subtracted from $c(i, j)$. Equation 5.92, which can be regarded in matrix form as the equation $\mathbf{Q}\phi = \mathbf{c}$, is then solved by means of the PCG or the weighted multigrid algorithm with the current solution $\phi_{i,j}$ taken to be the initial guess. The result becomes the new current solution $\phi_{i,j}$, and the process is repeated iteratively.

Although this algorithm highlights the correct iterative structure, it does not work well in practice for several reasons. We will address several modifications that will produce a working algorithm, but first we need to discuss the choice of the norm. Recall that the norm is based on minimizing the differences between the solution gradients and the data gradients. This is in essence a surface-fitting problem, where the local slopes are made to agree in some sense.

Since we are free to select any norm we desire, what does it mean to select a specific value for p? We know that for large p (i.e., $p \gg 2$), solution gradients that are far away from the corresponding measured gradients are penalized very severely. (Recall the simple line-fitting example of Figure 5.1.) All remaining solution gradients are adjusted so that the outliers have minimum deviation. Therefore norms where $p \gg 2$ provide solutions that tend to be very smooth globally, and outliers strongly pull the solution in their direction. This is not desirable for two-dimensional phase unwrapping because the resulting solution phase gradients will not equal the data gradients *anywhere*. Besides, $p \gg 2$ exaggerates all the potentially undesirable properties of least-*squares* solutions. We showed examples of the undesirable behavior of least-squares solutions in Section 5.3.4.

When $p < 2$, the solution tends to have local gradients that match the measured gradients *in as many places as possible*. We can see this effect through the data-dependent weights of Equations 5.29 and 5.30. For $p < 2$ the weights become large when the gradients of the current solution are close to the measured gradients. We can make this notion precise when $p \equiv 0$. In this case Equation 5.3 gives a value for J equal to the sum of the number of samples where the solution gradients *do not* exactly equal the measured gradients. This, it appears, is exactly what we desire. In other words, the minimum L^0-norm solution is such that the solution gradients exactly equal the measured gradients *in as many places as possible*. By comparison, if one unwraps a phase map by means of Goldstein's path-following algorithm in Chapter 4, the resulting phase gradients are identical to those of the measured data except across the branch cuts. These notions give credence to the desire to minimize the sum of cut lengths as described in [29]. Thus the minimum L^0-norm solution provides an interesting and thought-provoking link to branch cut methods where none existed previously. We will return to this issue in Chapter 6.

When $p = 0$, the solution gradients deviate from the measured gradients in as few places as possible, and the magnitude of the deviation is of no concern. It *is* of concern for $p \neq 0$, however. There appears to be a smooth transition of solution properties as p varies. Thus, for example, the $p = 1.9$ solution should be quite close to the $p = 2$ solution.

In this section we are concerned only with minimum L^p-norm solutions for $p < 2$. In practice we consider only the $p = 0$ solution because (1) the $p = 0$ solution has the optimality properties and the link to branch-cut methods described previously and (2) the obtained $p = 1$ solutions appear to provide no particular advantage. In fact we have observed slower and sometimes erratic convergence behavior. This may be due in part to the "algorithm-tuning" issues mentioned below. It appears that $p = 0$ computations are better behaved.

In theory, any values for weights are acceptable. In practice, weighted problems are better behaved numerically when the weights are normalized to lie in the range $(0, 1)$. The problem is to obtain some useful form of normalization of the weights given in Equations 5.29 and 5.30.

The following equations provide a means for normalizing the data-dependent weights:

$$U(i, j) = \frac{\epsilon_0}{|\phi_{i+1,j} - \phi_{i,j} - \Delta^x_{i,j}|^{2-p} + \epsilon_0}, \tag{5.95}$$

$$V(i, j) = \frac{\epsilon_0}{|\phi_{i,j+1} - \phi_{i,j} - \Delta^y_{i,j}|^{2-p} + \epsilon_0}. \tag{5.96}$$

We have found that setting $\epsilon_0 = 0.01$ radians, not to be confused with ϵ in Step 7 of the PCG algorithm (see Section 5.4.3), provides a reasonable compromise between how closely we require the solution gradients to match the measured gradients, on the one hand, and the rate of convergence of the PCG algorithm, on the other. For example, if ϵ_0 is too small, the PCG algorithm is slow to converge. We will discuss typical numbers for the iterations in Section 5.5.3.

There is another modification that can be made to speed convergence to a useful minimum L^0-norm solution that we will now describe. We mentioned before that any wrapped phase is uniquely unwrappable (to within an arbitrary constant) if it does not contain residues. This is true for the wrapped residual between the input wrapped phase and any partially converged estimate $\phi_k(i, j)$. Thus at the beginning of each iteration we compute a residual wrapped phase

$$R(i, j) = \mathcal{W}\{\psi(i, j) - \phi_k(i, j)\}. \tag{5.97}$$

Next we test $R(i, j)$ for residues by using the procedure outlined in Section 2.6. Note that the wrapped phase differences Δ^x and Δ^y in Equation 5.93 come from $R(i, j)$ rather than $\psi(i, j)$ in this case. If no residues exist, $R(i, j)$ can be unwrapped uniquely with a simple path-following method. The unwrapped residual can be added back to $\phi_k(i, j)$ to form the final estimate

$$\phi(i, j) = \phi_k(i, j) + \mathcal{W}^{-1}\{R(i, j)\}, \tag{5.98}$$

where $\mathcal{W}^{-1}$ represents a simple unwrapping operation.

Unwrapping the residual accomplishes two things: (1) It ensures that the final unwrapped phase differs from the measured values by only integral multiples of 2π, and (2) it speeds up the time to final solution by not having to wait for complete convergence. In practice, as long as the weighted least-squares solution is sufficiently converged, we always seem to obtain a residue-free residual at some point. As the algorithm converges to a minimum L^p-norm solution, the residual approaches zero and, as a result, becomes free of residues.

5.5.2 Pseudocode

We now incorporate these concepts into the following pseudocode for minimum L^p-norm phase unwrapping. While any value of p can be used, we suggest $p = 0$ for the best results. The algorithm is doubly iterative: The outer loop is controlled by the index k, while the inner loop occurs in the PCG computation, which is iterative. The purpose of the PCG computation is to compute the weighted least-squares solution.

L^p-Norm Phase Unwrapping Algorithm

```
Initialize the solution array φ_0 to zero (or an initial guess).
for (k=0 to MaxNumberOfIterations - 1)
  Compute the residual R(i, j) from Equation 5.97.
  Test R(i, j) for residues. If no residues exist, exit the loop.
  Compute the normalized data-dependent weights U_k(i, j) and V_k(i, j)
    from Equations 5.95 and 5.96.
  Compute the weighted phase Laplacian c_k(i, j) from Equation 5.93.
  Subtract the weighted Laplacian of the current solution φ_k (i.e., the
    left-hand side of Equation 5.92) from c_k.
  Solve Equation 5.94, Qφ_k = c_k, with the PCG algorithm. (Let φ_k be the
    initial guess at which the PCG algorithm begins.)
end
if (the residual contains no residues), then
  Unwrap R(i, j) with a simple path-following method and correct the
    current estimate with Equation 5.98.
else
  Apply the postprocessing congruence operation of Section 5.4.3 to the
    solution φ_k.
end
Compute the final set of weights U_k(i, j) and V_k(i, j) from Equations 5.95
  and 5.96.
```

The final computation of the weights from the converged solution allows us to view the spatial distribution of the weights, thereby helping us to assess the quality of the solution. In some of the examples to follow, we will show the sequence of weights as the iteration progresses.

Like Flynn's minimum discontinuity algorithm, the L^p-norm algorithm can easily be modified to accept an array of weights (not to be confused with the data-dependent weights) that indicate which phase values are more reliable than others. The discontinuities in the unwrapped surface can thus be constrained to the low-quality or zero-weighted regions. This modification is effected by multiplying the data-dependent weights $U_k(i,j)$ and $V_k(i,j)$ by the input weights. Results of the algorithm when run with and without the input weights will be shown in Section 5.5.4.

The ability of the PCG algorithm to smoothly interpolate through zero-weighted regions is a crucial element of the L^p-norm algorithm. If the solution were zero or "garbage"in the regions of low-valued weights, this solution would provide no assistance in obtaining a better set of weights for the next iteration. The iteration would then stagnate or even diverge.

5.5.3 Implementation

Our C language implementation of the L^p-norm algorithm can be found in Section A.13 of Appendix A. This implementation calls the PCG algorithm within each iteration. It gives the user the following choice of quality maps, which define the weights:

1. User-supplied correlation coefficients (or other quality map)
2. Pseudocorrelation
3. Variance of phase derivatives
4. Maximum phase gradient
5. No weights

The correlation coefficients can be averaged locally, and the other three quality maps can be computed in windows of any size. The user can also specify that a threshold be calculated and applied to the quality map to produce a mask that defines zero-weights. The algorithm used to compute the threshold is described in Section 3.4. As with the PCG and weighted multigrid algorithms, we recommend that a mask be used instead of a quality map because, in practice, the algorithm seems to converge faster with binary-valued weights. Like Flynn's minimum discontinuity algorithm of Section 4.5, the algorithm can be executed with or without a quality map.

The memory requirements are the same as those of the weighted multigrid algorithm. Seven arrays are required, all of the same size and data type as the input phase data. If a quality map or mask is supplied for defining the input weights (not to be confused with the data-dependent weights that evolve as the iteration progresses), an eighth array is required:

1. Phase data array
2. Array for residual (for the PCG algorithm)

Table 5.5 Execution times of the L^p-norm algorithm for various array sizes on an IBM RS/6000 workstation.

Array Size (pixels)	Execution Time (hr:min:s)
256 × 256	00:07:00
512 × 512	00:30:00
1024 × 1024	02:00:00
2048 × 2048	08:00:00

Note: The times are given for 10 outer iterations and 20 PCG iterations.

3. Array for preconditioner (for the PCG algorithm)
4. Array for conjugate gradient direction vector (for the PCG algorithm)
5. Array for row (x-direction) weights $U(i, j)$
6. Array for column (y-direction) weights $V(i, j)$
7. Solution array
8. Quality map or mask for input weights (optional)

There are three parameters the user needs to specify: (1) the number of iterations of the outer loop, (2) the number of iterations of the PCG algorithm, and (3) the weight normalization parameter ϵ_0 (see Equations 5.95 and 5.96). For most applications, 10 outer iterations and 20 PCG iterations are adequate for convergence. A value of 0.01 radians for the parameter ϵ_0 seems to work well. (This value worked for all of the phase unwrapping examples in this book.) If the algorithm converges slowly or not at all on a particular phase unwrapping problem, the user may need to adjust this parameter.

Table 5.5 lists the execution times of our implementation for various array sizes. The actual times depend on the number of iterations required for convergence. The times are given for 10 outer iterations and 20 PCG iterations. The algorithm is the slowest of all the algorithms we have considered in this book.[18]

5.5.4 Examples

In this section we examine the performance of the L^p-norm algorithm on a number of phase unwrapping examples. We fix $p = 0$ and use 30 iterations within the PCG algorithm, while the number of iterations in the outer loop of the L^p-norm algorithm varies for each example.

We begin our examples with the wrapped values of a 128 × 128-pixel surface

[18] The algorithm could perhaps be speeded up if a partially unwrapped solution were supplied as an initial guess, but we have not investigated this possibility.

(a) (b)

Figure 5.39 (a) Wrapped values of a 128 × 128-pixel two-dimensional phase surface with shear. (b) The five negative residues are located at the black dots. No other residues are present.

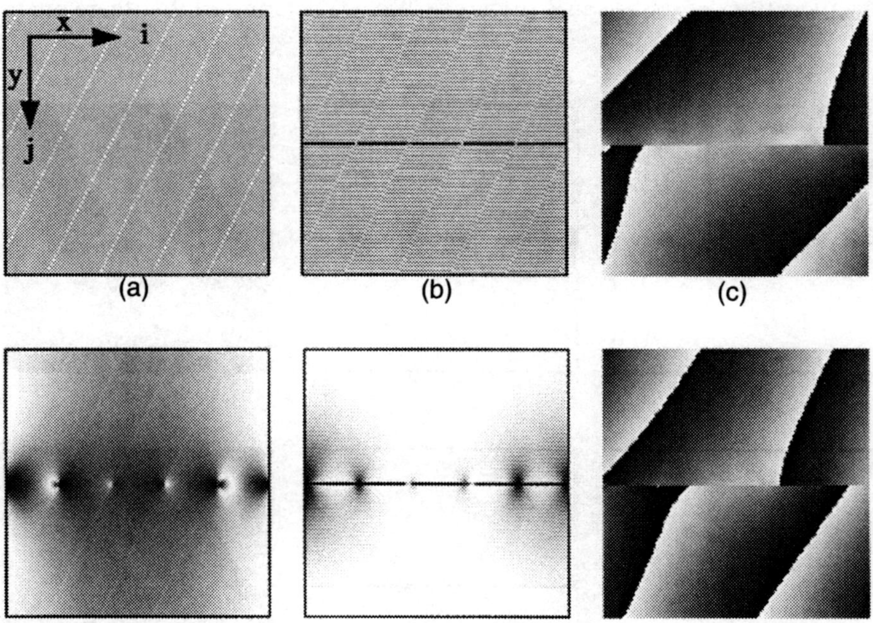

(a) (b) (c)

(d) (e) (f)

Figure 5.40 Progress of the L^p-norm algorithm on the phase-shear problem. In this, and all other examples in this section, we set $p = 0$. (a) Data-dependent x-gradient weights $U_0(i,j)$, which were computed from the initial guess $\phi(i,j) = 0$. (b) Data-dependent y-gradient weights $V_0(i,j)$ computed from the initial guess $\phi(i,j) = 0$. (c) Rewrapped solution after iteration $k = 0$. (d) $U_1(i,j)$. (e) $V_1(i,j)$. (f) Rewrapped solution after iteration $k = 1$.

with shear as shown in Figure 5.39(a). The residue locations are shown in Figure 5.39(b) as black dots on a uniform gray background. Our intuition (or preference) indicates that a horizontal shear line exists midway between the top and bottom of Figure 5.39(a). Thus we expect the "correct" unwrapped phase to "tear" along this shear line and nowhere else. This is such a strong notion that any other tear location would be considered incorrect or undesirable. It is interesting to watch the evolution of the data-dependent weights. Also note the unwrapped phase at each iteration toward the minimum L^p-norm solution. The unwrapped results are "rewrapped" to show the progress of the algorithm more clearly.

Figures 5.40 to 5.42 show the progress of the L^p-norm algorithm on the data depicted in Figure 5.39(a). Convergence on the shear problem was achieved after five iterations. The unwrapped phase, shown in Figure 5.42(f), "tears" along the perceived shear line as our intuition demanded. More important for this example, the unwrapped phase is indeed the minimum L^0-norm solution. Analysis of the y-gradient weights $V(i, j)$, shown in Figure 5.42(e), indicates that there are only 101 samples at which the gradients of the unwrapped phase differ from those of the original wrapped input data. With knowledge of the exact locations of the residues (Figure 5.39b), we determined that there are no

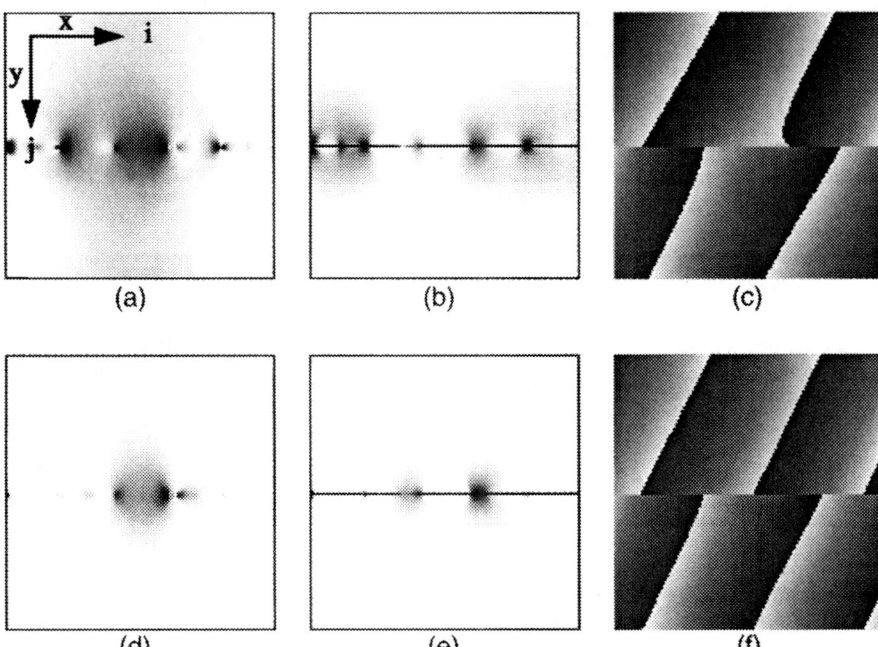

Figure 5.41 Continued progress of the L^p-norm algorithm on the phase-shear problem. (a) $U_2(i,j)$. (b) $V_2(i,j)$. (c) Rewrapped solution after iteration $k = 2$. (d) $U_3(i,j)$. (e) $V_3(i,j)$. (f) Rewrapped solution after iteration $k = 3$.

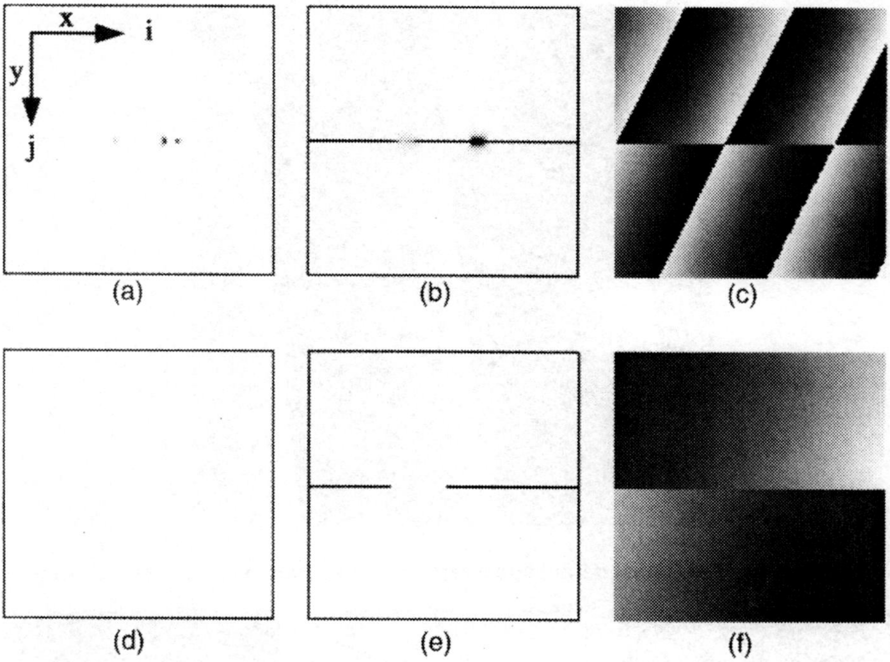

Figure 5.42 Convergence on the phase-shear problem. (a) $U_4(i,j)$. (b) $V_4(i,j)$. (c) Rewrapped solution after iteration $k = 4$. This is the result obtained (at convergence) after correcting the solution with the unwrapped residual. It agrees perfectly with the original input. (d) The x-gradient weights, $U_5(i,j)$, are all unity at convergence because there is no disagreement between the x-gradients of the solution and the x-gradients of the input data. (e) The y-gradient weights, $V_5(i,j)$, differ from unity at only 101 locations. Based on the locations of the residues, this is the minimum number of locations at which the gradients of the solution can differ from the gradients of the original data. Thus the result is a minimum L^0-norm. (f) The solution obtained after iteration $k = 4$.

other ways in which fewer gradient weights can be selected to yield an unwrapped result. In other words, the unwrapped phase has gradients that correspond exactly with the gradients of the input in as many places as possible.

Although we achieved a mathematically (and intuitively) correct result in this case, it is unfortunate that there appear to be many local minima into which the solution could fall. Some of the local minima could be close to an actual global minimum and could thereby yield a solution that is useful for the application at hand. In other cases a local minimum solution, while very good mathematically, might be counterintuitive and thus unacceptable for a particular application. This unfortunate circumstance stems, in part, from the fact that there is no unique global minimum for the L^0-norm problem. This differs from the uniquely solvable L^2-norm problem. While convergence to a local minimum is guaranteed, convergence to a global minimum is not.

Figure 5.43 Rotated and translated version of the wrapped phase-shear data.

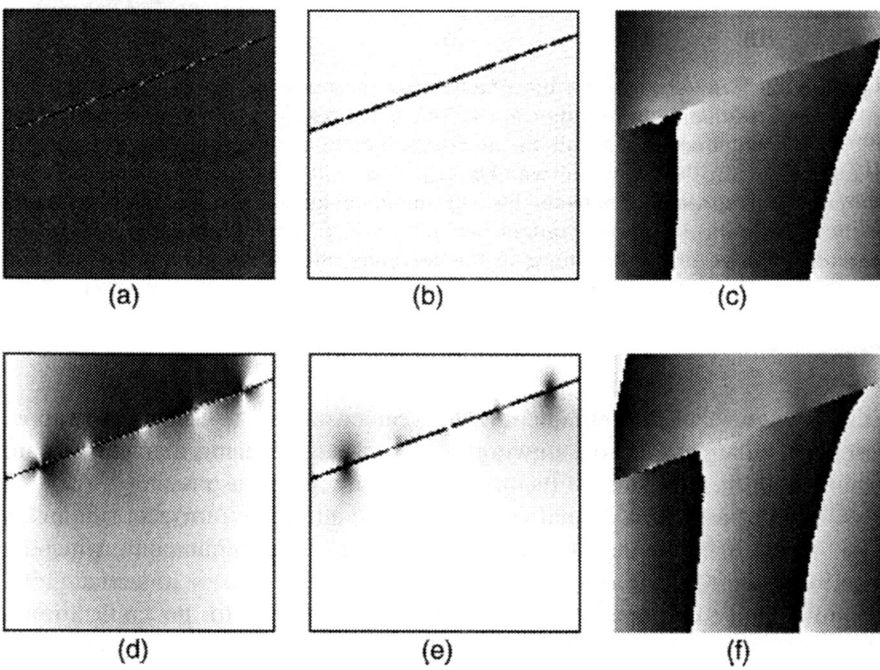

Figure 5.44 Progress of the L^p-norm algorithm on the rotated and translated phase-shear problem. (a) Data-dependent x-gradient weights $U_0(i,j)$ computed from the initial guess $\phi(i,j) = 0$. (b) Data-dependent y-gradient weights $V_0(i,j)$ computed from the initial guess $\phi(i,j) = 0$. (c) Rewrapped solution after iteration $k = 0$. (d) $U_1(i,j)$. (e) $V_1(i,j)$. (f) Rewrapped solution after iteration $k = 1$.

We now illustrate convergence to a local minimum where the resulting solution is intuitively undesirable. Figure 5.43 shows a rotated and translated wrapped phase shear. Figures 5.44 and 5.45 show a sequence similar to that of Figures 5.40 through 5.42. Because the input is nothing more than a rotated and translated version of the previous phase-shear problem, we would expect the solution to be a rotated and translated version of the previous solution shown in Figure 5.42(f).

Notice, however, that we did not obtain our intuitively correct answer, namely the rotated and translated version of Figure 5.42(f). Instead, we obtained the result shown in Figure 5.45(f). Even though the partially converged results shown in Figure 5.44(c), Figure 5.44(f), and Figure 5.45(c) looked like they were progressing to the intuitively correct solution, the data-dependent weight progression did not allow this to happen. Instead, the partially converged solution and the data-dependent weights "pinched off" the continuous solution along a nearly vertical shear line as shown in Figure 5.45(d). Thus the final converged result is simply the solution at a local minimum, not the minimum L^0-

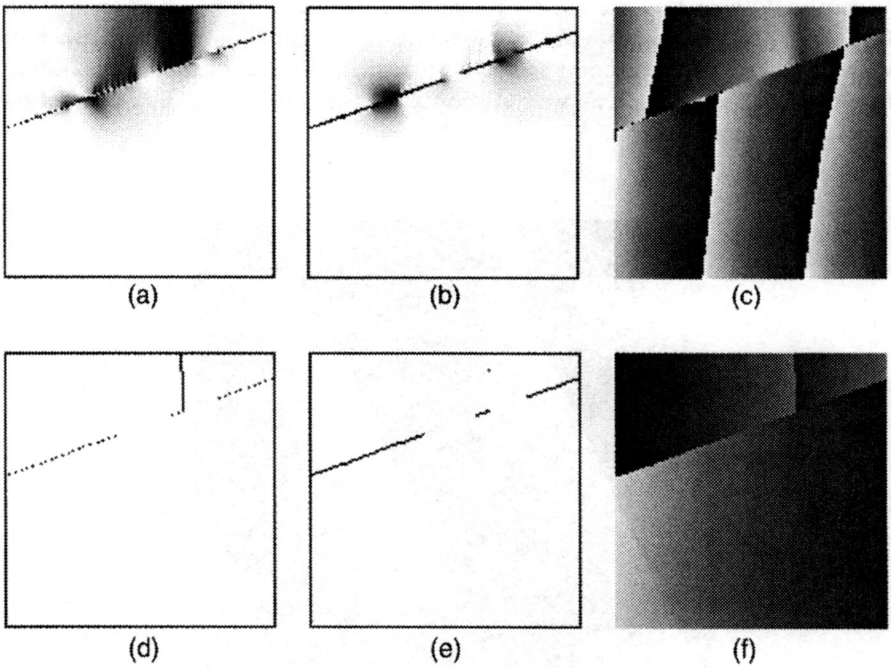

Figure 5.45 Further progress and final convergence on the rotated phase-shear problem. (a) $U_4(i,j)$. (b) $V_4(i,j)$. (c) Rewrapped solution after iteration $k = 4$. (d) The x-gradient weights $U_{17}(i,j)$ after convergence to a local minimum. (e) The y-gradient weights $V_{17}(i,j)$ after convergence to a local minimum. (f) Unwrapped result after 17 iterations. The rewrapped solution at this point agrees exactly with the original wrapped input.

norm solution as desired. The result shown in Figure 5.45(f), while not the global optimum, is still quite good in the mathematical sense. Specifically, this solution has gradients that differ from the input in only 115 locations. We would have preferred the final weights to lie entirely along the rotated shear line and to provide the globally optimal and intuitively correct solution, but this, of course, did not happen.

Even if the algorithm would converge to one of several identical (in norm) global minima, the solution might be undesirable. It might be undesirable because it could strongly violate our preconceived notion of correctness, or it could violate some other physical property required for the specific application, such as agreement with a quality map.

One such example whereby the true minimum L^0-norm solution might be undesirable is shown by unwrapping the shifted phase-shear data shown in Figure 5.46. Again we perceive the intuitively correct horizontal shear occurring along the line 13 samples below the upper border. The minimum L^0-norm solution and the data-dependent weights are shown at convergence (after nine iterations) in Figures 5.47(a)–(c). It is interesting that this solution, while mathematically optimal, does not support our preconceived notion that the shear should be maintained solely along the perceived horizontal shear line. In fact this is one of two possible mathematically optimal solutions, neither of which support the horizontal shear hypothesis. (The second optimal solution results from the x-gradient weights containing a zero-weight line emanating

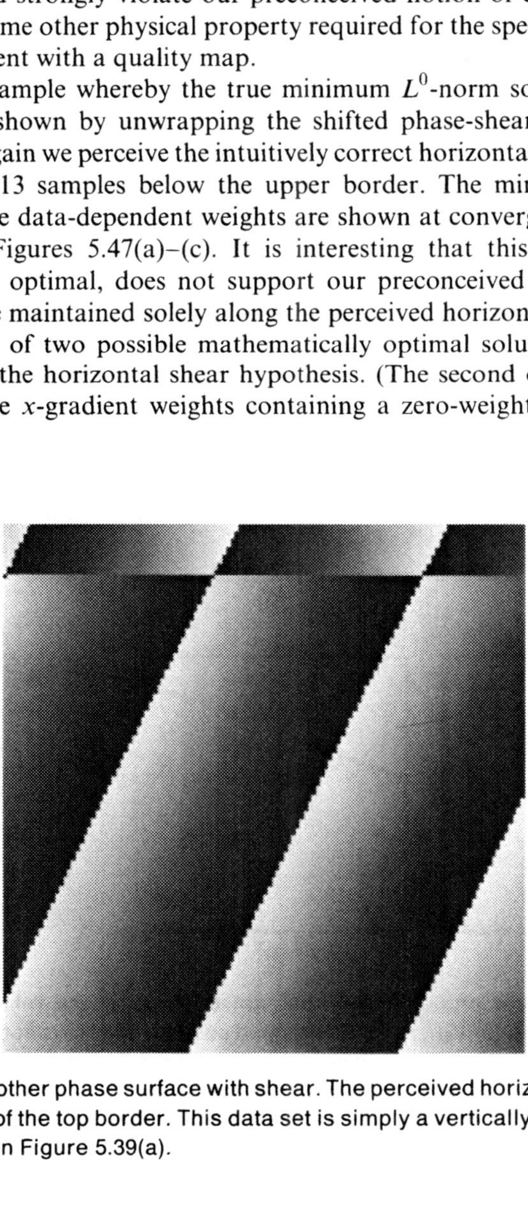

Figure 5.46 Another phase surface with shear. The perceived horizontal shear line is within 13 pixels of the top border. This data set is simply a vertically shifted version of the data shown in Figure 5.39(a).

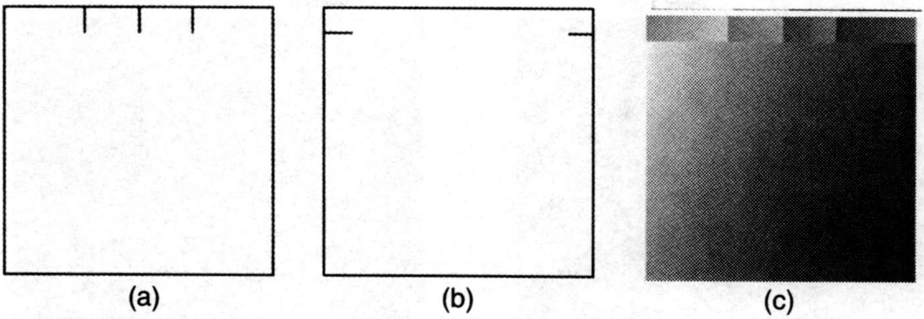

Figure 5.47 Converged solution of the vertically shifted phase-shear problem. (a) $U_9(i,j)$. (b) $V_9(i,j)$. (c) Minimum L^0-norm solution after nine iterations. This is one of two possible mathematically optimal solutions, neither of which maintains the phase shear solely along the horizontal shear line.

from the leftmost residue to the top border. The x-gradient weights consist of four equal-length vertical line segments instead of the three segments shown in Figure 5.47a. The y-gradient weights consist of only one rightmost horizontal line segment instead of the two segments shown in Figure 5.47b.) The 64 nonunity data-dependent weights begin at the residue locations and proceed in straight lines to the boundaries as shown in Figures 5.47(a) and (b). In order to maintain the intuitively correct horizontal shear, more than 64 nonunity weights would be required. If the horizontal shear line was desired or required for the particular application at hand, this minimum L^0-norm solution would be undesirable.

It should be emphasized that the raw wrapped data alone cannot support *any* preconceived notion of correctness. Because residues exist, many solutions are possible. Global or local minimum L^0-norm solutions sometimes give intuitively correct and usable results, at other times they do not. The same phenomenon is exhibited by any algorithm that does not use ancillary information to guide the phase unwrapping process. For example, Goldstein's algorithm generates vertical branch cuts similar to the zero-weights of Figure 5.47. Flynn's minimum (unweighted) discontinuity algorithm also generates a solution with vertical discontinuities similar to these zero-weights. The problem in all cases is the presence of a border near the shear line, permitting a solution with short discontinuities running to the border.

We now move on to the phase unwrapping examples of Chapter 3. The first such example is the IFSAR example of Figure 3.2(a), which is shown in Figure 5.48(a) for convenience. The unwrapped result (with the planar trend removed to allow easier visualization) is shown in Figure 5.48(b). The algorithm converged in 10 iterations. Notice the unwrapped phase estimates in the noisy regions. The data-dependent weights effectively isolate these noisy regions from the remainder of the unwrapped data. It is significant that the result was achieved without the explicit use of branch cuts or of externally supplied weights. The data-dependent weights at convergence are shown in Figures 5.48(c) and (d).

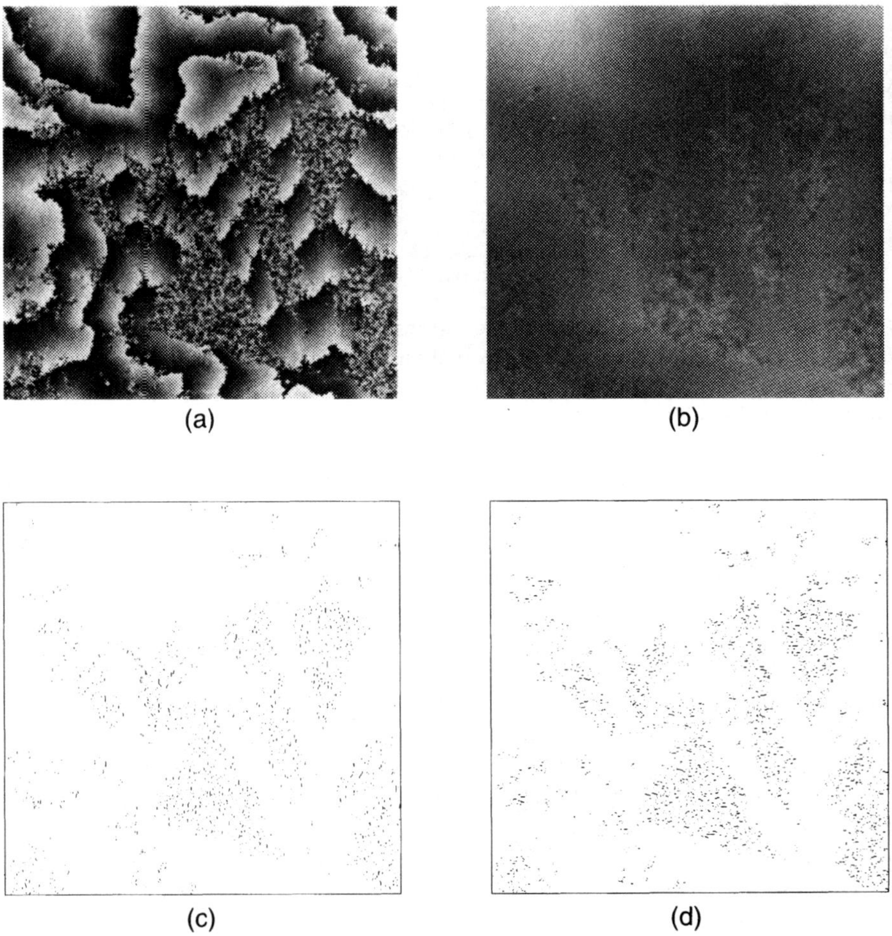

Figure 5.48 Results on the IFSAR example. (a) The wrapped phase. (b) The unwrapped result at convergence after 10 iterations with the planar trend removed to allow easier visualization. (c) The x-gradient weights at convergence. (d) The y-gradient weights at convergence.

The differences between the surface produced by the L^p-norm algorithm and that produced by Goldstein's algorithm are shown by the error map of Figure 5.49. The black regions are those in which the two surfaces differ. Notice that the differences are confined to the patches of corrupted phase. Exclusive of these corrupted regions, the algorithms produced the same result.

We now consider the MRI examples. The unwrapped phase for the MRI head data of Figure 3.3(a) is shown in Figures 5.50(a) and (b). This example provides one of the most visually interesting sequence of weight progressions of the experiments performed thus far. The data-dependent weights for the first four

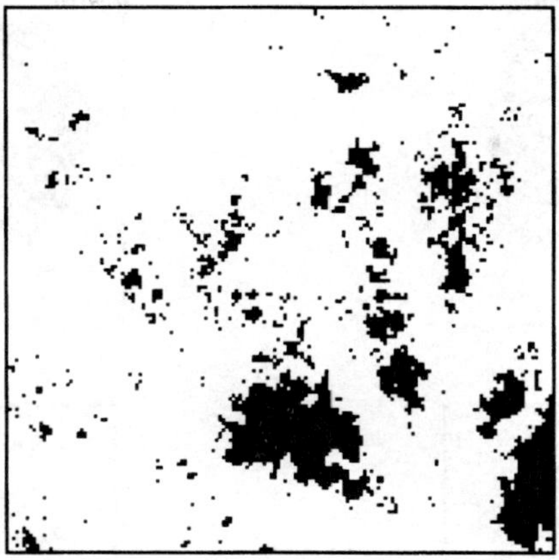

Figure 5.49 Differences between the surfaces produced by the L^p-norm algorithm and Goldstein's algorithm on the IFSAR example. The black pixels indicate where the two surfaces differ.

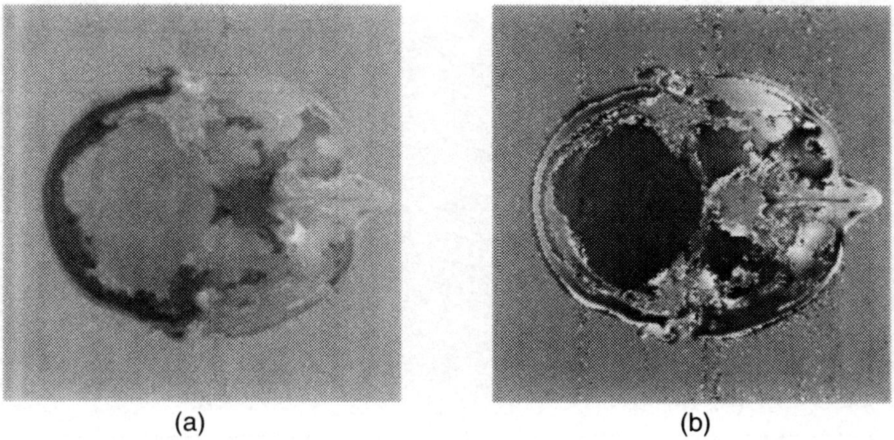

Figure 5.50 Result of the L^p-norm algorithm on the MRI head data. (a) The unwrapped result at convergence after eight iterations. (b) The rewrapped phase is identical to the wrapped input phase as expected.

iterations are shown in Figures 5.51(a)–(h). The weights at convergence after eight iterations are shown in Figures 5.52(a) and (b). The differences between the surface produced by the L^p-norm algorithm and that produced by Goldstein's algorithm are shown by the error map of Figure 5.53(a). Exclusive of the isolated regions within the head, the algorithms produced the same result.

We now consider the MRI knee example of Figure 3.4(a). Eleven iterations are required for convergence on this example. The differences between the surface produced by the L^p-norm algorithm and that produced by Goldstein's

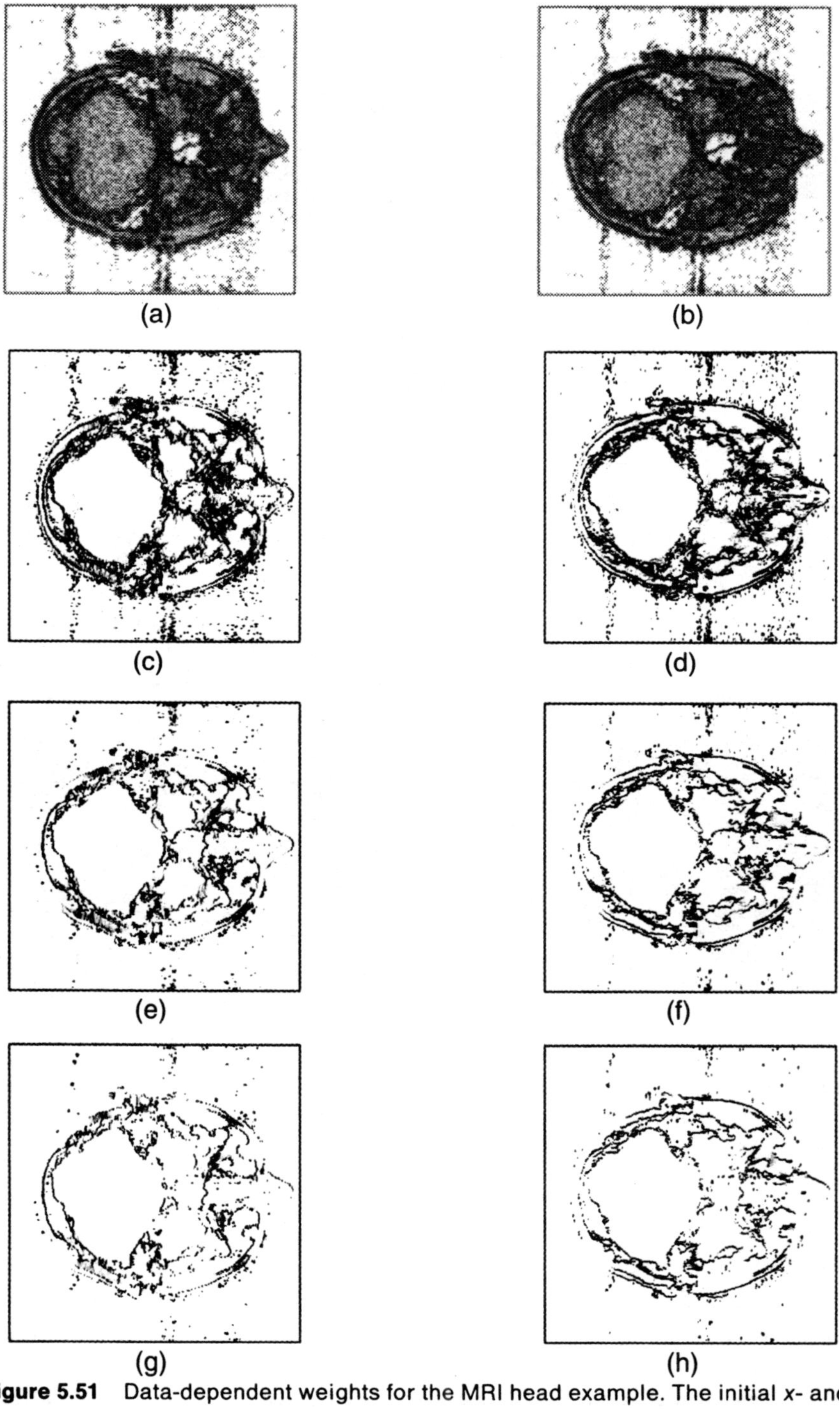

Figure 5.51 Data-dependent weights for the MRI head example. The initial x- and y-gradient weights are shown in (a) and (b), respectively. The x-gradient weights after the first, second, and third iterations are shown in (c), (e) and (g), respectively. The corresponding y-gradient weights are shown in (d), (f), and (h).

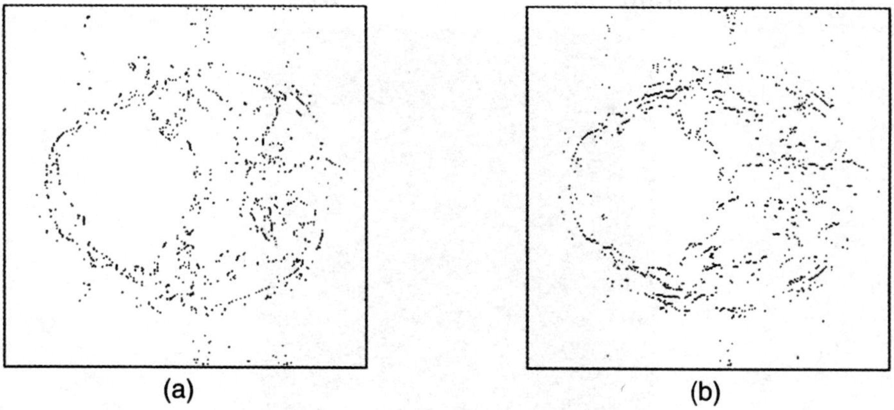

Figure 5.52 Data-dependent weights after convergence on the MRI head example. (a) The *x*-gradient weights. (b) The *y*-gradient weights.

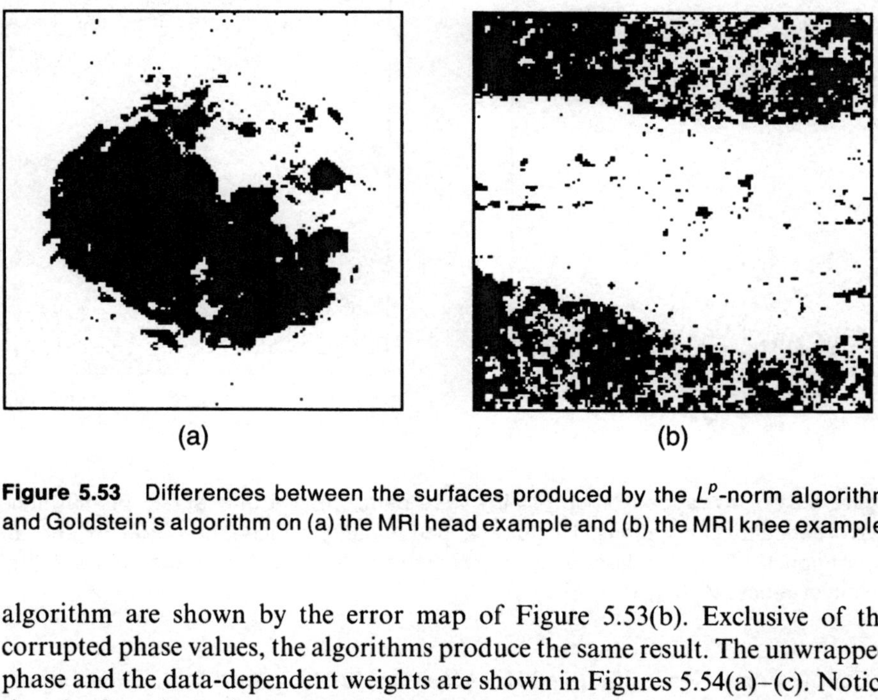

Figure 5.53 Differences between the surfaces produced by the L^p-norm algorithm and Goldstein's algorithm on (a) the MRI head example and (b) the MRI knee example.

algorithm are shown by the error map of Figure 5.53(b). Exclusive of the corrupted phase values, the algorithms produce the same result. The unwrapped phase and the data-dependent weights are shown in Figures 5.54(a)–(c). Notice that the data-dependent weights have low values in the noise-corrupted regions.

We now turn to the simulated IFSAR examples. As we did with Flynn's minimum discontinuity algorithm in Section 4.5, we will show the unwrapped results for both the unweighted and weighted versions of the L^p-norm algorithm. While the unweighted version was sufficient for unwrapping the preceding examples, the weighted version will be needed on the simulated IFSAR examples.

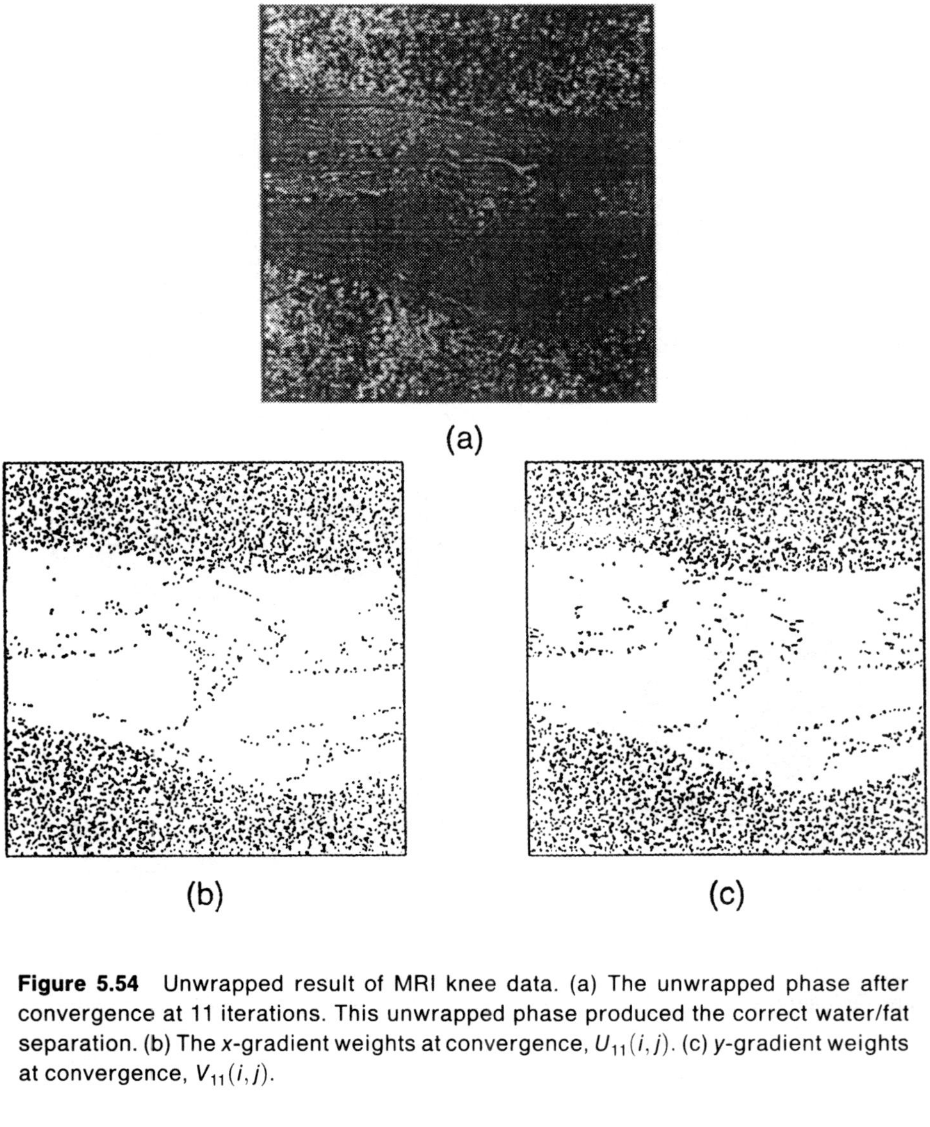

Figure 5.54 Unwrapped result of MRI knee data. (a) The unwrapped phase after convergence at 11 iterations. This unwrapped phase produced the correct water/fat separation. (b) The x-gradient weights at convergence, $U_{11}(i, j)$. (c) y-gradient weights at convergence, $V_{11}(i, j)$.

On the Long's Peak simulated IFSAR example of Figure 3.7, the unweighted algorithm produces incorrect results. The error map is shown in Figure 5.55(a). The entire lower half of the image has been unwrapped incorrectly. It is instructive to examine the discontinuities (i.e., the nonzero jump counts defined in Section 4.5), whose locations are determined by the data-dependent weights that have low values. The discontinuities, which are shown in Figure 5.55(b), are analogous to the branch cuts of Goldstein's algorithm, which are shown in Figure 4.11(a). To minimize the L^0 errors, the algorithm was forced to place the

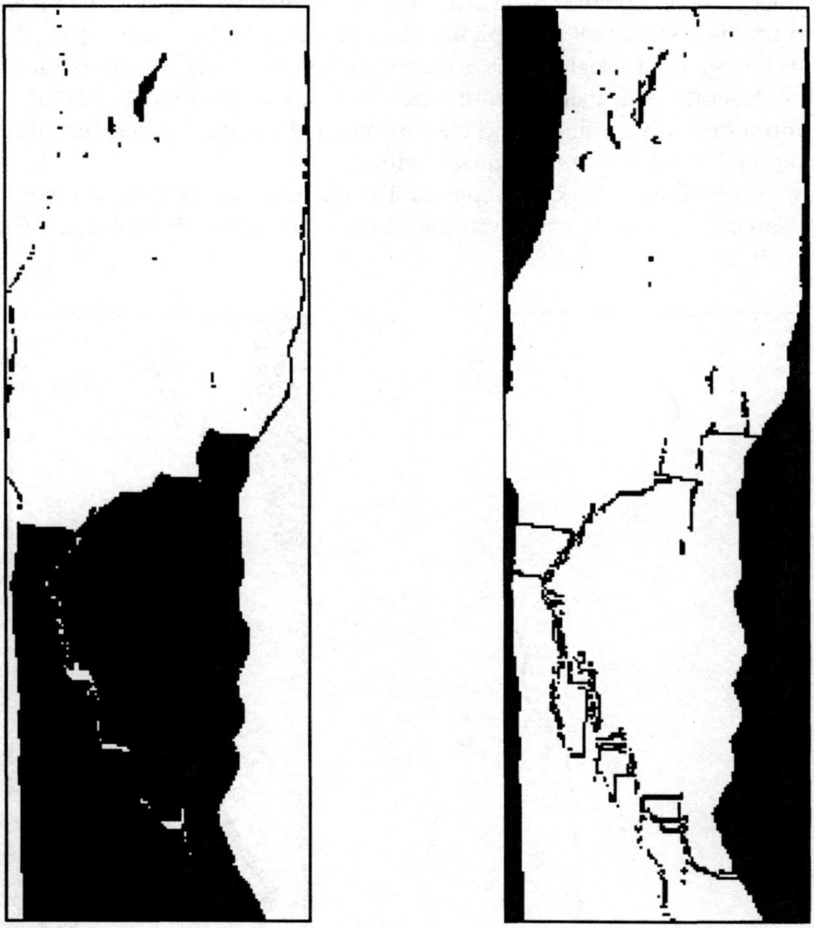

(a) (b)

Figure 5.55 The results of the *unweighted* L^p-norm algorithm on the Long's Peak example of Figure 3.7. (a) The error map. (b) The discontinuities produced by the algorithm. This example illustrates the need for externally supplied weights to guide the placement of the low-valued data-dependent weights (and hence the discontinuities).

low-valued data-dependent weights (and hence the discontinuities) across the image, thereby isolating the lower half of the image from the upper half. Interestingly this solution is almost identical to that produced by Flynn's minimum unweighted discontinuity algorithm, which is shown in Figure 4.52.

Like Flynn's minimum discontinuity algorithm, the L^p-norm algorithm can accept an array of weights that indicate which phase values are more reliable than others. These weights constrain the placement of the low-valued data-dependent weights to the low-quality (or zero-weighted) regions. In the algorithm this is effected by multiplying the data-dependent weights $U(i, j)$ and $V(i, j)$ by the input weights. The error map of the resulting surface is shown in Figure 5.56(a), and the discontinuities are shown in Figure 5.56(b). The discontinuities are confined to the zero-weighted regions of layover, and the unwrapping is flawless outside these regions.

On the Isolation Peak example of Figure 3.8, the story is similar. The *unweighted* L^p-norm algorithm produced an incorrect result whose error map

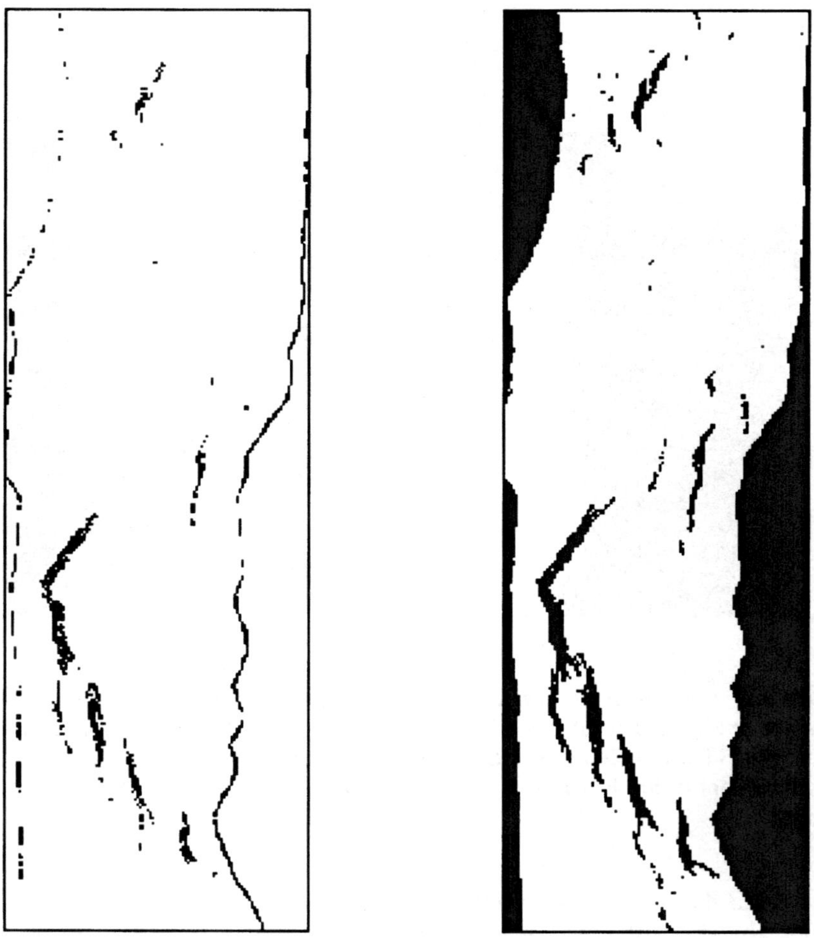

(a) (b)

Figure 5.56 Results of the *weighted* L^p-norm algorithm on the Long's Peak example of Figure 3.7. Figure 3.30(c) shows the mask that was used to define the zero-weights. (a) The error map. (b) The discontinuities in the unwrapped surface, which are restricted to the zero-weighted regions.

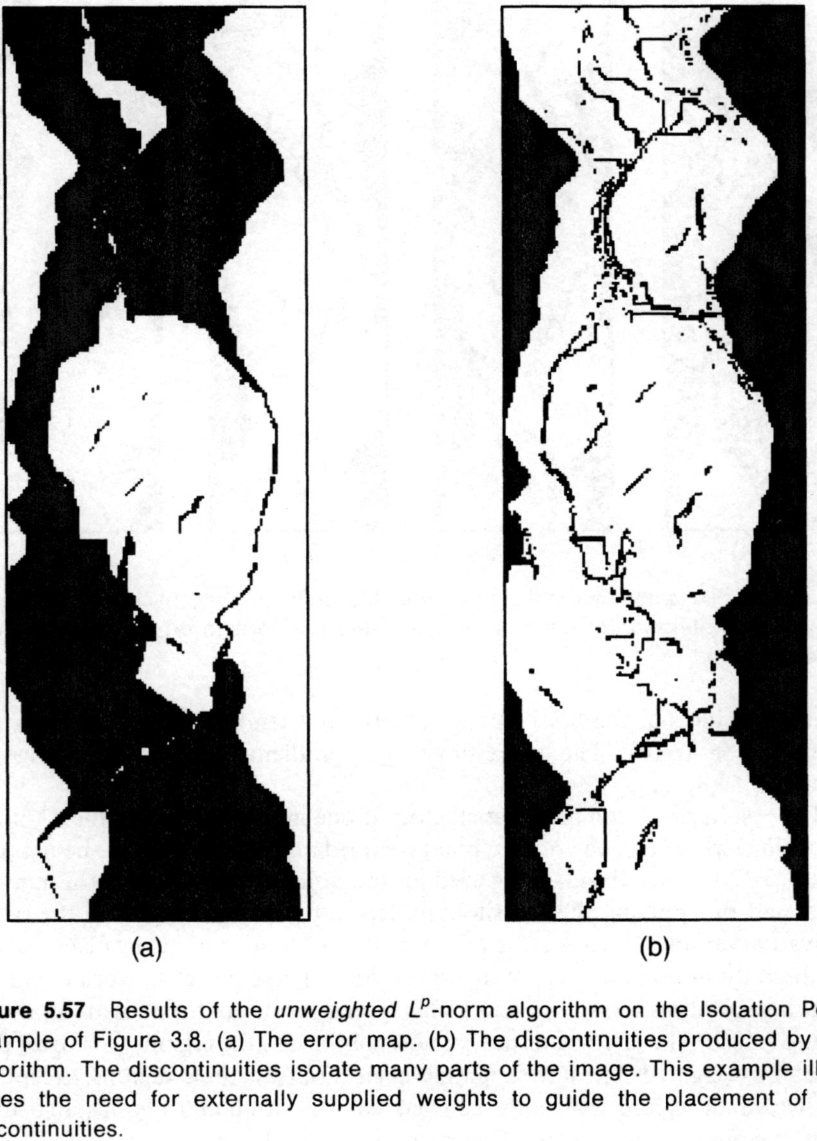

(a) (b)

Figure 5.57 Results of the *unweighted* L^p-norm algorithm on the Isolation Peak example of Figure 3.8. (a) The error map. (b) The discontinuities produced by the algorithm. The discontinuities isolate many parts of the image. This example illustrates the need for externally supplied weights to guide the placement of the discontinuities.

is shown in Figure 5.57(a). The locations of the discontinuities are shown in Figure 5.57(b). Many regions of the image have been isolated by the discontinuities, thereby resulting in incorrect wrap counts. Comparison of these results with those of Flynn's minimum unweighted discontinuity algorithm in Figure 4.54 shows that they are similar.

It is interesting to examine the progression of the data-dependent weights (not to be confused with the input weights, which are fixed) on this example. The x-

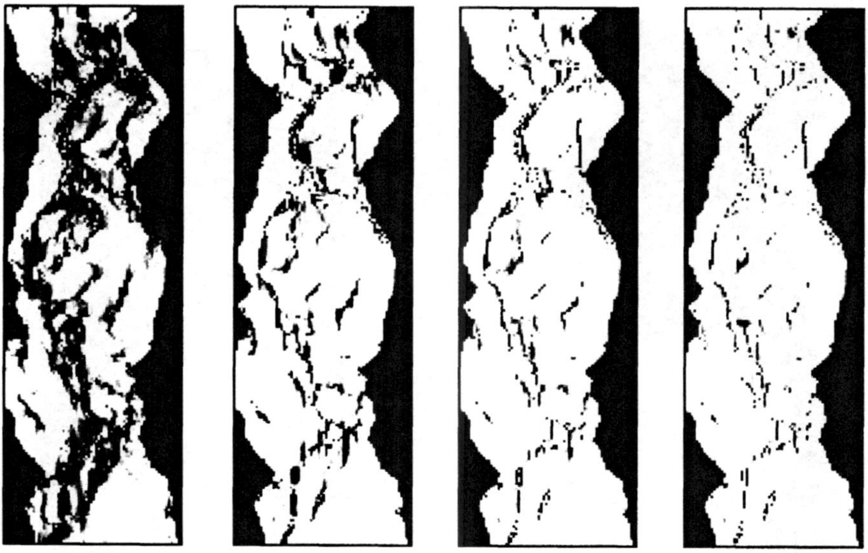

Figure 5.58 Data-dependent *x*-gradient weights for the (a) first, (b) second, (c) fourth, and (d) tenth iteration of the *unweighted* L^p-norm algorithm on the Isolated Peak example.

gradient weights at the first, second, fourth, and tenth iteration are shown in Figures 5.58(a)–(d). The corresponding y-gradient weights are shown in Figures 5.59(a)–(d).

The results are much more satisfactory if one inputs weights to the L^p-norm algorithm, where the weights are binary-valued and are defined by the mask of Figure 3.32(b). Like the mask we used for the Long's Peak example, this mask is generated by applying the thresholding technique of Section 3.4 to the phase derivative variance. It causes the L^p-norm algorithm to confine the placement of the discontinuities to the zero-weighted regions of layover. The error map of the resulting surface is shown in Figure 5.60(a), and the discontinuities are shown in Figure 5.60(b) The discontinuities are now confined to the regions of layover that were zero-weighted. The unwrapping is correct outside these regions, except for the two small regions that both the mask cut algorithm and Flynn's algorithm also unwrapped incorrectly. Compare these results with those of Flynn's minimum weighted discontinuity algorithm shown in Figure 4.55. The x-gradient weights at the first, second, fourth, and tenth iteration are shown in Figures 5.61(a)–(d).

We have already shown that the L^p-norm algorithm unwraps the sheared planes example. In this case there is no need to use weights. The algorithm also unwraps the noise-corrupted shear example of Figure 4.15, albeit with some minor errors. The error map and discontinuity map are shown in Figures 5.62(a) and (b), respectively.

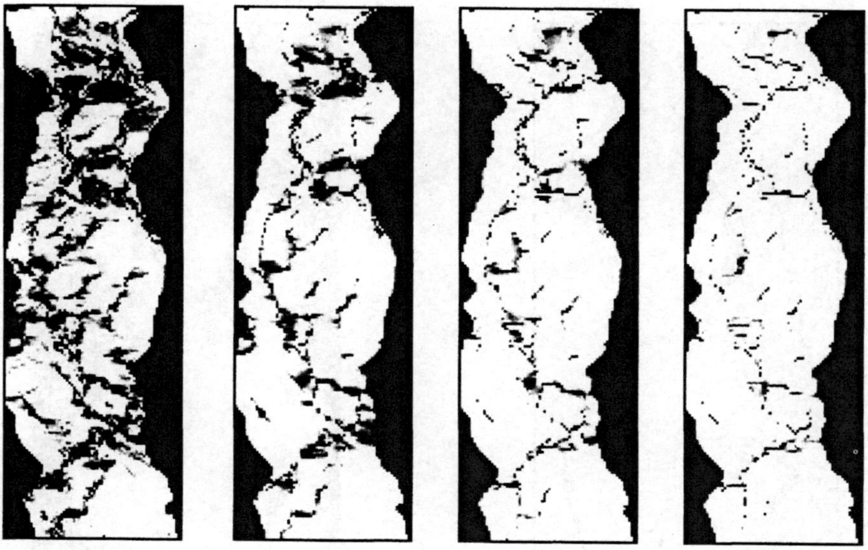

Figure 5.59 Data-dependent *y*-gradient weights for the (a) first, (b) second, (c) fourth, and (d) tenth iteration of the *unweighted* L^p-norm algorithm on the Isolated Peak example.

The L^p-norm also unwraps the spiral shear example of Figure 3.10, albeit extremely slowly like the PCG and the weighted multigrid algorithms. In this case weights are required to restrict the discontinuities to the spiral lines of shear. If these weights are not supplied, the algorithm fails and produces discontinuities similar to those of the unweighted version of Flynn's minimum discontinuity algorithm, which are shown in Figure 4.56(a).

Thus the L^p-norm algorithm shares with Flynn's minimum discontinuity algorithm the distinction of being able to unwrap both the noise-corrupted sheared planes example and the spiral shear example.

5.5.5 Conclusion

In this section we presented an algorithm for minimum L^p-norm two-dimensional phase unwrapping. This approach is a generalization of weighted least-squares phase unwrapping. The resulting algorithm is computationally intensive due to its doubly iterative structure. It iterates on the PCG algorithm, which in turn iterates on an FFT-based (or DCT-based) unweighted least-squares algorithm. Like Flynn's minimum discontinuity algorithm, however, the L^p-norm algorithm can be used with or without input weights. In contrast to the weighted least-squares algorithms, which require that explicit weights be provided, the L^p-norm algorithm generates its own weights from the gradients of the input data

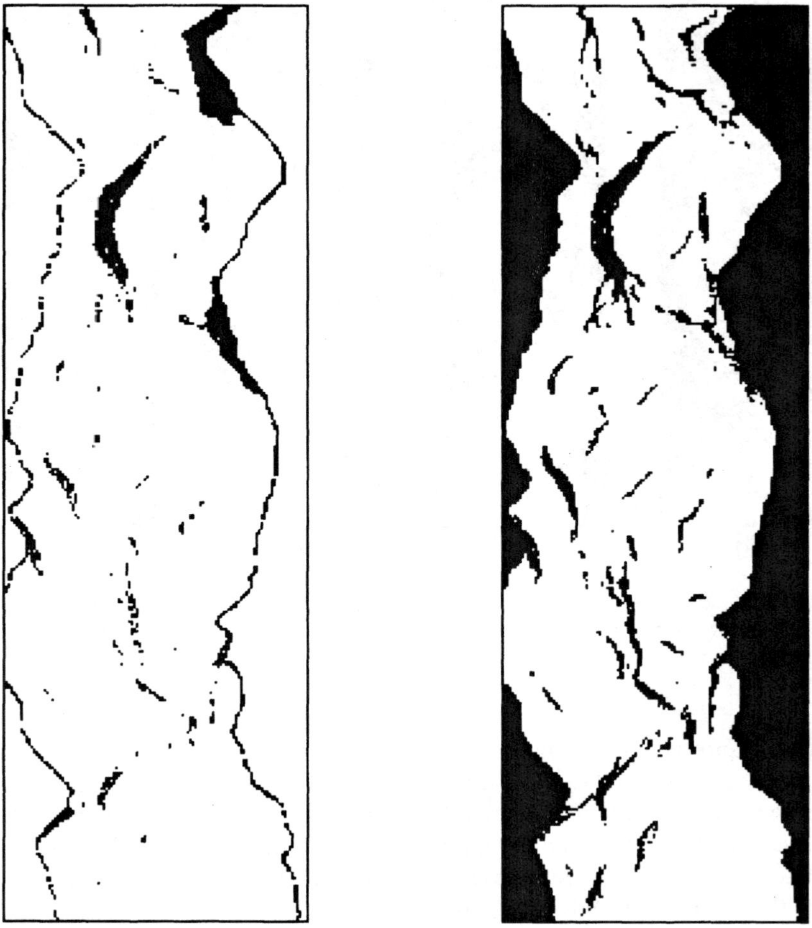

(a) (b)

Figure 5.60 Results of the weighted version of the L^p-norm algorithm on the Isolation Peak example of Figure 3.8. Figure 3.32(b) shows the mask that defines the zero-weights. (a) The error map. (b) The discontinuities in the unwrapped surface. The mask restricts the discontinuities to the regions of layover as desired.

(when $p \neq 2$). However, the performance of the algorithm can be enhanced with the inclusion of user-supplied weights or a quality map.

The L^p-norm algorithm correctly unwrapped all of the phase unwrapping examples, albeit with minor errors on the Isolation Peak example. The results were very similar to those of Flynn's minimum discontinuity algorithm, even including the minor unwrapping errors and the placement of the discontinuities. Thus the two algorithms may be regarded as roughly equivalent. It is interesting that the two algorithms produce such similar results because their methods of

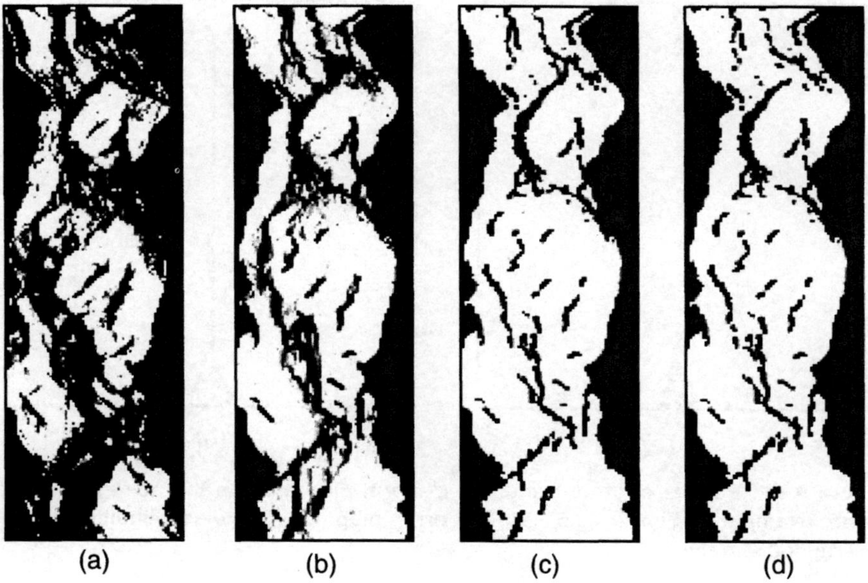

Figure 5.61 Data-dependent *x*-gradient weights for the Isolated Peak example after the (a) first, (b) second, (c) fourth, and (d) tenth iteration of the *weighted* L^p-norm algorithm.

solving the phase unwrapping problem are completely different. Flynn's algorithm is a path-following algorithm that depends on the local examination of the phase data, while the L^p-norm algorithm is a gradient-fitting algorithm that treats the phase data in a global rather than local fashion.

5.6 SUMMARY AND CONCLUSION

We have presented several minimum-norm algorithms that span a rather large range of requirements for two-dimensional phase unwrapping. In particular, we have examined the following four algorithms for *unweighted* least-squares phase unwrapping:

1. FFT method 1
2. FFT method 2
3. DCT algorithm
4. Unweighted multigrid algorithm

These are computationally efficient algorithms for solving the discretized Poisson equation, which is the equation that describes the least-squares phase unwrapping problem. These algorithms are fast and relatively simple to imple-

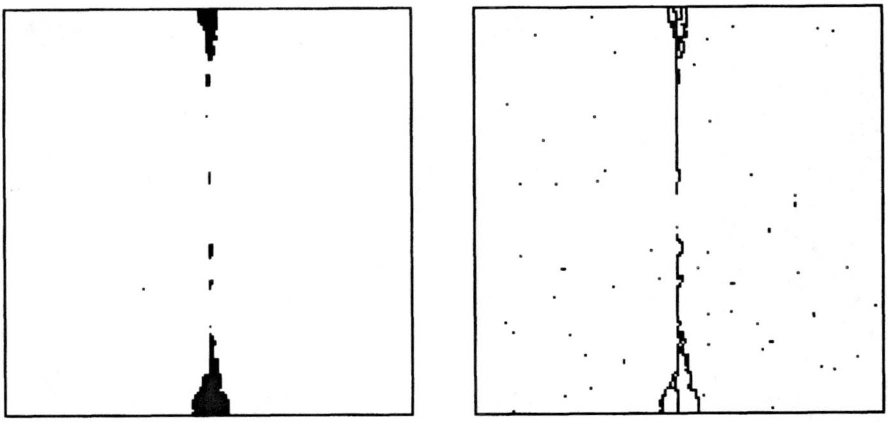

(a) (b)

Figure 5.62 Results of the unweighted L^p-norm algorithm on the noise-corrupted shear example of Figure 4.15. (a) The error map. (b) The discontinuities in the unwrapped surface.

ment, and they yield essentially identical results. There are differences in the boundary conditions (or mirror reflections), but these differences are insignificant. The algorithms are robust in that no user intervention is required to obtain a solution, and the algorithms never fail to obtain a solution. However, we also showed that unweighted least-squares solutions are not appropriate for many applications. Because these algorithms unwrap *through* rather than *around* the residues, the resulting solution is not congruent to the input data and typically represents a smoother solution than the data can conceivably support.

To mitigate these deficiencies we presented several algorithms that solve the *weighted* least-squares problem, which is a natural extension of the unweighted problem. We studied three algorithms in particular:

1. Picard algorithm
2. PCG algorithm
3. Weighted multigrid

Unlike the unweighted algorithms, these algorithms are iterative and therefore slower. The Picard algorithm is particularly easy to implement but is not guaranteed to converge. If it does converge, however, it converges to the correct solution. In contrast to this, the PCG algorithm has excellent convergence properties. For some applications, however, such as the sheared planes example, the weighted multigrid algorithm can be significantly faster. Because the multigrid algorithm can fail if the coarse grid zero-weights isolate portions of the phase, we recommend that the PCG algorithm be used in general applications.

As with the path-following algorithms that require quality maps, weighted least-squares algorithms require judicious weight selection. Chapter 3 discussed these issues in depth. It remains that weight selection, both for path-following and least-squares algorithms, is as much an art as it is a science and cannot be divorced from the application at hand.

We also studied an important postprocessing operation that renders any least-squares result congruent to the wrapped input phase. This operation reduces the errors in the result significantly.

As a generalization of weighted least squares, we presented an algorithm for finding a minimum L^p-norm solution. Although it is computationally intensive due to its doubly iterative structure, it can and does produce useful solutions for a variety of phase unwrapping problems. In contrast to weighted least squares, where explicit weights must be provided, the L^p-norm algorithm generates its own weights from the input data when $p \neq 2$. This algorithm draws heavily upon, and uses the methods devised for, unweighted and weighted least-squares algorithms. The algorithm is dominated by the computation of forward and inverse transforms (FFTs or DCTs), which are necessary for the PCG solution and in turn is necessary for the L^p-norm solution. The total number of forward and inverse transforms is equal to the product of the number of inner (i.e., PCG) and outer iterations.

Table 5.6 summarizes several characteristics of the four algorithms. In this table, "FFT/DCT" denotes any of the FFT- or DCT-based algorithms for unweighted least-squares phase unwrapping, "PCG" denotes the PCG algorithm and "Multigrid" denotes the weighted multigrid algorithm. For each algorithm the table lists the memory requirements and the average execution times for a 1024 × 1024-pixel phase array. The memory requirements are expressed in terms of the number of 1024 × 1024 arrays for the phase data, the unwrapped surface, the input weights and other data. The table also indicates whether or not the algorithm requires input weights.

Table 5.6 Characteristics of the minimum-norm algorithms.

Algorithm	Memory Requirements (arrays)	Execution Time	Input Weights Required?
FFT/DCT	1	23 second	*n*
PCG	5	10 minute	*y*
Multigrid	7	1.5 minute	*y*
L^p-norm	7 or 8	2 hour	*y*/*n*

Note: The memory requirements are the number of arrays required. The execution times are given in minutes for a 1024 × 1024-pixel phase array on an IBM RS/6000 workstation. The last column indicates whether or not the algorithm requires input weights. The L^p-norm algorithm works with or without input weights.

Table 5.7 Results of the minimum-norm algorithms on the phase unwrapping examples.

Algorithm	Examples: IFSAR	MRI Head	MRI Knee	Long's Peak
FFT/DCT	*n*	*n*	*n*	*n*
PCG	*y*	*y*	*y*	*y*
Multigrid	*y*	*y*	*y*	*y*
L^p norm	*y*	*y*	*y*	*y*

Algorithm	Isolation Peak	Sheared Planes	Noisy Shear	Spiral Shear
FFT/DCT	*n*	*n*	*n*	*n*
PCG	(*y*)	*y*	*n*	*y*
Multigrid	*n*	*y*	*n*	*y*
L^p norm	(*y*)	*y*	*y*	*y*

Note: An entry of *y* indicates the algorithm unwrapped the example correctly, *n* indicates the algorithm did not unwrap the example correctly, and (*y*) indicates the algorithm had minor errors in its unwrapped solution.

Table 5.7 summarizes the successes and failures of the algorithms on the seven phase-unwrapping examples of Chapter 3 and the noise-corrupted sheared planes example of Figure 4.15. An entry of *y* indicates the algorithm was successful, *n* indicates the algorithm failed, and *(y)* indicates the algorithm unwrapped the example successfully except for minor errors.[19] The L^p-norm algorithm was the only algorithm to unwrap all of the examples successfully. The FFT/DCT (unweighted least-squares) algorithm could not unwrap any of the examples correctly, but we have already noted that the primary importance of this algorithm is the foundation it lays for weighted least squares. (In addition the FFT/DCT algorithm is called iteratively by the PCG and the L^p-norm algorithms.)

Because the phase unwrapping problem was formulated as an optimization problem, the quantity being optimized may or may not be compatible with the desired outcome of a specific application. The general L^p-norm algorithm can converge to a local minimum instead of a global minimum. In some cases the local minimum can provide a useful solution; in other cases it does not. The same can be said for the global minimum. This is true for the path-following

[19] Just as we did for the path-following algorithms, we must make the following caveat: Because the correct unwrapped solution was unknown for the first three examples, an algorithm was judged a success or failure based on comparison of its result with the input phase and the results of the other phase unwrapping algorithms, as well as the authors' estimation of what constituted success on each example. The results for the other five examples were judged by comparison with the known solutions.

algorithms (e.g., Flynn's minimum discontinuity algorithm) as well. More important, perhaps, is that the minimum-norm formulations provide additional insight into the phase unwrapping problem in general. Specifically, the minimum L^0- and L^1-norm analysis provided a rather surprising link to some path-following methods in terms of their optimization criteria. We will explore these ideas further in Chapter 6.

REFERENCES

[1] D. C. Ghiglia and L. A. Romero, "Robust two-dimensional weighted and unweighted phase unwrapping that uses fast transforms and iterative methods," *Journal of the Optical Society of America A*, Vol. 11, pp. 107–117, November 1994.

[2] M. D. Pritt and J. S. Shipman, "Least-squares two-dimensional phase unwrapping using FFTs," *IEEE Transactions on Geoscience and Remote Sensing*, Vol. 32, No. 3, pp. 706–708, May 1994.

[3] D. C. Ghiglia and L. A. Romero, "Minimum L^p-norm two-dimensional phase unwrapping," *Journal of the Optical Society of America A*, Vol. 13, No. 10, pp. 1–15, October 1996.

[4] H. Tarayre, D. Massonnet, and J. A. Sirat, "Noise-robust phase unwrapping method in radar interferometry," *Synthetic Aperture Radar and Passive Microwave Sensing, Proceedings of the SPIE*, Vol. 2584, Society of Photo-Optical Instrumentation Engineers, Bellingham, WA, pp. 310–318, 1995.

[5] D. Labrousse, S. Dupont, and M. Berthod, "SAR interferometry: A Markovian approach to phase unwrapping with a discontinuity model," *Proceedings of the 1995 International Geoscience and Remote Sensing Symposium*, Tokyo, Japan, August 18–21, 1995, IEEE, Piscataway, NJ, pp. 556–558, 1995.

[6] U. Spagnolini, "2-D phase unwrapping and instantaneous frequency estimation," *IEEE Transactions on Geoscience and Remote Sensing*, Vol. 33, No. 3, pp. 579–589, May 1995.

[7] D. L. Fried, "Least-squares fitting a wave-front distortion estimate to an array of phase-difference measurements," *Journal of the Optical Society of America*, Vol. 67, No. 3, pp. 370–375, March 1977.

[8] R. H. Hudgin, "Wave-front reconstruction for compensated imaging," *Journal of the Optical Society of America*, Vol. 67, No. 3, pp. 375–378, March 1977.

[9] G. Fornaro, G. Franceschetti, and R. Lanari, "Interferometric SAR phase unwrapping using Green's formulation," *IEEE Transactions on Geoscience and Remote Sensing*, Vol. 34, No. 3, pp. 720–727, May 1996.

[10] S. M. Song, S. Napel, N. J. Pelc, and G. H. Glover, "Phase unwrapping of MR phase images using Poisson equation," *IEEE Transactions on Image Processing*, Vol. 4, No. 5, pp. 667–676, May 1995.

[11] M. D. Pritt, "Phase unwrapping by means of multigrid techniques for interferometric SAR," *IEEE Transactions on Geoscience and Remote Sensing*, Vol. 34, No. 3, pp. 728–738, May 1996.

[12] J. R. Rice, *The Approximation of Functions*, Addison-Wesley, Reading, MA, 1964.

[13] G. B. Arfken and H. J. Weber, *Mathematical Methods for Physicists*, 4th ed., Academic Press, San Diego, 1995, ch. 17.

[14] G. M. Ewing, *Calculus of Variations with Applications*, Dover, Inc., New York, 1985.

[15] B. R. Hunt, "Matrix formulation of the reconstruction of phase values from phase differences," *Journal of the Optical Society of America A*, Vol. 69, No. 3, pp. 393–399, March 1979.

[16] H. Takajo and T. Takahashi, "Least-squares phase estimation from phase differences," *Journal of the Optical Society of America A*, Vol. 5, No. 3, pp. 416–425, March 1988.

[17] H. Takajo and T. Takahashi, "Noniterative method for obtaining the exact solution for the normal equation in least-squares phase estimation from the phase difference," *Journal of the Optical Society of America A*, Vol. 5, No. 11, pp. 1818–1827, November 1988.

[18] D. C. Ghiglia and L. A. Romero, "Direct phase estimation from phase differences using fast elliptic partial differential equation solvers," *Optics Letters*, Vol. 14, pp. 1107–1109, Oct. 15, 1989.

[19] B. L. Busbee, G. H. Gollub, and C. W. Nielson, "On direct methods for solving Poisson's equations," *SIAM Journal of Numerical Analysis*, Vol. 7, pp. 627–656, 1970.

[20] J. L. Marroquin and M. Rivera, "Quadratic regularization functionals for phase unwrapping," *Journal of the Optical Society of America A*, Vol. 12, pp. 2393–2400, 1995.

[21] W. H. Press, B. P. Flannery, S. A. Teukolsky, and W. T. Vetterling, *Numerical Recipes: The Art of Scientific Computing*, 2nd ed., University of Cambridge Press, Cambridge, 1992, ch. 18.

[22] W. H. Press, B. P. Flannery, S. A. Teukolsky, and W. T. Vetterling, *Numerical Recipes in C: The Art of Scientific Computing*, 2nd ed., University of Cambridge Press, Cambridge, 1992, p. 864.

[23] H. Silverman, *Complex Variables*, Houghton-Mifflin, Boston, 1975, p. 270.

[24] W. H. Press, B. P. Flannery, S. A. Teukolsky, and W. T. Vetterling, *Numerical Recipes in C: the Art of Scientific Computing*, 2nd ed., University of Cambridge Press, Cambridge, 1992, pp. 517–519.

[25] J. S. Lim, "The discrete cosine transform," in *Two-dimensional Signal and Image Processing*, Prentice-Hall, Englewood Cliffs, NJ, 1990, pp. 148–157.

[26] W. L. Briggs, *A Multigrid Tutorial*, SIAM, Philadelphia, 1987.

[27] G. H. Gollub and C. F. Van Loan, "Iterative methods for linear systems," in *Matrix Computations*, 2nd ed., Johns Hopkins University Press, Baltimore, 1990, pp. 516–538.

[28] J. A. Scales and A. Gersztenkorn, "Robust methods in inverse theory," *Inverse Problems*, Vol. 4, pp. 1071–1091, 1988.

[29] J. M. Huntley and J. R. Buckland, "Characterization of sources of phase discontinuity in speckle interferograms," *Journal of the Optical Society of America A*, Vol. 12, No. 9, pp. 1990–1996, September 1995.

6

COMPARISONS AND CONCLUSION

In Chapters 1 and 2 we introduced the phase unwrapping problem and analyzed it in terms of differential forms and path integrals. We also elucidated the concept of the residue, which in many ways is the key to understanding the phase unwrapping problem. In Chapter 3 we presented the concept of a quality map for guiding the phase unwrapping process. We also presented a number of phase unwrapping examples from SAR interferometry and magnetic resonance imaging, along with methods for defining quality maps, quality masks, and filters for wrapped phase data. In Chapters 4 and 5 we presented a number of phase unwrapping algorithms, which we organized into two main classes: path-following algorithms and minimum-norm algorithms. The path-following algorithms use localized operations that, in some sense, follow paths through the wrapped phase, while the minimum-norm algorithms take a more global approach that seeks to minimize some measure of the difference between the gradients of the wrapped phase and those of the unwrapped solution.

In this chapter we attempt to summarize and consolidate what we have learned. In Section 6.1 we summarize the distinctive features of each algorithm and rank the algorithms in terms of their execution times, memory requirements, and successes and failures on the phase unwrapping examples. We also suggest a few guidelines for deciding which algorithm to apply to a given problem.

In Section 6.2 we explore the links between the path-following and the minimum-norm algorithms. At first glance the two classes of algorithms appear to have little in common. It turns out, however, that one of the path-following algorithms and one of the minimum-norm algorithms produce the *same* solution under certain general conditions. This observation is very surprising, given that their approaches to finding the solution are completely different.

The second link between the path-following and the minimum-norm algorithms is provided by the congruence operation, which we described in the section on "Postprocessing Congruence Operation" in Chapter 5. This operation minimizes the discontinuities in the solutions produced by the least-squares (minimum L^2-norm) algorithms. In this respect it is analogous to path-following algorithms like Goldstein's that seek to minimize the discontinuities by other means, namely branch cuts.

The third link is a theorem that states that the solutions produced by the least-squares algorithms are in some sense the averages of certain path-following solutions. The value of the least-squares solution at any point is shown to be the average of the path-following solutions obtained by integrating the phase values along radial paths from the given point to the boundary of the phase data.

In Section 6.3 we turn our attention to the problem of evaluating the solutions produced by the phase unwrapping algorithms. We define several measures of the success of an algorithm and then apply these measures to the solutions produced by the algorithms described in this book. These measures are generally dependent on a quality map that defines which phase values are valid and which are invalid. We then examine a number of images of these measures, which do not depend on a quality map. Last we propose a method for detecting incorrectly unwrapped phase values. This method works with any algorithm but depends on the choice of a quality map.

In Section 6.4 we discuss the directions in which future research in phase unwrapping is likely to go. We outline three main areas in which we believe further research is needed: (1) generation of quality maps, (2) evaluation of unwrapped results, and (3) development of a minimum L^0-norm algorithm. We conclude with a brief survey of recent efforts to circumvent difficult phase unwrapping problems altogether. In Section 6.5 we conclude the chapter and the book with some final comments and observations.

6.1 COMPARISONS OF PHASE UNWRAPPING ALGORITHMS

In this book we have studied eight phase unwrapping algorithms in detail. Four of these are path-following algorithms, and four are minimum-norm algorithms:

1. Goldstein's algorithm (Section 4.2)
2. Quality-guided algorithm (Section 4.3)
3. Mask cut algorithm (Section 4.4)
4. Flynn's minimum discontinuity algorithm (Section 4.5)
5. Unweighted least-squares algorithm (Section 5.3)
6. PCG algorithm (Section 5.4.3)
7. Multigrid algorithm (Section 5.4.4)
8. Minimum L^p-norm algorithm (Section 5.5)

Except for Goldstein's algorithm, each of these algorithms actually defines an entire class of algorithms. For example, the quality-guided algorithm can be used with a wide variety of quality maps, and different quality maps often yield different results on a given phase unwrapping problem. If the quality map is defined by the phase gradients as in Section 3.3.4, then the algorithm may be regarded as a gradient-following algorithm. If the quality map is defined by the correlation coefficients, then the algorithm may be regarded as a "maximum correlation path" algorithm. The same holds true for all of the algorithms except Goldstein's algorithm and the unweighted least-squares algorithm. In addition to this, Flynn's algorithm, the PCG algorithm, the multigrid algorithm, and the minimum L^p-norm algorithm can use either quality maps or quality masks (i.e., thresholded quality maps), and Flynn's algorithm and the minimum L^p-norm algorithm can be used with or without quality maps. The L^p-norm algorithm can be used with any value of p, although the values $p = 0$ and $p = 1$ are the most useful in practice. The unweighted least-squares algorithm, for its part, represents either direct solution approaches based on the FFT or DCT, or iterative solution approaches such as multigrid. We will continue to refer to them as *algorithms*, even though each algorithm actually represents a *class* of algorithms.

In this section we briefly review the distinctive features of each algorithm. We also summarize the execution time and memory requirements of each algorithm, as well as the performance on each of the phase unwrapping examples. We then present a few guidelines for choosing which algorithm to apply to a given phase unwrapping problem.

6.1.1 Goldstein's Algorithm

Goldstein's algorithm identifies the residues in the wrapped phase data and balances them with branch cuts. The algorithm uses a nearest-neighbor heuristic to find a configuration of branch cuts, the sum of whose lengths is approximately minimal. Because the branch cuts balance the residues, the discontinuities (i.e., adjoining pixels whose values differ by more than π radians) in the unwrapped solution must be confined to the branch cuts. In this way the algorithm approximately minimizes the discontinuities, and it thus approximately minimizes the L^0-norm. Goldstein's algorithm is very fast, requires little memory, and generally unwraps the phase correctly. Its shortcoming is that it does not use information in the phase data, other than the locations of the residues, to guide the placement of the branch cuts. In particular, it does not use a quality map. However, it generally produces a solution that is correct, unlike other algorithms that do not use quality maps such as the unweighted version of Flynn's algorithm, the unweighted version of the L^p-norm algorithm, and the unweighted least-squares algorithms.

6.1.2 Quality-Guided Algorithm

The quality-guided algorithm does not identify residues at all. Rather than depending on branch cuts, it relies completely on a quality map to guide the unwrapping paths. It begins at a pixel and "grows" a region of unwrapped pixels, beginning with the highest-quality pixels and ending with the lowest-quality pixels. The paths followed depend on the quality map. The phase derivative variance and the correlation coefficients yield the best unwrapping paths in practice. The algorithm is very successful, which may be a surprising fact given the lack of a guarantee that the unwrapping paths will not encircle unbalanced residues and thereby introduce spurious discontinuities. The algorithm cannot be used if a good quality map is not available. (However, a quality map can be constructed from phase data alone.)

6.1.3 Mask Cut Algorithm

The mask cut algorithm can be regarded as a cross between Goldstein's algorithm and the quality-guided algorithm. Like Goldstein's algorithm, it identifies residues and joins them with "branch cut" structures that interdict the unwrapping path to prevent the encircling of unbalanced residues. Like the quality-guided algorithm, it grows masks based on the quality values of each pixel as defined by a quality map. Rather than following the highest-quality pixels, however, it follows the lowest-quality pixels from one residue to another until the residues are balanced by regions of low-quality pixels called *mask cuts*. The algorithm does not attempt to minimize the lengths of the mask cuts, although it does perform a "thinning" operation. It relies on the quality map to ensure that the masks are placed and sized properly. Like the quality-guided algorithm, the mask cut algorithm cannot be used if a good quality map is lacking.

6.1.4 Flynn's Minimum Discontinuity Algorithm

Flynn's minimum discontinuity algorithm works with or without a quality map. This algorithm finds the surface that is congruent to the wrapped phase data and whose discontinuities (called *jump counts*) are minimal in some sense. The discontinuities are defined by the error measure of Equation 4.4 in the unweighted case or 4.9 in the weighted case. In Section 6.2 we will see that this is equivalent to minimizing the L^1-norm of the congruent solution. Flynn's algorithm finds the truly minimal solution—that is, it is not merely a *local* minimum. With the exception of the minimum L^p-norm algorithm, the memory and execution time requirements of Flynn's algorithm are higher than those of the other algorithms.

6.1.5 Unweighted Least-Squares Algorithm

The unweighted least-squares algorithm actually consists of several algorithms. There are four algorithms that find a direct solution by means of FFTs or DCTs, each algorithm depending on how the mirror reflection or the boundary conditions are defined. There is also an iterative algorithm based on multigrid techniques that is almost as fast as the direct algorithms. These five algorithms are listed below:

1. FFT method 1 (Section 5.3.1)
2. FFT method 2 (Section 5.3.1)
3. DCT method 1 (Section 5.3.2)
4. DCT method 2 (Section 5.3.2)
5. Multigrid algorithm (Section 5.2.3)

The solutions produced by these algorithms are equivalent to the solution of Poisson's equation with Neumann boundary conditions. The solution can be obtained by FFT techniques, which apply a mirror reflection and then solve the resulting equation by means of a Fourier transform, or by DCT techniques, which define "boundary conditions" and then solve the resulting equation by means of a discrete cosine transform. Poisson's equation can also be solved by Gauss-Seidel relaxation on the residual equations. Although this iterative method is too slow to be useful in practice, it can be sped up tremendously by the use of multigrid techniques. A disadvantage of the FFT/DCT approaches is that the dimensions of the array that holds the wrapped phase data must be powers of two.[1] The multigrid approach does not have this restriction.

The unweighted least-squares algorithms are similar to Goldstein's algorithm in that they are fast, require little memory, and do not need a quality map. However, because the least-squares algorithms unwrap *through* rather than *around* the residues in the data, the solution can be corrupted by these residues. This corruption, which is strongest near the residues, also has a significant effect far away from them. For this reason unweighted least-squares solutions are inappropriate for many applications. However, some applications require smooth solutions that do not need to match the wrapped phase gradients exactly. Examples of such applications are local strain tests in optical interferometry and mirror deformations in adaptive optics. Unweighted least-squares solutions are also suitable when one desires a "quick look" at the terrain elevations produced by SAR interferometry data. The global structure of the terrain is usually distorted by the residues in the wrapped phase data, making the results unsuitable for mensuration, but the local structure of the terrain is accurate enough for 3-D shaded views. We use the least-squares algorithms quite often for this purpose. The main significance of the least-squares algorithms for the purposes of this book is that they form the basis of the weighted least-squares algorithms and the minimum L^p-norm algorithm.

6.1.6 PCG Algorithm

The preconditioned conjugate gradient (PCG) algorithm overcomes the limitations of the unweighted least-squares algorithms by introducing weights. The weights are defined by a user-supplied quality map, and their purpose is to zero-weight the regions of corrupt phase (where the residues tend to lie) to prevent

[1] Or highly composite numbers (i.e., products of small primes) due to the use of the FFT or DCT.

them from corrupting the unwrapped solution. This weighted problem requires iterative techniques for its solution. The simplest algorithm is the Picard algorithm, which has a simple algorithmic structure but converges extremely slowly. The PCG algorithm is much faster because it uses unweighted least-squares solutions as "preconditioners" to speed the convergence. In practice, the algorithm works best with binary-valued weights (i.e, 0's and 1's). Thus it requires the quality map to be thresholded to produce a quality mask. We presented an algorithm for computing a good threshold in Chapter 3. The PCG algorithm seems to work best when the regions of zero-weights are thickened by a few pixels. If a congruent solution (i.e., an unwrapped solution that is congruent to the wrapped input phase) is required, the algorithm requires a postprocessing operation to accomplish this step. We presented such an operation in the section in "Postprocessing Congruence Operation" in Chapter 5.

6.1.7 Weighted Multigrid Algorithm

The weighted multigrid algorithm also obtains the weighted least-squares solution. It is a generalization of the multigrid algorithm for obtaining the unweighted least-squares solution. It applies multigrid techniques to speed the convergence of Gauss-Seidel relaxation on the weighted residual equations. The algorithm includes a somewhat complicated *ad hoc* procedure for transferring the zero-weights from the fine to the coarse grids. The need to define the weights on the coarser grids seems to cause the algorithm to fail on some types of data. We saw an example of this on the phase unwrapping example of Figure 5.36(b) where the coarse weights seemed to isolate portions of the phase. On some problems, however, the algorithm is superior to the PCG algorithm. In the presence of large discontinuities, as in the sheared planes example of Figure 3.9, the algorithm converges extremely quickly. Overall the algorithm converges faster than the PCG algorithm on average.

6.1.8 Minimum L^p-Norm Algorithm

The minimum L^p-norm algorithm finds a minimum L^p-norm solution to a phase unwrapping problem. In other words, it finds a solution whose gradients agree best with those of the wrapped phase in the sense of the L^p-norm. When $p = 2$, the algorithm is equivalent to the (weighted or unweighted) least-squares algorithms. When $p = 1$, the algorithm is in some sense equivalent to Flynn's minimum discontinuity algorithm, as we will see in Section 6.2. When $p = 0$, the algorithm minimizes the number of discontinuities in the unwrapped solution. One drawback of the algorithm is that it finds solutions that are only *locally* minimum.

In contrast to the PCG algorithm and the multigrid algorithm, which require explicit weights to be provided, the minimum L^p-norm algorithm generates its own data-dependent weights when $p \neq 2$. Like Flynn's minimum discontinuity algorithm, it can be used with or without user-supplied weights. If weights are supplied, they are combined with the data-dependent weights. Like the PCG and multigrid algorithms, the minimum L^p-norm algorithm seems to work best with binary-valued weights. The memory and computational requirements are very high due to the doubly iterative structure of the algorithm. It iterates the PCG algorithm, which in turn iterates a FFT/DCT algorithm.

6.1.9 Comparisons

In Table 6.1 we rank the eight algorithms in terms of their memory requirements and average execution times. The two rankings are approximately identical. The FFT/DCT algorithm has minimal memory requirements: All of the computations are performed in place in one array. Goldstein's algorithm is next, requiring only three arrays. The minimum L^p-norm algorithm requires the most memory: eight arrays. In terms of execution time, the FFT/DCT algorithm is the fastest, requiring only 23 seconds to unwrap a 1024×1024 array of data on our IBM RS/6000 workstation. Goldstein's algorithm requires only 30 seconds, while the minimum L^p-norm algorithm requires about two hours.

In Table 6.2 we rank the algorithms in terms of their successes and failures on the phase unwrapping examples. Flynn's algorithm and the minimum L^p-norm algorithm were the only algorithms that correctly unwrapped all of the examples, although they did have minor errors on the Isolation Peak simulated IFSAR example of Figure 3.8. The quality-guided algorithm was the only algorithm that unwrapped this example flawlessly, but it could not unwrap the noise-corrupted sheared planes example of Figure 4.15 due to the lack of a good quality map. The mask cut and the PCG algorithms yielded results comparable to those of the quality-guided algorithm except for minor errors on the Isolation Peak example. Goldstein's algorithm failed on the Long's Peak and Isolation Peak examples,

Table 6.1 Memory requirements and average execution times of the eight phase unwrapping algorithms for an array size of 1024 × 1024 on an IBM RS/6000 workstation.

Algorithm	Memory Requirements	Algorithm	Execution Time
FFT/DCT	1	FFT/DCT	23 seconds
Goldstein	3	Goldstein	30 seconds
Quality	4	Multigrid	1.5 minutes
Mask Cut	4	Quality	3 minutes
PCG	5	Mask Cut	4 minutes
Flynn	5	PCG	10 minutes
Multigrid	7	Flynn	20 minutes
L^p-norm	8	L^p-norm	2 hours

Note: The memory requirements are expressed as number of arrays. The algorithms are ranked in order from lowest- to highest-memory requirements, and again from fastest to slowest.

Table 6.2 Results of the algorithms on the eight phase unwrapping examples of Chapters 3 and 4.

	Examples							
Algorithm	*a*	*b*	*c*	*d*	*e*	*f*	*g*	*h*
Flynn	*y*	*y*	*y*	*y*	(*y*)	*y*	*y*	*y*
L^p-norm	*y*	*y*	*y*	*y*	(*y*)	*y*	*y*	*y*
Quality	*y*	*y*	*y*	*y*	*y*	*y*	*n*	*y*
Mask cut	*y*	*y*	*y*	*y*	(*y*)	*y*	*n*	*y*
PCG	*y*	*y*	*y*	*y*	(*y*)	*y*	*n*	*y*
Goldstein	*y*	*y*	*y*	*n*	*n*	*y*	*y*	*n*
Multigrid	*y*	*y*	*y*	*y*	*n*	*y*	*n*	*y*
FFT/DCT	*n*	*n*	*n*	*n*	*n*	*n*	*n*	*n*

Note: The examples are labeled as follows: (*a*) IFSAR, (*b*) MRI head, (*c*) MRI knee, (*d*) Long's Peak, (*e*) Isolation Peak, (*f*) Sheared planes, (*g*) Noisy shear and (*h*) Spiral shear. An entry of *y* signifies a successful unwrapping, (*y*) indicates that the unwrapping errors were only minor, while *n* indicates failure. The algorithms are ranked in order from most successful to least successful.

Table 6.3 Characteristics of the algorithms.

		Requires	
Algorithm	Identifies Residues?	Quality Map?	Map or Mask?
Goldstein	*y*	*n*	–
Quality	*n*	*y*	map
Mask Cut	*y*	*y*	map
Flynn	*n*	*y*/*n*	mask
FFT/DCT	*n*	*n*	–
PCG	*n*	*y*	mask
Multigrid	*n*	*y*	mask
L^p-norm	*n*	*y*/*n*	mask

Note: The first column indicates the algorithms that identify residues. The next column indicates which ones require a quality map. The entry *y*/*n* indicates the two algorithms that work with or without a quality map. The last column indicates which of the algorithms work best with quality masks (binary weights) as opposed to quality maps.

while the multigrid algorithm unwrapped Long's Peak correctly but failed on the noise-corrupted sheared planes example. The FFT/DCT algorithms failed on all of the examples.

In Table 6.3 we indicate which algorithms explicitly identify residues, which ones use quality maps or weights, and which ones work best with quality masks or binary weights. Only two algorithms identify residues: Goldstein's and the mask cut algorithm. (The L^p-norm algorithm also identifies residues, but only to determine if the iterations can be terminated sooner.) All of the algorithms use quality maps, except Goldstein's and the FFT/DCT (unweighted least-squares) algorithms. Of these, only the quality-guided and mask cut algorithms actually use quality *maps*, while the others seem to work better with quality *masks* in practice. Flynn's algorithm and the minimum L^p-norm algorithm can be used with or without a quality map.

Which algorithm should be used in practice? The answer to this question depends on the particular phase unwrapping problem to be solved. We will repeat and stress a few of the points we made in Chapter 4.

We recommend that Goldstein's algorithm should always be tried first. This algorithm has simple input requirements (i.e., a quality map is not needed), and it is very fast. When it fails, the user is usually warned by an inspection of the branch cuts to determine if any regions were isolated from the rest of the image. We will describe other techniques for detecting failure in Section 6.3. If Goldstein's algorithm fails and a good quality map is available, then the quality-guided, mask cut, or PCG algorithm should be tried. The multigrid algorithm can also be tried, although it may fail if the zero-weights isolate regions of the phase data on the coarse grids. (The techniques of Section 6.3 can be used to detect such a failure.) If the results are not satisfactory, then the user can "tinker" with the quality map to try to obtain a successful unwrapping. Different quality maps can yield different results on a given problem. The user can choose not only the measure used to compute the quality map but also the type and amount of filtering to be applied to the quality map, the threshold to be applied if a quality *mask* is desired, and whether any morphological operations such as thinning or thickening are to be applied to the mask. There are many ways to generate a quality map!

If these five algorithms fail, then we recommend the use of Flynn's minimum discontinuity algorithm. Although this algorithm takes longer to execute than the others, it can sometimes "crack" tough phase unwrapping problems on which the other algorithms fail. The same is true for the L^p-norm algorithm.

The following list ranks the algorithms in the order in which we recommend they be tried: (1) Goldstein, (2) quality-guided, (3) PCG, (4) mask cut, (5) multigrid (weighted), (6) Flynn's minimum discontinuity, (7) minimum L^p-norm, and (8) FFT/DCT.

After an algorithm has unwrapped a given phase unwrapping problem, how should the results be evaluated? Can one determine where the phase may have been unwrapped incorrectly? Comparison of the results from several algorithms usually provides insight into whether the problem was correctly unwrapped. For example, the fact that seven of the eight algorithms of this book produced exactly the same result on the IFSAR example (exclusive of the patches of noise) suggests quite strongly that this result is correct. Because we have supplied C language implementations for the algorithms in this book, there is no reason why *all* the algorithms should not be run and the results compared. We will provide another method for evaluating unwrapped solutions and detecting regions of errors in Section 6.3 after examining the links between the path-following and the minimum-norm algorithms in the next section.

6.2 LINKS BETWEEN PATH-FOLLOWING AND MINIMUM-NORM ALGORITHMS

When we began writing this book we did not expect it to turn out this way, but it appears to be true: Minimizing the lengths of branch cuts, minimizing the number of discontinuities, minimizing the number of places where the gradients do not match, and so forth, all appear to converge on the same theme. Good solutions to phase unwrapping problems appear to minimize some error norm. In Section 6.3 we will use this theme to motivate the definition of some measures that can be used to evaluate phase-unwrapping solutions.

In this section we use this theme to explore the links between the path-following algorithms and the minimum-norm algorithms. The most obvious link is that between Flynn's minimum discontinuity algorithm and the minimum L^p-norm algorithm. After exploring this connection, we examine in more detail the congruence operation, which was introduced in Chapter 5. This operation provides a connection between the least-squares algorithms and the path-following algorithms. Finally we explore yet another link, a theoretical result that states that the least-squares solutions are essentially the averages of path-following solutions.

6.2.1 First Link

We begin with the link between Flynn's minimum discontinuity algorithm and the minimum L^p-norm algorithm. Given an array of wrapped phase data to unwrap, Flynn's algorithm finds the solution $\phi_{m,n}$ that minimizes the error measure

$$E = \sum |v_{m,n}| + \sum |z_{m,n}|, \tag{6.1}$$

where $v_{m,n}$ and $z_{m,n}$ are the horizontal and vertical jump counts, respectively, and the summation is evaluated over all the indexes m and n. The jump counts express the number of 2π-multiples by which adjoining pixels of the solution differ. Because the solution is required to be congruent to the wrapped phase data, minimizing Equation 6.1 is equivalent to minimizing the error measure

$$\epsilon = \sum_{i=0}^{M-2} \sum_{j=0}^{N-1} |\phi_{i+1,j} - \phi_{i,j} - \Delta_{i,j}^{x}| + \sum_{i=0}^{M-1} \sum_{j=0}^{N-2} |\phi_{i,j+1} - \phi_{i,j} - \Delta_{i,j}^{y}|, \tag{6.2}$$

where $\Delta^x{}_{i,j}$ and $\Delta^y{}_{i,j}$ are the partial derivatives of the wrapped phase data. (The requirement of congruence makes the "quantized" jump count measure equal to the "continuous" measure of Equation 6.2.)

The minimum L^p-norm algorithm, on the other hand, finds a solution $\phi_{m,n}$ that minimizes the error measure

$$\epsilon = \sum_{i=0}^{M-2}\sum_{j=0}^{N-1} |\phi_{i+1,j} - \phi_{i,j} - \Delta^x_{i,j}|^p$$

$$+ \sum_{i=0}^{M-1}\sum_{j=0}^{N-2} |\phi_{i,j+1} - \phi_{i,j} - \Delta^y_{i,j}|^p. \qquad (6.3)$$

Comparison of Equations 6.2 and 6.3 shows that Flynn's algorithm and the L^p-norm algorithm find the *same* solution when $p = 1$! This conclusion is surprising because the two algorithms use completely different methods to find the solution. Flynn's algorithm approaches the problem as a "discontinuity minimization" problem akin to the branch cut minimization approach of Goldstein's algorithm. It utilizes a tree-growing approach that traces paths of discontinuities, detects the paths that form loops, and adds multiples of 2π to the phase values enclosed by the loops. The L^p-norm algorithm, on the other hand, generates data-dependent weights from the discontinuities and solves the weighted least-squares problem that minimizes the squared differences between the gradients of the solution and those of the wrapped phase. The only feature the algorithms seem to have in common is that they exploit the discontinuity information.

This connection between the two algorithms remains true in the weighted case as well. The weighted forms of Equations 6.2 and 6.3 remain equivalent when $p = 1$.

The advantages and disadvantages of each algorithm need to be pointed out. The primary disadvantage of the L^p-norm algorithm is that it is guaranteed to find a solution that is only *locally* minimal. Flynn's algorithm is guaranteed to find the *globally* minimal solution. The advantage of the L^p-norm algorithm, on the other hand, is that it can find minimum L^p-norm solutions for any $p \geq 0$. In particular, it can find (locally) minimal L^0-norm solutions—solutions that minimize the total number of discontinuities.

6.2.2 Second Link

We now move on to the second link between the path-following algorithms and the minimum-norm algorithms. When used with the postprocessing congruence operation, the least-square algorithms find solutions that are approximately minimal L^0-norm solutions. In this sense they find approximately the "same" solutions Goldstein's algorithm finds because, in minimizing the branch cut lengths, Goldstein's algorithm is essentially minimizing the number of discontinuities and hence the L^0-norm. Of course the solutions will differ on many phase-unwrapping problems because they only *approximately* minimize the L^0-norm. We now investigate in what sense the least-squares algorithms approximately minimize the L^0-norm.

We showed in Section 2.4 that the gradient field of any wrapped phase function can be decomposed into two parts by means of the Helmholtz decomposition theorem (see Equation 2.33). The conservative or irrotational part is residue-free and can be unwrapped easily by an unweighted least-squares algorithm. The nonconservative or rotational part contains the curl components, which lead to residues and path dependencies in phase unwrapping. This part can be obtained by subtracting the irrotational part (i.e., the least-squares solution) from the original wrapped phase and then rewrapping the result into the interval $(-\pi, \pi]$. In other words, the rotational part is defined by the equation $\phi_R = \mathcal{W}\{\psi - \phi\}$, where ψ is the wrapped phase, ϕ is the least-squares unwrapped phase, and $\mathcal{W}$ is the wrapping operator. Note that the right-hand side of this equation is simply the rightmost term of the congruence operation defined by Equation 5.88.

Because the phase gradient is unchanged if a constant is added to the wrapped phase, the least-squares solution ϕ can contain an arbitrary constant bias. Therefore the rotational component ϕ_R, which is derived from ϕ, also can be influenced by this arbitrary constant. In fact the *only* arbitrariness in the phase decomposition is this arbitrary constant. We are therefore free to adjust the bias on ϕ when computing ϕ_R. This one free parameter allows us to change the locations and shapes of the fringe lines contained in ϕ_R as illustrated in Figure 6.1. This figure shows the rotational components of Figure 2.7(d). Notice that the fringe lines originate and terminate on the residues and the boundary. By varying the bias, the fringe lines move around and change their shapes.

The postprocessing congruence algorithm in Chapter 5 selects the bias that minimizes the number of discontinuities in the unwrapped phase. Because the irrotational component has no discontinuities, the selected bias actually minimizes the number of discontinuities in the residual component ϕ_R, which is equivalent to minimizing its L^0-norm. Because the phase discontinuities across the fringe lines are (in the continuous limit) exactly 2π radians in magnitude, minimizing the overall length of the residual fringe lines is also equivalent to minimizing the L^1-norm. However, after adding ϕ_R to ϕ to force congruence with ψ, the resulting unwrapped phase will no longer have a minimum L^1-norm measure but simply a minimum L^0-norm measure.

An example where the unweighted least-squares solution is forced to be congruent to the original wrapped phase is shown in Figure 6.2. This leads to an interesting question. Does the congruence operation define another simple way to do phase unwrapping? That is, could one simply compute the least-squares solution and force congruence with the minimum L^0-norm measure? The answer is no. The unweighted least-squares algorithm obtains the conservative solution without preferential guidance from the weights or quality mask. The residual, or nonconservative solution, contains all the residues and therefore all the pro-

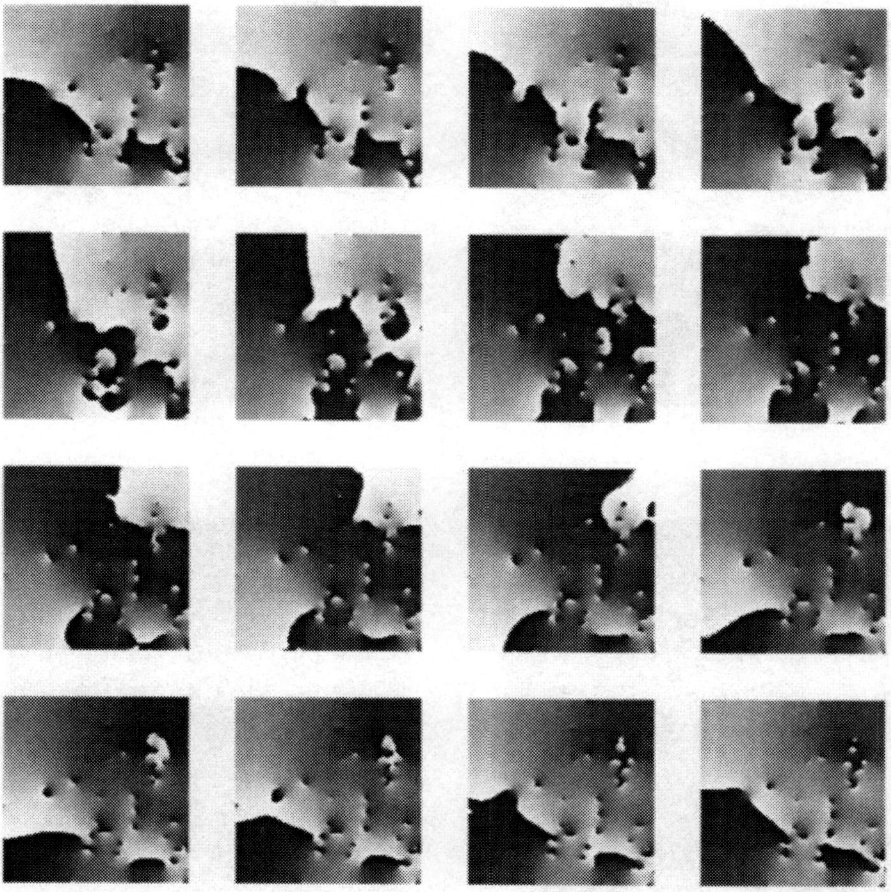

Figure 6.1 Illustration of how the fringe lines of a rotational phase component change under the influence of an additive constant. The sequence, which progresses from left-to-right, top-to-bottom, represents the rewrapped rotational phase component of Figure 2.7(d) with the "bias" or additive term $c_k = 2\pi k/16$, $k = 0, 1, 2, \ldots, 15$. The phase in the lower left corner ($k = 12$) has the minimum number of discontinuities.

blems. By adjusting the bias, one cannot do much to the residual at this point. If a more "intelligent" conservative solution were obtained by a weighted least-squares algorithm, fewer problems would be transferred to the residual. Forcing congruence at this stage would lead to a more desirable result. This was convincingly demonstrated by the PCG examples of Section 5.4.3.

6.2.3 Third Link

We now describe a third link between the path-following algorithms and the minimum-norm algorithms. Fornaro et al. [1] have shown that the solution ϕ

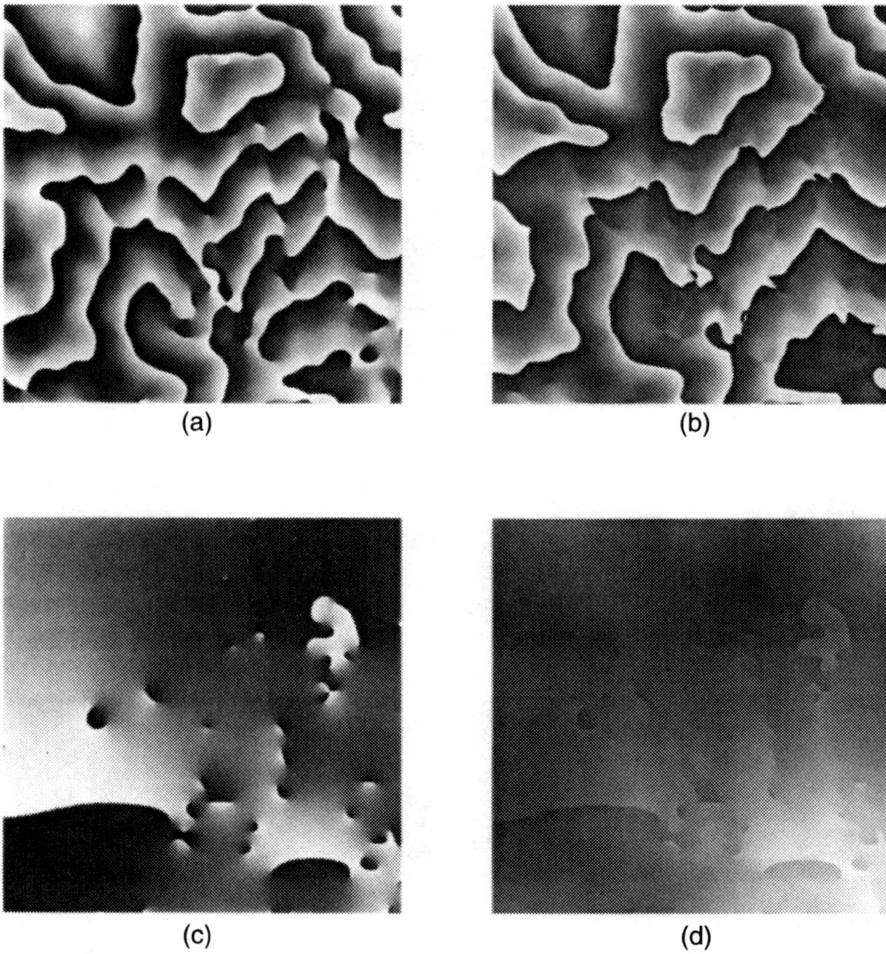

Figure 6.2 Forcing congruence between the least-squares phase and the original wrapped phase. (a) Original wrapped phase. (b) Rewrapped least-squares phase. (c) Residual phase (rotational component) that has the minimum number of discontinuities. (d) Congruent phase (with minimum number of discontinuities) obtained by adding the unwrapped phase of (b) to that of (c).

produced by an unweighted least-squares algorithm satisfies the equation

$$\phi(\mathbf{r}_0) = -\iint_D \mathbf{a}(\mathbf{r}) \cdot \nabla g(\mathbf{r}_0 - \mathbf{r})\, dS + \oint_C \phi(\mathbf{r}_C) \frac{\partial g(\mathbf{r}_0 - \mathbf{r}_C)}{\partial n}\, dc, \tag{6.4}$$

where $\phi(\mathbf{r}_0)$ is the value of the least-squares solution at any given point $\mathbf{r}_0$ of the domain D, C is the boundary of D, $\mathbf{a}$ is the wrapped phase gradient field, g is a special function called Green's function, and $\partial g/\partial n$ is the normal derivative of g along the boundary. Fornaro et al. [2] have also shown that the solution satisfies the equation

$$\phi(\mathbf{r}) = \frac{1}{2\pi}\int_0^{2\pi}\left[\phi(\mathbf{p}) + \int_C \mathbf{a}(\xi)\cdot d\xi\right] d\theta, \tag{6.5}$$

where $C = C(\mathbf{r}, \theta)$ is the radial path from the point $\mathbf{r}$ in the domain D to the point $\mathbf{p} = \mathbf{p}(\mathbf{r}, \theta)$ where it intersects the boundary. In other words, the least-squares solution at a given point is the average of all the solutions obtained by simple path-following on the radial lines from the point to the boundary. (Note that the terminal values of each path must be determined from the least-squares solution on the boundary.) This is a very interesting and somewhat surprising connection between the least-squares and path-following approaches. This approach also extends naturally to the case of weighted least squares. In this case the solution at any given point is also the average of the solutions obtained by path-integration along the radial paths. However, the value of the solution on the boundaries of the zero-weighted regions must be known *a priori.* Wherever a path intersects a zero-weighted region, the correct phase jump across the region, which comes from the knowledge of the solution along the boundary, must be used in the definition of the path integral. These results probably have little practical value, but they are very interesting from a theoretical standpoint.

6.3 MEASURES OF SUCCESS

We now turn our attention to the problem of evaluating the solutions produced by the phase-unwrapping algorithms. We will confine our attention to solutions that are congruent to the wrapped input phase. (This is always the case for the path-following algorithms, and the postprocessing congruence operation described in Section 5.4.3 ensures that it is the case for the minimum-norm algorithms as well.) After proposing several measures based on the L^p norm for determining how good an unwrapped solution is, and after examining the distributions of the discontinuities (called "discontinuity maps") on a number of examples, we will describe a method for identifying the regions of phase, if any, that were unwrapped incorrectly.

6.3.1 Measures

Because all of the algorithms, with the possible exception of the quality-guided algorithm and the mask cut algorithm, minimize or tend to minimize some error norm, the key to the effective evaluation of an unwrapped solution seems to be the formulation of a suitable measure. Flynn's algorithm minimizes the L^1-norm, Goldstein's algorithm approximately minimizes the L^0-norm, and the minimum L^p-norm algorithm locally minimizes the L^p-norm. A natural measure for evaluating solutions then appears to be the L^p-norm. We define the following measure for evaluating the quality of an unwrapped solution:

$$\epsilon = \frac{1}{MN}\sum_{i=0}^{M-2}\sum_{j=0}^{N-1} w^x{}_{i,j}|\phi_{i+1,j} - \phi_{i,j} - \Delta^x_{i,j}|^p$$

$$+ \frac{1}{MN}\sum_{i=0}^{M-1}\sum_{j=0}^{N-2} w^y{}_{i,j}|\phi_{i,j+1} - \phi_{i,j} - \Delta^y_{i,j}|^p, \qquad (6.6)$$

where $w^x{}_{i,j}$ and $w^y{}_{i,j}$ are user-defined weights that define the quality of the wrapped phase gradients $\Delta^x_{i,j}$ and $\Delta^y_{i,j}$, respectively. These weights, which are normally derived from a quality map, may take values between 0 and 1, or they may be binary-valued with the values 0 and 1. They may also be omitted altogether (i.e., $w^x{}_{i,j} = w^y{}_{i,j} = 1$) if there are no weights to supply. Given a quality map with the values $\{w_{i,j}\}$, the weights are normally defined by the formulas $w^x{}_{i,j} = min(w_{i+1,j}, w_{i,j})$ and $w^y{}_{i,j} = min(w_{i,j+1}, w_{i,j})$.

It is important to note that this measure depends on the quality map that defines the weights. In addition to the choice of a quality map, there is also a choice of the value p. We will consider only three cases: $p = 0$, $p = 1$ and $p = 2$. (There may be no real benefit when $p > 2$.) These values cause the measure to signify different qualities about the unwrapped solution:

1. $p = 0$ produces a *count* of the number of samples at which the gradients of the unwrapped solution *do not match* those of the wrapped phase. For congruent solutions, this is equivalent to the number of discontinuities—that is, samples that differ from a neighboring sample by more than π radians. In practice, we modify this measure so that each sample is counted at most once, since the two summations in Equation 6.6 may count the same sample twice.
2. $p = 1$ measures the average absolute deviation of the gradients. This is the measure that Flynn's minimum discontinuity algorithm minimizes (for congruent solutions).
3. $p = 2$ measures the average squared deviation of the gradients. This is the measure that the least-squares algorithms of Chapter 5 minimize (without the postprocessing congruent operation).

We will now restrict our attention to the case $p = 0$, since this seems to be the error measure that one most often seeks to minimize in practice. The reader may object that the $p = 0$ measure will always be minimized by the minimum L^p-norm algorithm, but this is not true, since this algorithm can find only *locally* minimal solutions.

Figure 6.3(a) shows the discontinuities (marked in black) for the unwrapped result produced by Goldstein's algorithm on the Long's Peak example of

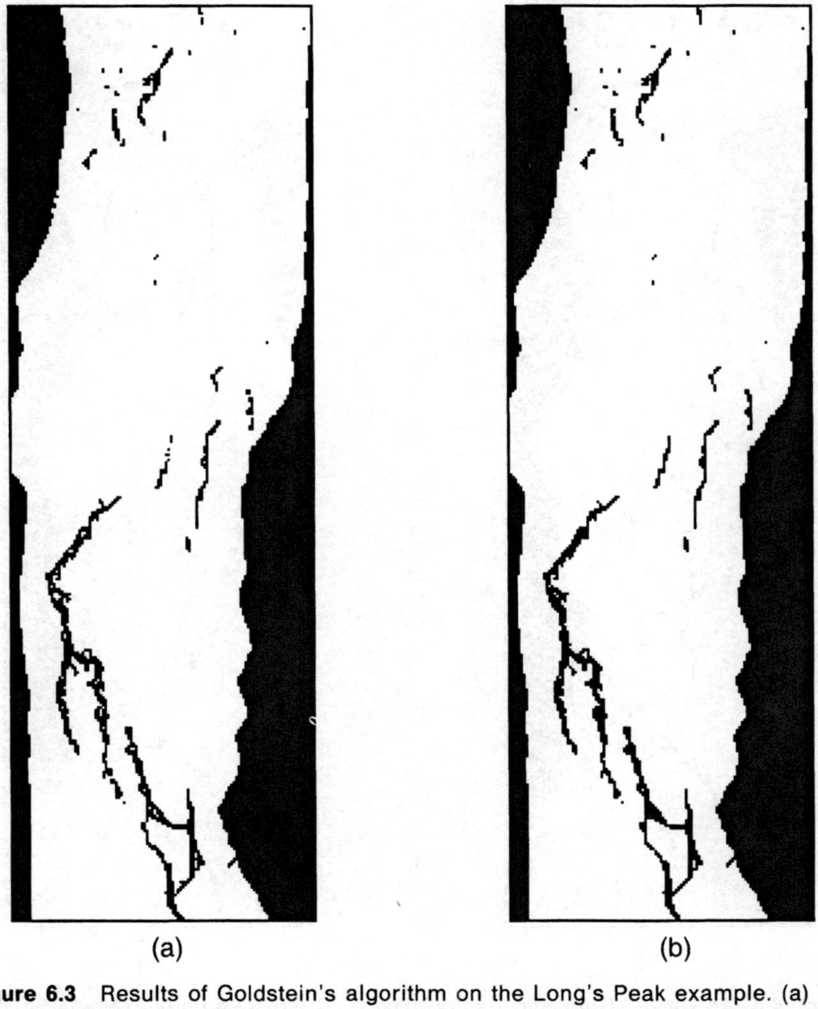

(a) (b)

Figure 6.3 Results of Goldstein's algorithm on the Long's Peak example. (a) The discontinuities in the unwrapped result. (b) The branch cuts produced by Goldstein's algorithm.

Figure 3.7. Goldstein's branch cuts are shown in Figure 6.3(b) for comparison. The discontinuities and the branch cuts coincide almost everywhere. In fact it is easy to see that the discontinuities must form a subset of the branch cuts. That is, if there is a discontinuity in Goldstein's result, it must lie on a branch cut. This is due to the fact that the branch cuts balance the residues, as well as the fact that the path integrals that unwrap the phase do not cross any branch cuts.

The L^0 measure for this example is 0.02505. That is, approximately 2.5 percent, or 1321 of the 52,733 nonmasked pixels, are discontinuities, or pixels

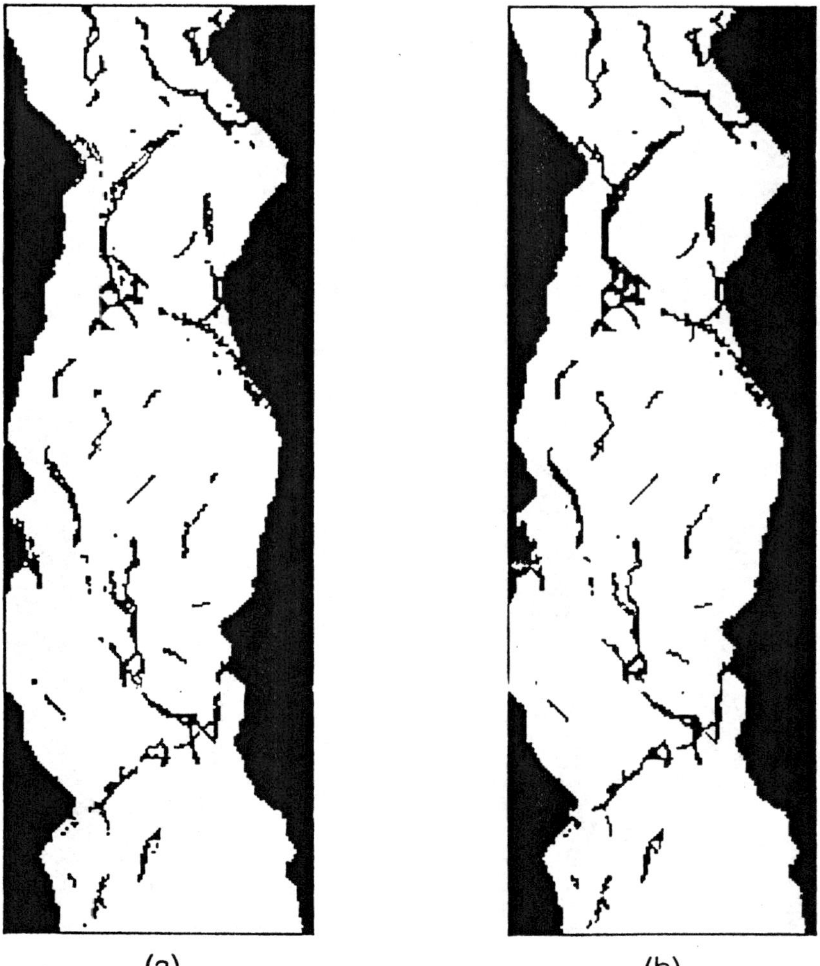

Figure 6.4 Results of Goldstein's algorithm on the Isolation Peak example. (a) The discontinuities in the unwrapped result. (b) The branch cuts produced by Goldstein's algorithm.

whose values differ by more than π radians from those of a neighbor. When the mask of Figure 3.30(c) is used to zero-weight the regions of corrupted phase, the L^0 measure drops to 0.003265. In this case approximately 0.3 percent, or 172, of the pixels are discontinuities. Interestingly, these discontinuities lie on the "misplaced" branch cuts discussed in Chapter 4.

The discontinuities and the branch cuts produced by Goldstein's algorithm on the Isolation Peak example of Figure 3.8 are shown in Figures 6.4(a) and (b). Once again the discontinuities are seen to be a subset of the branch cuts, and the two maps are nearly identical.

Table 6.4 Unweighted and weighted L^0 measures for a number of unwrapped solutions of the Long's Peak problem.

Algorithm	Unweighted L^0 Measure	Weighted L^0 Measure
Goldstein	0.02505	0.003265
Quality–variance	0.02114	0.000146
Quality–correlation	0.02312	0.000582
Quality–gradient	0.02107	0.000811
Quality–pseudocorrelation	0.08338	0.067945
Mask cut–variance	0.02596	0.000104 .
Mask cut–correlation	0.02996	0.003972
Mask cut–gradient	0.02822	0.001456
Mask cut–pseudocorrelation	0.11105	0.099141
PCG	0.03330	0.005761
Multigrid (weighted)	0.03055	0.000270
Flynn's (weighted)	0.03108	0.005407
L^p-norm (weighted)	0.02983	0.000083
Flynn's (unweighted)	0.02327	0.009172
L^p-norm (unweighted)	0.02018	0.005499

Note: The quality-guided and mask cut algorithms were run with four different quality maps.

Table 6.4 lists the values of the L^0 measure for a number of unwrapped solutions for the Long's Peak example. The first column of numbers lists the unweighted measures, while the second column lists the weighted measures when the mask of Figure 3.30(c) is used. As can be seen, the unweighted measures are not useful for ranking the algorithms. For example, the measure of the unweighted version of Flynn's algorithm is 0.02327, which is less that that of the weighted version (0.03108). However, we saw from Figures 4.52 and 4.54 that the unweighted version places its discontinuities in the "wrong" places, while the weighted version unwraps the problem correctly. Thus the unweighted measure merely indicates that the unweighted version of Flynn's algorithms has fewer discontinuities than the weighted version, but it says nothing at all about the placement of these discontinuities.

The weighted measures are much more useful for ranking the solutions. The unweighted version of Flynn's algorithm has a higher measure (0.009172) than that of the weighted version (0.005407) as one would expect. The "best" algorithm is the weighted L^p-norm algorithm with a measure of 0.000083, followed by the mask cut algorithm and the quality-guided algorithm (where the quality map is defined by the phase derivative variance) with measures of 0.000104 and 0.000146, respectively. However, the PCG algorithm has a higher measure (0.005761) than that of the unweighted L^p-norm algorithm (0.005499), even though the former algorithm, unlike the latter, unwrapped the example

Table 6.5 Unweighted and weighted L^0 measures for the Isolation Peak example.

Algorithm	Unweighted L^0 Measure	Weighted L^0 Measure
Goldstein	0.04517	0.006160
Quality–variance	0.04136	0.001294
Quality–correlation	0.04132	0.001414
Quality–gradient	0.04046	0.001750
Quality–pseudocorrelation	0.13370	0.106882
Mask cut–variance	0.05424	0.003020
Mask cut–correlation	0.04811	0.001462
Mask cut–gradient	0.05995	0.006040
Mask cut–pseudocorrelation	0.12543	0.099163
PCG	0.06107	0.011553
Multigrid (weighted)	0.05286	0.002852
Flynn's (weighted)	0.04734	0.007891
L^p-norm (weighted)	0.05182	0.001031
Flynn's (unweighted)	0.04144	0.018984
L^p-norm (unweighted)	0.03888	0.012225

Note: The quality mask of Figure 3.32(b) was used to calculate the weighted measures. (The quality-guided and mask cut algorithms were run with four different quality maps.)

correctly. These results are essentially the same for the Isolation Peak example, as Table 6.5 shows. What is the problem here?

It turns out that while the PCG algorithm unwraps the examples correctly, it has many discontinuities just outside the borders of the zero-weights. We noticed in Chapter 5 that the errors in the unwrapped results of the PCG algorithm are located at the correct regions of corrupted phase, but they tend to be thick. See, for example, Figures 5.23(b) and 5.25(b). The problem with the measures in Tables 6.4 and 6.5 can be overcome by thickening the masks used to calculate the weighted measures. In any case, the measures are useful to some degree for indicating whether an algorithm succeeded or failed. For example, one can see from the tables that the use of the pseudocorrelation quality map (with either the quality-guided or the mask cut algorithm) resulted in relatively high L^0 measures, an observation that points to the presence of errors in the results.

6.3.2 Discontinuity Maps

The L^p-norm measures are much more useful for indicating where the discontinuities are located as in Figures 6.3 and 6.4. We will now examine a number of these "discontinuity maps." Figure 6.5(a) shows the discontinuity maps for the quality-guided algorithm on the Long's Peak example. The phase derivative variance was used to define the quality map. Figure 6.5(b) shows the discontinuities when the correlation coefficients are used instead to define the quality

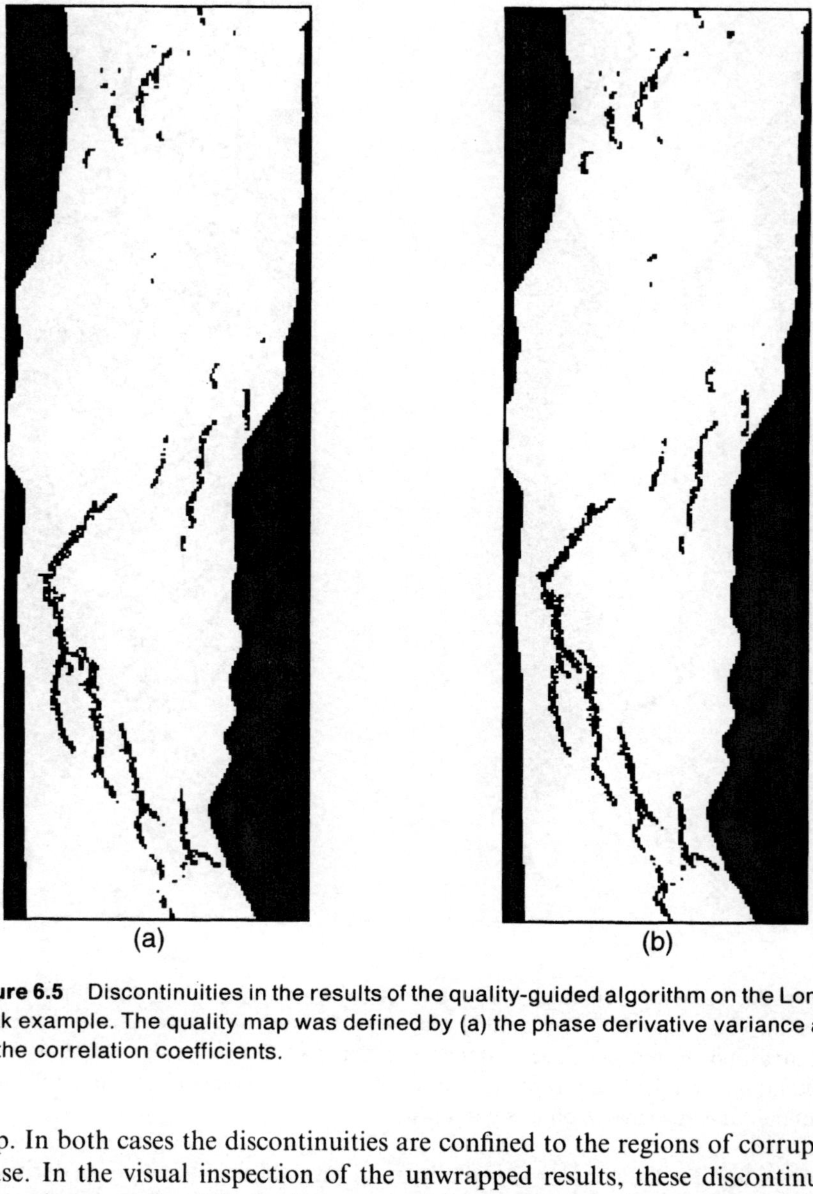

Figure 6.5 Discontinuities in the results of the quality-guided algorithm on the Long's Peak example. The quality map was defined by (a) the phase derivative variance and (b) the correlation coefficients.

map. In both cases the discontinuities are confined to the regions of corrupted phase. In the visual inspection of the unwrapped results, these discontinuity maps play the role of "branch cut maps" for algorithms that do not generate branch cuts. Comparison with Figure 6.3(a) shows that there are differences between the quality-guided algorithm and Goldstein's algorithm in regard to where the "branch cuts" (discontinuities) were placed and hence which paths were followed to unwrap the phase. Refer to the discussions of the various results on the Long's Peak example in Chapter 4 for more details.

Figure 6.6(a) shows the discontinuities for the quality-guided algorithm on

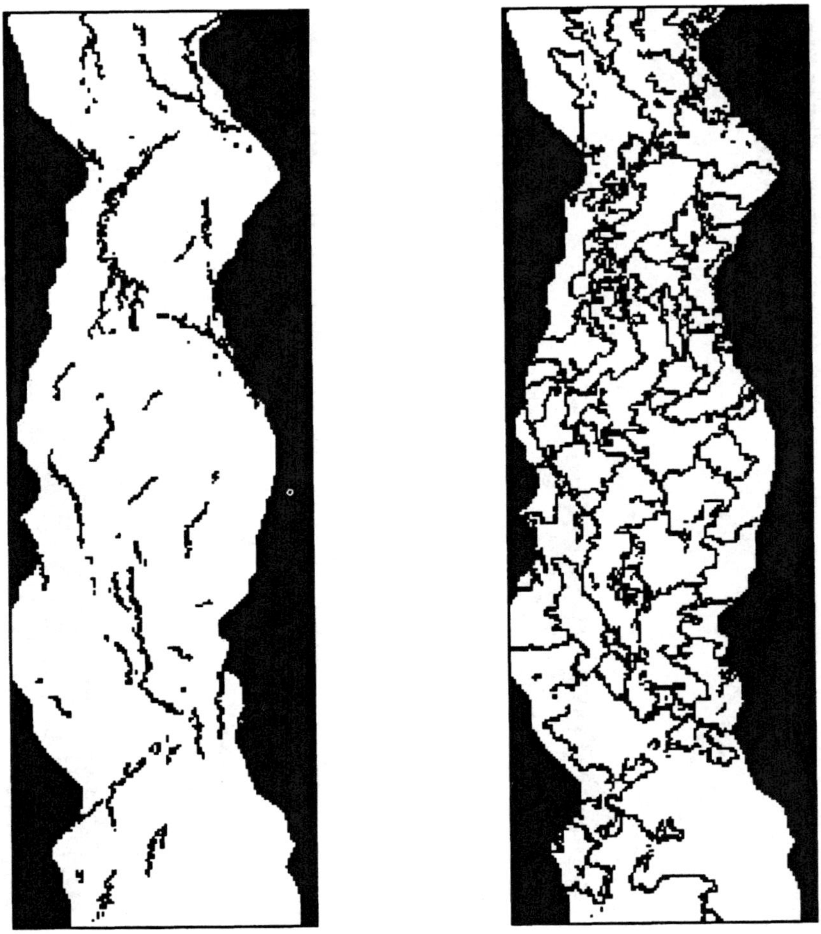

(a) (b)

Figure 6.6 Discontinuities in the results of the quality-guided algorithm on the Isolation Peak example. The quality map was defined by (a) the phase derivative variance and (b) the pseudocorrelation coefficients. The excessive number of discontinuities and isolated regions shows that the pseudocorrelation coefficients produced useless results on this example.

the Isolation Peak example. The quality map was defined by the phase derivative variance. The result is essentially correct as the error map of Figure 4.28(a) shows. Comparison of the discontinuity map with Figure 6.4(a) shows where this algorithm and Goldstein's algorithm differed in their placement of branch cuts (i.e., discontinuities) and hence where they differed in the unwrapping paths. Figure 6.6(b) shows the discontinuity map of a "catastrophic failure." In this case the quality map was defined by pseudocorrelation.

Figures 6.7(a) and (b) show the discontinuity maps for the mask cut algorithm

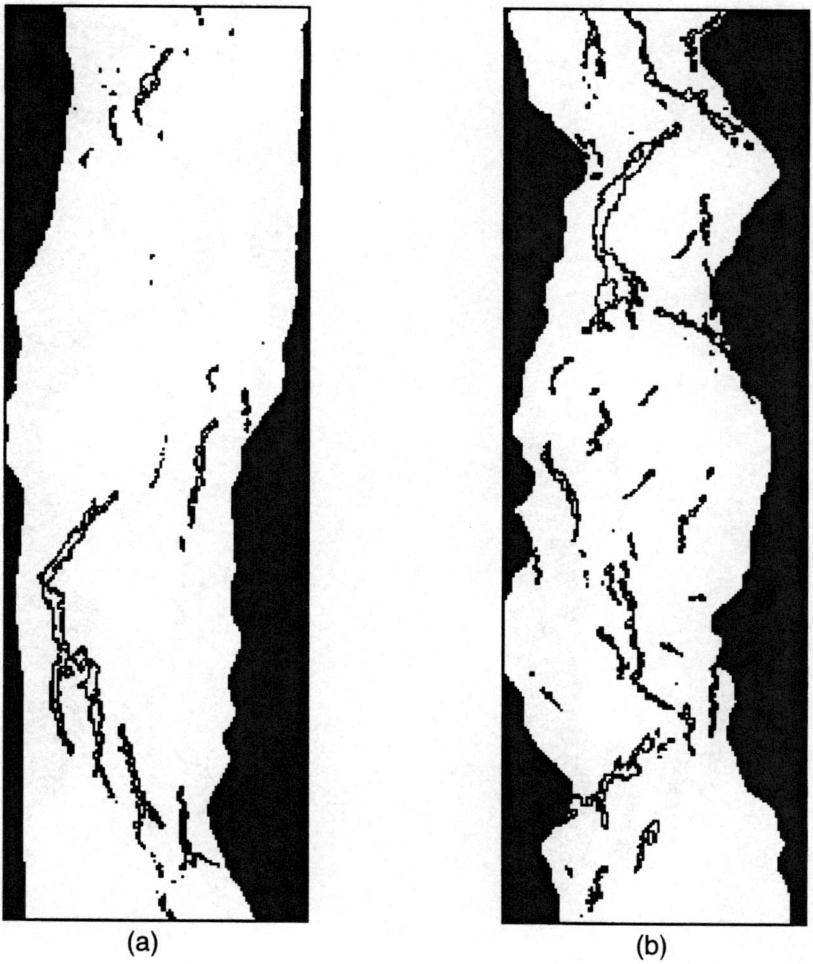

Figure 6.7 Discontinuities in the solutions produced by the mask cut algorithm on (a) the Long's Peak example and (b) the Isolation Peak example.

on the Long's Peak and Isolation Peak examples. The quality map was defined by the phase derivative variance. Compare these figures with the mask cuts shown in Figures 4.36(a) and 4.38(a). The unusual "outline" effect in the figures is due to the fact that the regions within the mask cuts are defined have an unwrapped value of zero.

Figures 6.8(a) and (b) show the discontinuity maps for the PCG algorithm. The discontinuities are very thick, as we noted in the earlier discussion of the L^0 measures listed in the tables. Compare these figures with the error maps shown in Figures 5.23(b) and 5.25(b). Figures 6.9(a) and (b) show the discontinuity maps for the multigrid algorithm.

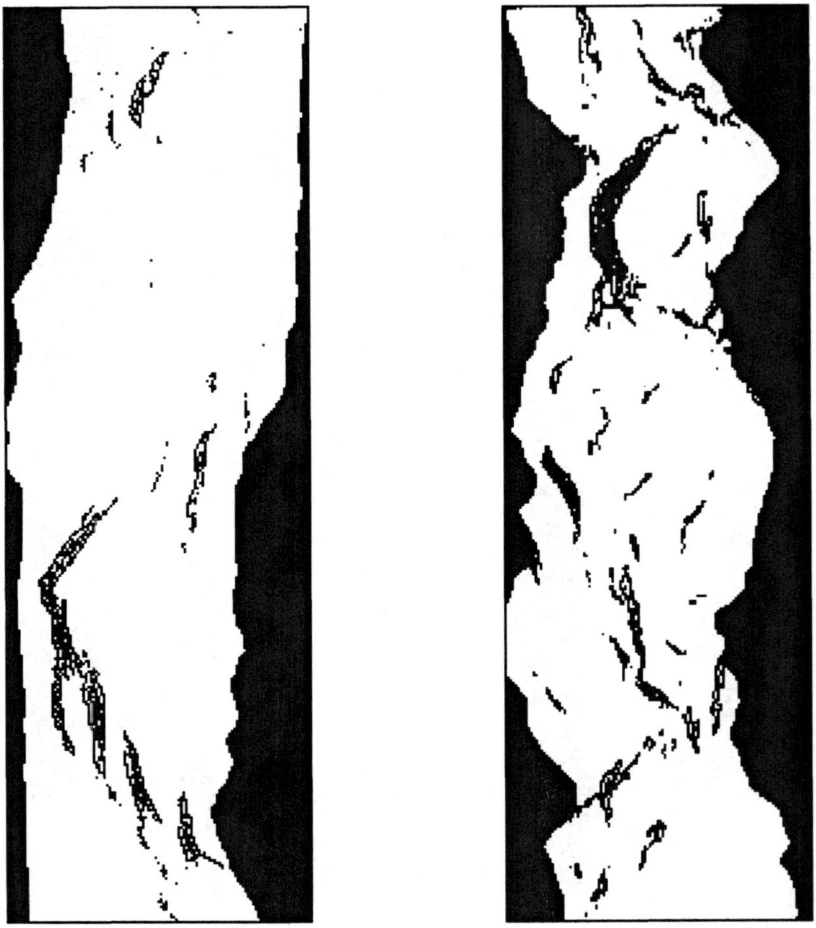

(a) (b)

Figure 6.8 Discontinuities in the solutions produced by the PCG algorithm on (a) the Long's Peak example and (b) the Isolation Peak example. Note the thickness of the regions of discontinuities.

The discontinuity maps for Flynn's minimum discontinuity algorithm and the minimum L^p-norm algorithm can be found in Sections 4.5.4 and 5.5.4, respectively.

As we noted before, these discontinuity maps play the role of "branch cut maps" in the evaluation of the unwrapped results. That is, they indicate where the phase unwrapping paths did *not* go. In Chapter 4 we illustrated the importance of the branch cuts produced by Goldstein's algorithm for helping one determine why phase-unwrapping errors occurred. However, these branch cuts were available only for two algorithms (Goldstein's and the mask cut algorithm). Since the discontinuity maps can be computed for the results of *any*

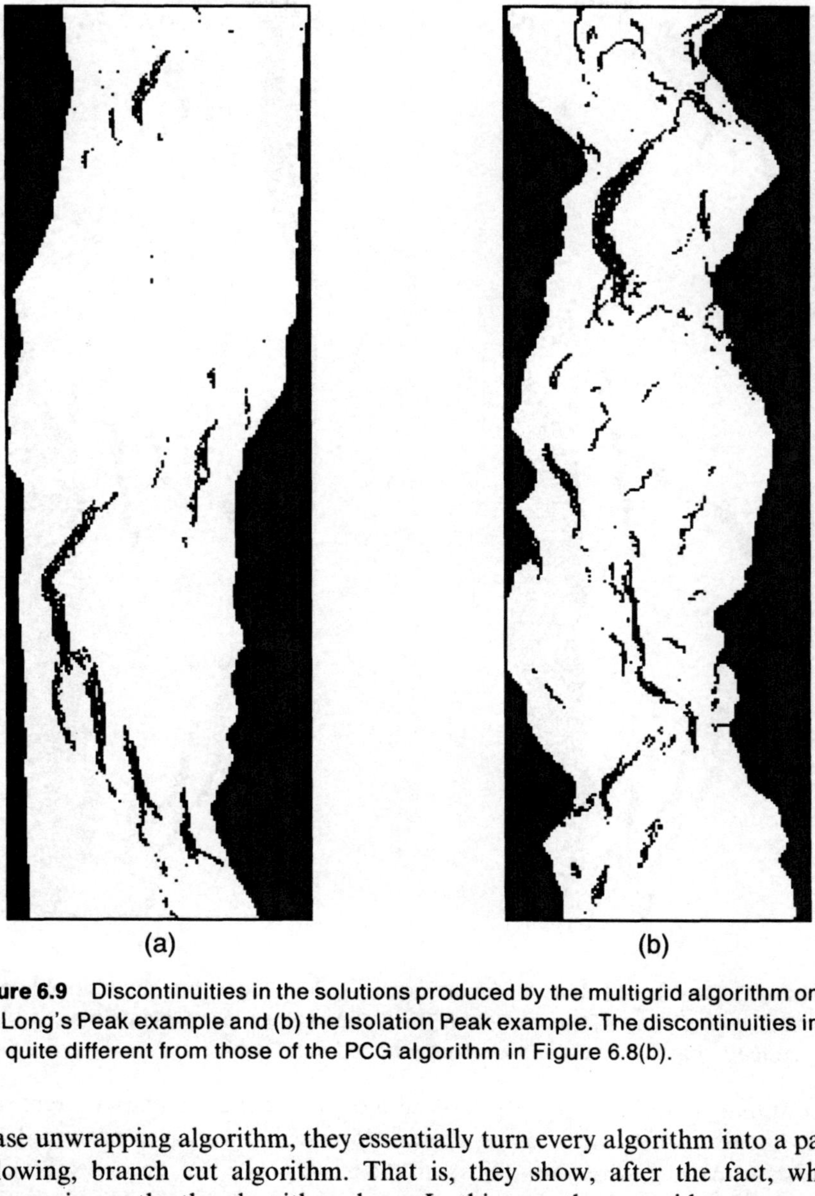

Figure 6.9 Discontinuities in the solutions produced by the multigrid algorithm on (a) the Long's Peak example and (b) the Isolation Peak example. The discontinuities in (b) are quite different from those of the PCG algorithm in Figure 6.8(b).

phase unwrapping algorithm, they essentially turn every algorithm into a path-following, branch cut algorithm. That is, they show, after the fact, which unwrapping paths the algorithm chose. In this way they provide a means for comparing the results of very different algorithms.

6.3.3 Evaluation Method

We now describe a method for detecting where phase unwrapping errors may have occurred in an unwrapped result. The method consists of combining the

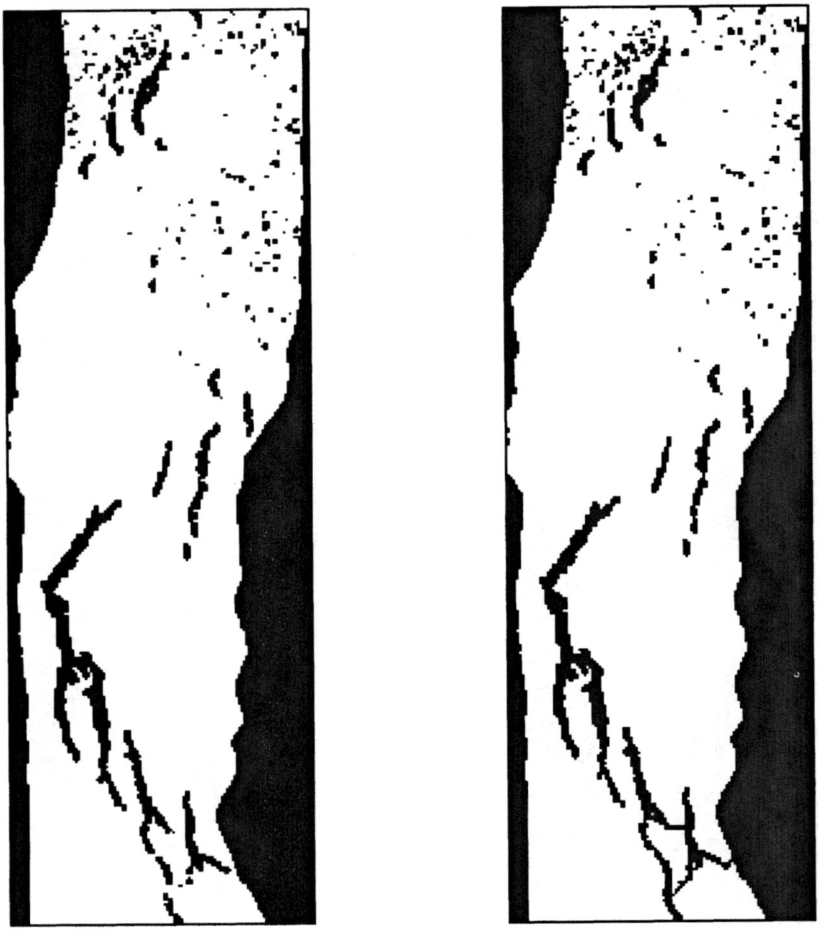

(a) (b)

Figure 6.10 Combined maps for (a) the quality-guided algorithm and (b) Goldstein's algorithm on the Long's Peak example. The isolated regions in Goldstein's combined map indicate that these regions were unwrapped incorrectly.

discontinuity map with a quality mask to show where the two maps disagree and where there may be unwrapping errors. Specifically, a map is produced that consists of both the discontinuities and the mask pixels (i.e., zero-weights). We will name this map (for lack of a better term) a *combined map*. Discontinuities that fall outside the mask and that are induced by wrong-multiple-of-2π errors (the only types of errors in congruent solutions) will isolate portions of the image in the combined map. This isolation may be total or partial. By examining the combined map, the user can reasonably conclude that these regions were unwrapped incorrectly. A few examples will suffice to illustrate the effectiveness of this technique.

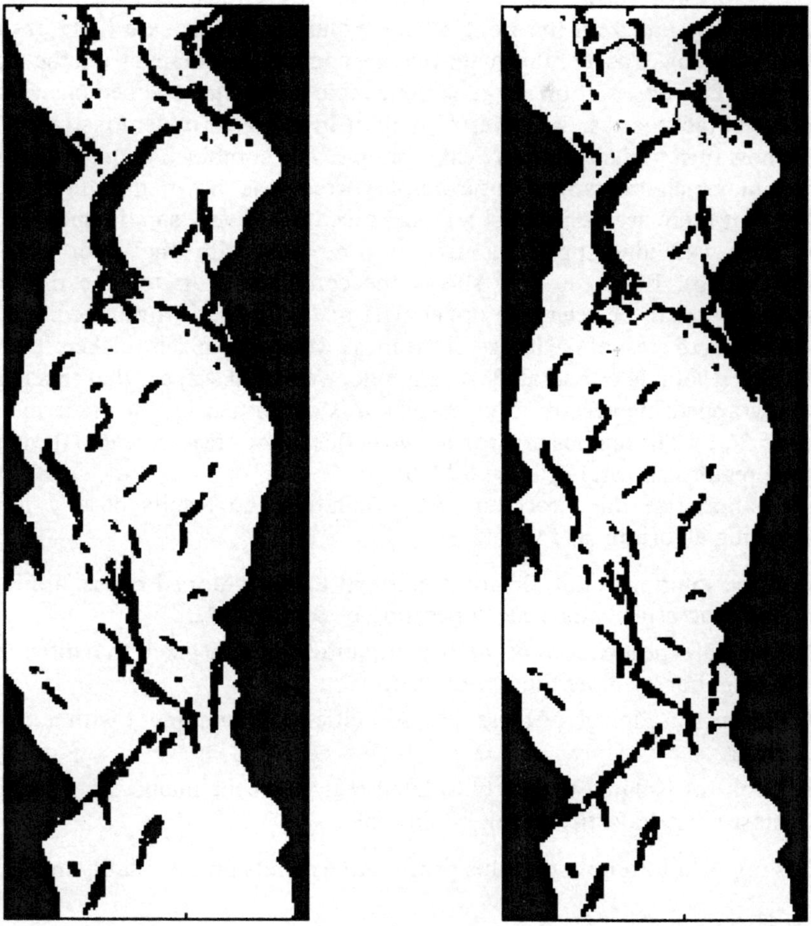

Figure 6.11 Combined maps for (a) the PCG algorithm and (b) the multigrid algorithm on the Isolation Peak example. The nearly isolated regions in (b) indicate that these regions were unwrapped incorrectly.

Figure 6.10(a) shows the combined map for the quality-guided algorithm (with the phase derivative variance as the quality map) on the Long's Peak example. There are no isolated regions, which implies that the discontinuities are confined to the mask pixels (zero-weights). This in turn indicates that the unwrapped result is correct, at least insofar as the mask is "correct." The error map of Figure 4.26(a), which was generated from the true surface, substantiates this conclusion. In practice, it is of course impossible to produce an error map, because the true surface is unknown. This is why the combined map is useful.

Figure 6.10(b) shows the combined map for Goldstein's algorithm, an algorithm that failed on the Long's Peak example. In this case there are two regions in the lower part of the image that are completely isolated from the rest of the image. This observation suggests these regions were unwrapped incorrectly. This was in fact the case, as the error map of Figure 4.10(b) demonstrates.

We now turn to the Isolation Peak example. The combined map for the PCG algorithm, which unwrapped this example correctly, is shown in Figure 6.11(a). Notice that there are no isolated regions, except for a very small region on the right that was unwrapped incorrectly. Compare with the error map of Figure 5.25(b). Figure 6.11(b) shows the combined map for the multigrid algorithm. As can be seen, the upper part of the image is almost completely isolated from the rest of the image. If the mask were a few pixels thicker, it would probably be completely isolated. At any rate, we would suspect that this region was unwrapped incorrectly. This suspicion is confirmed by the error map of Figure 5.36(a). The regions of error shown in this figure are identical to the nearly isolated regions shown in Figure 6.11(b).

We summarize this procedure for evaluating the results of any phase-unwrapping algorithm as follows:

1. If the solution is not already congruent to the wrapped phase, apply the postprocessing congruence operation of Section 5.4.3.
2. Compute the locations of the discontinuities (i.e., the pixels that differ from a neighbor by more that π radians.)
3. Generate an image of these discontinuities and combine it with a quality mask.
4. Look for isolated or nearly isolated regions in the image. These regions most likely have the wrong 2π-multiple.

It is important to note that this procedure depends on the quality mask.

6.4 FUTURE DIRECTIONS

Where should the research in phase unwrapping be directed next? When we survey the results that have been achieved over the past decade or so, we see three areas where further work needs to be done. These areas are the generation of good quality maps, the evaluation of phase unwrapping results, and the development of an algorithm that finds the *global* minimal L^0-norm solution. After describing these areas, we will conclude the section with a brief discussion of recent techniques for circumventing the phase unwrapping problem altogether.

6.4.1 Quality Maps

Much work still needs to be done in the area of generating quality maps. The success or failure of a phase unwrapping algorithm seems to hinge on the availability of a good quality map. For example, if one corrupts the spiral shear example of Figure 3.10 with Gaussian noise, then no phase unwrapping algorithm (as far as we know) can produce a correct or even nearly correct solution. This remains true for even simpler examples consisting of a curved line of shear corrupted with noise. (Some algorithms can unwrap the noise-corrupted sheared planes example of Figure 4.15, but only because the line of shear is a straight line.) These failures are due to the fact that there is no way to produce a quality map to guide the unwrapping process on these types of examples.

Quality maps are necessary not only for guiding phase unwrapping algorithms but also for evaluating the results of these algorithms. In Chapter 3 we presented several techniques for producing quality maps. We also presented an algorithm for automatically thresholding the quality map to produce a quality mask. There is still much work that can be done to improve these methods.

6.4.2 Evaluation

A second area where we believe research needs to be concentrated is the evaluation of phase unwrapping results. In the preceding section we described a technique for detecting regions of incorrectly unwrapped phase. This technique depends on the choice of a quality mask. Can this dependency be removed?

6.4.3 Minimum L^0-Norm Solution

A third area of future research is the development of an algorithm for finding the truly minimal L^0-norm solution to any given phase unwrapping problem. The L^0-norm is what one most often wishes to minimize in practice. We have yet to find an algorithm that produces the true L^0-norm solution to every phase unwrapping problem. Flynn's minimum discontinuity algorithm finds the minimum L^1-norm solution, while the minimum L^p-norm algorithm finds only a *local* minimum to the L^0 problem. Other algorithms find solutions that are only *approximately* minimal. For example, Goldstein's algorithm finds an approximately minimal set of branch cuts, which produces an approximately minimum L^0-norm solution. The algorithm based on the Hungarian algorithm [3], which was mentioned in Section 4.2, finds a truly minimal set of dipole cuts, but because the minimum L^0-norm solution does not necessarily join the residues in pairs, the unwrapped result produced by the algorithm is not guaranteed to be the minimum L^0-norm solution.

6.4.4 Avoiding Phase Unwrapping

In some application areas, such as optical interferometry and SAR interferometry, the phase unwrapping problem can be avoided altogether if additional data are available. In optical interferometry the technique of "temporal phase unwrapping" avoids the need for two-dimensional *spatial* phase unwrapping [4], [5]. The basic idea is to measure the phase of each pixel as a function of time and then carry out one-dimensional phase unwrapping along the time axis. Phase errors are thereby constrained within the high-noise regions of the phase, and they are not able to propagate to the low-noise regions. The phase must not alias in the time direction for this method to work. In SAR interferometry, "multi-baseline" techniques have been proposed to avoid the phase unwrapping problem [6]–[10]. At least three (instead of two) SAR images are collected. In the three-image case, one of the baselines (i.e., the distance between two of the sensor positions) should be large and the other small. The small-baseline interferogram has its fringes spaced widely apart—ideally containing only one cycle of phase—while the large-baseline interferogram has closely spaced fringes. The terrain elevation at each point of the scene is then computed by deriving a coarse estimate from the first interferogram and a more precise estimate from the second interferogram [10]. Because the function of the first interferogram is to provide a coarse estimate, a rough digital elevation model (DEM) may be substituted instead [8]. More elaborate multi-baseline schemes utilize a maximum-likelihood formulation for the terrain-height estimate [6]. These methods impose no restrictions on the ratios of baseline lengths. Multi-baseline SAR methods, like temporal phase unwrapping methods, are pointwise estimators of terrain height. No spatial integration of phase gradients is required, and unwrapping errors cannot propagate. However, incorrect 2π multiples of phase can occur on a pixel-by-pixel basis. The severity of this problem is related to the local phase noise of the SAR imagery.

6.5 CONCLUSION

In this chapter we have summarized and consolidated the material covered in this book. We began the section by summarizing the distinctive features of each phase unwrapping algorithm, and we provided several comparative rankings of the algorithms. We then explored the links between the path-following algorithms and the minimum-norm algorithms. We showed that while the two approaches differ in methodology, they essentially produce the same results, namely minimum (or approximately minimum) L^p-norm solutions. We then examined the use of the L^p-norm measures for the evaluation of phase unwrapping results, and we proposed a method for detecting regions of incorrectly unwrapped phase. We then discussed the future directions in which phase unwrapping research is likely to head.

One of our motivations for writing this book was the observation that the theory and principles of phase unwrapping are poorly understood. Technical journals and conferences are filled with a bewildering array of algorithms that, while capable of solving some problems, fail miserably on others. It is quite difficult, if not impossible, to assess how an algorithm designed for one particular problem will perform on another.

After reading this book, it should be apparent that phase unwrapping is a difficult task. Strictly speaking, it is an unsolvable problem! The main difficulty stems from the fact that there is only so much information available in the wrapped phase data (or in a pair of mutually coherent observations). For example, although the locations and polarities of the residues are determined by the wrapped phase data, the configuration of the optimal set of branch cuts or discontinuities is not. Furthermore the optimum quality map for guiding the phase unwrapping process is not given a priori. There are many choices that must be made about the source of the quality values, the window size to use for computations or filtering, the threshold to apply, and so on. Thus additional information must be injected to obtain the "correct" (i.e., desired) unwrapped solution. The researcher must have a thorough understanding of phase unwrapping in order to inject the right information and obtain the desired result. We hope that this book has removed some of the mystery of phase unwrapping. We also hope that the algorithms, examples, and software not only provide newcomers to the field with a rapid and high-level entry but also prove useful to researchers already working in the field.

Phase unwrapping is an exciting and growing field. In SAR interferometry alone, one of the many areas in which the need for phase unwrapping arises, there are many airborne collection systems and several satellite systems. The satellites include Canada's RADARSAT, Europe's ERS-1 and ERS-2, Japan's JERS-1, and the United States' NASA Space Shuttle Imaging Radar. Many other nations are currently planning to launch SAR satellites in the near future. The application of phase-unwrapping techniques to the data from these systems will lead to many new accomplishments, from accurate topographic mapping on a global scale to millimeter deformation measurements of the earth's crust for earthquake prediction. And this does not even take into account the many other applications of phase unwrapping, including magnetic resonance imaging, optical interferometry, and adaptive optics. New applications of phase unwrapping are still appearing as researchers continue to exploit the phase properties of coherent signals. The future of phase unwrapping is bright indeed, and it will be interesting to see how it unfolds.

REFERENCES

[1] G. Fornaro, G. Franceschetti, R. Lanari, and E. Sansosti, "Robust phase-unwrapping techniques: A comparison," *Journal of the Optical Society of America A*, Vol. 12, pp. 2355–2366, December 1996.

[2] G. Fornaro, G. Franceschetti, R. Lanari, E. Sansosti, and M. Tesauro, "Global and local phase unwrapping techniques: A comparison," *Journal of the Optical Society of America A*, Vol. 14, No. 10, January 1997.

[3] J. R. Buckland, J. M. Huntley, and S. R. E. Turner, "Unwrapping noisy phase maps by use of a minimum cost matching algorithm," *Applied Optics*, Vol. 34, No. 23, pp. 5100–5108, 1995.

[4] J. M. Huntley and H. O. Saldner, "Temporal phase-unwrapping algorithm for automated interferogram analysis," *Applied Optics*, Vol. 32, No. 17, pp. 3047–3052, 10 June 1993.

[5] H. O. Saldner and J. M. Huntley, "Profilometry using temporal phase unwrapping and a spatial light modulator-based fringe projector," *Optical Engineering*, Vol. 36, No. 2, pp. 610–615, 1997.

[6] D. C. Ghiglia and D. E. Wahl, "Interferometric synthetic aperture radar terrain elevation mapping from multiple observations," *Proceedings of the 1994 Sixth IEEE Digital Signal Processing Workshop,* Yosemite National Park, CA, October 2–5, 1994, IEEE, Piscataway, NJ, pp. 33–36, 1994.

[7] W. Xu, E. C. Chang, L. K. Kwoh, H. Lim, and W. C. A. Heng, "Phase-unwrapping of SAR interferogram with multi-frequency or multi-baseline," *Proceedings of the 1994 International Geoscience and Remote Sensing Symposium, Pasadena, Cal., Aug. 8–12, 1994*, IEEE, Piscataway, New Jersey, pp. 730–732, 1994.

[8] D. Massonnet, H. Vadon, and M. Rossi, "Reduction of the need for phase unwrapping in radar interferometry," *IEEE Transactions on Geoscience and Remote Sensing*, Vol. 34, No. 2, pp. 489–497, 1996.

[9] J. Homer, I. D. Longstaff, and G. Callaghan, "High resolution 3-D SAR via multi-baseline interferometry," *Proceedings of the 1996 International Geoscience and Remote Sensing Symposium*, Lincoln, NE, May 27–31, 1996, IEEE, Piscataway, NJ, pp. 796–798, 1996.

[10] C. V. Jakowatz Jr., D. E. Wahl, and P. A. Thompson, "Ambiguity resolution in SAR interferometry by use of three phase centers," *Proceedings of the SPIE*, Vol. 2757, Society of Photo-Optical Instrumentation Engineers, Bellingham, WA, pp. 82–91, 1996.

APPENDIX A

SOFTWARE

In this appendix we present the `C` language computer programs for the algorithms discussed in this book. See Appendix C for directions on downloading the software. Before proceeding, we clarify the licensing and warranty issues.

License: If you are the individual owner of a copy of this book, we hereby authorize you to type into your computer one copy of each program for your noncommercial use. You are not authorized to transfer or distribute a machine-readable copy to any other person.

Disclaimer of Warranty: The computer programs in this appendix are provided "AS-IS". We make no representation or warranties, express or implied. By way of example, but not of limitation, we make no representations or warranties of merchantability or fitness for any particular purpose, or that the use of the computer programs will not infringe any patents, copyrights or any other rights of any third party. The authors, the publisher, Lockheed Martin Corp., and Sandia National Laboratories shall not be held to any liability with respect to any claim by a third party on account of, or arising from, the use of these computer programs, including, but not limited to liability for lost profits, loss of use, and consequential or similar damage.

We chose the `C` programming language in which to implement the algorithms because it is widely available on many different computers from PCs to UNIX workstations to mainframes. The universality, portability, and flexibility of the language have attracted increasing numbers of scientists and engineers, a trend that is expected to continue. We developed, compiled, and tested all of the software on an IBM RS/6000 workstation with an `ANSI C` compiler. We also compiled and tested the software on Sun and Silicon Graphics workstations. The size of the `int` data type on our computer system is four bytes. If the size of this data type on your system differs, you may have to change the int's to long's.

Before proceeding, we must clarify our convention for array indexing. Two-dimensional arrays in C can be confusing because there are two ways to define them. A two-dimensional array can be defined as a one-dimensional array of pointers to the rows of the array, in which case the declaration would be

```
float *array[ysize];
```

or as a true two-dimensional array

```
float array[ysize][xsize];
```

In the latter case the array is stored by rows (i.e., the rightmost subscript changes the fastest as the array elements are accessed in memory), in contrast to other languages such as FORTRAN which store arrays by columns. To avoid these complications, we have chosen to define all arrays as one-dimensional arrays. We define an xsize×ysize array for storing the phase data, where xsize is the number of columns and ysize is the number of rows, by declaring float *phase and then allocating the memory for the array with the statement

```
phase = (float *) malloc(xsize*ysize*sizeof(float));
```

We access the pixel location (i,j), which is the intersection of the ith column and the jth row, with the statement

```
phi = phase[j*xsize + i];
```

We use this convention consistently in all of the software in this appendix. We also follow the usual C convention that all arrays are indexed starting at 0 (rather than 1).

For an excellent introduction to these and other C language issues, we encourage the reader to consult the classic book *The* C *Programming Language*, second edition, by Brian W. Kernighan and Dennis M. Ritchie, published by Prentice-Hall, Englewood Cliffs, New Jersey, 1988.

We now explain the organization of the appendix. The software is organized in 47 source code files and 38 "header" files. The source code files, which are identified by the suffix .c, contain the main programs and the functions (subroutines). The header files, which are identified by the suffix .h, contain the function declarations and the definitions of the constants, data types, and function macros. Not all of the source code files contain programs; most contain only functions. All of the source code files that contain programs begin with the prefix main. For example, the file mainpcg.c contains the main program for running the PCG algorithm, while pcg.c contains functions required by this program. Associated with every source code file (except the main files) is a header file with the same name. For example, corresponding to the file pcg.c is the header file pcg.h, but the file mainpcg.c has no corresponding header file. Conversely, all the header files except files.h and pi.h have corresponding source code files. The files are organized in this appendix in 13 sections:

SECTION	TITLE
A.1	General-purpose utilities
A.2	Derivatives and residues
A.3	Quality maps, masks, and weights
A.4	Least squares, Laplacians, and congruence
A.5	Visualization and evaluation
A.6	Goldstein's algorithm
A.7	Quality-guided algorithm
A.8	Mask cut algorithm
A.9	Flynn's minimum discontinuity algorithm
A.10	Unweighted least-squares algorithms
A.11	PCG algorithm
A.12	Weighted multigrid algorithm
A.13	L^p-norm Algorithm

The main programs for the phase-unwrapping algorithms are found in Sections A.6–A.13, while the functions needed by these programs for tasks such as memory allocation, residue extraction, and list management are in Sections A.1–A.4. Programs and functions for visualizing the unwrapped surfaces are in Section A.5. The functions in Sections A.1–A.4 are used and re-used in the later sections. For example, in Section A.1 one finds the file util.c, which contains the basic file I/O (input and output) and memory management functions. This file is required by nearly all of the programs in the later sections. The file residues.c of Section A.2 contains the function Residues, which detects the residues in an array of phase data and is required by the programs of Section A.6 (Goldstein's algorithm), A.8 (mask cut algorithm), and A.13 (L^p-norm algorithm).

It is a straightforward matter to determine the list of files required to compile, link and execute a given program. Every source code file has "include" statements at the beginning of the file that cause the header files to be included at compile time. The corresponding source code files are required at link time. For example, suppose that one wishes to compile and link the software for Goldstein's algorithm. Turning to Section A.6 on Goldstein's algorithm, one finds the files maingold.c, gold.c and gold.h. The file maingold.c contains the following "include" statements:

```
# include "file.h"
# include "pi.h"
# include "residues.h"
# include "util.h"
```

```
# include "extract.h"
# include "list.h"
# include "gold.h"
# include "dipole.h"
```

The header files (`file.h`, `pi.h`, `residues.h`, etc.) are required for compilation, along with the corresponding source code files for linking. Additional files are also needed, a fact one determines by examining the source code files. For example, turning to Section A.2, one sees that `dipole.c` includes the header file `brcut.h`, which means that this file and the corresponding source code file `brcut.c` also are needed. If one of the files is inadvertently omitted, the program will be impossible to link due to a missing function. The error message from the linker will indicate the name of the missing function. The omitted file can then be determined from the software index in Appendix B. (To make this process easier, we have also listed all of the required files in each `main` source code file.)

A complete index of the software can be found in Appendix B. It lists all of the files along with a description of the file contents and the appendix section in which each file can found. It also lists all the functions, constants, data types, and macros along with the files in which they are defined.

A.1 GENERAL-PURPOSE UTILITIES

The following files are listed in this section: `util.c`, `util.h`, `extract.c`, `extract.h`, `file.h`, and `pi.h`.

We present below a number of general-purpose functions. Most of the file I/O and memory management functions are contained in the file `util.c`:

```
/*
 * util.c -- utility Functions: I/O, memory, etc.
 */
#include <stdlib.h>
#include <stdio.h>
#include <string.h>
#include <math.h>
#include "pi.h"
#include "file.h"
#include "util.h"

/* Print error message and exit */
void ErrorHandler(char *msg, char *item, int code)
{
  fprintf(stderr, "%s\nItem: %s\n", msg, item);
  exit(code);
}
```

```
/* open the file */
void OpenFile(FILE **fp, char *name, char *mode)
{
  if ((*fp = fopen(name, mode))==NULL)
    ErrorHandler("Cannot open file", name, FILE_OPEN_ERROR);
}

/* Search for the keyword in the string. If present, return 1. */
int Keyword(char *string, char *keyword)
{
  int  i, lenstr, lenkey;
  char str[500], key[500];
  lenstr = strlen(string);
  lenkey = strlen(keyword);
  if (lenstr != lenkey) return 0;
  str[lenstr] = key[lenstr] = 0;
  for (i=0; i<=lenstr; i++)
    str[i] = tolower(string[i]);
  for (i=0; i<=lenstr; i++)
    key[i] = tolower(keyword[i]);
  return (!strcmp(str, key));
}

/* Allocate array of bytes and initialize to zero. */
void AllocateByte(unsigned char **ptr, int len, char *name)
{
  int i;
  *ptr = (unsigned char *) malloc(len*sizeof(unsigned char));
  if ((*ptr)==NULL)
    ErrorHandler("Cannot allocate memory", name,
                 MEMORY_ALLOCATION_ERROR);
  for (i=0; i<len; i++) (*ptr)[i] = 0;
}

/* Allocate array of short integers  and initialize to zero. */
void AllocateShort(short **ptr, int len, char *name)
{
  int i;
  *ptr = (short *) malloc(len*sizeof(short));
  if ((*ptr)==NULL)
    ErrorHandler("Cannot allocate memory", name,
                 MEMORY_ALLOCATION_ERROR);
  for (i=0; i<len; i++) (*ptr)[i] = 0;
}
```

```
/* Allocate array of integers and initialize to zero. */
void AllocateInt(int **ptr, int len, char *name)
{
  int i;
  *ptr = (int *) malloc(len*sizeof(int));
  if ((*ptr)==NULL)
    ErrorHandler("Cannot allocate memory", name,
                 MEMORY_ALLOCATION_ERROR);
  for (i=0; i<len; i++) (*ptr)[i] = 0;
}

/* Allocate array of floats and initialize to zero. */
void AllocateFloat(float **ptr, int len, char *name)
{
  int i;
  *ptr = (float *) malloc(len*sizeof(float));
  if ((*ptr)==NULL)
    ErrorHandler("Cannot allocate memory", name,
                 MEMORY_ALLOCATION_ERROR);
  for (i=0; i<len; i++) (*ptr)[i] = 0.0;
}

/* Allocate array of doubles and initialize to zero. */
void AllocateDouble(double **ptr, int len, char *name)
{
  int i;
  *ptr = (double *) malloc(len*sizeof(double));
  if ((*ptr)==NULL)
    ErrorHandler("Cannot allocate memory", name,
                 MEMORY_ALLOCATION_ERROR);
  for (i=0; i<len; i++) (*ptr)[i] = 0.0;
}

/* Read array of bytes */
void ReadByte(FILE *fp, unsigned char *data, int len, char *name)
{
  if (len != fread(data, sizeof(unsigned char), len, fp))
    ErrorHandler("File read error", name, FILE_READ_ERROR);
  fclose(fp);
}

/* Read array of short integers */
void ReadShort(FILE *fp, short *data, int len, char *name)
{
```

```
  if (len != fread(data, sizeof(short), len, fp))
    ErrorHandler("File read error", name, FILE_READ_ERROR);
  fclose(fp);
}

/* Read array of integers */
void ReadInt(FILE *fp, int *data, int len, char *name)
{
  if (len != fread(data, sizeof(int), len, fp))
    ErrorHandler("File read error", name, FILE_READ_ERROR);
  fclose(fp);
}

/* Read array of floats */
void ReadFloat(FILE *fp, float *data, int len, char *name)
{
  if (len != fread(data, sizeof(float), len, fp))
    ErrorHandler("File read error", name, FILE_READ_ERROR);
  fclose(fp);
}

/* Read array of double-precision floats */
void ReadDouble(FILE *fp, double *data, int len, char *name)
{
  if (len != fread(data, sizeof(double), len, fp))
    ErrorHandler("File read error", name, FILE_READ_ERROR);
  fclose(fp);
}

/* Write array of bytes */
void WriteByte(FILE *fp, unsigned char *data, int len, char *name)
{
  if (len != fwrite(data, sizeof(unsigned char), len, fp))
    ErrorHandler("File write error", name, FILE_WRITE_ERROR);
  fclose(fp);
}

/* Write array of short integers */
void WriteShort(FILE *fp, short *data, int len, char *name)
{
  if (len != fwrite(data, sizeof(short), len, fp))
    ErrorHandler("File write error", name, FILE_WRITE_ERROR);
  fclose(fp);
}
```

```
/* Write array of integers */
void WriteInt(FILE *fp, int *data, int len, char *name)
{
  if (len != fwrite(data, sizeof(int), len, fp))
    ErrorHandler("File write error", name, FILE_WRITE_ERROR);
  fclose(fp);
}

/* Write array of floats */
void WriteFloat(FILE *fp, float *data, int len, char *name)
{
  if (len != fwrite(data, sizeof(float), len, fp))
    ErrorHandler("File write error", name, FILE_WRITE_ERROR);
  fclose(fp);
}

/* Write array of doubles */
void WriteDouble(FILE *fp, double *data, int len, char *name)
{
  if (len != fwrite(data, sizeof(double), len, fp))
    ErrorHandler("File write error", name, FILE_WRITE_ERROR);
  fclose(fp);
}

/* Save an array of bytes in a file.  If neg = 1, reverse    */
/* the values (like photographic negative).  If binary = 1, */
/* save values as 0's and 255's (black and white binary     */
/* image).  If mask_code is not 0, then ignore the pixels    */
/* that are marked with the bits defined by mask_code.       */
void SaveByteToImage(unsigned char *im, char *what, char *filename,
           int xsize, int ysize, int neg, int binary, int mask_code)
{
  int  k;
  FILE *fp;
  unsigned char *out, mask;
  printf("Saving %s to file %s\n", what, filename);
  AllocateByte(&out, xsize*ysize, "byte array");
  mask = (mask_code) ? mask_code : 0xFF;     /* bits all 1's */
  for (k=0; k<xsize*ysize; k++) {
    if (binary)
      out[k] = ((neg && !(im[k]&mask))
                 || (!neg && (im[k]&mask))) ? 255 : 0;
    else
      out[k] = (neg) ? 255 - (im[k]&mask) : (im[k]&mask);
  }
```

```
  OpenFile(&fp, filename, "w");
  WriteByte(fp, out, xsize*ysize, filename);
  free(out);
}

/* Scale the floating-point array to 0-255 (byte array),    */
/* and then save values in a file.  If neg = 1, reverse     */
/* the values (like photographic negative).  If binary = 1, */
/* save values as 0's and 255's (black and white binary     */
/* image).  If logflag = 1, then perform a log-linear       */
/* scaling on the data (useful for "brightening" images).   */
void SaveFloatToImage(float *data, char *what, char *filename,
           int xsize, int ysize, int neg, int binary, int logflag)
{
  int  k;
  unsigned char *im;
  double  r, rmin, rmax, rscale;
  AllocateByte(&im, xsize*ysize, "byte array");
  for (rmin=1.0e+10, rmax=-1.0e+10, k=0; k<xsize*ysize; k++) {
    if (rmin > data[k]) rmin = data[k];
    if (rmax < data[k]) rmax = data[k];
  }
  if (logflag)
    r = (rmin==rmax) ? 1.0 : 255.0/log(1.0 + rmax - rmin);
  else
    r = (rmin==rmax) ? 1.0 : 255.0/(rmax - rmin);
  printf("Min & max of %s = %lf %lf\n", what, rmin, rmax);
  for (k=0; k<xsize*ysize; k++)
    im[k] = (logflag) ? r*log(1.0 + data[k] - rmin)
                                        : r*(data[k] - rmin);
  SaveByteToImage(im, what, filename, xsize, ysize, neg, binary, 0);
  free(im);
}

/* Averages byte values and scale from 0-255 to 0-1.  Store      */
/* results in array "out".  tsize is size of averaging template. */
void AverageByteToFloat(unsigned char *in, float *out, int tsize,
                        int xsize, int ysize)
{
  int  hs, i, j, n, ii, jj, c, sum;
  double  scale=1.0/255.0;
  hs = tsize/2;
  for (j=0; j<ysize; j++) {
    for (i=0; i<xsize; i++) {
      for (n=0, sum=0.0, jj=j-hs; jj<=j+hs; jj++) {
```

```
        if (jj < 0 || jj > ysize - 1) continue;
        for (ii=i-hs; ii<=i+hs; ii++) {
          if (ii < 0 || ii > xsize - 1) continue;
          c = in[jj*xsize + ii];
          sum += c;
          ++n;
        }
      }
      out[j*xsize + i] = scale*sum/n;
    }
  }
}

/* Search for the keyword "key" in the command line arguments, */
/* then read the value into the variable pointed to by ptr.    */
/* If required = 1, exit if the parameter is not found.  Print */
/* the string "usage" in this case before exiting.             */
int CommandLineParm(int argc, char *argv[], char *key,
  CommandLineParmType type, void *ptr, int required, char *usage)
{
  int  i, found=0;
  for (i=1; i<argc; i++) {
    if (Keyword(argv[i], key)) {
      if (i < argc - 1) {
        found = 1;
        if (type==IntegerParm)
          sscanf(argv[i+1], "%d", (int *)ptr);
        else if (type==FloatParm)
          sscanf(argv[i+1], "%f", (float *)ptr);
        else if (type==DoubleParm)
          sscanf(argv[i+1], "%lf", (double *)ptr);
        else if (type==StringParm)
          sscanf(argv[i+1], "%s", (char *)ptr);
        break;
      }
      else {
        fprintf(stderr, "Missing parameter value for %s\n",
                argv[i]);
        fprintf(stderr, "%s", usage);
        exit(BAD_USAGE);
      }
    }
  }
  if (required && !found) {
    fprintf(stderr, "Required parameter missing: %s\n", key);
```

```
    fprintf(stderr, "%s", usage);
    exit(BAD_USAGE);
  }
  return found;
}

/* Compute and print the min & max of the array "soln". */
void PrintMinAndMax(int w, int h, float *soln, char *keyword)
{
  int   i, j;
  float rmin, rmax;
  printf("Finding min, max of %s...", keyword);
  rmin = rmax = soln[0];
  for (j=0; j<w*h; j++) {
    if (soln[j] > rmax) rmax = soln[j];
    if (soln[j] < rmin) rmin = soln[j];
  }
  printf("Min %f, max %f\n", rmin, rmax);
}
```

The header file `util.h` is as follows:

```
#ifndef __UTIL
#define __UTIL
typedef enum {IntegerParm, FloatParm, DoubleParm, StringParm}
  CommandLineParmType;
void ErrorHandler(char *msg, char *item, int code);
void OpenFile(FILE **fp, char *name, char *mode);
void AllocateByte(unsigned char **ptr, int len, char *name);
void AllocateShort(short **ptr, int len, char *name);
void AllocateInt(int **ptr, int len, char *name);
void AllocateFloat(float **ptr, int len, char *name);
void AllocateDouble(double **ptr, int len, char *name);
void ReadByte(FILE *fp, unsigned char *data, int len, char *name);
void ReadShort(FILE *fp, short *data, int len, char *name);
void ReadInt(FILE *fp, int *data, int len, char *name);
void ReadFloat(FILE *fp, float *data, int len, char *name);
void ReadDouble(FILE *fp, double *data, int len, char *name);
void WriteByte(FILE *fp, unsigned char *data, int len, char *name);
void WriteShort(FILE *fp, short *data, int len, char *name);
void WriteInt(FILE *fp, int *data, int len, char *name);
void WriteFloat(FILE *fp, float *data, int len, char *name);
void WriteDouble(FILE *fp, double *data, int len, char *name);
void AverageByteToFloat(unsigned char *in, float *out, int tsize,
                        int xsize, int ysize);
```

```
void SaveFloatToImage(float *data, char *what, char *filename,
          int xsize, int ysize, int neg, int binary, int logflag);
void SaveByteToImage(unsigned char *im, char *what, char *filename,
          int xsize, int ysize, int neg, int binary, int mask_code);
int Keyword(char *string, char *keyword);
int CommandLineParm(int argc, char *argv[], char *key,
  CommandLineParmType type, void *ptr, int required, char *usage);
void PrintMinAndMax(int w, int h, float *soln, char *keyword);
#endif
```

The routines for reading the input data and extracting the phase are contained in the file `extract.c`. The input data can be in four different formats: (1) 8-byte complex values, in which the real and imaginary parts are 4-byte single-precision floating-point values stored consecutively in the file, (2) 4-byte complex values, in which the real and imaginary parts are 2-byte integers, (3) 1-byte phase values that are scaled to the range 0–255 and quantized to the nearest integer, and (4) 4-byte floating-point phase values. After reading the data and calling function `ExtractPhase` to extract the phase, the function `GetPhase` scales the phase values to the range 0–1 and stores them in a `float` array.

```
/*
 * extract.c -- extract phase from input data
 */
#include <stdlib.h>
#include <stdio.h>
#include <string.h>
#include <math.h>
#include "extract.h"
#include "pi.h"
/*
 *  Get the phase data from the input data.  Allocate the temporary
 *  memory required and read the input file.
 *
 *  in_format = 0 for 8-byte complex data,
 *              1 for 4-byte complex data,
 *              2 for 1-byte phase data,
 *              3 for 4-byte float phase data
 *
 *  Output: normalized (0 - 1) phase values in array "phase"
 */
void GetPhase(int in_format, FILE *ifp, char *infile, float *phase,
              int xsize, int ysize)
{
  void  *in_data;
  if (in_format==0)  /* floating-pt real & imaginary parts */
```

```
    AllocateFloat((float**)(&in_data), 2*xsize*ysize, "input data");
  else if (in_format==1)  /* short integer real & imaginary parts */
    AllocateShort((short**)(&in_data), 2*xsize*ysize, "input data");
  else if (in_format==2)  /* 1-byte (quantized) phase values */
    AllocateByte((unsigned char **)(&in_data), xsize*ysize,
                 "input data");
  else                    /* floating-pt phase values */
    AllocateFloat((float **)(&in_data), xsize*ysize, "input data");

  printf("Reading input data...\n");
  if (in_format==1)       /* short int real and imaginary parts */
    ReadShort(ifp, (short *)in_data, 2*xsize*ysize, infile);
  else if (in_format==0) /* floating-pt real and imag parts */
    ReadFloat(ifp, (float *)in_data, 2*xsize*ysize, infile);
  else if (in_format==3) /* floating-pt phase */
    ReadFloat(ifp, (float *)in_data, xsize*ysize, infile);
  else                   /* 1-byte quantized phase */
    ReadByte(ifp, (unsigned char *)in_data, xsize*ysize, infile);

  ExtractPhase(in_format, in_data, phase, xsize, ysize, 1);
  free(in_data);
}

/* Extract the phase from the input data.  If status_flag is 0 */
/* then do not print any status.                               */
void ExtractPhase(int in_format, void *in_data, float *phase,
                  int xsize, int ysize, int status_flag)
{
  int            i, j;
  double         x, y, r, angle;
  static double  one_over_twopi = 1.0/TWOPI;
  static double  scale;
  float          *in8_data = in_data;
  short          *in4_data = in_data;
  unsigned char  *quantized_phase = in_data;
  float          *float_phase = in_data;
  if (in_format==0 || in_format==1) {
    for (j=0; j<ysize; j++) {
      if (status_flag && j%100==0) {
        printf("%d ", j);
        fflush(stdout);
        if (j+100 >= ysize) printf("\n");
      }
      for (i=0; i<xsize; i++) {
        if (in_format==1) {
```

```
          x = in4_data[2*(j*xsize + i)];
          y = in4_data[2*(j*xsize + i) + 1];
        }
        else {
          x = in8_data[2*(j*xsize + i)];
          y = in8_data[2*(j*xsize + i) + 1];
        }
        r = sqrt((double)(x*x + y*y));
        r = (r==0.0) ? 0.0 : 1.0/r;
        x *= r;
        y *= r;
        angle = atan2((double)y, (double)x);
        if (angle < 0) angle += TWOPI;
        if (angle >= TWOPI) angle -= TWOPI;
        angle *= one_over_twopi;
        phase[j*xsize + i] = angle;
      }
    }
  }
  else if (in_format==2) {      /* quantized phase */
    scale = 1.0/256.0;
    for (j=0; j<xsize*ysize; j++) {
      phase[j] = quantized_phase[j]*scale;
    }
  }
  else {    /* 4-byte float phase */
    scale = one_over_twopi;
    for (j=0; j<xsize*ysize; j++) {
      /* re-scale phase to interval (0,1) */
      phase[j] = float_phase[j]*scale;
    }
  }
}
```

The header file `extract.h` is as follows:

```
#ifndef __EXTRACT
#define __EXTRACT
void GetPhase(int in_format, FILE *ifp, char *in_file, float *phase,
              int xsize, int ysize);
void  ExtractPhase(int in_format, void *in_data, float *phase,
                   int xsize, int ysize, int status_flag);
#endif
```

The header file `file.h` contains the definitions of the error codes that are returned when a program must terminate prematurely:

```
/*  exit codes  */
#define  BAD_USAGE                  10
#define  MEMORY_ALLOCATION_ERROR    20
#define  FILE_OPEN_ERROR            30
#define  FILE_READ_ERROR            31
#define  FILE_WRITE_ERROR           32
#define  FILE_CLOSE_ERROR           33
#define  BAD_PARAMETER             100
#define  BAD_DIMENSIONS            110
```

The header file pi.h contains the definitions of π and 2π:

```
#define PI              3.141592653589793
#define TWOPI           6.28318530717959
```

A.2 DERIVATIVES AND RESIDUES

The following files are listed in this section: grad.c, grad.h, raster.c, raster.h, dxdygrad.c, dxdygrad.h, residues.c, residues.h, brcut.c, brcut.h, dipole.c, dipole.h, path.c, path.h, list.c, and list.h.

We now present the software for a number of functions that compute phase derivatives, extract residues from phase data, generate branch cuts, detect and remove dipole residues, and manage the pixels to be unwrapped.

The first file grad.c contains the function Gradient that computes the phase derivative (i.e., wrapped phase difference) of the parameters p1 and p2:

```
/*
 *   grad.c -- function for computing phase derivative
 *             (wrapped phase difference)
 */
#include <stdio.h>
#include <math.h>
#include "grad.h"
/* return wrapped phase difference */
float Gradient(float p1, float p2)
{
  float  r;
  r = p1 - p2;
  if (r > 0.5) r -= 1.0;
  if (r < -0.5) r += 1.0;
  return r;
```

```
}
```

The header file `grad.h` is as follows:

```
#ifndef __GRAD
#define __GRAD
float Gradient(float, float);
#endif
```

The second file `raster.c` contains the function `RasterUnwrap` that unwraps phase data by means of the modified Itoh's method discussed in Section 2.8. The phase data must be residue-free.

```
/*
 *  raster.c -- simple raster function for unwrapping residue-free
 *              phase data by means of Itoh's method.
 */
#include <stdio.h>
#include <math.h>
#include "raster.h"
#include "grad.h"
/* Unwrap phase using simple raster technique based on Itoh's */
/* method.  This method works inly if there are no residues.  */
void RasterUnwrap(float *phase, float *soln, int xsize, int ysize)
{
  int    i, j, k;
  soln[0] = phase[0];
  for (j=1; j<ysize; j++) {  /* unwrap first column */
    k = j*xsize + 0;
    soln[k] = soln[k-xsize] + Gradient(phase[k], phase[k-xsize]);
  }
  for (j=0; j<ysize; j++) {   /* unwrap all the rows */
    for (i=1; i<xsize; i++) {
      k = j*xsize + i;
      soln[k] = soln[k-1] + Gradient(phase[k], phase[k-1]);
    }
  }
}
```

The header file `raster.h` is as follows:

```
#ifndef __RASTER
#define __RASTER
void RasterUnwrap(float *phase, float *soln, int xsize, int ysize);
#endif
```

The file dxdygrad.c contains the function DxPhaseGradient and the function DyPhaseGradient, which compute the d/dx and d/dy phase gradients or partial derivatives (i.e., wrapped row and column differences) and store them in arrays. The xsize×ysize input array phase contains the phase data, and the results are stored in the array dx or dy.

```
/*
 * dxdygrad.c -- functions for computing phase derivatives in arrays
 */
#include <stdio.h>
#include <math.h>
#include "file.h"
#include "pi.h"
#include "grad.h"
#include "dxdygrad.h"

/* Compute the wrapped row differences (d/dx partial derivatives */
/* of the phase) and store them in the array "dx".              */
void DxPhaseGradient(float *phase, float *dx, int xsize, int ysize)
{
  int  i, j, k;
  float  grad;
  for (j=0; j<ysize; j++) {
    for (i=0; i<xsize; i++) {
      k = j*xsize + i;
      if (i < xsize - 1)
        grad = Gradient(phase[k+1], phase[k]);
      else
        grad = Gradient(phase[k-1], phase[k]);
      dx[k] = grad;
    }
  }
}

/* Compute the wrapped column differences (d/dy partial  */
/* derivatives) and store them in the array "dy".         */
void DyPhaseGradient(float *phase, float *dy, int xsize, int ysize)
{
  int  i, j, k;
  float  grad;
  for (j=0; j<ysize; j++) {
    for (i=0; i<xsize; i++) {
      k = j*xsize + i;
      if (j < ysize - 1)
        grad = Gradient(phase[k+xsize], phase[k]);
```

```
      else
        grad = Gradient(phase[k-xsize], phase[k]);
      dy[k] = grad;
    }
  }
}
```

The header file is `dxdygrad.h`:

```
#ifndef __DXDYGRAD
#define __DXDYGRAD
void DxPhaseGradient(float *phase, float *dx, int xsize, int ysize);
void DyPhaseGradient(float *phase, float *dy, int xsize, int ysize);
#endif
```

The file `residues.c` defines the function `Residues` that identifies the residues in the array `phase` of phase data and labels them in the `bitflags` array. The bit flags for the positive and negative residues are defined by `posres_code` and `negres_code`, respectively, which should be powers of two (see file `list.h`).

```
/*
 *   residues.c -- function for extracting residues from phase
 */
#include <stdio.h>
#include <math.h>
#include "residues.h"
#include "grad.h"
/* Detect residues in phase data and mark them as positive or  */
/* negative residues in the bitflags array.  Ignore the pixels */
/* marked with avoid_code in the bitflags araay.               */
int Residues(float *phase, unsigned char *bitflags, int posres_code,
             int negres_code, int avoid_code, int xsize, int ysize)
{
  int  i, j, k, NumRes=0;
  double  r;
  for (j=0; j<ysize - 1; j++) {
    for (i=0; i<xsize - 1; i++) {
      k = j*xsize + i;
      if (bitflags && ((bitflags[k] & avoid_code)
           || (bitflags[k+1] & avoid_code)
             || (bitflags[k+1+xsize] & avoid_code)
                 || (bitflags[k+xsize] & avoid_code))) {
        continue; /* masked region: don't unwrap */
      }
      r = Gradient(phase[k+1], phase[k])
```

```
            + Gradient(phase[k+1+xsize], phase[k+1])
              + Gradient(phase[k+xsize], phase[k+1+xsize])
                + Gradient(phase[k], phase[k+xsize]);
      if (bitflags) {
        if (r > 0.01) bitflags[k] |= posres_code;
        else if (r < -0.01) bitflags[k] |= negres_code;
      }
      if (r*r > 0.01) ++NumRes;
    }
  }
  return NumRes;
}
```

The header file is `residues.h`:

```
#ifndef __RESIDUES
#define __RESIDUES
int Residues(float *phase, unsigned char *bitflags, int pos_code,
             int neg_code, int avoid_code, int xsize, int ysize);
#endif
```

The file `brcut.c` contains two functions for placing branch cuts in an array. These functions are used by only one phase unwrapping algorithm: Goldstein's. Function `PlaceCut` places a branch cut from the pixel (a,b) to the pixel (c,d) by labeling the appropriate pixels in the bitflags array `array`. The bit position of the branch cut pixels are defined by `code`, which should be a power of two (see file `list.h`). Function `DistToBorder` computes the distance from the pixel (a,b) to the nearest border pixel (`ra`,`rb`), which is computed and returned by the function. The border pixels are those that border the edges of the array or those marked `border_code` in the `bitflags` array. Border pixels are pixels that are masked out and intended to be ignored.

```
/*
 *   brcut.c -- functions for branch cutting
 */
#include <stdio.h>
#include <math.h>
#include "brcut.h"

/* Place a branch cut in the bitflags array from pixel (a,b) */
/* to pixel (c,d).  The bit for the branch cut pixels is     */
/* given by the value of "code".                             */
void PlaceCut(unsigned char *array, int a, int b, int c, int d,
              int xsize, int ysize, int code)
{
```

```
  int  i, j, k, ii, jj, m, n, istep, jstep;
  double  r;

  /* residue location is upper-left corner of 4-square */
  if (c > a && a > 0) a++;
  else if (c < a && c > 0) c++;
  if (d > b && b > 0) b++;
  else if (d < b && d > 0) d++;

  if (a==c && b==d) {
    array[b*xsize + a] |= code;
    return;
  }
  m = (a < c) ? c - a : a - c;
  n = (b < d) ? d - b : b - d;
  if (m > n) {
    istep = (a < c) ? +1 : -1;
    r = ((double)(d - b))/((double)(c - a));
    for (i=a; i!=c+istep; i+=istep) {
      j = b + (i - a)*r + 0.5;
      array[j*xsize + i] |= code;
    }
  }
  else {   /* n < m */
    jstep = (b < d) ? +1 : -1;
    r = ((double)(c - a))/((double)(d - b));
    for (j=b; j!=d+jstep; j+=jstep) {
      i = a + (j - b)*r + 0.5;
      array[j*xsize + i] |= code;
    }
  }
  return;
}

/* Return the squared distance between the pixel (a,b) and the */
/* nearest border pixel.  The border pixels are encoded in the */
/* bitflags array by the value of "border_code".               */
int DistToBorder(unsigned char *bitflags, int border_code,
               int a, int b, int *ra, int *rb, int xsize, int ysize)
{
  int  besta, bestb, found, dist2, best_dist2;
  int  i, j, k, bs;
  *ra = *rb = 0;
  for (bs=0; bs<xsize + ysize; bs++) {
    found = 0;
```

```
    best_dist2 = 1000000;
    /* search boxes of increasing size until border pixel found */
    for (j=b - bs; j<=b + bs; j++) {
      for (i=a - bs; i<=a + bs; i++) {
        k = j*xsize + i;
        if (i<=0 || i>=xsize - 1 || j<=0 || j>=ysize - 1
              || (bitflags[k] & border_code)) {
          found = 1;
          dist2 = (j - b)*(j - b) + (i - a)*(i - a);
          if (dist2 < best_dist2) {
            best_dist2 = dist2;
            besta = i;
            bestb = j;
          }
        }
      }
    }
    if (found) {
      *ra = besta;
      *rb = bestb;
      break;
    }
  }
  return best_dist2;
}
```

The header file `brcut.h` is as follows:

```
#ifndef __BRCUT
#define __BRCUT
void PlaceCut(unsigned char *array, int a, int b, int c, int d,
              int xsize, int ysize, int code);
int DistToBorder(unsigned char *bitflags, int border_code,
            int a, int b, int *ra, int *rb, int xsize, int ysize);
#endif
```

The function `Dipole` in file `dipole.c` identifies dipole residues (pairs of adjoining residues of opposite sign), places a branch cut between them in the `bitflags` array, and then removes them from the array. This function comprises the "dipole preprocessing" step of Section 4.2.5 that enhances Goldstein's algorithm.

```
/*
 *   dipole.c -- function for finding dipoles and eliminating them
```

```
 */
#include <stdio.h>
#include <math.h>
#include "util.h"
#include "brcut.h"
#include "list.h"

/* Find the dipoles (adjoining residues of opposite sign) in    */
/* the bitflags array, remove them, and place branch cut.       */
/* The bits for the positive and negative residues are defined */
/* by POS_RES and NEG_RES (defined in brcut.h), while the       */
/* branch cut bits are defined by "branchcut_code".             */
void Dipole(unsigned char *bitflags, int xsize, int ysize,
            int branchcut_code)
{
  int  i, j, k, kk;
  printf("Dipoles...\n");
  for (j=0; j<ysize; j++) {
    for (i=0; i<xsize; i++) {
      k = j*xsize + i;
      kk = 0;
      if ((bitflags[k] & POS_RES)) {
        if (i<xsize-1 && (bitflags[k+1] & NEG_RES)) kk=k+1;
        else if (j<ysize-1) {
          if ((bitflags[k+xsize] & NEG_RES)) kk=k+xsize;
        }
      }
      else if ((bitflags[k] & NEG_RES)) {
        if (i<xsize-1 && (bitflags[k+1] & POS_RES)) kk=k+1;
        else if (j<ysize-1) {
          if ((bitflags[k+xsize] & POS_RES)) kk=k+xsize;
        }
      }
      if (kk) {
        printf("Connecting dipoles %d,%d - %d,%d\n", i, j, kk%xsize,
               kk/xsize);
        PlaceCut(bitflags, i, j, kk%xsize, kk/xsize,
                 xsize, ysize, branchcut_code);
        bitflags[k] &= (~(RESIDUE));
        bitflags[kk] &= (~(RESIDUE));
      }
    }
  }
}
```

The header file is dipole.h:

```
#ifndef __DIPOLE
#define __DIPOLE
void Dipole(unsigned char *bitflags, int xsize, int ysize,
            int cut_code);
#endif
```

The file path.c contains the function UnwrapAroundCuts that manages the unwrapping of the phase data around the branch cuts. This function is used by Goldstein's algorithm and the quality-guided algorithm. The phase array stores the phase data to be unwrapped, the bitflags array stores the bit flags of the pixels that are to be avoided (masked pixels and branch cut pixels), and the soln array stores the unwrapped result. The bit flag of the branch cuts is defined by cut_code, which should be a power of 2. If the debug_flag is 1, images of the intermediate stages of the unwrapping process are saved to files. The character string infile defines the prefix of the file names.

```
/*
 *   path.c -- function for unwrapping around cuts
 */
#include <stdio.h>
#include <math.h>
#include "list.h"
#include "grad.h"
#include "path.h"

/* Unwrap the phase data (by Itoh's method) without crossing
 * any branch cuts.  Return number of disconnected pieces.
 */
int UnwrapAroundCuts(float *phase, unsigned char *bitflags,
              float *soln, int xsize, int ysize, int cut_code,
              int debug_flag, char *infile)
{
  int  i, j, k, a, b, c, n, num_pieces=0;
  float  value;
  float  min_qual, small_val = -1.0E+10;
  int    num_index, max_list_size, bench, benchout;
  int    *index_list;
  int    unwrapped_code=UNWRAPPED, postponed_code = POSTPONED;
  int    avoid_code;

  bench = xsize*ysize/100;
  if (bench < 1) bench = 1;
  benchout = 10*bench;
```

```
min_qual = small_val;
max_list_size = xsize*ysize; /* this size may be reduced */
AllocateInt(&index_list, max_list_size + 1,
            "bookkeeping list (index)");

avoid_code = cut_code | unwrapped_code | BORDER;

/* find starting point */
n = 0;
num_index = 0;
for (j=0; j<ysize; j++) {
  for (i=0; i<xsize; i++) {
    k = j*xsize + i;
    if (!(bitflags[k] & avoid_code)) {
      bitflags[k] |= unwrapped_code;
      if (bitflags[k] & postponed_code)
        /* soln[k] already stores the unwrapped value */
        value = soln[k];
      else {
        ++num_pieces;
        value = soln[k] = phase[k];
      }
      UpdateList(NULL, i, j, value, phase, soln, bitflags, xsize,
                 ysize, index_list, &num_index, avoid_code,
                 unwrapped_code, postponed_code, max_list_size,
                 0, &min_qual);
      while (num_index > 0) {
        if (n%bench==0) {
          printf("%d ", n/bench);
          fflush(stdout);
          if (0 && debug_flag && n%benchout==0 && n>0) {
            char filename[300];
            static char file_num = 0;
            sprintf(filename, "%s.fill-%d", infile, ++file_num);
            SaveByteToImage(bitflags, "fill path", filename,
                            xsize, ysize, 1, 1, avoid_code);
          }
        }
        ++n;
        if (!GetNextOneToUnwrap(&a, &b, index_list,
                                &num_index, xsize, ysize))
          break;   /* no more to unwrap */
        c = b*xsize + a;
        bitflags[c] |= unwrapped_code;
        value = soln[c];
```

```
          UpdateList(NULL, a, b, value, phase, soln, bitflags,
                     xsize, ysize, index_list, &num_index,
                     avoid_code, unwrapped_code, postponed_code,
                     max_list_size, 0, &min_qual);
        }
      }
    }
  }
  printf("\n");
  free(index_list);
  /* unwrap branch cut pixels */
  for (j=1; j<ysize; j++) {
    for (i=1; i<xsize; i++) {
      k = j*xsize + i;
      if (bitflags[k] & cut_code) {
        if (!(bitflags[k-1] & cut_code))
          soln[k] = soln[k-1] + Gradient(phase[k], phase[k-1]);
        else if (!(bitflags[k-xsize] & cut_code))
          soln[k]
            = soln[k-xsize] + Gradient(phase[k], phase[k-xsize]);
      }
    }
  }
  return num_pieces;
}
```

The header file is `path.h`:

```
#ifndef __PATH
#define __PATH
int UnwrapAroundCuts(float *phase, unsigned char *bitflags,
                float *soln, int xsize, int ysize, int cut_code,
                int debug_flag, char *infile);
#endif
```

The file `list.c` contains the list management functions that keep track of the unwrapped pixels. See Section 4.3.1. The function `GetNextOneToUnwrap` fetches the next pixel to be unwrapped, passes back the pixel location (a,b), and decrements the number of pixels `num_index` in the list. The `index_list` stores the indexes of the pixels on the list. The function `InsertList` inserts a new pixel with the unwrapped value `val` in the list, stores `val` in the `soln` array, and maintains the pixels in sorted order. This function also manages the list-trimming procedure described in Section 4.3.1. The function `UpdateList` calls `InsertList` to insert the four neighbors of a given pixel in the list.

```
/*
 *   list.c -- functions for managing list of pixels to unwrap
 */
#include <stdio.h>
#include <math.h>
#include "file.h"
#include "pi.h"
#include "list.h"
#include "grad.h"

/* Returns 0 if no pixels left, 1 otherwise */
int GetNextOneToUnwrap(int *a, int *b, int *index_list,
                       int *num_index, int xsize, int ysize)
{
  int index;
  if (*num_index < 1)
    return 0;   /* return if list empty */
  index = index_list[*num_index - 1];
  *a = index%xsize;
  *b = index/xsize;
  --(*num_index);
  return 1;
}

/* Insert new pixel into the list.  */
/* Note: qual_map can be NULL       */
void InsertList(float *soln, float val, float *qual_map,
           unsigned char *bitflags, int a, int b, int *index_list,
           int *num_index, int xsize, int ysize, int processed_code,
           int postponed_code, float *min_qual, int max_list_size)
{
  int i, n, index, top, bot, mid;
  double  quality;

  index = b*xsize + a;
  quality = (qual_map) ? qual_map[index] : 0.0;
  if (~(bitflags[index] & postponed_code)) {
    /* if not postponed, store new unwrapped value */
    if (soln) soln[index] = val;
  }
  else {
    /* if postponed, leave old value */
  }
  /* if quality is too low, postpone it */
  if (qual_map && min_qual && quality < *min_qual) {
```

```
    bitflags[index] |= postponed_code;
    return;
  }
  /* otherwise, add to list */
  if (!qual_map) {   /* don't order if qual_map is NULL */
    index_list[*num_index] = index;
    ++(*num_index);
  }
  else {
    /* insert in list in order from lowest to highest quality */
    n = *num_index;
    if (n < 1) {
      (*num_index) = 0;   /* will be incremented below */
      index_list[0] = index;
    }
    else {
      if (quality <= qual_map[index_list[0]]) {
        /* insert at top of list */
        for (i=n; i>0; i--)
          index_list[i] = index_list[i-1];
        index_list[0] = index;
      }
      else if (quality > qual_map[index_list[n - 1]]) {
        /* insert at bottom */
        index_list[n] = index;
      }
      else {   /* insert in middle */
        top = 0;
        bot = n - 1;
        while (bot - top > 1) {
          mid = (top + bot)/2;
          if (quality <= qual_map[index_list[mid]])  bot = mid;
          else  top = mid;
        }
        for (i=n; i>top+1; i--)
          index_list[i] = index_list[i-1];
        index_list[top+1] = index;
      }
    }
    ++(*num_index);
  }
  bitflags[index] |= processed_code;
  bitflags[index] &= (~postponed_code);

  /* trim list if it's too big, and increase the quality */
```

```
  if (qual_map && min_qual
         && max_list_size > 0 && *num_index >= max_list_size) {
    n = 0.50*(*num_index);  /* discard 50% */
    for (i=0; i<n; i++) {
      bitflags[index_list[i]] |= postponed_code;
      bitflags[index_list[i]] &= (~processed_code);
    }
    for (i=0; i<*num_index - n; i++)
      index_list[i] = index_list[i + n];
    *num_index -= n;
    *min_qual = qual_map[index_list[0]];
  }

  return;
}

/* Insert the four neighboring pixels of the given pixel */
/* (x,y) into the list.  The quality value of the given  */
/* pixel is "val".                                        */
void UpdateList(float *qual_map, int x, int y, float val,
         float *phase, float *soln, unsigned char *bitflags,
         int xsize, int ysize, int *index_list, int *num_index,
         int ignore_code, int processed_code, int postponed_code,
         int max_list_size, int dxdy_flag, float *min_qual)
{
  int    i, a, b, k, w;
  float  grad;
  a = x - 1;
  b = y;
  k = b*xsize + a;

  if (a >= 0
       && !(bitflags[k] & (ignore_code | processed_code))) {
    w = y*xsize + x-1;
    grad = Gradient(phase[w], phase[w+1]);
    if (dxdy_flag && qual_map)
      qual_map[k] = -fabs(grad);
    InsertList(soln, val + grad, qual_map, bitflags, a, b,
              index_list, num_index, xsize, ysize, processed_code,
              postponed_code, min_qual, max_list_size);
  }

  a = x + 1;
  b = y;
  k = b*xsize + a;
```

```
  if (a < xsize
        && !(bitflags[k] & (ignore_code | processed_code))) {
    w = y*xsize + x;
    grad = - Gradient(phase[w], phase[w+1]);
    if (dxdy_flag && qual_map)
      qual_map[k] = -fabs(grad);
    InsertList(soln, val + grad, qual_map, bitflags, a, b,
               index_list, num_index, xsize, ysize, processed_code,
               postponed_code, min_qual, max_list_size);
  }

  a = x;
  b = y - 1;
  k = b*xsize + a;
  if (b >= 0
        && !(bitflags[k] & (ignore_code | processed_code))) {
    w = (y-1)*xsize + x;
    grad = Gradient(phase[w], phase[w+xsize]);
    if (dxdy_flag && qual_map)
      qual_map[k] = -fabs(grad);
    InsertList(soln, val + grad, qual_map, bitflags, a, b,
               index_list, num_index, xsize, ysize, processed_code,
               postponed_code, min_qual, max_list_size);
  }

  a = x;
  b = y + 1;
  k = b*xsize + a;
  if (b < ysize
        && !(bitflags[k] & (ignore_code | processed_code))) {
    w = y*xsize + x;
    grad = - Gradient(phase[w], phase[w+xsize]);
    if (dxdy_flag && qual_map) qual_map[k] = -fabs(grad);
    InsertList(soln, val + grad, qual_map, bitflags, a, b,
               index_list, num_index, xsize, ysize, processed_code,
               postponed_code, min_qual, max_list_size);
  }
}
```

The header file is `list.h`. It includes the definitions of the bit flags `POS_RES`, `NEG_RES`, `VISITED`, etc., which are used to label the pixels in the `bitflags` array.

```
#ifndef __LIST
#define __LIST
#define POS_RES      0x01   /* 1st bit */
```

```
#define NEG_RES      0x02    /* 2nd bit */
#define VISITED      0x04    /* 3rd bit */
#define ACTIVE       0x08    /* 4th bit */
#define BRANCH_CUT   0x10    /* 5th bit */
#define BORDER       0x20    /* 6th bit */
#define UNWRAPPED    0x40    /* 7th bit */
#define POSTPONED    0x80    /* 8th bit */
#define RESIDUE       (POS_RES | NEG_RES)
#define AVOID         (BRANCH_CUT | BORDER)
int GetNextOneToUnwrap(int *a, int *b, int *index_list,
                       int *num_index, int xsize, int ysize);
void UpdateList(float *qual_map, int x, int y, float val,
                float *phase, float *soln, unsigned char *bitflags,
                int xsize, int ysize, int *index_list,
                int *num_index, int ignore_code, int processed_code,
                int postponed_code, int max_list_size,
                int dxdy_flag, float *min_qual);
void InsertList(float *soln, float val, float *qual_map,
                unsigned char *bitflags, int a, int b,
                int *index_list, int *num_index, int xsize,
                int ysize, int processed_code, int postponed_code,
                float *min_qual, int max_list_size);
#endif
```

A.3 QUALITY MAPS, MASKS, AND WEIGHTS

The following files for generating and managing quality maps and masks are contained in this section: qualvar.c, qualvar.h, qualgrad.c, qualgrad.h, qualpseu.c, qualpseu.h, getqual.c, getqual.h, histo.c, histo.h, maskfat.c, maskfat.h, maskthin.c, maskthin.h, and mainmap.c.

We present below a number of functions for producing quality maps and masks. See Chapter 3 for a discussion of these topics. The file qualvar.c contains functions for producing the phase derivative variance quality map. The function PhaseVariance computes the quality map from the phase data and stores the quality values in the array result. The bitflags array indicates the pixels to ignore (defined by the bit flag ignore_code). The array temp1 is required for scratch calculations. The quality values are computed in tsize×tsize-pixel windows. The two functions DxGradVar and DyGradVar compute the variances of the d/dx and d/dy gradients in these windows.

```
/*
 *  qualvar.c -- functions for computing phase derivative variance
 *               quality map
 */
```

```
#include <stdio.h>
#include <math.h>
#include "qualvar.h"
#include "dxdygrad.h"

/* Compute variance of phase derivatives in array "result".    */
/* The size of the averaging template is tsize x tsize pizels. */
/* The array temp1 is required for temporary data.             */
void PhaseVariance(float *phase, float *result,
                   unsigned char *bitflags, int ignore_code,
                   float *temp1, int tsize, int xsize, int ysize)
{
  int  add_flag;
  printf("Extracting dx gradients\n");
  DxPhaseGradient(phase, temp1, xsize, ysize);
  add_flag = 0;
  printf("Extracting dx variances\n");
  DxGradVar(temp1, result, xsize, ysize, tsize, bitflags,
            ignore_code, add_flag);
  printf("Extracting dy gradients\n");
  DyPhaseGradient(phase, temp1, xsize, ysize);
  add_flag = 1;
  printf("Extracting dy variances\n");
  /* add to dx variances */
  DyGradVar(temp1, result, xsize, ysize, tsize, bitflags,
            ignore_code, add_flag);
}

/* Compute variance of dx phase derivs in array "dxvar" in */
/* tsize x tsize windows.  If add_code is 1, the results   */
/* are added to the dxvar array, otherwise they overwrite. */
void DxGradVar(float *dx, float *dxvar, int xsize, int ysize,
  int tsize, unsigned char *bitflags, int avoid_code, int add_code)
{
  int    i, j, k, a, b, aa, bb, cc, n, hs;
  float  r, avg, avgsqr;
  if (tsize < 3 && !add_code) {
    for (i=0; i<xsize*ysize; i++)
      dxvar[i] = 0.0;
  }
  else {
    hs = tsize/2;
    for (j=0; j<ysize; j++) {
      for (i=0; i<xsize; i++) {
        avg = avgsqr = 0.0;
```

```
        for (n=0, b=j-hs; b<=j+hs; b++) {
          if ((bb = b) < 0) bb = -bb;
          else if (bb >= ysize) bb = 2*ysize - 2 - bb;
          for (a=i-hs; a < i+hs; a++) {   /* a < to match dx */
            if ((aa = a) < 0) aa = -aa;
            else if (aa >= xsize) aa = 2*xsize - 2 - aa;
            cc = bb*xsize + aa;
            if (bitflags && (bitflags[cc]&avoid_code)) continue;
            r = dx[cc];
            avg += r;
            avgsqr += r*r;
            ++n;
          }
        }
        r = (n>0) ? 1.0/n : 0.0;
        avg *= r;
        avgsqr *= r;
        if (add_code)
          dxvar[j*xsize + i] += avgsqr - avg*avg;   /* variance */
        else
          dxvar[j*xsize + i] = avgsqr - avg*avg;   /* variance */
      }
    }
  }
}

/* Compute variance of dy phase derivs in array "dyvar" in */
/* tsize x tsize windows.  If add_code is 1, the results   */
/* are added to the dyvar array, otherwise they overwrite. */
void DyGradVar(float *dy, float *dyvar, int xsize, int ysize,
  int tsize, unsigned char *bitflags, int avoid_code, int add_code)
{
  int    i, j, k, a, b, aa, bb, cc, n, hs;
  float  r, avg, avgsqr;
  if (tsize < 3 && !add_code) {
    for (i=0; i<xsize*ysize; i++)
      dyvar[i] = 0.0;
  }
  else {
    hs = tsize/2;
    for (j=0; j<ysize; j++) {
      for (i=0; i<xsize; i++) {
        avg = avgsqr = 0.0;
        for (n=0, b=j-hs; b < j+hs; b++) { /* b < to match dy */
          if ((bb = b) < 0) bb = -bb;
```

```
          else if (bb >= ysize) bb = 2*ysize - 2 - bb;
          for (a=i-hs; a<=i+hs; a++) {
            if ((aa = a) < 0) aa = -aa;
            else if (aa >= xsize) aa = 2*xsize - 2 - aa;
            cc = bb*xsize + aa;
            if (bitflags && (bitflags[cc]&avoid_code)) continue;
            r = dy[cc];
            avg += r;
            avgsqr += r*r;
            ++n;
          }
        }
        r = (n>0) ? 1.0/n : 0.0;
        avg *= r;
        avgsqr *= r;
        if (add_code)
          dyvar[j*xsize + i] += avgsqr - avg*avg;   /* variance */
        else
          dyvar[j*xsize + i] = avgsqr - avg*avg;   /* variance */
      }
    }
  }
}
```

The header file is qualvar.h:

```
#ifndef __QUALVAR
#define __QUALVAR
void PhaseVariance(float *phase, float *result,
                   unsigned char *bitflags, int ignore_code,
                   float *soln, int tsize, int xsize, int ysize);
void DxGradVar(float *dx, float *dxvar, int xsize, int ysize,
  int size, unsigned char *bitflags, int avoid_code, int add_flag);
void DyGradVar(float *dy, float *dyvar, int xsize, int ysize,
  int size, unsigned char *bitflags, int avoid_code, int add_flag);
#endif
```

The functions for producing the maximum phase gradient quality maps are in file qualgrad.c. The function MaxPhaseGradients computes the quality map from the phase data and stores the quality values in the array result. The bitflags array indicates the pixels to ignore (defined by the bit flag ignore_code). The array temp1 is required for scratch calculations. The maximum phase gradients are computed in tsize×tsize-pixel windows. The two functions DxGradMax and DyGradMax compute the maxima of the d/dx and d/dy gradients in these windows.

```
/*
 *  qualgrad.c -- functions for computing max gradient quality map
 */
#include <stdio.h>
#include <math.h>
#include "qualgrad.h"
#include "dxdygrad.h"

/* Compute max gradient quality map in array "result".  Size */
/* of averaging template is tsize x tsize pizels.  The array */
/* temp1 is required for temporary data.                     */
void MaxPhaseGradients(float *phase, float *result,
                    unsigned char *bitflags, int ignore_code,
                    float *temp1, int tsize, int xsize, int ysize)
{
  int  add_flag;
  printf("Extracting dx gradients\n");
  DxPhaseGradient(phase, temp1, xsize, ysize);
  add_flag = 0;
  printf("Extracting dx max's\n");
  DxGradMax(temp1, result, xsize, ysize, tsize, bitflags,
            ignore_code, add_flag);
  printf("Extracting dy gradients\n");
  DyPhaseGradient(phase, temp1, xsize, ysize);
  add_flag = 1;
  printf("Extracting dy max's\n");
  /* add to dx max's */
  DxGradMax(temp1, result, xsize, ysize, tsize, bitflags,
            ignore_code, add_flag);
}

/* Compute max dx gradients in array "dxmax". If add_code */
/* is 1, the results are added to the dxmax array,        */
/* otherwise they overwrite.                              */
void DxGradMax(float *dx, float *dxmax, int xsize, int ysize,
   int tsize, unsigned char *bitflags, int avoid_code, int add_code)
{
  int    i, j, a, b, c, hs;
  float  r, dmax;
  if (tsize < 3) {
    for (i=0; i<xsize*ysize; i++)
      dxmax[i] = (add_code) ? dxmax[i] + dx[i] : dx[i];
  }
  else {
    hs = tsize/2;
```

```
    for (j=0; j<ysize; j++) {
      for (i=0; i<xsize; i++) {
        for (dmax = 0.0, b=j-hs; b<=j+hs; b++) {
          if (b < 0 || b >= ysize) continue;
          for (a=i-hs; a < i+hs; a++) {  /* a < to match dx */
            if (a < 0 || a >= xsize) continue;
            c = b*xsize + a;
            if (bitflags && (bitflags[c]&avoid_code)) continue;
            r = dx[c];
            if (r < 0) r = -r;
            if (dmax < r) dmax = r;
          }
        }
        if (add_code)
          dxmax[j*xsize + i] += dmax;
        else
          dxmax[j*xsize + i] = dmax;
      }
    }
  }
}

/* Compute max dy gradients in array "dymax". If add_code */
/* is 1, the results are added to the dymax array,        */
/* otherwise they overwrite.                              */
void DyGradMax(float *dy, float *dymax, int xsize, int ysize,
   int tsize, unsigned char *bitflags, int avoid_code, int add_code)
{
  int    i, j, a, b, c, hs;
  float  r, dmax;
  if (tsize < 3) {
    for (i=0; i<xsize*ysize; i++)
      dymax[i] = (add_code) ? dymax[i] + dy[i] : dy[i];
  }
  else {
    hs = tsize/2;
    for (j=0; j<ysize; j++) {
      for (i=0; i<xsize; i++) {
        for (dmax = 0.0, b=j-hs; b < j+hs; b++) {
          /* b < j+hs to match dy */
          if (b < 0 || b >= ysize) continue;
          for (a=i-hs; a<=i+hs; a++) {
            if (a < 0 || a >= xsize) continue;
            c = b*xsize + a;
            if (bitflags && (bitflags[c] & avoid_code)) continue;
```

```
          r = dy[c];
          if (r < 0) r = -r;
          if (dmax < r) dmax = r;
        }
      }
      if (add_code)
        dymax[j*xsize + i] += dmax;
      else
        dymax[j*xsize + i] = dmax;
    }
  }
 }
}
```

The header file is qualgrad.h:

```
#ifndef __QUALGRAD
#define __QUALGRAD
void MaxPhaseGradients(float *phase, float *result,
                      unsigned char *bitflags, int ignore_code,
                      float *soln, int tsize, int xsize, int ysize);
void DxGradMax(float *dx, float *dxvar, int xsize, int ysize,
   int size, unsigned char *bitflags, int avoid_code, int add_flag);
void DyGradMax(float *dy, float *dyvar, int xsize, int ysize,
   int size, unsigned char *bitflags, int avoid_code, int add_flag);
#endif
```

The functions for generating the pseudocorrelation quality maps are in file qualpseu.c. The function PseudoCorrelation computes the quality map from the phase data and stores the quality values in the array result. The bitflags array indicates the pixels to ignore (defined by the bit flag ignore_code). The array temp1 is required for scratch calculations. The pseudocorrelation values are computed in tsize×tsize-pixel windows by calling the function SqrAvg-Filter, which averages the values in these windows and squares the result.

```
/*
 *   qualpseu.c -- functions for computing pseudocorrelation
 *                 quality map
 */
#include <stdio.h>
#include <math.h>
#include "pi.h"
#include "qualpseu.h"
#include "dxdygrad.h"
```

```
/* Compute pseudocorrelation quality map in array "result".    */
/* The size of the averaging template is tsize x tsize pizels. */
/* The array temp1 is required for temporary data.             */
void PseudoCorrelation(float *phase, float *result,
                     unsigned char *bitflags, int ignore_code,
                     float *temp1, int tsize, int xsize, int ysize)
{
  int  i, j, k, add_flag;
  printf("Extracting cos's\n");
  for (k=0; k<xsize*ysize; k++)
    temp1[k] = cos(TWOPI*phase[k]);
  add_flag = 0;
  printf("Filtering cos's\n");
  SqrAvgFilter(temp1, result, xsize, ysize, tsize,
               bitflags, ignore_code, add_flag);
  printf("Extracting sin's\n");
  for (k=0; k<xsize*ysize; k++)
    temp1[k] = sin(TWOPI*phase[k]);
  add_flag = 1;
  printf("Filtering sin's\n");
  SqrAvgFilter(temp1, result, xsize, ysize, tsize,
               bitflags, ignore_code, add_flag);
  printf("Square root\n");
  for (k=0; k<xsize*ysize; k++)
    result[k] = 1.0 - sqrt(result[k]);
}

/* Computes average of array 'in' in tsize x tsize windows,  */
/* then squares the values.  If add_code is 1, the results   */
/* are added to the values in out, otherwise they overwrite. */
void SqrAvgFilter(float *in, float *out, int xsize, int ysize,
   int tsize, unsigned char *bitflags, int avoid_code, int add_code)
{
  int    i, j, k, a, b, aa, bb, cc, n, hs;
  float  r, avg;
  if (tsize < 3 && !add_code) {
    for (i=0; i<xsize*ysize; i++)
      out[i] = 0.0;
  }
  else {
    hs = tsize/2;
    for (j=0; j<ysize; j++) {
      for (i=0; i<xsize; i++) {
        avg = 0.0;
        for (n=0, b=j-hs; b<=j+hs; b++) {
```

```
          if ((bb = b) < 0) bb = -bb;
          else if (bb >= ysize) bb = 2*ysize - 2 - bb;
          for (a=i-hs; a <= i+hs; a++) {
            if ((aa = a) < 0) aa = -aa;
            else if (aa >= xsize) aa = 2*xsize - 2 - aa;
            cc = bb*xsize + aa;
            if (aa>=0 && aa<xsize-1 && bb>=0 && bb<ysize-1) {
              r = in[cc];
              avg += r;
              ++n;
            }
          }
        }
        r = (n>0) ? 1.0/n : 0.0;
        avg *= r;
        if (add_code)
          out[j*xsize + i] += avg*avg;
        else
          out[j*xsize + i] = avg*avg;
      }
    }
  }
}
```

The header file is qualpseu.h:

```
#ifndef __QUALPSEU
#define __QUALPSEU
void PseudoCorrelation(float *phase, float *result,
                     unsigned char *bitflags, int ignore_code,
                     float *temp1, int tsize, int xsize, int ysize);
void SqrAvgFilter(float *in, float *out, int xsize, int ysize,
                  int size, unsigned char *bitflags, int avoid_code,
                  int add_flag);
#endif
```

The functions for coordinating the generation of the quality maps are contained in file getqual.c. The function GetQualityMap generates the quality map by calling one of the functions in qualvar.c, qualgrad.c, or qualpseu.c and rescaling the quality values to the range 0–1. The function SetQualityMode parses the modekey string and returns a code that indicates the type of quality map to use.

```
/*
 * getqual.c -- generate or process quality map and set quality mode
```

```
 */
#include <stdio.h>
#include <math.h>
#include "getqual.h"
#include "file.h"
#include "util.h"
#include "pi.h"
#include "qualvar.h"
#include "qualpseu.h"
#include "qualgrad.h"

/* Generate a new quality map from the phase data, or process */
/* the quality map that was input.                            */
void GetQualityMap(int mode, float *qual_map, float *phase,
                   unsigned char *bitflags, int border_code,
                   int tsize, int xsize, int ysize)
{
  float  *temp;
  double rmin, rmax, rscale;
  int    i, j, k;
  /* process phase gradients */
  if (mode==variance) {
    AllocateFloat(&temp, xsize*ysize, "temp data");
    PhaseVariance(phase, qual_map, bitflags, border_code, temp,
                  tsize, xsize, ysize);
    free(temp);
    /* convert from cost to quality, and scale to interval (0,1) */
    for (rmin = rmax = qual_map[0], k=0; k<xsize*ysize; k++) {
      if (rmin > qual_map[k]) rmin = qual_map[k];
      if (rmax < qual_map[k]) rmax = qual_map[k];
    }
    printf("Min & max phase derivative variance = %lf, %lf\n",
           rmin, rmax);
    rscale = (rmin != rmax) ? 1.0/(rmax - rmin) : 0.0;
    for (k=0; k<xsize*ysize; k++) {
      qual_map[k] = (rmax - qual_map[k])*rscale;
      if (bitflags && (bitflags[k]&border_code)) qual_map[k] = 0.0;
    }
  }
  else if (mode==gradient) {
    AllocateFloat(&temp, xsize*ysize, "temp data");
    MaxPhaseGradients(phase, qual_map, bitflags, border_code, temp,
                      tsize, xsize, ysize);
    free(temp);
    /* convert from cost to quality, and scale to interval (0,1) */
```

```
    for (rmin = rmax = qual_map[0], k=0; k<xsize*ysize; k++) {
      if (rmin > qual_map[k]) rmin = qual_map[k];
      if (rmax < qual_map[k]) rmax = qual_map[k];
    }
    printf("Min&max 'max phase gradient' = %lf %lf\n", rmin,rmax);
    rscale = (rmin != rmax) ? 1.0/(rmax - rmin) : 0.0;
    for (k=0; k<xsize*ysize; k++) {
      qual_map[k] = (rmax - qual_map[k])*rscale;
      if (bitflags && (bitflags[k]&border_code)) qual_map[k] = 0.0;
    }
  }
  else if (mode==pseudocorr) {
    AllocateFloat(&temp, xsize*ysize, "temp data");
    PseudoCorrelation(phase, qual_map, bitflags, border_code, temp,
                      tsize, xsize, ysize);
    for (k=0; k<xsize*ysize; k++) {
      if (bitflags && (bitflags[k]&border_code)) qual_map[k] = 0.0;
    }
    free(temp);
  }
  else if (mode==none) {
    for (k=0; k<xsize*ysize; k++) {
      qual_map[k] = 1.0;
      if (bitflags && (bitflags[k]&border_code)) qual_map[k] = 0.0;
    }
  }
  else { /* quality map was input */
    /* scale to interval (0,1) */
    for (rmin = rmax = qual_map[0], k=0; k<xsize*ysize; k++) {
      if (rmin > qual_map[k]) rmin = qual_map[k];
      if (rmax < qual_map[k]) rmax = qual_map[k];
    }
    printf("Min & max corr. coeff. = %lf, %lf\n", rmin, rmax);
    rscale = (rmin != rmax) ? 1.0/(rmax - rmin) : 0.0;
    for (k=0; k<xsize*ysize; k++) {
      qual_map[k] = (qual_map[k] - rmin)*rscale;
      if (bitflags && (bitflags[k]&border_code)) qual_map[k] = 0.0;
    }
  }
}

/* Determine the quality mode based on a keyword, and return a */
/* corresponding integer.  If keyword unrecognized, return -1.  */
/* Allow the keyword "none" only if allow_none = 1.             */
int SetQualityMode(char *modekey, char *qualfile, int allow_none)
```

```
{
  int mode;
  if (Keyword(modekey, "none")) {
    mode = none;
    printf("No weights supplied\n");
    if (!allow_none) {
      fprintf(stderr, "Invalid mode: %s\n", modekey);
      mode = -1;  /* "none" not allowed */
    }
  }
  else if (Keyword(modekey, "max_corr")) {
    mode = corr_coeffs;
    if (Keyword(qualfile, "none")) {
      fprintf(stderr, "Must supply quality image for this mode\n");
      mode = -1;
    }
  }
  else if (Keyword(modekey, "min_grad")) {
    if (!Keyword(qualfile, "none"))
      printf("Correlation image file will be ignored\n");
    mode = gradient;
  }
  else if (Keyword(modekey, "min_var")) {
    if (!Keyword(qualfile, "none"))
      printf("Correlation image file will be ignored\n");
    mode = variance;
  }
  else if (Keyword(modekey, "max_pseu")) {
    if (!Keyword(qualfile, "none"))
      printf("Correlation image file will be ignored\n");
    mode = pseudocorr;
  }
  else {
    fprintf(stderr, "Invalid mode = '%s'.  Must be\n", modekey);
    fprintf(stderr, "max_corr, min_var, max_grad, max_pseu "
                    "or none.\n");
    mode = -1;  /* invalid */
  }
  return mode;
}
```

The header file is getqual.h:

```
#ifndef __GETQUAL
#define __GETQUAL
```

```
typedef enum {none, gradient, variance, pseudocorr,
              corr_coeffs, dxdygrad
} UnwrapMode;
void GetQualityMap(int mode, float *qual_map, float *phase,
                   unsigned char *bitflags, int border_code,
                   int tsize, int xsize, int ysize);
int SetQualityMode(char *mode_key, char *qual_file, int allow_none);
#endif
```

The functions for computing histograms and performing histogram equalization are in file `histo.c`. Function `HistoAndThresh` executes the algorithm for computing and applying a threshold to the quality map to produce a quality mask. This algorithm was presented in Section 3.4. Function `Histogram` computes the histogram in `num_bins` histogram bins. The pixels defined by the bit flag `ignore_code` in the bit flag array `ignore` are not included in the computations.

There are several ways to control the thresholding process. If the parameters `thresh`, `percent_flag`, and `percentage` are all zero, then the threshold is computed automatically. If `thresh` is nonzero, then it defines the threshold. If `thresh` is zero but `percent_flag` is nonzero, then the threshold is calculated so that `percentage` percent of the pixels fall below the threshold. In any case the quality values are first "stretched" so that about 5 percent of the pixels fall in the first histogram bin and 5 percent fall in the last bin.

```
/*
 *  histo.c -- functions for histogramming and thresholding
 */
#include <stdio.h>
#include <math.h>
#include "histo.h"
#define NUM_BINS           10000
#define NUM_BINS_TO_PRINT  10

/*
 * Compute histogram, apply histogram equalization and
 * apply threshold.  If thresh=0, the threshold is determined
 * automatically.  If percent_flag is 1, the threshold is
 * selected to threshold the percentage of the pixels
 * defined by "percentage".  The bitflags array is "ignore",
 * and ignore_code defines those pixels to ignore in the
 * array "qual" of values to be histogrammed and thresholded.
 * The results overwrite the values in "qual".
 */
void HistoAndThresh(float *qual, int xsize, int ysize,
                double thresh, int percent_flag, double percentage,
                unsigned char *ignore, int ignore_code)
```

```
{
  int     i, j, k;
  int     num_zero=0, sum, num;
  int     histo[NUM_BINS + 1];
  double  first_val, last_val, rmult;
  printf("Computing histogram, equalizing & applying threshold\n");
  /* compute and print histogram */
  Histogram(qual, xsize, ysize, histo, NUM_BINS,
            ignore, ignore_code);
  /* stretch histogram so at least 5% in first & last bins */
  for (num=0.0, i=0; i<NUM_BINS; i++) { /* find number of pixels */
    num += histo[i];
  }
  for (sum=0.0, i=0; i<NUM_BINS; i++) {
    sum += histo[i];
    if (sum >= 0.05*num) break;
  }
  first_val = (1.0*i)/NUM_BINS;
  for (sum=0.0, i=0; i<NUM_BINS; i++) {
    sum += histo[i];
    if (sum >= 0.95*num) break;
  }
  last_val = (1.0*i)/NUM_BINS;
  if (last_val <= first_val) last_val = first_val + 1.0/NUM_BINS;
  rmult = 1.0/(last_val - first_val);
  /* stretch weights */
  for (j=0; j<xsize*ysize; j++) {
    if (ignore[j] & ignore_code) continue;   /* ignore these */
    if (qual[j] < first_val)
      qual[j] = 0.0;
    else if (qual[j] > last_val)
      qual[j] = 1.0;
    else
      qual[j] = (qual[j] - first_val)*rmult;
  }
  /* compute histogram again */
  printf("Stretching histogram...\n");
  Histogram(qual, xsize, ysize, histo,
            NUM_BINS, ignore, ignore_code);
  for (num=0.0, i=0; i<NUM_BINS; i++) {  /* find number of pixels */
    num += histo[i];
  }
  if (thresh==0.0) {
    /* Find optimal threshold by finding the trough of the */
    /* histogram.  Failing that, choose percentage = 50.   */
```

```
int  downhill=0, uphill=0, num_coarse_bins = 10;
double  coarse_histo[NUM_BINS];
printf("Looking for trough of histogram to set threshold...\n");
while (num_coarse_bins <= 100) {
  for (k=0; k<num_coarse_bins; k++)
    coarse_histo[k] = 0.0;
  for (i=0, k=0, sum=0.0; i<NUM_BINS; i++) {
    if (i==NUM_BINS - 1) sum += histo[i];
    if ((int)((num_coarse_bins*1.0*i)/NUM_BINS) > k
                                  || i==NUM_BINS - 1) {
      /* coarse_histo holds the percentages */
      coarse_histo[k] = (100.0*sum)/(num);
      sum = 0.0;
      ++k;
    }
    sum += histo[i];
  }
  /* find first minima */
  for (i=1, downhill=0; i<num_coarse_bins; i++) {
    if (downhill) {
      if (coarse_histo[i] > coarse_histo[i-1] + 0.5) {
        printf("Uphill at %d\n", i);
        break;
      }
    }
    else {
      if (coarse_histo[i] < coarse_histo[i-1] - 1.0) {
        downhill = 1;
        printf("Downhill at %d\n", i);
      }
    }
  }
  if (i < num_coarse_bins) {
    /* thresh = (i - 1.0)/num_coarse_bins; */
    for (j=0, percentage=0.0; j<i; j++) {
      percentage += coarse_histo[j];
    }
    for (sum=0.0, i=0; i<NUM_BINS; i++) {
      sum += histo[i];
      if (sum >= percentage*num*0.01) break;
    }
    if (i<NUM_BINS - 1) thresh = (i + 1.0)/NUM_BINS;
    else thresh = (NUM_BINS - 1.0)/NUM_BINS;
    printf("Found trough.  Setting threshold to %g (%g%%).\n",
           thresh, percentage);
```

```
        break;
      }
      else {
        /* repeat with more bins in histogram */
        num_coarse_bins += 10;
      }
    }
    if (thresh==0.0) {
      percentage = 50.0;
      for (sum=0.0, i=0; i<NUM_BINS; i++) {
        sum += histo[i];
        if (sum >= percentage*num*0.01) break;
      }
      if (i<NUM_BINS - 1) thresh = (i + 1.0)/NUM_BINS;
      else thresh = (NUM_BINS - 1.0)/NUM_BINS;
      printf("Could not find min of histogram.\n");
      printf("Setting thresh at 50%%.\n");
    }
  }

  /* compute threshold from desired percentage */
  if (percent_flag) {
    for (sum=0.0, i=0; i<NUM_BINS; i++) {
      sum += histo[i];
      if (sum >= percentage*num*0.01) break;
    }
    if (i<NUM_BINS - 1) thresh = (i + 1.0)/NUM_BINS;
    else thresh = (NUM_BINS - 1.0)/NUM_BINS;
  }
  /* apply threshold */
  printf("Threshold = %lf\n", thresh);
  for (j=0, num_zero=0, num=0; j<xsize*ysize; j++) {
    if (ignore[j] & ignore_code) continue;   /* ignore these */
    ++num;
    if (qual[j]<thresh) ++num_zero;
    qual[j] = (qual[j]<thresh) ? 0.0 : 1.0;
  }
  printf("%lg percent of the weights are now zero-weights\n",
         100.0*((double)num_zero)/((double)(num)));
}

/* Generate histogram (in array histo) of the values in array */
/* "qual" using num_bins bins.  The bigflags array is array   */
/* "ignore", and the pixels marked with the ignore_code must  */
```

```
/* be ignored (these are mask pixels, for example).            */
void Histogram(float *qual, int xsize, int ysize, int *histo,
               int num_bins, unsigned char *ignore, int ignore_code)
{
  double min_qual, max_qual;
  int   sum, num, i, j, k, off=0;
  /* zero histogram bins */
  min_qual = 1.0e+10;
  max_qual = -1.0e+10;
  for (i=0; i<=num_bins; i++) histo[i] = 0;
  /* find min & max weight, and histogram of weights */
  for (j=0, num=0; j<xsize*ysize; j++) {
    if (ignore[j] & ignore_code) continue;   /* ignore these */
    ++num;
    if (qual[j] < min_qual) min_qual = qual[j];
    if (qual[j] > max_qual) max_qual = qual[j];
    if ((qual[j] < -1.0e-5 || qual[j] > 1.0 + 1.0e-5) && !off) {
      off = 1;
      printf("WARNING: Quality values must be between 0 and 1!\n");
      printf("(Found a magnitude of %f.)\n", qual[j]);
      printf("Results of this run may be invalid!\n\n");
    }
    if (qual[j] < 0.0) qual[j] = 0.0;
    if (qual[j] > 1.0) qual[j] = 1.0;
    ++histo[(int)(qual[j]*num_bins)];
  }
  histo[NUM_BINS - 1] += histo[NUM_BINS];  /* ignore last bin */
  /* print histogram */
  printf("Min & max qual = %lf, %lf\nHistogram (percentage) = \n",
         min_qual, max_qual);
  /* quantize histogram to fewer bins and print */
  for (i=0, k=0, sum=0.0; i<NUM_BINS; i++) {
    if (i==NUM_BINS - 1) sum += histo[i];
    if ((int)((NUM_BINS_TO_PRINT*1.0*i)/NUM_BINS) > k
                                        || i==NUM_BINS - 1) {
      printf("Bins %.3f-%.3f:  %d\n", (k + 0.0)/NUM_BINS_TO_PRINT,
         (k + 1.0)/NUM_BINS_TO_PRINT, (int)((100.0*sum)/(num)+0.5));
      sum = 0.0;
      ++k;
    }
    sum += histo[i];
  }
}
```

The header file is histo.h:

```
#ifndef __HISTO
#define __HISTO
void HistoAndThresh(float *qual, int xdim, int ydim,
               double thresh, int percent_flag, double percentage,
               unsigned char *ignore, int ignore_code);
void Histogram(float *qual, int xdim, int ydim, int *histo,
               int num_bins, unsigned char *ignore, int ignore_code);
#endif
```

File maskfat.c contains functions for thickening masks or zero-weights. Function FattenMask thickens the mask pixels (identified by the bit flag maskcode) in the bitflags array array by thickness pixels. For example, if thickness = 2, then all pixels that lie within two pixels of any masked pixel will be masked (i.e., the maskcode bit in the bitflags array will be set). Similarly the function FattenQual thickens the zero-weights in the quality map qual_map by thickness pixels.

```
/*
 * maskfat.c -- fatten mask
 */
#include <stdlib.h>
#include <stdio.h>
#include <string.h>
#include <math.h>
#include "util.h"

/* Fattens mask and border by "thickness" pixels */
void FattenMask(unsigned char *array, int mask_code, int thickness,
                int xsize, int ysize)
{
  int i, j, k, ii, jj, kk, t2=thickness;
  unsigned char *temp;
  if (t2 < 1) return;
  AllocateByte(&temp, xsize*ysize, "temp array");
  for (j=0; j<ysize; j++) {
    for (i=0; i<xsize; i++) {
      k = j*xsize + i;
      temp[k] = 0;
      for (jj=j-t2; jj<=j+t2 && !temp[k]; jj++) {
        for (ii=i-t2; ii<=i+t2 && !temp[k]; ii++) {
          kk = jj*xsize + ii;
          if (ii>=0 && ii<xsize && jj>=0 && jj<ysize
                                    && (array[kk] & mask_code))
```

```
            temp[k] = 1;
          }
        }
      }
    }
  for (k=0; k<xsize*ysize; k++) {
    if (temp[k]) array[k] |= mask_code;
  }
  free(temp);
}

/* Fattens zero values in quality map by "thickness" pixels */
void FattenQual(float *qual_map, int thickness,
                int xsize, int ysize)
{
  int i, j, k, ii, jj, kk, t2=thickness;
  unsigned char *temp;
  if (t2 < 1) return;
  AllocateByte(&temp, xsize*ysize, "temp array");
  for (j=0; j<ysize; j++) {
    for (i=0; i<xsize; i++) {
      k = j*xsize + i;
      temp[k] = 0;
      for (jj=j-t2; jj<=j+t2 && !temp[k]; jj++) {
        for (ii=i-t2; ii<=i+t2 && !temp[k]; ii++) {
          kk = jj*xsize + ii;
          if (ii>=0 && ii<xsize && jj>=0 && jj<ysize
                       && qual_map[kk]==0.0)
            temp[k] = 1;
        }
      }
    }
  }
  for (k=0; k<xsize*ysize; k++) {
    if (temp[k]) qual_map[k] = 0.0;
  }
  free(temp);
}
```

The header file is maskfat.h:

```
#ifndef __MASKFAT
#define __MASKFAT
void FattenMask(unsigned char *mask, int mask_code, int thickness,
                int xsize, int ysize);
```

```
void FattenQual(float *qualmap, int thickness, int xsize,
                int ysize);
#endif
```

File `maskthin.c` contains the function `ThinMask` for thinning the mask in the `bitflags` array. The algorithm, which was presented in Section 4.4.2, is used to thin the mask cuts produced by the mask cut algorithm. The mask pixels are those identified by the bit flag `cut_code` in the `bitflags` array. This function is used to trim the mask cuts produced by the mask cut algorithm of Section 4.4.1.

```
/*
 *   maskthin.c -- function for thinning a mask
 */
#include <stdio.h>
#include <math.h>
#include "list.h"
#include "maskthin.h"

/* Thins mask without disconnecting it. */
void ThinMask(unsigned char *bitflags, int xsize, int ysize,
              int cut_code)
{
  int   i, j, k, n, ii, jj, kk, k1, k2, k3;
  int   numflips, removed_some=1, avoid_code;
  avoid_code = RESIDUE | BORDER;
  while (removed_some) {
    removed_some = 0;
    for (j=1; j<ysize - 1; j++) {
      for (i=1; i<xsize - 1; i++) {
        k = j*xsize + i;
        /* only consider branch cut pixels for removal */
        if (!(bitflags[k] & cut_code)) continue;
        /* don't remove pixel k if it borders a residue or mask */
        if (bitflags[k-xsize-1] & avoid_code) continue;
        if (bitflags[k-xsize] & avoid_code) continue;
        if (bitflags[k-xsize+1] & avoid_code) continue;
        if (bitflags[k-1] & avoid_code) continue;
        if (bitflags[k] & avoid_code) continue;
        if (bitflags[k+1] & avoid_code) continue;
        if (bitflags[k+xsize-1] & avoid_code) continue;
        if (bitflags[k+xsize] & avoid_code) continue;
        if (bitflags[k+xsize+1] & avoid_code) continue;
        /* walk around border of 3x3 nbhd; make sure removal */
        /* does not leave branchcut pixels disconnected in  */
        /* 3x3 neighborhood  */
```

```
        kk = k - xsize;   /* start at upper middle pixel */
        for (n=0, numflips=0; n<4; n++) {
          if (n==0) { k1 = k-xsize; k2 = k1+1; k3 = k2+xsize; }
          else if (n==1) { k1 = k+1; k2 = k1+xsize; k3 = k2-1; }
          else if (n==2) { k1 = k+xsize; k2 = k1-1; k3 = k2-xsize; }
          else if (n==3) { k1 = k-1; k2 = k1-xsize; k3 = k2+1; }
          if (bitflags[k1] & cut_code) {
            if (!(bitflags[k3] & cut_code))
              ++numflips;
          }
          else {
            if (bitflags[k2] & cut_code) {
              ++numflips;
              if (!(bitflags[k3] & cut_code)) ++numflips;
            }
            else if (bitflags[k3] & cut_code) ++numflips;
          }
        }
        if (numflips < 3) { /* remove pixel from branch cut */
          if (!(bitflags[k-xsize]&cut_code)
                || !(bitflags[k+1]&cut_code)
                     || !(bitflags[k+xsize]&cut_code)
                          || !(bitflags[k-1]&cut_code)) {
            bitflags[k] &= (~cut_code);
            ++removed_some;
          }
        }
      }
    }
    printf("Trimmed %d mask pixels\n", removed_some);
  }
}
```

The header file is maskthin.h:

```
#ifndef __MASKTHIN
#define __MASKTHIN
void ThinMask(unsigned char *bitflags, int xsize, int ysize,
              int cut_code);
#endif
```

A.4 LEAST SQUARES, LAPLACIANS, AND CONGRUENCE

The following files are contained in this section: solncos.c, solncos.h, dct.c, dct.h, laplace.c, laplace.h, congruen.c, and congruen.h.

We now present the software for solving Poisson's equation by means of a fast cosine transform. We also present functions for computing Laplacians and making surfaces congruent to the wrapped phase data.

File `solncos.c` contains the function `DirectSolnByCosineTranform` that solves Poisson's equation by means of a fast cosine transform. See Sections 5.3.1 and 5.3.2. This function is required by the PCG algorithm. It is also required by the program in file `mainunwt.c` of Section A.10 to solve the unweighted least-squares phase-unwrapping problem. The small arrays `xcos` and `ycos`, which are of lengths `xsize` and `ysize`, respectively, are required for scratch calculations.

```
/*
 *  solncos.c -- function for solving Poisson's equation by means of
 *               fast cosine transform
 */
#include <stdio.h>
#include <math.h>
#include "dct.h"
#include "pi.h"

/* Main function for direct solution of Poisson's equation for      */
/* unweighted least squares phase unwrapping by means of discrete   */
/* cosine transform (which actually uses FFTs to compute the DCTs) */
void DirectSolnByCosineTransform(float *array, int xsize, int ysize,
                                 double *xcos, double *ycos)
{
  int    i, j, m, n;
  /*  Transform  */
  printf("Forward transform...\n");
  FastCosineTransform(array, xsize, ysize, 1);

  /*  Divide  */
  printf("Scaling...\n");
  for (i=0; i<xsize; i++) {
    xcos[i] = cos(i*PI/(xsize - 1.0));
  }
  for (j=0; j<ysize; j++) {
    ycos[j] = cos(j*PI/(ysize - 1.0));
  }
  array[0] = 0.0;
  for (j=0; j<ysize; j++) {
    for (i=0; i<xsize; i++) {
      if (i==0 && j==0) {
        array[0] = 0.0;
      }
      else {
```

```
        array[j*xsize + i] /= (4.0 - 2.0*xcos[i] - 2.0*ycos[j]);
      }
    }
  }
  /*  Transform  */
  printf("Inverse transform...\n");
  FastCosineTransform(array, xsize, ysize, -1);
}
```

The header file is `solncos.h`:

```
#ifndef __SOLNCOS
#define __SOLNCOS
void DirectSolnByCosineTransform(float *laplace, int xsize,
                            int ysize, double *xcos, double *ycos);
#endif
```

The functions for computing the two-dimensional fast cosine transform are in file `dct.c`. Function `FastCosineTransform` computes the two-dimensional DCT (discrete cosine transform) of the data in `array`. The `sign` parameter should be positive for the forward DCT and negative for the inverse DCT. The functions `RowDcts` and `ColDcts` compute the one-dimensional DCT's of the rows and the columns, respectively. These functions call `FastDct` to compute the one-dimensional DCT. This function in turn calls the functions `RealFft` and `Fft`, which compute the one-dimensional FFT of real and complex data, respectively. The DCT and FFT functions are based on those in the book *Numerical Recipes in C*, second edition. (See the References of Chapter 5 for the author and publisher information.)

```
#include <malloc.h>
#include <stdio.h>
#include <math.h>
#include "file.h"
#include "dct.h"


/*
 * Two-dimensional cosine tranform of array.
 * Array must be dimensioned from 0 to xsize * ysize - 1.
 * (xsize & ysize each must be one more than a power of two)
 * For forward transform, sign = 1.  For inverse transform,
 * sign = -1.
 */
void FastCosineTransform(float *array, int xsize, int ysize,
                          int sign)
```

```
{
  int           i, j;
  double        tmpr;
  float         *vector, *buffer;
  vector = (float*) malloc(ysize*sizeof(float));
  if (!vector) {
    fprintf(stderr, "Cannot allocate memory for cosine transf.\n");
    exit(MEMORY_ALLOCATION_ERROR);
  }
  buffer = (float *) malloc(BLOCKSIZE*ysize*sizeof(float));
  if (!buffer) {
    fprintf(stderr, "Cannot allocate buffer for cosine transf.\n");
    exit(MEMORY_ALLOCATION_ERROR);
  }
  RowDcts(array, xsize, ysize, sign);
  ColDcts(array, xsize, ysize, sign, vector, buffer);
  free(vector);
  free(buffer);
}

/* Perform 1-dim fast DCT's on rows of array */
void RowDcts(float *array, int xsize, int ysize, int sign)
{
  int  i, j;
  for (j=0; j<ysize; j++) {
    FastDct(&array[j*xsize], xsize, sign);
  }
}

/* Perform 1-dim fast DCT's on columns of array */
void ColDcts(float *array, int xsize, int ysize, int sign,
             float *vector, float *buffer)
{
  int  i, j;
  int  ii, istop, blocksize=BLOCKSIZE;
  if (blocksize<=0) {
    for (i=0; i<xsize; i++) {
      for (j=0; j<ysize; j++) {
        vector[j] = array[j*xsize + i];
      }
      FastDct(vector, ysize, sign);
      for (j=0; j<ysize; j++) {
        array[j*xsize + i] = vector[j];
      }
    }
```

```
  }
  else {
    for (ii=0; ii<xsize; ii+=blocksize) {
      istop = (ii + blocksize <= xsize) ? ii + blocksize : xsize;
      for (j=0; j<ysize; j++) {
        for (i=ii; i<istop; i++) {
          buffer[(i - ii)*ysize + j] = array[j*xsize + i];
        }
      }
      for (i=ii; i<istop; i++) {
        FastDct(&buffer[(i - ii)*ysize], ysize, sign);
      }
      for (j=0; j<ysize; j++) {
        for (i=ii; i<istop; i++) {
          array[j*xsize + i] = buffer[(i - ii)*ysize + j];
        }
      }
    }
  }
}

/* Fast Discrete Cosine Transform
 *   The array must be dimensioned from 0 to size-1
 *   (size must be one more than a power of two).
 *   If sign = -1, the inverse transform is computed.
 */
void FastDct(float *array, int size, int sign)
{
  int     j;
  double  rmult, sum, p1, p2;
  double  angle, zr, zi, zpi, zpr, r;
  --array;              --size;
  angle = PI/size;      r = sin(0.5*angle);
  zpr = -2.0*r*r;       zpi = sin(angle);
  zr = 1.0;             zi = 0.0;
  sum =       0.5*(array[1] - array[size + 1]);
  array[1] = 0.5*(array[1] + array[size + 1]);
  for (j=2; j<=size/2; j++) {
    r = zr;
    zr =  r*zpr - zi*zpi + zr;
    zi = zi*zpr +  r*zpi + zi;
    p1 = 0.5*(array[j] + array[size + 2 - j]);
    p2 =     (array[j] - array[size + 2 - j]);
    array[j] = p1 - zi*p2;
    array[size + 2 - j] = p1 + zi*p2;
```

```
    sum += zr*p2;
  }
  RealFft(array, size, 1);
  array[size + 1] = array[2];
  array[2] = sum;
  for (j=4; j<=size; j+=2) {
    sum += array[j];
    array[j] = sum;
  }
  if (sign < 0) rmult = 1.0/size;
  else rmult = 2.0;
  for (j=1; j<=size+1; j++) array[j] *= rmult;
}

/* FFT of real function (size must be a power of two) */
void RealFft(float *array, int size, int sign)
{
  int     i, i1, i2, i3, i4;
  double  s1, s2, qr, qi, rr, ri;
  double  zr, zi, zpr, zpi, r, angle;

  angle = PI/(size/2);
  s1 = 0.5;
  if (sign > 0) {
    s2 = -0.5;
    Fft(array, size/2, 1);
  }
  else {
    s2 = 0.5;
    angle = -angle;
  }
  r = sin(0.5*angle);
  zpr = -2.0*r*r;     zpi = sin(angle);
  zr = 1.0 + zpr;     zi = zpi;
  for (i=2; i<= size/4; i++) {
    i1 = 2*i - 1;
    i2 = i1 + 1;
    i3 = size + 3 - i2;
    i4 = i3 + 1;
    qr = s1*(array[i1] + array[i3]);
    qi = s1*(array[i2] - array[i4]);
    rr = -s2*(array[i2] + array[i4]);
    ri =  s2*(array[i1] - array[i3]);
    array[i1]  =  qr + zr*rr - zi*ri;
    array[i2]  =  qi + zr*ri + zi*rr;
```

```
    array[i3]  =  qr - zr*rr + zi*ri;
    array[i4]  = -qi + zr*ri + zi*rr;
    r = zr;
    zr = r*zpr - zi*zpi + zr;
    zi = zi*zpr + r*zpi + zi;
  }
  qr = array[1];
  if (sign > 0) {
    array[1] = qr + array[2];
    array[2] = qr - array[2];
  }
  else {
    array[1] = s1*(qr + array[2]);
    array[2] = s1*(qr - array[2]);
    Fft(array, size/2, -1);
  }
}

/* FFT (size must be a power of two) */
void Fft(float *array, int size, int sign)
{
  int            n, nmax, m, j, istep, i;
  double         r, zr, zpr, zpi, zi, angle;
  double         tmp, tmpr, tmpi;
  j = 1;   n = 2*size;
  for (i=1; i<n; i+=2) {
    if (j > i) {
      /* perform swaps */
      tmp = array[j]; array[j] = array[i]; array[i] = tmp;
      tmp = array[j+1]; array[j+1] = array[i+1]; array[i+1] = tmp;
    }
    m = n/2;
    while (j > m && m > 1) {
      j -= m;    m /= 2;
    }
    j += m;
  }
  nmax = 2;
  while (n > nmax) {
    istep = 2*nmax;
    angle = TWOPI/nmax;
    if (sign < 0) angle = -angle;
    r = sin(0.5*angle);
    zpr = -2.0*r*r;     zpi = sin(angle);
    zr = 1.0;           zi = 0.0;
```

```
    for (m=1; m<nmax; m+=2) {
      for (i=m; i<=n; i+=istep) {
        j = i + nmax;
        tmpr = zr*array[j] - zi*array[j+1];
        tmpi = zr*array[j+1] + zi*array[j];
        array[j]   = array[i]   - tmpr;
        array[j+1] = array[i+1] - tmpi;
        array[i]   += tmpr;
        array[i+1] += tmpi;
      }
      r = zr;
      zr = r*zpr - zi*zpi + zr;
      zi = zi*zpr + r*zpi + zi;
    }
    nmax = istep;
  }
}
```

The header file is `dct.h`:

```
#ifndef __DCT
#define __DCT
#include "pi.h"
#define BLOCKSIZE     50
#define MAX_NUM_ROWS  2050
void FastCosineTransform(float *array, int xsize, int ysize,
                         int sign);
void RowDcts(float *array, int xsize, int ysize, int sign);
void ColDcts(float *array, int xsize, int ysize, int sign,
             float *vector, float *buffer);
void FastDct(float *array, int size, int sign);
void RealFft(float *array, int size, int sign);
void Fft(float *array, int size, int sign);
#endif
```

The functions for computing the Laplacians are in `laplace.c`. The function `ComputeLaplacian` computes the weighted or unweighted Laplacian of wrapped or unwrapped data. Zero, one, or two sets of weights may be supplied. For example, to compute the unweighted Laplacian, the arrays `dxwts` and `dywts` should be null pointers. If the parameter `laptype` is zero, the usual (unwrapped) Laplacian of the data in the `input` array is computed. Otherwise, the (wrapped) phase Laplacian is computed. The function `ComputeDerivWts` computes the d/dx and d/dy weights based on the L^0-norm and stores the results in the arrays `dxwts` and `dywts`. This function is called by the L^p-norm algorithm of Section A.13 to compute the data-dependent weights $U(i,j)$ and $V(i,j)$. These weights are multiplied by the user-supplied weights in the array qual_map. The mask pixels that are to be ignored should be indicated by zero-weights in this array. If there are no user-supplied weights or mask pixels, qual_map should be a null pointer. The parameter e0 is the normalization term ϵ_0 that was discussed in Section 5.5.1.

```
/*
 *  laplace.c -- compute weighted (or unweighted) Laplacian
 */
#include <stdio.h>
#include <math.h>
#include "grad.h"
#include "laplace.h"
#define SIMIN(x,y) (((x)*(x) < (y)*(y)) ? (x)*(x) : (y)*(y))

/* Compute the dx and dy weighted (or unweighted) Laplacian.      */
/* laptype=1 for wrapped phase Laplacian, =0 for usual Laplacian. */
void ComputeLaplacian(float *input, float *laplacian, float *dxwts,
                      float *dywts, int xsize, int ysize, int laptype)
{
  double w1, w2, w3, w4;
  int    i, j, k1, k2, k3, k4, k;
  float  *wts;
  for (j=0; j<ysize; j++) {
    for (i=0; i<xsize; i++) {
      k = j*xsize + i;
      k1 = (i<xsize-1) ? k + 1 : k - 1;
      k2 = (i>0) ? k - 1 : k + 1;
      k3 = (j<ysize-1) ? k + xsize : k - xsize;
      k4 = (j>0) ? k - xsize : k + xsize;
      if (dxwts==NULL && dywts==NULL) {  /* unweighted */
        w1 = w2 = w3 = w4 = 1.0;
      }
      else if (dxwts==NULL || dywts==NULL) {  /* one set of wts */
        wts = (dxwts) ? dxwts : dywts;
        w1 = SIMIN(wts[k], wts[k1]);
        w2 = SIMIN(wts[k], wts[k2]);
        w3 = SIMIN(wts[k], wts[k3]);
        w4 = SIMIN(wts[k], wts[k4]);
      }
      else {    /* dxwts and dywts are both supplied */
        w1 = dxwts[k];
        w2 = (i>0) ? dxwts[k-1] : dxwts[k];
        w3 = dywts[k];
        w4 = (j>0) ? dywts[k-xsize] : dywts[k];
```

```
      }
      if (laptype) {   /* compute wrapped phase Laplacian */
        float *phase = input;
        laplacian[k]  = w1*Gradient(phase[k], phase[k1])
                          + w2*Gradient(phase[k], phase[k2])
                             + w3*Gradient(phase[k], phase[k3])
                                + w4*Gradient(phase[k], phase[k4]);
      }
      else {   /* compute usual (not wrapped) Laplacian */
        float *surface = input;
        laplacian[k]  = w1*(surface[k] - surface[k1])
                          + w2*(surface[k] - surface[k2])
                             + w3*(surface[k] - surface[k3])
                                + w4*(surface[k] - surface[k4]);
      }
    }
  }
}

/* Compute the dx and dy weighted laplacian.  The parameter */
/* e0 is deiscussed in the text.                            */
void ComputeDerivWts(float *phase, float *soln, float *dxwts,
                     float *dywts, float *qual_map,
                     double e0, int xsize, int ysize)
{
  double w, r, eps=1.0e-06;
  int    i, j, k, k1, k2, k3, k4;
  w = 1.0;  /* w must be 1 if qual_map is NULL */
  for (j=0; j<ysize; j++) {
    for (i=0; i<xsize; i++) {
      k = j*xsize + i;
      k1 = (i < xsize - 1) ? k + 1 : k - 1;
      r = soln[k1] - soln[k] - Gradient(phase[k1], phase[k]);
      dxwts[k] = e0/(r*r + e0);
      if (qual_map) {
        w = qual_map[k];
        if (w > qual_map[k1]) w = qual_map[k1];
      }
      dxwts[k] *= w;  /* must have 0 <= w <= 1 */
    }
  }
  for (j=0; j<ysize; j++) {
    for (i=0; i<xsize; i++) {
      k = j*xsize + i;
      k3 = (j < ysize - 1) ? k + xsize : k - xsize;
```

```
      r = soln[k3] - soln[k] - Gradient(phase[k3], phase[k]);
      dywts[k] = e0/(r*r + e0);
      if (qual_map) {
        w = qual_map[k];
        if (w > qual_map[k3]) w = qual_map[k3];
      }
      dywts[k] *= w;  /* must have 0 <= w <= 1 */
    }
  }
}
```

The header file is laplace.h:

```
#ifndef __LAPLACE
#define __LAPLACE
void ComputeLaplacian(float *phase, float *laplacian, float *dxwts,
                      float *dywts, int xsize, int ysize, int laptype);
void ComputeDerivWts(float *phase, float *soln, float *dxwts,
                      float *dywts, float *qual_map, double e0,
                      int xsize, int ysize);
#endif
```

File congruen.c contains the function CongruentSoln for making the values of the surface surf congruent to the wrapped phase data phase by the method described in Section 5.3. The zero-weighted pixels (indicated by the zero values in the array qual_map, which may be a null pointer if there are no weights) are ignored.

```
/*
 *  congruen.c -- make surface congruent to phase
 */
#include <stdio.h>
#include <math.h>
#include "grad.h"
#include "util.h"

/* Make the surface in the array "surf" congruent to the wrapped */
/* phase (scaled to 0-1) in the array "phase".  Ignore the       */
/* zero-weights in the quality map array qual_map.               */
void CongruentSoln(float *surf, float *phase, float *qual_map,
                   int xsize, int ysize)
{
  int     i, j, k, n, npos;
  int     kk, num, rnum, rnum_min;
  double  rmin, rr, rr_min, r;
```

```
float   *ss;
AllocateFloat(&ss, xsize*ysize, "temp array");
/* ensure surface is positive */
rmin = surf[0];
for (k=0; k<xsize*ysize; k++) {
  if (rmin > surf[k]) rmin = surf[k];
}
npos = (rmin < 0) ? -rmin + 2.0 : 0.0;
for (k=0; k<xsize*ysize; k++) {
  surf[k] += npos;
}
/* find best additive offset (to minimize discontinuities) */
for (kk=0; kk<=10; kk++) {
  rr = 0.1*kk;
  /* create temporary surface by adding a constant rr */
  for (k=0; k<xsize*ysize; k++) {
    if (qual_map && qual_map[k]==0.0) continue; /* ignore 0-wts */
    n = (int)(surf[k] + rr);
    r = (surf[k] + rr) - n;
    r -= phase[k];
    if (r < -0.5) ss[k] = n - 1 + phase[k];
    else if (r > 0.5) ss[k] = n + 1 + phase[k];
    else  ss[k] = n + phase[k];
  }
  /* measure the discontinuities in the temporary surface */
  for (k=0, rnum=0; k<xsize*ysize; k++) {
    if (qual_map && qual_map[k]==0.0) continue; /* ignore 0-wts */
    i = k%xsize;
    j = k/xsize;
    if (i>0) {
      if (qual_map && qual_map[k-1]==0.0) continue; /* ignore */
      r = ss[k] - ss[k-1];
      if (r > 0.5 || r < -0.5) rnum++;
    }
    if (j>0) {
      if (qual_map && qual_map[k-xsize]==0.0) continue; /*ignore*/
      r = ss[k] - ss[k-xsize];
      if (r > 0.5 || r < -0.5) rnum++;
    }
  }
  /* save the rr that gives the minimum error (rnum) */
  if (kk==0 || rnum < rnum_min) {
    rnum_min = rnum;
    rr_min = rr;
  }
```

```
    printf("Offset %lf yields %d discontinuities\n", rr, rnum);
  }
  free(ss);
  printf("Adding offset %lf to surface\n", rr_min);
  for (k=0; k<xsize*ysize; k++) {
    surf[k] += rr_min;
  }
  for (k=0; k<xsize*ysize; k++) {
    n = (int)surf[k];
    r = surf[k] - n;
    r -= phase[k];
    if (r < -0.5) surf[k] = n - 1 + phase[k];
    else if (r > 0.5) surf[k] = n + 1 + phase[k];
    else  surf[k] = n + phase[k];
    surf[k] -= npos;  /* also un-do addition of npos */
  }
}
```

The header file is `congruen.h`:

```
#ifndef __CONGRUEN
#define __CONGRUEN
void CongruentSoln(float *surf, float *phase, float *qual_map,
                   int xsize, int ysize);
#endif
```

A.5 VISUALIZATION AND EVALUATION

The following files are listed in this section: `view.c`, `view.h`, `maindiff.c`, and `mainjump.c`.

We now present several functions and programs for visualizing and evaluating an unwrapped surface. File `view.c` contains functions for producing grayscale images from surfaces. Function `Shade` produces a bird's-eye shaded view of the surface, an example of which is shown in Figure 3.6(a). `SurfImage` converts the surface elevations to gray scales. An example of the latter type of image is shown in Figure 5.48(b). `Rewrap` produces the phase pattern that results from rewrapping the values of the surface. All of these functions produce simple 1-byte arrays whose values represent pixel brightness, ranging from 0 for black to 255 for white. These arrays may be saved to files for viewing with graphics software (which is not supplied).

```
/*
 * view.c -- utility functions for viewing surface data
 */
#include <stdlib.h>
```

```
#include <stdio.h>
#include <string.h>
#include <math.h>
#include "pi.h"
#include "view.h"

/* Generate shaded bird's eye view of surface.               */
/* elev_angle = elevation angle of light vector in radians   */
/* (elev_angle=0.0 results in a default angle of 45 deg.)    */
void Shade(float *surf, char *image, int xsize, int ysize,
           double elev_angle)
{
  int    i, j, k;
  double angle, ht, r, ca, sa, ch, sh, lightness;
  ca = cos(elev_angle);
  sa = sin(elev_angle);
  for (j=0; j<ysize; j++) {
    for (i=0; i<xsize; i++) {
      k = j*xsize + i;
      if (i==xsize - 1) {
        image[k] = 0;
      }
      else {
        ht = surf[k] - surf[k+1];
        if (ht < 0.0) {
          image[k] = 0;
        }
        else {
          /* Dot product of surface normal with light vector. */
          /* Assumes light vector perpendicular to array columns. */
          r = sqrt(1.0 + ht*ht);
          ch = 1.0/r;
          sh = ht/r;
          lightness = ch*ca + sh*sa;
          image[k] = 255.0*lightness;
        }
      }
    }
  }
}

/* Generate image of surface elevations.                     */
void SurfImage(float *surf, char *image, int xsize, int ysize)
{
  int     i, j, k;
```

```
  double  rmin, rmax, rscale;
  for (k=0, rmax = -(rmin = 1.0e+20); k<xsize*ysize; k++) {
    if (rmin > surf[k]) rmin = surf[k];
    if (rmax < surf[k]) rmax = surf[k];
  }
  rscale = (rmin < rmax) ? 1.0/(rmax - rmin) : 0.0;
  for (k=0, rmax = -(rmin = 1.0e+20); k<xsize*ysize; k++) {
    image[k] = 255.0*rscale*(surf[k] - rmin);
  }
}

/* Generate image of rewrapped phase.              */
/* Assumes surface values are scaled so that the   */
/* interval 0-TWOPI is scaled to the interval 0-1. */
void Rewrap(float *surf, char *image, int xsize, int ysize)
{
  int     i, j, k;
  double  r;
  for (k=0; k<xsize*ysize; k++) {
    r = surf[k];
    if (r < 0) r += (int)(-r) + 2;   /* ensure r > 0 */
    r -= (int) r;
    image[k] = 256.0*r;
  }
}
```

The header file is view.h:

```
#ifndef __VIEW
#define __VIEW
void Shade(float *surf, char *image, int xsize, int ysize,
           double elev_angle);
void SurfImage(float *surf, char *image, int xsize, int ysize);
void Rewrap(float *surf, char *image, int xsize, int ysize);
#endif
```

File maindiff.c contains a program for comparing two surfaces. The differences between the surfaces are stored in a file as a surface, and the rms (root mean square) and mean absolute values of the differences are computed after subtracting the average difference. The "usage statement" in the program listing below explains how to run the program.

```
/*
 * maindiff.c - difference between two surfaces &
 *              RMS and absolute difference measures
```

```
 *
 * Source code files required:
 *   maindiff.c        util.c
 */
#include <stdio.h>
#include <math.h>
#include "file.h"
#include "pi.h"
#include "util.h"

main (int argc, char *argv[])
{
  int             k, num;
  FILE            *ifp1, *ifp2, *ofp, *mfp;
  float           *surf1, *surf2;    /* array */
  float           *outsurf;     /* array */
  unsigned char   *mask;
  char            ifile1[200], ifile2[200], outfile[200];
  char            maskfile[200];
  int             xsize, ysize;   /* dimensions of arrays */
  double          sum, avg, sumsqr, avgsqr, rms, sumabs, avgabs;
  double          r, r1, r2;

  printf("Surface difference and RMS measure\n");
  if (argc!=7) {
    fprintf(stderr, "Usage: %s surf1 surf2 outfile xsize ysize"
                    " maskfile\n", argv[0]);
    fprintf(stderr, "\nwhere surf1 = first surface file\n");
    fprintf(stderr, "       surf2 = second surface file\n");
    fprintf(stderr, "       outfile = output surface file\n");
    fprintf(stderr, "       xsize, ysize = array dimensions\n");
    fprintf(stderr, "       maskfile = mask file (0=masked).\n");
    fprintf(stderr, "       (If no mask, enter 'none')\n");
    exit(BAD_USAGE);
  }
  sscanf(argv[1], "%s", ifile1);
  sscanf(argv[2], "%s", ifile2);
  sscanf(argv[3], "%s", outfile);
  sscanf(argv[4], "%d", &xsize);
  sscanf(argv[5], "%d", &ysize);
  sscanf(argv[6], "%s", maskfile);
  printf("Input file 1 = %s\n", ifile1);
  printf("Input file 2 = %s\n", ifile2);
  printf("Output file = %s\n", outfile);
  printf("Mask file = %s\n", maskfile);
```

```
printf("Array size = %d x %d\n", xsize, ysize);
if (Keyword(maskfile, "none")) printf("No mask file\n");

/*  OPEN FILES, ALLOCATE MEMORY       */
mfp = ifp1 = ifp2 = ofp = NULL;
OpenFile(&ifp1, ifile1, "r");
OpenFile(&ifp2, ifile2, "r");
OpenFile(&ofp, outfile, "w");
if (!Keyword(maskfile, "none")) OpenFile(&mfp, maskfile, "r");

surf1 = surf2 = outsurf = NULL;
mask = NULL;
AllocateFloat(&surf1, xsize*ysize, "surface 1 data");
AllocateFloat(&surf2, xsize*ysize, "surface 2 data");
AllocateFloat(&outsurf, xsize*ysize, "output data");
if (mfp) AllocateByte(&mask, xsize*ysize, "mask data");

/*  READ DATA  */
printf("Reading data...\n");
ReadFloat(ifp1, surf1, xsize*ysize, ifile1);
ReadFloat(ifp2, surf2, xsize*ysize, ifile2);
if (mfp) ReadByte(mfp, mask, xsize*ysize, maskfile);

/*  PROCESS  */
for (k=0, num=0, sum=sumsqr=0.0; k<xsize*ysize; k++) {
  r1 = (surf1) ? surf1[k] : 0.0;
  r2 = (surf2) ? surf2[k] : 0.0;
  r = r1 - r2;
  if (mask && !mask[k]) {
    r = 0;
  }
  else {
    sum += r;
    sumsqr += r*r;
    ++num;
  }
}
avg = (num > 0) ? sum/num : 0.0;
avgsqr = (num > 0) ? sumsqr/num : 0.0;
rms = sqrt(avgsqr - avg*avg);
for (k=0, sumabs=0.0; k<xsize*ysize; k++) {
  r1 = (surf1) ? surf1[k] : 0.0;
  r2 = (surf2) ? surf2[k] : 0.0;
  r = (mask && !mask[k]) ? 0.0 : r1 - r2;
  outsurf[k] = r - avg;  /* subtract average */
```

```
    if (!mask || mask[k])  sumabs += (r > avg) ? r - avg : avg - r;
  }
  avgabs = (num > 0) ? sumabs/num : 0.0;
  printf("Subtracting average value %lf from differences.\n", avg);
  printf("RMS value of differences = %lf.\n", rms);
  printf("Absolute difference = %lf.\n", avgabs);
  printf("Saving unwrapped surface to file '%s'\n", outfile);
  WriteFloat(ofp, outsurf, xsize*ysize, outfile);
  free(surf1);
  free(surf2);
  free(outsurf);
  if (mask) free(mask);
  fclose(ifp1);
  fclose(ifp2);
  fclose(ofp);
  if (mfp) fclose(mfp);
}
```

File mainjump.c contains a program that detects the locations of the discontinuities (i.e., adjoining pixels whose values differ by more than π radians) in an unwrapped surface. The program produces an image "map" of these locations as shown in Figure 4.52(b). This image map consists of a simple byte array whose values represent pixel brightness. This array may be saved to a file and viewed with graphics software (which is not supplied). The usage statement in the program listing below explains how to run the program.

```
/*
 * mainjump.c - Generate image of discontinuities
 *
 * Source code files required:
 *      maskfat.c    mainjump.c       util.c
 */
#include <stdio.h>
#include <math.h>
#include "file.h"
#include "pi.h"
#include "util.h"
#include "maskfat.h"

main (int argc, char *argv[])
{
  int            i, j, k, m, n, a, b;
  int            ii, jj, kk;
  FILE           *ifp, *ofp, *mfp=NULL;
  float          *surf;     /* array */
```

```
unsigned char  *mask=NULL, *out;
char           infile[200], outfile[200];
char           maskfile[200];
int            xsize, ysize;    /* dimensions of arrays */
double         r, r1, r2, sum, xsum;

printf("Generate discontinuity map of surface data\n");
if (argc!=6) {
  fprintf(stderr,"Usage: %s surf image xsize ysize mask\n",
              argv[0]);
  fprintf(stderr,"\nwhere surf = input surface (float) file\n");
  fprintf(stderr,"        image = output image (byte) file\n");
  fprintf(stderr,"        xsize, ysize = array dimensions\n");
  fprintf(stderr,"        mask = mask file (0=mask).\n");
  fprintf(stderr,"        (If no mask, enter 'none'.)\n");
  exit(BAD_USAGE);
}
sscanf(argv[1], "%s", infile);
sscanf(argv[2], "%s", outfile);
sscanf(argv[3], "%d", &xsize);
sscanf(argv[4], "%d", &ysize);
sscanf(argv[5], "%s", maskfile);

printf("Input file = %s\n", infile);
printf("Output file = %s\n", outfile);
printf("Mask file = %s\n", maskfile);
printf("Array size = %d x %d\n", xsize, ysize);
if (Keyword(maskfile, "none")) printf("No mask file\n");

/*  OPEN FILES, ALLOCATE MEMORY      */
OpenFile(&ifp, infile, "r");
OpenFile(&ofp, outfile, "w");
if (!Keyword(maskfile, "none")) OpenFile(&mfp, maskfile, "r");
AllocateFloat(&surf, xsize*ysize, "surface data");
AllocateByte(&out, xsize*ysize, "binary image");
if (mfp) AllocateByte(&mask, xsize*ysize, "mask data");

/*  READ DATA  */
printf("Reading data...\n");
ReadFloat(ifp, surf, xsize*ysize, infile);
if (mfp) ReadByte(mfp, mask, xsize*ysize, maskfile);

/* FATTEN MASK BY 1 PIXEL (OPTIONAL) */
if (mask) {
```

```
    for (k=0; k<xsize*ysize; k++) mask[k] = !mask[k];
    FattenMask(mask, 1, 1, xsize, ysize);
    for (k=0; k<xsize*ysize; k++) mask[k] = !mask[k];
  }

  /* Generate image of discontinuities (where derivs > PI) */
  for (j=0; j<ysize-1; j++) {
    for (i=0; i<xsize-1; i++) {
      k = j*xsize + i;
      out[k] = 255;
      if (mask && (!mask[k] || !mask[k+1] || !mask[k+xsize])) {
        out[k] = 0;
      }
      else {
        r = surf[k] - surf[k+1];
        if (r > PI || r < -PI) out[k] = 0;
        r = surf[k] - surf[k+xsize];
        if (r > PI || r < -PI) out[k] = 0;
      }
    }
  }
  /* Compute measure (sum) of discontinuities */
  for (j=0, sum=xsum=0.0; j<ysize-1; j++) {
    for (i=0; i<xsize-1; i++) {
      k = j*xsize + i;
      if (mask && (!mask[k] || !mask[k+1] || !mask[k+xsize]))
        continue;
      if (!out[k]) ++sum;
      ++xsum;
    }
  }
  printf("Saving image to %s\n", outfile);
  printf("L0-measure (normalized to 0-1) of discontinuities %lf\n",
         sum/xsum);
  printf("(%lf%% of %d non-masked pixels)\n",
         100.0*sum/xsum, (int)xsum);
  WriteByte(ofp, out, xsize*ysize, outfile);
  free(surf);
  fclose(ifp);
  free(out);
  fclose(ofp);
  if (mfp) free(mask);
  if (mfp) fclose(mfp);
}
```

A.6 GOLDSTEIN'S ALGORITHM

The following files are contained in this section: `maingold.c`, `gold.c`, and `gold.h`.

We can now present the software for Goldstein's phase-unwrapping algorithm of Section 4.2. The main program is contained in file `maingold.c`. Among other parameters, the user can specify the maximum branch cut length `-cutlen` and whether or not to perform the dipole-preprocessing step (`-dipole`). A typical command to run the program would be

```
gold -input mri.phase -format byte -output mri.surf
        -bmask mri.bmask -xsize 256 -ysize 256 -dipole yes
```

where `gold` is the name of the executable file, `mri.phase` is the name of the byte file that contains the 256 × 256 array of quantized phase data, `mri.surf` is the output file, and `mri.bmask` is the "border mask" file (i.e., a byte file that identifies with 0's the pixels to mask out and ignore). The border mask file and dipole parameters are optional.

The format of the input phase file is specified by the `-format` parameter. Its argument can be `complex8`, `complex4`, `byte`, or `float`. These arguments specify that the format of the input data is 8-byte complex (i.e., four bytes for the real part followed by four bytes for the imaginary part), 4-byte complex, 1-byte unsigned integer (the quantized phase values scaled from 0–2π to 0–255), or 4-byte floating-point, respectively. These data formats are discussed in the file `extract.c` in Section A.1.

The program operates as follows: The command line arguments are parsed by the functions `CommandLineParm` and `Keyword`. The files are opened and the memory for the arrays is allocated. The phase data and the (optional) border mask data are then read from files and processed. The border mask is fattened slightly by a call to the function `FattenMask`. The residues are identified and labeled in the `bitflags` array by function `Residues`, and the optional dipole preprocessing step is performed by a call to `Dipole`. The branch cuts are then generated and labeled in the `bitflags` array with a call to `GoldsteinBranchCuts`. The phase data are then unwrapped around the branch cuts by function `UnwrapAroundCuts`, and the results are written to the output file.

```
/*
 * maingold.c - phase unwrapping by means of residues & branch-cuts
 *
 * Source code files required:
 *        brcut.c      dipole.c     extract.c        gold.c
 *         grad.c        list.c    maingold.c     maskfat.c
 *         path.c    residues.c        util.c
 */
```

```
#include <stdio.h>
#include <math.h>
#include "file.h"
#include "pi.h"
#include "residues.h"
#include "util.h"
#include "extract.h"
#include "list.h"
#include "gold.h"
#include "dipole.h"

main (int argc, char *argv[])
{
  int            i, j, k, m, n, a, b;
  FILE           *ifp, *ofp, *mfp;
  float          *phase;     /* array */
  float          *soln;      /* array */
  unsigned char  *bitflags;
  char           buffer[200], string[200];
  char           infile[200], outfile[200];
  char           maskfile[200], tempstr[200], format[200];
  int            in_format, debug_flag, dipole_flag;
  int            xsize, ysize;   /* dimensions of arrays */
  int            NumRes, MaxCutLen;
  char           use[] =     /* define usage statement */
    "Usage: prog-name -input file -format fkey -output file\n"
    "  -xsize x -ysize y [ -mask file -cutlen c -debug yes/no\n"
    "  -dipole yes/no ]\n"
    "where 'fkey' is a keyword designating the input file type\n"
    "(key = complex8, complex4, float or byte), 'x' and 'y' are\n"
    "the dimensions of the file, mask is an optional byte-file\n"
    "of masks for masking out undefined phase values and cutlen\n"
    "is the max branchcut length allowed.  All files are simple\n"
    "raster files, and the output file consists of floating\n"
    "point numbers that define the heights of the unwrapped\n"
    "surface.  If the 'debug' parm is 'yes', then intermediate\n"
    "byte-files are saved (residues, branch cuts, etc.)  If\n"
    "the 'dipole' parm is 'yes', then dipole-residues are\n"
    "eliminated before unwrapping.\n";

  printf("Phase Unwrapping By Means of Goldstein's Algorithm\n");

  /* GET COMMAND LINE PARAMETERS AND CHECK */
  CommandLineParm(argc,argv, "-input", StringParm, infile, 1, use);
  CommandLineParm(argc,argv, "-format", StringParm, format, 1, use);
```

```
CommandLineParm(argc,argv, "-output", StringParm, outfile, 1,use);
CommandLineParm(argc, argv, "-xsize", IntegerParm, &xsize, 1,use);
CommandLineParm(argc, argv, "-ysize", IntegerParm, &ysize, 1,use);
if (!CommandLineParm(argc, argv, "-mask", StringParm, maskfile,
      0, use)) strcpy(maskfile, "none");
if (!CommandLineParm(argc, argv, "-cutlen", IntegerParm,
   &MaxCutLen, 0, use)) MaxCutLen = 0; /* default defined below */
if (!CommandLineParm(argc, argv, "-debug", StringParm, tempstr,
      0, use)) debug_flag = 0;
else debug_flag = Keyword(tempstr, "yes");
if (!CommandLineParm(argc, argv, "-dipole", StringParm, tempstr,
      0, use)) dipole_flag=0;
else dipole_flag = Keyword(tempstr, "yes");

if (Keyword(format, "complex8"))  in_format = 0;
else if (Keyword(format, "complex4"))  in_format = 1;
else if (Keyword(format, "byte"))  in_format = 2;
else if (Keyword(format, "float"))  in_format = 3;
else {
  fprintf(stderr, "Unrecognized format: %s\n", format);
  exit(BAD_PARAMETER);
}

printf("Input file =  %s\n", infile);
printf("Input file type = %s\n", format);
printf("Output file =  %s\n", outfile);
printf("File dimensions = %dx%d (cols x rows).\n", xsize, ysize);

if (Keyword(maskfile, "none")) printf("No mask file.\n");
else printf("Mask file = %s\n", maskfile);

if (dipole_flag) printf("Dipole-residues will be eliminated.\n");
if (debug_flag)
  printf("Intermediate files will be saved (i.e. debug on).\n");

/*  OPEN FILES, ALLOCATE MEMORY       */
OpenFile(&ifp, infile, "r");
OpenFile(&ofp, outfile, "w");
mfp = NULL;
if (!Keyword(maskfile, "none"))
  OpenFile(&mfp, maskfile, "r");

AllocateFloat(&phase, xsize*ysize, "phase data");
AllocateFloat(&soln, xsize*ysize, "unwrapped data");
AllocateByte(&bitflags, xsize*ysize, "bitflag data");
```

```
/*  READ AND PROCESS DATA  */
printf("Reading phase data...\n");
GetPhase(in_format, ifp, infile, phase, xsize, ysize);

/* mask data */
printf("Processing mask data...\n");
if (mfp) {
  ReadByte(mfp, bitflags, xsize*ysize, maskfile);
}
else {
  for (k=0; k<xsize*ysize; k++)
    bitflags[k] = 255;
}
for (k=0; k<xsize*ysize; k++)
  bitflags[k] = (!bitflags[k]) ? BORDER : 0;

FattenMask(bitflags, BORDER, 1, xsize, ysize);

/*  LOCATE AND PROCESS RESIDUES  */
/* compute residues and store in bitflags array */
NumRes = Residues(phase, bitflags, POS_RES, NEG_RES,
                  BORDER, xsize, ysize);
printf("%d Residues\n", NumRes);

if (debug_flag) {
  char filename[300];
  sprintf(filename, "%s.res", outfile);
  SaveByteToImage(bitflags, "residues", filename, xsize, ysize,
                  1, 1, 0);
}

/*  GENERATE BRANCH CUTS  */
if (dipole_flag) {  /* elimate dipole-residues first */
  Dipole(bitflags, xsize, ysize, BRANCH_CUT);
  i = Residues(phase, bitflags, 0, 0, BORDER | BRANCH_CUT,
               xsize, ysize);
  printf("%d Residues are left\n", i);
}

if (MaxCutLen==0) MaxCutLen = (xsize + ysize)/2;
GoldsteinBranchCuts(bitflags, MaxCutLen, NumRes, xsize, ysize,
                    BRANCH_CUT);
if (debug_flag) {
  char filename[300];
  sprintf(filename, "%s.brc", outfile);
```

```
    SaveByteToImage(bitflags, "branch cuts", filename,
                    xsize, ysize, 1, 1, BRANCH_CUT | BORDER);
  }

  /* UNWRAP AROUND CUTS */
  for (k=0, i=0; i<xsize*ysize; i++)
    if (bitflags[i] & BRANCH_CUT) k++;
  printf("%d BRANCH CUT PIXELS\n", k);
  printf("Unwrapping around branch cuts\n");
  k = UnwrapAroundCuts(phase, bitflags, soln, xsize, ysize, AVOID,
                       0, NULL);
  if (k > 1) printf("%d disconnected pieces.\n", k);
  else printf("%d piece\n", k);
  printf("\nFinished\n");
  PrintMinAndMax(xsize, ysize, soln, "solution");

  /* SAVE RESULT */
  for (k=0; k<xsize*ysize; k++)
    soln[k] *= TWOPI; /* scale output */
  printf("Saving unwrapped surface to file '%s'\n", outfile);
  WriteFloat(ofp, soln, xsize*ysize, outfile);
  free(soln);
  free(phase);
  free(bitflags);
}
```

The function `GoldsteinBranchCuts` that generates the branch cuts is in file `gold.c`. Rather than accepting as input the phase data, the function accepts the `bitflags` array, in which the positive and negative residues have been labeled. The elements of this array are labeled with bit flags, which are represented by numbers that are powers of two. For example, border pixels (those pixels that lie along the border or are masked out) are labeled with the bit flag whose binary value is 00100000. (See file `list.h` in Section A.2 for a list of the bit flags.) This means that any `bitflags` array entry whose sixth bit is set is a border pixel. Similarly the positive and negative residues are marked with the bit flags 00000001 and 00000010, respectively, and the branch cuts are marked with 00010000. Thus, if an entry in the `bitflags` array has the value 17 (decimal) = 00010001 (binary), then it is a positive residue and also a branch cut pixel. The bit flags are tested and manipulated with the `C` language bit operators. For example, `&` is the bitwise `AND` operator and `~` is bitwise complement. In the function below the branch cuts are generated and labeled with the bit flag `branchcut_code`, which should be a power of two distinct from the other bit flags.

```
/*
 * gold.c -- generate branch cuts by Goldstein's algorithm
 */
#include <stdio.h>
#include <math.h>
#include "util.h"
#include "brcut.h"
#include "list.h"

/* Goldstein's phase unwrapping algorithm.  The bitflags store */
/* the masked pixels (to be ignored) and the residues and      */
/* accumulates other info such as the branch cut pixels.       */
void GoldsteinBranchCuts(unsigned char *bitflags, int MaxCutLen,
               int NumRes, int xsize, int ysize, int branchcut_code)
{
  int            i, j, k, ii, jj, kk, m, n, ri, rj;
  int            charge, boxctr_i, boxctr_j, boxsize, bs2;
  int            dist, min_dist, rim_i, rim_j, near_i, near_j;
  int            ka, num_active, max_active, *active_list;
  int            bench;
  int            draw_cut_line;
  double         r;
  bench = ysize/100;
  if (bench < 1) bench = 1;
  if (MaxCutLen < 2) MaxCutLen = 2;
  max_active = NumRes + 10;
  AllocateInt(&active_list, max_active + 1, "book keeping data");
  /* branch cuts */
  printf("Computing branch cuts\n");
  for (j=0; j<ysize; j++) {
    if (j%bench==0) {
      printf("%d ", j/bench);
      fflush(stdout);
    }
    for (i=0; i<xsize; i++) {
      k = j*xsize + i;
      if ((bitflags[k] & (POS_RES | NEG_RES))
               && !(bitflags[k] & VISITED)) {
        bitflags[k] |= VISITED;  /* turn on visited flag */
        bitflags[k] |= ACTIVE;   /* turn on active flag */
        charge = (bitflags[k] & POS_RES) ? 1 : -1;
        num_active = 0;
        active_list[num_active++] = k;
        if (num_active > max_active) num_active = max_active;
        for (boxsize = 3; boxsize<=2*MaxCutLen; boxsize += 2) {
```

```
    bs2 = boxsize/2;
    for (ka=0; ka<num_active; ka++) {
      boxctr_i = active_list[ka]%xsize;
      boxctr_j = active_list[ka]/xsize;
      for (jj=boxctr_j - bs2; jj<=boxctr_j + bs2; jj++) {
        for (ii=boxctr_i - bs2; ii<=boxctr_i + bs2; ii++) {
          kk = jj*xsize + ii;
          if (ii<0 || ii>=xsize || jj<0 || jj>=ysize) {
            continue;
          }
          else {
            if (ii==0 || ii==xsize-1 || jj==0 || jj==ysize-1
                  || (bitflags[kk] & BORDER)) {
              charge = 0;
              DistToBorder(bitflags, BORDER, boxctr_i,
                      boxctr_j, &ri, &rj, xsize, ysize);
              PlaceCut(bitflags, ri, rj, boxctr_i, boxctr_j,
                      xsize, ysize, branchcut_code);
            }
            else if ((bitflags[kk] & (POS_RES | NEG_RES))
                          && !(bitflags[kk] & ACTIVE)) {
              if (!(bitflags[kk] & VISITED)) {
                charge += (bitflags[kk] & POS_RES) ? 1 : -1;
                bitflags[kk] |= VISITED;   /* set flag */
              }
              active_list[num_active++] = kk;
              if (num_active > max_active)
                    num_active = max_active;
              bitflags[kk] |= ACTIVE;  /* set active flag */
              PlaceCut(bitflags, ii, jj, boxctr_i, boxctr_j,
                      xsize, ysize, branchcut_code);
            }
            if (charge==0)
              goto continue_scan;
          } /* else */
        }  /* for (ii ... */
      }  /* for (jj ... */
    } /* for (ka ... */
  }  /* for (boxsize ... */

  if (charge != 0) {   /* connect branch cuts to rim */
    min_dist = xsize + ysize;  /* large value */
    for (ka=0; ka<num_active; ka++) {
      ii = active_list[ka]%xsize;
      jj = active_list[ka]/xsize;
```

```
          if ((dist = DistToBorder(bitflags, BORDER,
                      ii, jj, &ri, &rj, xsize, ysize))<min_dist) {
            min_dist = dist;
            near_i = ii;
            near_j = jj;
            rim_i = ri;
            rim_j = rj;
          }
        }
        PlaceCut(bitflags, near_i, near_j, rim_i, rim_j,
                 xsize, ysize, branchcut_code);
      }
      continue_scan :
      /* mark all active pixels inactive */
      for (ka=0; ka<num_active; ka++)
        bitflags[active_list[ka]] &= ~ACTIVE;  /* turn flag off */
    } /* if (bitflags ... */
  } /* for (i ... */
} /* for (j ... */
printf("\n");
free(active_list);
return;
}
```

The header file is `gold.h`:

```
#ifndef __GOLD
#define __GOLD
void GoldsteinBranchCuts(unsigned char *bitflags, int MaxCutLen,
                int NumRes, int xsize, int ysize, int mask_code);
#endif
```

A.7 QUALITY-GUIDED ALGORITHM

The following files are contained in this section: `mainqual.c`, `quality.c`, and `quality.h`.

We now present the program and functions for the quality-guided phase-unwrapping algorithm of Section 4.3. The main program is contained in the file `mainqual.c`. The usage statement in the program listing below explains how to run the program. A typical example would be

```
qual -input mri.phase -format byte -output mri.surf -bmask
    mri.bmask -xsize 256 -ysize 256 -mode min_var -tsize 3
```

where qual is the name of the executable file, mri.phase is the name of the byte file that contains the 256 × 256 array of quantized phase data, mri.surf is the output file, and mri.bmask is the "border mask" file. The border mask file, which masks out the pixels that should be ignored, such as the border pixels of the simulated IFSAR data shown in Figure 3.8(b), is optional. See Section A.6 for an explanation of the -format parameter. The usage is very similar to that of Goldstein's algorithm of Section A.6, except for the new parameter -mode, which specifies the source of the quality map. The modes allowed are min_var for minimum phase variance, min_grad for maximum phase gradient, max_pseu for maximum pseudocorrelation, and max_corr for maximum correlation. The quality maps are computed or averaged in small windows, and the parameter -tsize specifies how large this window should be (in this case 3 × 3 pixels). See Section 3.3 for a discussion of the quality maps and how they are computed. If the max_corr quality map is specified, then the user must specify the name of a byte file that contains the correlation coefficients, quantized to the range 0–255. The correlation file may contain any quality map, however.

The program works as follows: The command line arguments are parsed by the functions CommandLineParm, Keyword, and SetQualityMode. The files are opened, the memory for the arrays is allocated, and the phase data, the quality map, and the (optional) border mask data are then read from the input files and processed. The border mask is fattened slightly by a call to the function FattenMask. The phase is then unwrapped with a call to the function QualityGuidedPathFollower, and the unwrapped surface is written to the output file.

```
/*
 * mainqual.c -- phase unwrapping by quality-guided path following
 *
 * Source code files required:
 *      dxdygrad.c      extract.c      getqual.c        grad.c
 *          list.c      mainqual.c     maskfat.c     quality.c
 *      qualgrad.c      qualpseu.c     qualvar.c        util.c
 */
#include <stdio.h>
#include <math.h>
#include "file.h"
#include "pi.h"
#include "quality.h"
#include "util.h"
#include "extract.h"
#include "getqual.h"

#define BORDER (0x20)
main (int argc, char *argv[])
{
```

```
int             i, j, k;
FILE            *ifp, *ofp, *mfp=0, *qfp=0;
float           *phase;     /* array */
float           *soln;      /* array */
float           *qual_map;  /* array */
unsigned char   *bitflags;  /* array */
char            buffer[200], string[200];
char            infile[200], outfile[200];
char            bmaskfile[200], qualfile[200];
char            tempstr[200], format[200], modekey[200];
int             xsize, ysize;   /* dimensions of arrays */
int             tsize, in_format, debug_flag, avoid_code;
double          rmin, rmax, rscale;
UnwrapMode      mode;
char            use[] =     /* define usage statement */
  "Usage: prog-name -input file -format fkey -output file\n"
  "  -xsize x -ysize y -mode mkey [ -bmask file -corr file\n"
  "  -tsize size -debug yes/no]\n"
  "where 'fkey' is a keyword designating the input file type\n"
  "(key = complex8, complex4, float or byte), 'x' and 'y' are\n"
  "the dimensions of the file, bmask is an optional byte-file\n"
  "of masks for masking out undefined phase values, corr\n"
  "is an optional byte-file of cross-correlation values,\n"
  "tsize is the size of the square template for averaging the\n"
  "corr file or quality values (default = 1), and 'mkey' is\n"
  "a keyword designating the source of the quality values\n"
  "that guide the unwrapping path.  The value of mkey must be\n"
  "'min_grad' for Minimum Gradient unwrapping, 'min_var' for\n"
  "Minimum Variance unwrapping, 'max_corr' for Maximum\n"
  "Correlation unwrapping, or 'max_pseu' for Maximum\n"
  "Pseudocorrelation unwrapping.  All files are simple\n"
  "raster files, and the output file consists of floating\n"
  "point numbers that define the heights of the unwrapped\n"
  "surface.  If the 'debug' parm is 'yes', then intermediate\n"
  "byte-files are saved (quality map, unwrapping paths, etc.)\n";

printf("Phase Unwrapping by Quality Map Guided Path Following\n");

/* GET COMMAND LINE PARAMETERS AND CHECK */
CommandLineParm(argc,argv, "-input", StringParm, infile, 1, use);
CommandLineParm(argc,argv, "-format", StringParm, format, 1, use);
CommandLineParm(argc,argv, "-output", StringParm, outfile, 1,use);
CommandLineParm(argc,argv, "-xsize", IntegerParm, &xsize, 1, use);
CommandLineParm(argc,argv, "-ysize", IntegerParm, &ysize, 1, use);
CommandLineParm(argc,argv, "-mode", StringParm, modekey, 1, use);
```

```
if (!CommandLineParm(argc, argv, "-bmask", StringParm,
      bmaskfile, 0, use)) strcpy(bmaskfile, "none");
if (!CommandLineParm(argc, argv, "-corr", StringParm,
      qualfile, 0, use)) strcpy(qualfile, "none");
if (!CommandLineParm(argc, argv, "-tsize", IntegerParm,
      &tsize, 0, use)) tsize = 1;
if (!CommandLineParm(argc, argv, "-debug", StringParm,
      tempstr, 0, use)) debug_flag = 0;
else debug_flag = Keyword(tempstr, "yes");

if (Keyword(format, "complex8"))  in_format = 0;
else if (Keyword(format, "complex4"))  in_format = 1;
else if (Keyword(format, "byte"))  in_format = 2;
else if (Keyword(format, "float"))  in_format = 3;
else {
  fprintf(stderr, "Unrecognized format: %s\n", format);
  exit(BAD_PARAMETER);
}

printf("Input file =  %s\n", infile);
printf("Input file type = %s\n", format);
printf("Output file =  %s\n", outfile);
printf("File dimensions = %dx%d (cols x rows).\n", xsize, ysize);

if (Keyword(bmaskfile, "none")) printf("No border mask file.\n");
else printf("Border mask file = %s\n", bmaskfile);
if (Keyword(qualfile, "none")) printf("No quality file.\n");
else printf("Correlation image file = %s\n", qualfile);

printf("Averaging template size = %d\n", tsize);
if (tsize < 0 || tsize > 30) {
  fprintf(stderr, "Illegal size: must be between 0 and 30\n");
  exit(BAD_PARAMETER);
}

printf("Quality mode = %s\n", modekey);
mode = SetQualityMode(modekey, qualfile, 0);
if (mode < 0) exit(BAD_PARAMETER);  /* error msg already printed */

/*  OPEN FILES, ALLOCATE MEMORY   */
OpenFile(&ifp, infile, "r");
OpenFile(&ofp, outfile, "w");
if (!Keyword(bmaskfile, "none"))
  OpenFile(&mfp, bmaskfile, "r");
if (mode==corr_coeffs)
```

```
    OpenFile(&qfp, qualfile, "r");

  AllocateFloat(&phase, xsize*ysize, "phase data");
  AllocateFloat(&soln, xsize*ysize, "unwrapped data");
  AllocateByte(&bitflags, xsize*ysize, "bitflag data");
  AllocateFloat(&qual_map, xsize*ysize, "quality map");

  /*  READ AND PROCESS DATA  */
  printf("Reading phase data...\n");
  GetPhase(in_format, ifp, infile, phase, xsize, ysize);

  if (qfp) {
    printf("Reading quality data...\n");
    /* borrow the bitflags array temporarily */
    ReadByte(qfp, bitflags, xsize*ysize, qualfile);
    /* process data and store in quality map array */
    AverageByteToFloat(bitflags, qual_map, tsize, xsize, ysize);
  }

  /* border mask data */
  printf("Processing border mask data...\n");
  if (mfp) {
    ReadByte(mfp, bitflags, xsize*ysize, bmaskfile);
  }
  else {
    for (k=0; k<xsize*ysize; k++)
      bitflags[k] = 255;
  }
  for (k=0; k<xsize*ysize; k++)
    bitflags[k] = (!bitflags[k]) ? BORDER : 0;
  FattenMask(bitflags, BORDER, 1, xsize, ysize);

  GetQualityMap(mode, qual_map, phase, bitflags, BORDER,
                tsize, xsize, ysize);
  if (debug_flag) {
    char filename[300];
    sprintf(filename, "%s.qual", outfile);
    SaveFloatToImage(qual_map, "quality", filename, xsize, ysize,
                     0, 0, 0);
  }

  /* initialize soln array */
  for (k=0; k<xsize*ysize; k++)
    soln[k] = 0;
```

```
  /*  UNWRAP  */
  avoid_code = BORDER;
  if (mode==gradient && tsize==1) mode = dxdygrad;
  k = QualityGuidedPathFollower(phase, qual_map, bitflags, soln,
            xsize, ysize, avoid_code, mode, debug_flag, outfile);
  if (k > 1) printf("\n%d pieces\n", k);

  printf("\nFinished\n");
  PrintMinAndMax(xsize, ysize, soln, "solution");

  /*  SAVE RESULT  */
  /* scale output */
  for (k=0; k<xsize*ysize; k++)
    soln[k] *= TWOPI;

  printf("Saving unwrapped surface to file '%s'\n", outfile);
  WriteFloat(ofp, soln, xsize*ysize, outfile);
  free(soln);
  free(phase);
  free(bitflags);
  free(qual_map);
}
```

The function QualityGuidedPathFollower that performs the quality-guided unwrapping is in file quality.c. The pseudocode for this function is in Section 4.3.2.

```
/*
 *  quality.c -- function for quality-guided path-following
 */
#include <stdio.h>
#include <math.h>
#include "quality.h"
#include "list.h"
#include "util.h"

/* Main function for quality-guided algorithm for phase unwrapping */
int QualityGuidedPathFollower(float *phase, float *qual_map,
                              unsigned char *bitflags, float *soln,
                              int xsize, int ysize, int avoid_code,
                              UnwrapMode unwrap_mode, int debug_flag,
                              char *infile)
{
  int  i, j, k, a, b, c, n, num_pieces=0;
  float  value;
```

```
float  min_qual, small_val = -1.0E+10;
int    num_index, max_list_size, bench, benchout;
int    *index_list;
int    postponed_code=POSTPONED, unwrapped_code=UNWRAPPED;

bench = xsize*ysize/100;
if (bench < 1) bench = 1;
benchout = 10*bench;
min_qual = small_val;
max_list_size = (xsize + ysize);
AllocateInt(&index_list, max_list_size + 1,
            "book-keeping list (index)");
avoid_code |= unwrapped_code;
/* find starting point */
n = 0;
num_index = 0;
for (j=0; j<ysize; j++) {
  for (i=0; i<xsize; i++) {
    k = j*xsize + i;
    if (!(bitflags[k] & avoid_code)) {
      if (bitflags[k] & postponed_code)
        /* soln[k] already stores the unwrapped value */
        value = soln[k];
      else {
        ++num_pieces;
        value = soln[k] = phase[k];
      }
      bitflags[k] |= unwrapped_code;
      bitflags[k] &= ~postponed_code;
      UpdateList(qual_map, i, j, value, phase, soln, bitflags,
                 xsize, ysize, index_list, &num_index, avoid_code,
                 unwrapped_code, postponed_code, max_list_size,
                 unwrap_mode==dxdygrad, &min_qual);
      while (num_index > 0) {
        if (n%bench==0) {
          printf("%d ", n/bench);
          fflush(stdout);
          if (0 && debug_flag && n%benchout==0 && n>0) {
            char filename[300];
            static char file_num = 0;
            sprintf(filename, "%s.path-%d", infile, ++file_num);
            SaveByteToImage(bitflags, "unwrapping path", filename,
                            xsize, ysize, 1, 1, avoid_code);
          }
        }
```

```
        ++n;
        if (!GetNextOneToUnwrap(&a, &b, index_list, &num_index,
              xsize, ysize)) break;   /* no more to unwrap */
        c = b*xsize + a;
        bitflags[c] |= unwrapped_code;
        bitflags[c] &= ~postponed_code;
        value = soln[c];
        UpdateList(qual_map, a, b, value, phase, soln, bitflags,
                 xsize, ysize, index_list, &num_index, avoid_code,
                 unwrapped_code, postponed_code, max_list_size,
                 unwrap_mode==dxdygrad, &min_qual);
        if (num_index <= 0) {
          min_qual = small_val;
          for (c=0; c<xsize*ysize; c++) {
            if ((bitflags[c] & postponed_code)
                  && !(bitflags[c] & unwrapped_code)) {
              a = c%xsize;
              b = c/xsize;
              InsertList(soln, soln[c], qual_map, bitflags, a, b,
                         index_list, &num_index, xsize, ysize,
                         unwrapped_code, postponed_code,
                         &min_qual, max_list_size);
            }
          }
          if (num_index > 0) min_qual = qual_map[index_list[0]];
        }
      }  /* while ... */
    }
  }   /* for (b ...  */
 }   /* for (a ... */
 printf("\n");
 free(index_list);
 return num_pieces;
}
```

The header file is quality.h:

```
#ifndef __QUALITY
#define __QUALITY
#include "getqual.h"
int QualityGuidedPathFollower(float *phase, float *qual_map,
      unsigned char *bitflags, float *soln, int xsize, int ysize,
      int avoid_code, UnwrapMode unwrap_mode, int debug_mode,
      char *infile);
#endif
```

A.8 MASK CUT ALGORITHM

The following files are listed in this section: `mainmcut.c`, `maskcut.c`, and `maskcut.h`.

We now present the program and functions for the mask cut algorithm discussed in Section 4.4. The main program is in file `mainmcut.c`. The usage statement in the program listing below explains how to run the program. It is identical to that of the quality guided algorithm of Section A.7. A typical command to execute the program would be

```
mcut -input mri.phase -format byte -output mri.surf -bmask
    mri.bmask -xsize 256 -ysize 256 -mode min_var -tsize 3
```

where `mcut` is the name of the executable file.

The program flow is very similar to that of the quality-guided algorithm. The command line arguments are parsed by the functions `CommandLineParm`, `Keyword`, and `SetQualityMode`. The files are opened, the memory for the arrays is allocated, and the phase data, the quality map, and the (optional) border mask data are then read from the input files and processed. The border mask is fattened slightly by a call to the function `FattenMask`. (The border mask consists of pixels that should be ignored, such as the border pixels of the simulated IFSAR data shown in Figure 3.8b). The residues are located by the function `Residues` and labeled in the `bitflags` array. The residues and border pixels are assigned the lowest possible quality value, and the values in the quality map are reversed because the algorithm seeks the lowest-quality pixels as it "grows" the mask cuts. The mask cuts are generated by the region-growing algorithm performed by `QualityGuidedMask`. After the mask cuts are thinned by `ThinMask`, the phase is unwrapped by `UnwrapAroundCuts` and written to the output file.

```
/*
 * mainmcut.c - phase unwrapping with quality-guided mask cuts
 *
 * Source code files required:
 *     dxdygrad.c     extract.c     getqual.c        grad.c
 *         list.c    mainmcut.c     maskcut.c     maskfat.c
 *     maskthin.c        path.c    qualgrad.c    qualpseu.c
 *      qualvar.c    residues.c        util.c
 */
#include <stdio.h>
#include <math.h>
#include "file.h"
#include "pi.h"
#include "list.h"
#include "util.h"
```

```
#include "extract.h"
#include "getqual.h"
#include "maskcut.h"
#include "residues.h"
#include "path.h"

main (int argc, char *argv[])
{
  int            i, j, k, NumRes, mask_code;
  FILE           *ifp, *ofp, *mfp=0, *qfp=0;
  float          *phase;     /* array */
  float          *soln;      /* array */
  float          *qual_map;  /* array */
  unsigned char  *bitflags;  /* array */
  char           buffer[200], string[200], infile[200];
  char           outfile[200], bmaskfile[200], qualfile[200];
  char           tempstr[200], format[200], modekey[200];
  int            xsize, ysize;   /* dimensions of arrays */
  int            in_format, debug_flag, avoid_code, tsize;
  double         minqual, maxqual, rmin, rmax, rscale;
  UnwrapMode     mode;
  char           use[] =    /* define usage statement */
    "Usage: prog-name -input file -format fkey -output file\n"
    "  -xsize x -ysize y -mode mkey [ -bmask file -corr file\n"
    "  -tsize size -debug yes/no]\n"
    "where 'fkey' is a keyword designating the input file type\n"
    "(key = complex8, complex4, float or byte), 'x' and 'y' are\n"
    "the dimensions of the file, bmask is an optional byte-file\n"
    "of masks for masking out undefined phase values, corr\n"
    "is an optional byte-file of cross-correlation values,\n"
    "tsize is the size of the square template for averaging the\n"
    "corr file or quality values (default = 1), and 'mkey' is\n"
    "a keyword designating the source of the quality values\n"
    "that guide the masking path.  The value of mkey must be\n"
    "'min_grad' for minimum gradients, 'min_var' for minimum\n"
    "phase variances, 'max_corr' for maximum cross\n"
    "correlations (i.e., corr file), or 'max_pseu' for max-\n"
    "imum pseudocorrelations.  All of the files are simple\n"
    "raster files, and the output file consists of floating\n"
    "point numbers that define the heights of the unwrapped\n"
    "surface.  If the 'debug' parm is 'yes', then intermediate\n"
    "byte-files are saved (residues, unwrapping paths, etc.)\n";

  printf("Phase Unwrapping By Quality-Guided Mask Cuts\n");
```

```
/* GET COMMAND LINE PARAMETERS AND CHECK */
CommandLineParm(argc,argv, "-input", StringParm, infile, 1, use);
CommandLineParm(argc,argv, "-format", StringParm, format, 1, use);
CommandLineParm(argc,argv, "-output", StringParm, outfile, 1,use);
CommandLineParm(argc,argv, "-xsize", IntegerParm, &xsize, 1, use);
CommandLineParm(argc,argv, "-ysize", IntegerParm, &ysize, 1, use);
CommandLineParm(argc,argv, "-mode", StringParm, modekey, 1, use);
if (!CommandLineParm(argc, argv, "-bmask", StringParm, bmaskfile,
      0, use)) strcpy(bmaskfile, "none");
if (!CommandLineParm(argc, argv, "-corr", StringParm, qualfile,
      0, use)) strcpy(qualfile, "none");
if (!CommandLineParm(argc, argv, "-tsize", IntegerParm, &tsize,
      0, use)) tsize = 1;
if (!CommandLineParm(argc, argv, "-debug", StringParm, tempstr,
      0, use)) debug_flag = 0;
else debug_flag = Keyword(tempstr, "yes");

if (Keyword(format, "complex8"))  in_format = 0;
else if (Keyword(format, "complex4"))  in_format = 1;
else if (Keyword(format, "byte"))  in_format = 2;
else if (Keyword(format, "float"))  in_format = 3;
else {
  fprintf(stderr, "Unrecognized format: %s\n", format);
  exit(BAD_PARAMETER);
}

printf("Input file =  %s\n", infile);
printf("Input file type = %s\n", format);
printf("Output file =  %s\n", outfile);
printf("File dimensions = %dx%d (cols x rows).\n", xsize, ysize);

if (Keyword(bmaskfile, "none")) printf("No border mask file.\n");
else printf("Border mask file = %s\n", bmaskfile);
if (Keyword(qualfile, "none")) printf("No quality file.\n");
else printf("Correlation image file = %s\n", qualfile);

printf("Averaging template size = %d\n", tsize);
if (tsize < 0 || tsize > 10) {
  fprintf(stderr, "Illegal size: must be between 0 and 10\n");
  exit(BAD_PARAMETER);
}
printf("Quality mode = %s\n", modekey);
mode = SetQualityMode(modekey, qualfile, 1);
if (mode < 0) exit(BAD_PARAMETER);  /* error msg already printed */
```

```
/*  OPEN FILES, ALLOCATE MEMORY       */
OpenFile(&ifp, infile, "r");
OpenFile(&ofp, outfile, "w");
if (!Keyword(bmaskfile, "none"))
  OpenFile(&mfp, bmaskfile, "r");
if (mode==corr_coeffs)
  OpenFile(&qfp, qualfile, "r");

AllocateFloat(&phase, xsize*ysize, "phase data");
AllocateFloat(&soln, xsize*ysize, "unwrapped data");
AllocateByte(&bitflags, xsize*ysize, "bitflags array");
AllocateFloat(&qual_map, xsize*ysize, "quality map");

/*  READ AND PROCESS DATA  */
printf("Reading phase data...\n");
GetPhase(in_format, ifp, infile, phase, xsize, ysize);

if (qfp) {
  printf("Reading quality data...\n");
  /* borrow the bitflags array temporarily */
  ReadByte(qfp, bitflags, xsize*ysize, qualfile);
  /* process data and store in qual_map */
  AverageByteToFloat(bitflags, qual_map, tsize, xsize, ysize);
}

/* border mask data */
printf("Processing border mask data...\n");
if (mfp) {
  ReadByte(mfp, bitflags, xsize*ysize, bmaskfile);
}
else {
  for (k=0; k<xsize*ysize; k++)
    bitflags[k] = 255;
}
for (k=0; k<xsize*ysize; k++)
  bitflags[k] = (!bitflags[k]) ? BORDER : 0;
FattenMask(bitflags, BORDER, 1, xsize, ysize);

GetQualityMap(mode, qual_map, phase, bitflags, BORDER,
              tsize, xsize, ysize);

/* initialize soln array */
for (k=0; k<xsize*ysize; k++)
  soln[k] = 0;
```

```
/*  LOCATE AND PROCESS RESIDUES  */
/* compute residues and store in bitflags array */
NumRes = Residues(phase, bitflags, POS_RES, NEG_RES, BORDER,
                  xsize, ysize);
printf("%d Residues\n", NumRes);

if (debug_flag) {
  char filename[300];
  sprintf(filename, "%s.res", outfile);
  SaveByteToImage(bitflags, "residues", filename, xsize, ysize,
                  1, 1, 0);
}

/* mark residues and border in quality map array */
for (k=0, maxqual = - (minqual = 1.0E+10); k<xsize*ysize; k++) {
  if (maxqual < qual_map[k]) maxqual = qual_map[k];
  if (minqual > qual_map[k]) minqual = qual_map[k];
}
minqual -= 0.0001;
maxqual += 0.0001;
/* change quality values to cost values */
/* (to follow low-quality mask cuts) */
/* i.e., reverse high and low qualities */
for (k=0; k<xsize*ysize; k++)
  qual_map[k] = maxqual - qual_map[k];
/* mark residues and border as high quality */
for (j=0; j<ysize; j++) {
  for (i=0; i<xsize; i++) {
    k = j*xsize + i;
    if (i==0 || i==xsize-1 || j==0 || j==ysize-1)
      qual_map[k] = maxqual;
    else if (bitflags[k] & BORDER)
      qual_map[k] = maxqual;
    else if (bitflags[k] & RESIDUE)
      qual_map[k] = maxqual;
  }
}

if (debug_flag) {
  char filename[300];
  sprintf(filename, "%s.qual", outfile);
  SaveFloatToImage(qual_map, "quality map", filename,
                   xsize, ysize, 1, 0, 0);
}
```

```
/*  GENERATE MASK CUTS  */
if (mode==gradient && tsize==1) mode = dxdygrad;
mask_code = BRANCH_CUT;
k = QualityGuidedMask(phase, qual_map, bitflags, xsize, ysize,
                      mask_code, mode, debug_flag, outfile);
if (k > 1) printf("\n%d pieces\n", k);
if (debug_flag) {
  char filename[300];
  sprintf(filename, "%s.fatmask", outfile);
  SaveByteToImage(bitflags, "fat masks", filename, xsize, ysize,
                  1, 1, mask_code | BORDER);
}
for (k=0; k<xsize*ysize; k++) {
  bitflags[k] &= (mask_code | BORDER | RESIDUE);
}

/*  THIN MASK CUTS  */
ThinMask(bitflags, xsize, ysize, mask_code);

if (debug_flag) {
  char filename[300];
  sprintf(filename, "%s.thinmask", outfile);
  SaveByteToImage(bitflags, "thin masks", filename, xsize, ysize,
                  1, 1, mask_code | BORDER);
}

/*  UNWRAP AROUND CUTS */
for (k=0; k<xsize*ysize; k++)  /* initialize solution array */
  soln[k] = 0.0;
printf("Unwrapping around branch cuts\n");
k = UnwrapAroundCuts(phase, bitflags, soln, xsize, ysize,
                     mask_code, debug_flag, outfile);
if (k > 1) printf("%d disconnected pieces.\n", k);
else printf("%d piece\n", k);

printf("\nFinished\n");
PrintMinAndMax(xsize, ysize, soln, "solution");

/*  SAVE RESULT  */
/* scale output */
for (k=0; k<xsize*ysize; k++)
  soln[k] *= TWOPI;
printf("Saving unwrapped surface to file '%s'\n", outfile);
WriteFloat(ofp, soln, xsize*ysize, outfile);
```

```
  free(soln);
  free(phase);
  free(bitflags);
  free(qual_map);
}
```

The mask cuts that join and balance the residues are generated by the function `QualityGuidedMask` in file `maskcut.c`. This function is similar to the function `QualityGuidedPathFollower`, which performs the quality-guided algorithm of Section A.7, but there is no need to perform the list trimming procedure because mask cuts are generally small.

```
/*
 *   maskcut.c -- function for quality map guided mask cut generation
 */
#include <stdio.h>
#include <math.h>
#include "list.h"
#include "util.h"
#include "maskcut.h"

/* Main function for Mask Cut Algorithm for phase unwrapping */
int QualityGuidedMask(float *phase, float *qual_map,
                      unsigned char *bitflags, int xsize, int ysize,
                      int mask_code, UnwrapMode unwrap_mode,
                      int debug_flag, char *infile)
{
  int  i, j, k, a, b, c, n, num_pieces=0;
  float  value;
  float  min_qual, small_val = -1.0E+10;
  int    num_index, max_list_size, bench, benchout;
  int    *index_list;
  int    postponed_code=POSTPONED;
  int    avoid_code;
  int    charge, num_pixels;

  bench = xsize*ysize/100;
  if (bench < 1) bench = 1;
  benchout = 10*bench;
  max_list_size = xsize*ysize/2; /* this size can be reduced */
  AllocateInt(&index_list, max_list_size + 1,
            "book-keeping list (index)");
```

```
/* find starting point */
n = 0;
num_index = 0;
avoid_code = BORDER | mask_code;
for (j=0; j<ysize; j++) {
  for (i=0; i<xsize; i++) {
    min_qual = small_val;
    num_index = 0;
    k = j*xsize + i;
    if ((bitflags[k] & RESIDUE) && !(bitflags[k] & avoid_code)) {
      charge = (bitflags[k] & POS_RES) ? 1 : -1;
      ++num_pieces;
      num_pixels = 1;
      bitflags[k] |= mask_code;
      if ((bitflags[k] & BORDER)
          || i==0 || i==xsize-1 || j==0 || j==ysize-1) {
        continue;
      }
      UpdateList(qual_map, i, j, value, phase, NULL, bitflags,
              xsize, ysize, index_list, &num_index,
              postponed_code, VISITED, postponed_code,
              max_list_size, unwrap_mode==dxdygrad, &min_qual);
      while (charge != 0 && num_index > 0) {
        if (n%bench==0) {
          printf("%d ", n/bench);
          fflush(stdout);
        }
        ++n;
        if (!GetNextOneToUnwrap(&a, &b, index_list, &num_index,
                                 xsize, ysize))
          break;   /* no more to unwrap */
        c = b*xsize + a;
        ++num_pixels;
        if ((bitflags[c] & RESIDUE) && !(bitflags[c] & mask_code))
          charge += (bitflags[c] & POS_RES) ? 1 : -1;
        if ((bitflags[c] & BORDER)
            || a==0 || a==xsize-1 || b==0 || b==ysize-1) {
          charge = 0;
        }
        bitflags[c] |= mask_code;
        UpdateList(qual_map, a, b, value, phase, NULL, bitflags,
              xsize, ysize, index_list, &num_index,
              postponed_code, VISITED, postponed_code,
              max_list_size, unwrap_mode==dxdygrad, &min_qual);
      }
```

```
        /* reset pixels on list */
        for (k=0; k<num_index; k++)
          bitflags[index_list[k]] &= ~(VISITED);
        for (k=0; k<xsize*ysize; k++)
          bitflags[k] &= ~(VISITED);
      }
    }
  }
  printf("\n");
  free(index_list);
  return num_pieces;
}
```

The header file is `maskcut.h`:

```
#ifndef __MASKCUT
#define __MASKCUT
#include "getqual.h"
int QualityGuidedMask(float *phase, float *qual_map,
                      unsigned char *bitflags, int xsize, int ysize,
                      int mask_code, UnwrapMode unwrap_mode,
                      int debug_flag, char *infile);
#endif
```

A.9 FLYNN'S MINIMUM DISCONTINUITY ALGORITHM

The following files are found in this section: `mainflyn.c`, `flynn.c`, `flynn.h`, `trees.c`, and `trees.h`.

We present below the software for Flynn's minimum discontinuity algorithm of Section 4.5. The main program is in file `mainflyn.c`. The usage statement in the program listing below explains how to run the program. It is similar to that of the quality-guided algorithm of Section A.7, but there are several additional parameters. The `-mode` parameter, which specifies the source of the quality map, can take the argument `none` or be omitted altogether, in which case the unweighted version of the algorithm will be executed. The quality map, if it is supplied, can be thresholded by specifying `-thresh yes`, and the resulting zero-weights can be thickened by means of the `-fat` parameter if desired. (On the phase-unwrapping examples of this book, we found the algorithm worked best when the quality map was thresholded but the zero-weights not thickened.) Furthermore the input file can contain an intermediate (or partially unwrapped) surface instead of phase data. In this case `-guess yes` should be specified, and the surface must be congruent to the wrapped phase. This option is useful if a partially unwrapped surface is available, in which case the program may converge faster.

A typical command to execute the program would be

```
mwd -input mri.phase -format byte -output mri.surf -bmask
    mri.bmask -xsize 256 -ysize 256 -mode min_var -tsize 3
    -thresh yes -fat 0 -debug yes -guess no
```

where `mwd` is the name of the executable file.

In this program and those of the following sections, the `-bmask` parameter (i.e., the border mask file) should not be confused with the quality mask. The border mask file defines the pixels that should be ignored (e.g., the border pixels of the simulated IFSAR data shown in Figure 3.8b), while the quality mask, which is derived from the quality map, defines the low-quality pixels.

The program flow is as follows: The command line arguments are parsed by the functions `CommandLineParm`, `Keyword` and `SetQualityMode`. The files are opened, the memory for the arrays is allocated, and the phase data, the quality map, and the (optional) border mask data are then read from the input files and processed. (The border mask consists of pixels that should be ignored, e.g., the border pixels of the simulated IFSAR data shown in Figure 3.8b.) The border mask is thickened slightly by a call to the function `FattenMask`, the quality map is thresholded if `-thresh yes` was specified, and the zero-weights are thickened if the `-fat` argument was specified. The phase is unwrapped by the function `FlynnMinDisc`, which performs Flynn's algorithm, and saved to the output file.

```
/*
 * mainflyn.c -- phase unwrapping by Flynn's min discontinuity alg.
 *
 * Source code files required:
 *     dxdygrad.c      extract.c        flynn.c      getqual.c
 *         grad.c        histo.c     mainflyn.c      maskfat.c
 *     qualgrad.c     qualpseu.c      qualvar.c        trees.c
 *         util.c
 */
#include <stdio.h>
#include <math.h>
#include "file.h"
#include "histo.h"
#include "maskfat.h"
#include "pi.h"
#include "util.h"
#include "extract.h"
#include "getqual.h"
#include "flynn.h"
#define BORDER  (0x20)

main (int argc, char *argv[])
```

```
{
  int             i, j, k, tsize;
  FILE            *ifp, *ofp, *mfp=0, *qfp=0;
  float           *phase;     /* array */
  float           *qual_map;  /* array */
  unsigned char   *bitflags;  /* array */
  short           *hjump;     /* array */
  short           *vjump;     /* array */
  int             *value;     /* array */
  char            buffer[200], tempstr[200];
  char            infile[200], outfile[200];
  char            bmaskfile[200], qualfile[200];
  char            format[200], modekey[200];
  int             in_format, debug_flag;
  int             xsize, ysize;   /* dimensions of arrays */
  int             avoid_code, thresh_flag, fatten, guess_mode;
  UnwrapMode      mode;
  double          rmin, rmax, rscale, one_over_twopi = 1.0/TWOPI;
  /* define "use" statement */
  char            use[] =     /* define usage statement */
    "Usage: program-name -input file -format fkey -output file\n"
    "  -xsize x -ysize y [ -mode mkey -bmask file -corr file\n"
    "  -tsize size -debug yes/no -thresh yes/no -fat n\n"
    "  -guess yes/no]\n"
    "where 'fkey' is a keyword designating the input file type\n"
    "(key = complex8, complex4, float or byte), 'x' and 'y' are\n"
    "the dimensions of the file, bmask is an optional byte-file\n"
    "of masks for masking out undefined phase values, corr\n"
    "is an optional byte-file of cross-correlation values,\n"
    "tsize is the size of the square template for averaging the\n"
    "corr file or quality values (default = 1), and 'mkey' is\n"
    "a keyword designating the source of the quality values\n"
    "that guide the unwrapping path.  The value of mkey may be\n"
    "'min_grad' for Minimum Gradient unwrapping, 'min_var' for\n"
    "Minimum Variance unwrapping, 'max_corr' for Maximum\n"
    "Correlation unwrapping, 'max_pseu' for Maximum Pseudo-\n"
    "correlation unwrapping, or 'none'.  All files are simple\n"
    "raster files, and the output file consists of floating\n"
    "point numbers that define the heights of the unwrapped\n"
    "surface.  If the 'debug' parm is 'yes', then the\n"
    "intermediate byte-files are saved (quality map, etc.)\n"
    "To apply an automatic threshold to the quality map to make\n"
    "a quality mask, the 'thresh' parm should be yes.  To\n"
    "thicken the quality mask, the 'fat' parm should be the\n"
    "number of pixels by which to thicken.  If the 'guess'\n"
```

```
    "parameter is yes, then the input file must be a floating\n"
    "point array that represents an intermediate solution.\n"
    "This solution must be congruent to the wrapped phase.\n";

  printf("Phase Unwrapping by Flynn's Min. Discontinuity Method\n");

  /* GET COMMAND LINE PARAMETERS AND CHECK */
  CommandLineParm(argc,argv, "-input", StringParm, infile, 1, use);
  CommandLineParm(argc,argv, "-format", StringParm, format, 1, use);
  CommandLineParm(argc,argv, "-output", StringParm, outfile, 1,use);
  CommandLineParm(argc,argv, "-xsize", IntegerParm, &xsize, 1, use);
  CommandLineParm(argc,argv, "-ysize", IntegerParm, &ysize, 1, use);
  if (!CommandLineParm(argc, argv, "-mode", StringParm,
        modekey, 0, use))   strcpy(modekey, "none");
  if (!CommandLineParm(argc, argv, "-bmask", StringParm,
        bmaskfile, 0, use)) strcpy(bmaskfile, "none");
  if (!CommandLineParm(argc, argv, "-corr", StringParm,
        qualfile, 0, use)) strcpy(qualfile, "none");
  if (!CommandLineParm(argc, argv, "-tsize", IntegerParm,
        &tsize, 0, use)) tsize = 1;
  if (!CommandLineParm(argc, argv, "-debug", StringParm,
        tempstr, 0, use)) debug_flag = 0;
  else debug_flag = Keyword(tempstr, "yes");
  if (!CommandLineParm(argc, argv, "-thresh", StringParm,
        tempstr, 0, use)) thresh_flag = 0;
  else thresh_flag = Keyword(tempstr, "yes");
  if (!CommandLineParm(argc, argv, "-fat", IntegerParm,
        &fatten, 0, use)) fatten = 0;
  if (!CommandLineParm(argc, argv, "-guess", StringParm,
        tempstr, 0, use)) guess_mode = 0;
  else guess_mode = Keyword(tempstr, "yes");

  if (Keyword(format, "complex8"))  in_format = 0;
  else if (Keyword(format, "complex4"))  in_format = 1;
  else if (Keyword(format, "byte"))  in_format = 2;
  else if (Keyword(format, "float"))  in_format = 3;
  else {
    fprintf(stderr, "Unrecognized format: %s\n", format);
    exit(BAD_PARAMETER);
  }

  printf("Input file =  %s\n", infile);
  printf("Input file type = %s\n", format);
  printf("Output file =  %s\n", outfile);
  printf("File dimensions = %dx%d (cols x rows).\n", xsize, ysize);
```

```
if (Keyword(bmaskfile, "none")) printf("No border mask file.\n");
else printf("Border mask file = %s\n", bmaskfile);
printf("Quality mode = %s\n", modekey);
if (modekey==none) {
  printf("No quality map.\n");
  strcpy(qualfile, "none");
}
else {
  if (Keyword(qualfile, "none")) printf("No correlation file.\n");
  else printf("Correlation image file = %s\n", qualfile);
  printf("Averaging template size = %d\n", tsize);
  if (tsize < 0 || tsize > 30) {
    fprintf(stderr, "Illegal size: must be between 0 and 30\n");
    exit(BAD_PARAMETER);
  }
}
mode = SetQualityMode(modekey, qualfile, 1);
if (mode < 0) exit(BAD_PARAMETER);  /* error msg already printed */

/*  OPEN FILES, ALLOCATE MEMORY   */
OpenFile(&ifp, infile, "r");
OpenFile(&ofp, outfile, "w");
if (!Keyword(bmaskfile, "none")) OpenFile(&mfp, bmaskfile, "r");
if (mode==corr_coeffs) OpenFile(&qfp, qualfile, "r");
AllocateFloat(&phase, xsize*ysize, "phase data");
AllocateByte(&bitflags, (xsize+1)*(ysize+1), "bitflags array");
AllocateFloat(&qual_map, xsize*ysize, "quality map");
AllocateShort(&hjump, (xsize+1)*(ysize+1), "hjump array");
AllocateShort(&vjump, (xsize+1)*(ysize+1), "vjump array");
AllocateInt(&value, (xsize+1)*(ysize+1), "value array");

/*  READ AND PROCESS DATA  */
if (guess_mode) {
  printf("Guess Mode: Inputting initial guess...\n");
  ReadFloat(ifp, phase, xsize*ysize, infile);
  for (k=0; k<xsize*ysize; k++) phase[k] *= one_over_twopi;
}
else {
  printf("Reading phase data...\n");
  GetPhase(in_format, ifp, infile, phase, xsize, ysize);
}

if (qfp) {
  printf("Reading quality data...\n");
  /* borrow the bitflags array temporarily */
```

```
    ReadByte(qfp, bitflags, xsize*ysize, qualfile);
    /* process data and store in quality map array */
    AverageByteToFloat(bitflags, qual_map, tsize, xsize, ysize);
  }

  /* border mask data */
  printf("Processing border mask data...\n");
  if (mfp) {
    ReadByte(mfp, bitflags, xsize*ysize, bmaskfile);
  }
  else {
    for (k=0; k<xsize*ysize; k++)
      bitflags[k] = 255;
  }
  for (k=0; k<xsize*ysize; k++) {
    bitflags[k] = (!bitflags[k]) ? BORDER : 0;
  }
  if (mfp) FattenMask(bitflags, BORDER, 1, xsize, ysize);
  GetQualityMap(mode, qual_map, phase, bitflags, BORDER,
                tsize, xsize, ysize);
  if (thresh_flag) {
    HistoAndThresh(qual_map, xsize, ysize, 0.0, 0, 0.0,
                   bitflags, BORDER);
    if (fatten > 0)
      FattenQual(qual_map, fatten, xsize, ysize);
  }
  if (debug_flag) {
    char filename[300];
    sprintf(filename, "%s.qual", outfile);
    SaveFloatToImage(qual_map, "quality", filename, xsize, ysize,
                     0, 0, 0);
  }

  /* embed bitflags array in larger array for use by Flynn's routines
*/
  for (j=ysize - 1; j>=0; j--) {
    for (i=xsize - 1; i>=0; i--) {
      bitflags[j*(xsize+1) + i] = bitflags[j*xsize + i];
    }
  }
  for (j=0; j<=ysize; j++) bitflags[j*(xsize + 1) + xsize] = 0;
  for (i=0; i<=xsize; i++) bitflags[ysize*(xsize + 1) + i] = 0;

  /*  UNWRAP  */
  printf("Unwrapping...\n");
```

```
  FlynnMinDisc(phase, value, bitflags, qual_map, vjump, hjump,
               xsize, ysize);
  printf("\nFinished\n");
  PrintMinAndMax(xsize, ysize, phase, "solution");

  /*  SAVE RESULT  */
  for (k=0; k<xsize*ysize; k++)
    phase[k] *= TWOPI;
  printf("Saving unwrapped surface to file '%s'\n", outfile);
  WriteFloat(ofp, phase, xsize*ysize, outfile);
  free(phase);
  free(bitflags);
  if (qual_map) free(qual_map);
  free(hjump);
  free(vjump);
  free(value);
}
```

The main iterative loop for Flynn's algorithm is contained in function `FlynnMinDisc` in file `flynn.c`. After initializing the arrays `hjump` and `vjump`, the function executes the main loop and, when it terminates, computes the unwrapped phase from `hjump` and `vjump`.

```
/*
 *  flynn.c -- functions for Flynn's min. discontinuity algorithm
 */
#include <stdio.h>
#include <math.h>
#include "flynn.h"
#include "trees.h"
/* define large value-increase */
#define BIG  25500

/* Main iteration (and initialization steps) for Flynn's minimum
 * discontinuity phase unwrapping algorithm.  The quality values
 * (weights) are stored in the qual_map array.  This array and the
 * phase array are xsize x ysize pixels in size.  The bitflags array,
 * which stores the change and direction info, and the other arrays
 * (value, vjump and hjump) are working arrays for managing the data.
 * The dimensions of these arrays are (xsize + 1) x (ysize + 1).
 * If there are masked (border) pixels, they should be coded as
 * zero weights in qual_map.
 */
void FlynnMinDisc(float *phase, int *value, unsigned char *bitflags,
                  float *qual_map, short *vjump, short *hjump,
```

```
                  int xsize, int ysize)
{
  double          eps = 1.0e-06, phd;
  short           jmpold;
  int             j, i, jnext, inext;
  int             ilast, jlast;
  int             new_loops, new_edges;
  int             val, value_incr;
  int             loop_found;
  int             value_change;

  /* Compute phase jump counts.
   * If dx(i,j) and dy(i,j) represent the wrapped derivatives
   * phase(i+1,j) - phase(i,j) and phase(i,j+1) - phase(i,j),
   * respectively, then the jump count hjump(i,j) corresponds to
   * dy(i-1,j-1), and vjump(i,j) corresponds to dx(i-1,j-1).
   */
  for (j=1; j <= ysize-1; j++) {
    for (i=1; i <= xsize; i++) {
      phd = phase[j*xsize + i-1] - phase[(j-1)*xsize + i-1] + eps;
      hjump[j*(xsize+1) + i] = (short) nint(phd);
    }
  }
  for (j=1; j <= ysize; j++) {
    for (i=1; i <= xsize-1; i++) {
      phd = phase[(j-1)*xsize + i] - phase[(j-1)*xsize + i-1] + eps;
      vjump[j*(xsize+1) + i] = (short) nint(phd);
    }
  }
  /* Make add nodes bitflags initially */
  for (j=0; j <= ysize; j++) {
    for (i=0; i <= xsize; i++) {
      bitflags[j*(xsize+1) + i] |= THIS_TIME;
    }
  }
  /* Main iteration */
  do {
    new_loops = 0;
    new_edges = 0;
    /* Add left-to-right edges to tree */
    for (j=0; j <= ysize-1; j++) {
      for (i=0; i <= xsize; i++) {
        jnext = j+1;
        inext = i;
        if ((bitflags[j*(xsize+1) + i] & THIS_TIME)
```

```
          || (bitflags[jnext*(xsize+1) + inext] & THIS_TIME)) {
        if (jnext <= 1 || jnext >= ysize
             || inext <= 0 || inext >= xsize) {
          value_incr = -isign(1, -vjump[jnext*(xsize+1) + inext]);
        }
        else {
          /* quality map values must be between 0 and 1 */
          val = (int)(1.0 + BIG*qual_map[jnext*xsize + inext]);
          value_incr
             = -isign(val, -vjump[jnext*(xsize+1) + inext]);
        }
        value_change = value[j*(xsize+1) + i] + (int)value_incr
                                - value[jnext*(xsize+1) + inext];
        if (value_change > 0) {
          new_edges++;
          /* revise values in subtree of [jnext][inext] */
          /* and check for loop */
          ilast = i;  jlast = j;  loop_found = 0;
          ChangeExten(inext, jnext, ilast, jlast, &loop_found,
                      value_change, value, bitflags, xsize, ysize);
          if (loop_found) {
            RemoveLoop(inext, jnext, i, j, value, bitflags,
                       vjump, hjump, xsize, ysize);
            ChangeOrphan(inext, jnext, -value_change, value,
                         bitflags, xsize, ysize);
            new_loops++;
          }
          else {
            /* add edge and remove edges to [jnext][inext] */
            bitflags[j*(xsize+1) + i] |= RIGHT;
            if (jnext<ysize)
              bitflags[(jnext+1)*(xsize+1) + inext] &= ~LEFT;
            if (inext<xsize)
              bitflags[jnext*(xsize+1) + inext+1] &= ~UP;
            if (inext > 0)
              bitflags[jnext*(xsize+1) + inext-1] &= ~DOWN;
          }
        }
      }
    }
  }
  /* Add top-to-bottom edges to tree */
  for (j=0; j <= ysize; j++) {
    for (i=0; i <= xsize-1; i++) {
      jnext = j;
```

```
      inext = i+1;
      if ((bitflags[j*(xsize+1) + i] & THIS_TIME)
           || (bitflags[jnext*(xsize+1) + inext] & THIS_TIME)){
        if (jnext <= 0 || jnext >= ysize
             || inext <= 1 || inext >= xsize) {
          value_incr = -isign(1, hjump[jnext*(xsize+1) + inext]);
        }
        else {
          /* quality map values must be between 0 and 1 */
          val = (int)(1.0 + BIG*qual_map[jnext*xsize + inext]);
          value_incr
              = -isign(val, hjump[jnext*(xsize+1) + inext]);
        }
        value_change = value[j*(xsize+1) + i] + (int)value_incr
                                 - value[jnext*(xsize+1) + inext];
        if (value_change > 0) {
          new_edges++;
          ilast = i; jlast = j; loop_found = 0;
          ChangeExten(inext, jnext, ilast, jlast, &loop_found,
                      value_change, value, bitflags, xsize, ysize);
          if (loop_found){
            RemoveLoop(inext, jnext, i, j, value, bitflags,
                       vjump, hjump, xsize, ysize);
            ChangeOrphan(inext, jnext, -value_change, value,
                         bitflags, xsize, ysize);
            new_loops++;
          }
          else {
            bitflags[j*(xsize+1) + i] |= DOWN;
            if (jnext<ysize)
              bitflags[(jnext+1)*(xsize+1) + inext] &= ~LEFT;
            if (jnext > 0)
              bitflags[(jnext-1)*(xsize+1) + inext] &= ~RIGHT;
            if (inext<xsize)
              bitflags[jnext*(xsize+1) + inext+1] &= ~UP;
          }
        }
      }
    }
  }
  /* Add right-to-left edges to tree */
  for (j=ysize; j >=1 ; j--) {
    for (i=xsize; i >=0; i--) {
      jnext = j-1;
      inext = i;
```

```
      if ((bitflags[j*(xsize+1) + i] & THIS_TIME)
            || (bitflags[jnext*(xsize+1) + inext] & THIS_TIME)){
        if (j <= 1 || j >= ysize || i <= 0 || i >= xsize) {
          value_incr = -isign(1, vjump[j*(xsize+1) + i]);
        }
        else {
          /* quality map values must be between 0 and 1 */
          val = (int)(1.0 + BIG*qual_map[j*xsize + i]);
          value_incr = -isign(val, vjump[j*(xsize+1) + i]);
        }
        value_change = value[j*(xsize+1) + i] + (int)value_incr
                              - value[jnext*(xsize+1) + inext];
        if (value_change > 0) {
          new_edges++;
          ilast = i; jlast = j; loop_found = 0;
          ChangeExten(inext, jnext, ilast, jlast, &loop_found,
                      value_change, value, bitflags, xsize, ysize);
          if (loop_found){
            RemoveLoop(inext, jnext, i, j, value, bitflags,
                       vjump, hjump, xsize, ysize);
            ChangeOrphan(inext, jnext, -value_change, value,
                         bitflags, xsize, ysize);
            new_loops++;
          }
          else {
            bitflags[j*(xsize+1) + i] |= LEFT;
            if (jnext > 0)
              bitflags[(jnext-1)*(xsize+1) + inext] &= ~RIGHT;
            if (inext < xsize)
              bitflags[jnext*(xsize+1) + inext+1] &= ~UP;
            if (inext > 0)
              bitflags[jnext*(xsize+1) + inext-1] &= ~DOWN;
          }
        }
      }
    }
  }
  /* Add bottom-to-top edges to tree */
  for (j=ysize; j >=0; j--) {
    for (i=xsize; i >= 1; i--) {
      jnext = j;
      inext = i-1;
      if ((bitflags[j*(xsize+1) + i] & THIS_TIME)
            || (bitflags[jnext*(xsize+1) + inext] & THIS_TIME)){
        if (j <= 0 || j >= ysize || i <= 1 || i >= xsize) {
```

```
          value_incr = -isign(1, -hjump[j*(xsize+1) + i]);
        }
        else {
          /* quality map values must be between 0 and 1 */
          val = (int)(1.0 + BIG*qual_map[j*xsize + i]);
          value_incr = -isign(val, -hjump[j*(xsize+1) + i]);
        }
        value_change = value[j*(xsize+1) + i] + (int)value_incr
                              - value[jnext*(xsize+1) + inext];
        if (value_change > 0) {
          new_edges++;
          ilast = i; jlast = j; loop_found = 0;
          ChangeExten(inext, jnext, ilast, jlast, &loop_found,
                      value_change, value, bitflags, xsize, ysize);
          if (loop_found){
            RemoveLoop(inext, jnext, i, j, value, bitflags,
                       vjump, hjump, xsize, ysize);
            ChangeOrphan(inext, jnext, -value_change, value,
                         bitflags, xsize, ysize);
            new_loops++;
          }
          else {
            bitflags[j*(xsize+1) + i] |= UP;
            if (jnext < ysize)
              bitflags[(jnext+1)*(xsize+1) + inext] &= ~LEFT;
            if (jnext > 0)
              bitflags[(jnext-1)*(xsize+1) + inext] &= ~RIGHT;
            if (inext > 0)
              bitflags[jnext*(xsize+1) + inext-1] &= ~DOWN;
          }
        }
      }
    }
  }
  printf("New edges: %d  New loops: %d\n",new_edges,new_loops);
  /* Update testability matrix */
  for (j=0; j <= ysize; j++) {
    for (i=0; i <= xsize; i++) {
      if (bitflags[j*(xsize+1) + i] & NEXT_TIME) {
        bitflags[j*(xsize+1) + i] |= THIS_TIME;
      }
      else {
        bitflags[j*(xsize+1) + i] &= ~NEXT_TIME;
        bitflags[j*(xsize+1) + i] &= ~THIS_TIME;
      }
```

```
      }
    }
  } while (new_edges > 0);

  /* Compute new unwrapped phase and exit */
  printf("Computing revised unwrapped phase...\n");
  for (i = 0; i < xsize - 1; i++) {
    phd = phase[0*xsize + (i + 1)] - phase[0*xsize + i] + eps;
    jmpold = (short) nint(phd);
    phase[0*xsize + (i + 1)]
        += (float)(vjump[1*(xsize + 1) + (i + 1)] - jmpold);
  }
  for (j = 0; j < ysize - 1; j++) {
    for (i = 0; i < xsize; i++) {
      phd = phase[(j + 1)*xsize + i]
                   - phase[j*xsize + i] + eps;
      jmpold = (short) nint(phd);
      phase[(j + 1)*xsize + i]
        += (float)(hjump[(j + 1)*(xsize + 1) + (i + 1)] - jmpold);
    }
  }
}
```

The header file is `flynn.h`:

```
#ifndef __FLYNN
#define __FLYNN
#define isign(A,B) ((B) >= 0 ? (A) : (-(A)))
#define nint(A) ((A)>0. ? (int)((A)+0.5) : (int)((A)-0.5))
void FlynnMinDisc(float *phase, int *value, unsigned char *bitflags,
                  float *qual_map, short *vjump, short *hjump,
                  int xsize, int ysize);
#endif
```

The functions RemoveLoop, ChangeExten, and ChangeOrphan for managing the trees, edges, and loops are found in file `trees.c`:

```
/*
 *  trees.c -- functions for managing trees, loops, edges, etc.
 *             in Flynn's min. discontinuity algorithm
 */
#include <stdio.h>
#include <math.h>
#include "trees.h"
```

```
/* Remove the loop and update all of the edges. */
void RemoveLoop(int ibase, int jbase, int ilast, int jlast,
                int *value, unsigned char *bitflags, short *vjump,
                short *hjump, int xsize, int ysize)
{
  int   jtip,itip;
  int   value_shift;
  /* Remove edge from last node to base of loop */
  if (jbase>jlast)       --vjump[jbase*(xsize+1) + ibase];
  else if (jbase<jlast)  ++vjump[jlast*(xsize+1) + ilast];
  else if (ibase>ilast)  ++hjump[jbase*(xsize+1) + ibase];
  else if (ibase<ilast)  --hjump[jlast*(xsize+1) + ilast];
  /* Back up over loop: remove edges and change subtree values */
  jtip = jlast;
  itip = ilast;
  do {
    value_shift = -value[jtip*(xsize+1) + itip];
    ChangeOrphan(itip, jtip, value_shift, value, bitflags,
                 xsize, ysize);
    if (jtip > 0
          && (bitflags[(jtip-1)*(xsize+1) + itip] & RIGHT)) {
      vjump[jtip*(xsize+1) + itip]--;
      bitflags[(jtip-1)*(xsize+1) + itip] &= ~RIGHT;
      jtip--;
    }
    else if ((jtip<ysize) &&
        (bitflags[(jtip+1)*(xsize+1) + itip] & LEFT)) {
      vjump[(jtip+1)*(xsize+1) + itip]++;
      bitflags[(jtip+1)*(xsize+1) + itip] &= ~LEFT;
      jtip++;
    }
    else if (itip > 0
                && (bitflags[jtip*(xsize+1) + itip-1] & DOWN)) {
      hjump[jtip*(xsize+1) + itip]++;
      bitflags[jtip*(xsize+1) + itip-1] &= ~DOWN;
      itip--;
    }
    else if ((itip<xsize)
                && (bitflags[jtip*(xsize+1) + itip+1] & UP)) {
      hjump[jtip*(xsize+1) + itip+1]--;
      bitflags[jtip*(xsize+1) + itip+1] &= ~UP;
      itip++;
    }
    else {
      printf("Error: broken loop at col %d, row %d\n",itip,jtip);
```

```
      exit(-1);
    }
  } while ((jtip != jbase) || (itip != ibase));
}

/*
 * Add 'value_change' (which is > 0) to the values of all nodes
 * of the tree rooted at the node (i,j).  Set a flag if the tree
 * includes (ilast,jlast). Make (i,j) child in the next iteration.
 */
void ChangeExten(int i, int j, int ilast, int jlast,
                 int *loop_found, int value_change, int *value,
                 unsigned char *bitflags, int xsize, int ysize)
{
  unsigned char kids;
  *loop_found |= ((j == jlast) && (i == ilast));
  kids = bitflags[j*(xsize+1) + i];
  if (kids & LEFT)
    ChangeExten(i, j-1, ilast, jlast, loop_found, value_change,
                value, bitflags, xsize, ysize);
  if (kids & RIGHT)
    ChangeExten(i, j+1, ilast, jlast, loop_found, value_change,
                value, bitflags, xsize, ysize);
  if (kids & UP)
    ChangeExten(i-1, j, ilast, jlast, loop_found, value_change,
                value, bitflags, xsize, ysize);
  if (kids & DOWN)
    ChangeExten(i+1, j, ilast, jlast, loop_found, value_change,
                value, bitflags, xsize, ysize);
  value[j*(xsize+1) + i] += value_change;
  bitflags[j*(xsize+1) + i] |= (NEXT_TIME | THIS_TIME);
}

/*
 * Add 'value_shift' (which is < 0) to the values of all nodes
 * of the tree rooted at the node (i,j).  If the change would make
 * 'value' < 0, make 'value' 0 instead  and split the node from
 * its parent.  Make (i,j) child in the next iteration.
 */
void ChangeOrphan(int i, int j, int value_shift, int *value,
                  unsigned char *bitflags, int xsize, int ysize)
{
  unsigned char kids;
  if (value[j*(xsize+1) + i] + (value_shift) < 0) {
    value_shift = -value[j*(xsize+1) + i];
```

```
    if (j>0)      bitflags[(j-1)*(xsize+1) + i] &=  ~RIGHT;
    if (j<ysize) bitflags[(j+1)*(xsize+1) + i] &=  ~LEFT;
    if (i>0)      bitflags[j*(xsize+1) + i-1]   &=  ~DOWN;
    if (i<xsize) bitflags[j*(xsize+1) + i+1]   &=  ~UP;
  }
  kids = bitflags[j*(xsize+1) + i];
  if (kids & LEFT) ChangeOrphan(i, j-1, value_shift, value,
                                 bitflags, xsize, ysize);
  if (kids & RIGHT) ChangeOrphan(i, j+1, value_shift, value,
                                  bitflags, xsize, ysize);
  if (kids & UP) ChangeOrphan(i-1, j, value_shift, value,
                               bitflags, xsize, ysize);
  if (kids & DOWN) ChangeOrphan(i+1, j, value_shift, value,
                                 bitflags, xsize, ysize);
  bitflags[j*(xsize+1) + i] |= (NEXT_TIME | THIS_TIME);
  value[j*(xsize+1) + i] += value_shift;
}
```

The header file is `trees.h`. Much of the record keeping performed by Flynn's algorithm is accomplished by means of the bit flags `LEFT`, `RIGHT`, `UP`, `DOWN`, `THIS_TIME`, and `NEXT_TIME`, which are used to label the `bitflags` array.

```
#ifndef __TREES
#define __TREES
#define LEFT      (0x08)
#define RIGHT     (0x04)
#define UP        (0x02)
#define DOWN      (0x01)
#define THIS_TIME   (0x40)
#define NEXT_TIME   (0x80)
void RemoveLoop(int ibase, int jbase, int ilast, int jlast,
                int *value, unsigned char *bitflags, short *vjump,
                short *hjump, int xsize, int ysize);
void ChangeExten(int i, int j, int ilast, int jlast,
                 int *loop_found, int value_change, int *value,
                 unsigned char *bitflags, int xsize, int ysize);
void ChangeOrphan(int i, int j, int value_shift, int *value,
                  unsigned char *bitflags, int xsize, int ysize);
#endif
```

A.10 UNWEIGHTED LEAST-SQUARES ALGORITHMS

The following files are listed in this section: `mainunwt.c`, `mainunmg.c`, `unfmg.c`, `unfmg.h`, `ungrid.c`, `ungrid.h`, `gridops.c`, `gridops.h`, `gridmem.c`, `gridmem.h`, `relax.c`, and `relax.h`.

We now present the software for two unweighted least-squares algorithms for phase unwrapping. The first is the FFT/DCT based algorithm of Sections 5.3.1 and 5.3.2, and the second is the unweighted multigrid algorithm of Section 5.3.3. Both programs are equivalent to the weighted phase-unwrapping programs of Sections A.11 and A.12 when unity weights (i.e., weights that are all 1's) are supplied. For example, running one iteration of the PCG program of Section A.11 with unity weights will perform the DCT-based algorithm. Similarly running the weighted multigrid program of Section A.12 with unity weights will perform the unweighted multigrid algorithm. However, the execution times will be higher than necessary due to the overhead of managing the weights. Because of this fact, and because the unweighted programs are simpler and shorter than their weighted counterparts, we present the software for the unweighted algorithms.

The main program for the direct transform (i.e., DCT or FFT) approach to the unweighted least-squares phase-unwrapping problem is in file `mainunwt.c`. While this program uses DCTs to solve the Poisson equation, the DCTs are computed in turn by means of FFTs (see Section A.4). Because the program does not accept masks or quality maps, it is particularly simple to use. The array dimensions `xsize` and `ysize` must be of the form $2^n + 1$ due to the use of the DCTs.

The program works as follows: After parsing the command line arguments, opening the files, allocating the memory, and reading the phase data, the program calls the function `ComputeLaplacian` to compute the wrapped Laplacian of the phase data. It then allocates two small scratch arrays (`xcos` and `ycos`) and calls `DirectSolnbyCosineTransform` to solve the Poisson equation (see Section A.4). The solution is then written to the output file.

```
/*
 * mainunwt.c -- unweighted least-squares phase unwrapping
 *               by means of direct transforms
 *
 * Source code files required:
 *          dct.c      extract.c          grad.c      laplace.c
 *      mainunwt.c     solncos.c          util.c
 */
#include <stdio.h>
#include <math.h>
#include "solncos.h"
#include "laplace.h"
#include "file.h"
#include "pi.h"
#include "util.h"
#include "extract.h"

main (int argc, char *argv[])
```

```
{
  int           i, j, k, n;
  FILE          *ifp, *ofp;
  float         *phase;     /* array */
  float         *soln;      /* array */
  double        *xcos, *ycos;
  char          infile[200], outfile[200];
  char          format[200];
  int           in_format;
  int           xsize, ysize;   /* dimensions of arrays */
  int           xsize_dct, ysize_dct;
  double        rmin, rmax, rscale;
  double        one_over_twopi = 1.0/TWOPI;
  char          use[] =       /* define usage statement */
   "Usage: prog-name -input file -format fkey -output file\n"
   "  -xsize x -ysize y\n"
   "where 'fkey' is a keyword designating the input file type\n"
   "(key = complex8, complex4, float or byte).  The dimensions\n"
   "x and y must be powers of two plus 1 (e.g., 257, 513, etc.)\n";

  printf("Unweighted Phase Unwrapping by DCT/FFT\n");

  /* GET COMMAND LINE PARAMETERS AND CHECK */
  CommandLineParm(argc,argv, "-input", StringParm, infile, 1, use);
  CommandLineParm(argc,argv, "-format", StringParm, format, 1, use);
  CommandLineParm(argc,argv, "-output", StringParm, outfile, 1,use);
  CommandLineParm(argc,argv, "-xsize", IntegerParm, &xsize, 1, use);
  CommandLineParm(argc,argv, "-ysize", IntegerParm, &ysize, 1, use);

  if (Keyword(format, "complex8"))  in_format = 0;
  else if (Keyword(format, "complex4"))  in_format = 1;
  else if (Keyword(format, "byte"))  in_format = 2;
  else if (Keyword(format, "float"))  in_format = 3;
  else {
    fprintf(stderr, "Unrecognized format: %s\n", format);
    exit(BAD_PARAMETER);
  }

  printf("Input file =  %s\n", infile);
  printf("Input file type = %s\n", format);
  printf("Output file =  %s\n", outfile);
  printf("File dimensions = %dx%d (cols x rows).\n", xsize, ysize);

  /* Verify dimensions are of form 2**n + 1 */
```

```
  for (xsize_dct = 1; xsize_dct + 1 < xsize; xsize_dct*=2)
    ;
  xsize_dct = xsize_dct + 1;
  for (ysize_dct = 1; ysize_dct + 1 < ysize; ysize_dct*=2)
    ;
  ysize_dct = ysize_dct + 1;
  if (xsize_dct != xsize || ysize_dct != ysize) {
    fprintf(stderr, "Error: dims %dx%d must be 2**n + 1\n",
            xsize, ysize);
    exit(BAD_PARAMETER);
  }

  /*  OPEN FILES, ALLOCATE MEMORY    */
  OpenFile(&ifp, infile, "r");
  OpenFile(&ofp, outfile, "w");
  AllocateFloat(&phase, xsize*ysize, "phase data");
  AllocateFloat(&soln, xsize*ysize, "scratch data");

  /*  READ AND PROCESS DATA  */
  printf("Reading phase data...\n");
  GetPhase(in_format, ifp, infile, phase, xsize, ysize);

  /*  UNWRAP  */
  printf("Computing Laplacian...\n");
  ComputeLaplacian(phase, soln, NULL, NULL, xsize, ysize, 1);
  free(phase);
  AllocateDouble(&xcos, xsize, "cosine terms");
  AllocateDouble(&ycos, ysize, "cosine terms");
  printf("Performing direct transform...\n");
  DirectSolnByCosineTransform(soln, xsize, ysize, xcos, ycos);
  free(xcos);
  free(ycos);
  printf("\nFinished\n");
  PrintMinAndMax(xsize, ysize, soln, "solution");

  /*  SAVE RESULT  */
  for (k=0; k<xsize*ysize; k++)
    soln[k] *= TWOPI;
  printf("Saving unwrapped surface to file '%s'\n", outfile);
  WriteFloat(ofp, soln, xsize*ysize, outfile);
  free(soln);
}
```

The unweighted multigrid phase-unwrapping program is in file `mainunmg.c`. The usage is exactly like that of the preceding program, except that the user can also specify the number of multigrid cycles (`-cycles`) and the number of Gauss-Seidel iterations (`-iter`) to execute. If these parameters are omitted, the default values are two cycles and two iterations. There are no power-of-two restrictions on the array dimensions `xsize` and `ysize`. At a high level the program flow is simple. After parsing the command line arguments, opening the files, allocating the memory, and reading the phase data, the program allocates additional memory for the phase derivatives (`dx` and `dy`) and calls function `UnweightedMultigridUnwrap` to solve the Poisson equation. The resulting solution is written to the output file.

```
/*
 * mainunmg.c -- unweighted least-squares phase unwrapping
 *               by means of unweighted multigrid algorithm
 *
 * Source code files required:
 *     dxdygrad.c     extract.c        grad.c     gridmem.c
 *      gridops.c    mainunmg.c       relax.c        unfmg.c
 *       ungrid.c        util.c
 */
#include <stdio.h>
#include <math.h>
#include "file.h"
#include "pi.h"
#include "util.h"
#include "extract.h"
#include "unfmg.h"
#define DEFAULT_NUM_ITER    2
#define DEFAULT_NUM_CYCLES  2

main (int argc, char *argv[])
{
  int            i, j, k, n;
  FILE           *ifp, *ofp;
  float          *phase;     /* array */
  float          *soln;      /* array */
  float          *dx;        /* array */
  float          *dy;        /* array */
  char           infile[200], outfile[200];
  char           format[200];
  int            in_format;
  int            xsize, ysize;    /* dimensions of arrays */
  int            num_iter=DEFAULT_NUM_ITER;
  int            num_cycles=DEFAULT_NUM_CYCLES;
  char           use[] =   /* define usage statement */
    "Usage: prog-name -input file -format fkey -output file\n"
```

```
  "  -xsize x -ysize y [ -cycles numc -iter num]\n"
  "where 'fkey' is a keyword designating the input file type\n"
  "(key = complex8, complex4, float or byte) and 'x' and 'y'\n"
  "are the dimensions of the file.  All files are simple\n"
  "raster files, and the output file consists of floating\n"
  "point numbers that define the heights of the unwrapped\n"
  "surface.  The number of cycles is given by 'numc'.  The\n"
  "number of Gauss-Seidel iterations is given by 'num'.\n";

printf("Unweighted Phase Unwrapping by Multigrid Algorithm\n");


/* GET COMMAND LINE PARAMETERS AND CHECK */
CommandLineParm(argc,argv, "-input", StringParm, infile, 1, use);
CommandLineParm(argc,argv, "-format", StringParm, format, 1, use);
CommandLineParm(argc,argv, "-output", StringParm, outfile, 1,use);
CommandLineParm(argc,argv, "-xsize", IntegerParm, &xsize, 1, use);
CommandLineParm(argc,argv, "-ysize", IntegerParm, &ysize, 1, use);
CommandLineParm(argc, argv, "-iter", IntegerParm, &num_iter,
                0, use);
if (num_iter < 0) num_iter = 1;
CommandLineParm(argc, argv, "-cycles", IntegerParm, &num_cycles,
                0, use);
if (num_cycles < 0) num_cycles = 1;


if (Keyword(format, "complex8"))  in_format = 0;
else if (Keyword(format, "complex4"))  in_format = 1;
else if (Keyword(format, "byte"))  in_format = 2;
else if (Keyword(format, "float"))  in_format = 3;
else {
  fprintf(stderr, "Unrecognized format: %s\n", format);
  exit(BAD_PARAMETER);
}


printf("Input file =  %s\n", infile);
printf("Input file type = %s\n", format);
printf("Output file =  %s\n", outfile);
printf("File dimensions = %dx%d (cols x rows).\n", xsize, ysize);


/*  OPEN FILES, ALLOCATE MEMORY   */
OpenFile(&ifp, infile, "r");
OpenFile(&ofp, outfile, "w");
AllocateFloat(&phase, xsize*ysize, "phase data");
AllocateFloat(&soln, xsize*ysize, "solution array");
```

```
  /*  READ AND PROCESS DATA  */
  printf("Reading phase data...\n");
  GetPhase(in_format, ifp, infile, phase, xsize, ysize);

  /*  UNWRAP  */
  printf("Unwrapping...\n");
  AllocateFloat(&dx, xsize*ysize, "dx derivs");
  AllocateFloat(&dy, ysize*ysize, "dy derivs");
  DxPhaseGradient(phase, dx, xsize, ysize);
  DyPhaseGradient(phase, dy, xsize, ysize);
  for (k=0; k<xsize*ysize; k++)  soln[k] = 0.0;
  UnweightedMultigridUnwrap(soln, dx, dy, xsize, ysize,
                            num_cycles, num_iter);
  printf("\nFinished\n");
  PrintMinAndMax(xsize, ysize, soln, "solution");

  /*  SAVE RESULT  */
  for (k=0; k<xsize*ysize; k++)
    soln[k] *= TWOPI;
  printf("Saving unwrapped surface to file '%s'\n", outfile);
  WriteFloat(ofp, soln, xsize*ysize, outfile);
  free(dx);
  free(dy);
  free(soln);
}
```

The main program calls the function `UnweightedMultigridUnwrap` in file `unfmg.c` to perform the multigrid algorithm:

```
/*
 *  unfmg.c -- function for multigrid solution of UNWEIGHTED
 *             least-squares phase-unwrapping problem
 */
#include <stdio.h>
#include <math.h>
#include "unfmg.h"
#include "ungrid.h"
#include "dxdygrad.h"
/* Unweighted multigrid function.                  */
/* Initialize soln array to zero before calling. */
void UnweightedMultigridUnwrap(float *soln, float *dx, float *dy,
                int xsize, int ysize, int num_cycles, int num_iter)
{
  int  n, coarsest_dim=3;
  for (n=0; n<num_cycles; n++) {
```

```
    printf("\nFMG CYCLE %d\n", n+1);
    Ufmg(soln, dx, dy, xsize, ysize, num_iter, coarsest_dim);
  }
}
```

The header file is `unfmg.h`:

```
#ifndef __UNFMG
#define __UNFMG
void UnweightedMultigridUnwrap(float *soln, float *dx, float *dy,
              int xsize, int ysize, int num_cycles, int num_iter);
#endif
```

The main functions for the multigrid algorithm are in `ungrid.c`. `Ufmg` is a recursive function that performs the unweighted version of the full multigrid algorithm. It in turn calls the recursive function `Umv` to perform the unweighted multigrid V-cycle algorithm. The parameter `mindim`, which specifies the dimensions of the coarsest grid, is usually defined to be 3. The parameter `numit`, which specifies the number of Gauss-Seidel relaxation sweeps to perform on each grid, is usually defined to be 2. Additional iterations are performed on the coarsest grid, where relaxation is computationally inexpensive due to the small size, to ensure convergence.

```
/*
 * ungrid.c - functions for UNWEIGHTED multigrid phase
 *            unwrapping
 */
#include <stdio.h>
#include <math.h>
#include "ungrid.h"
#include "gridops.h"
#include "gridmem.h"
#include "relax.h"

/* Main function for unweighted multigrid phase unwrapping. */
/* (Called recursively.)                                     */
void Ufmg(float *soln, float *dx, float *dy,
          int w, int h, int numit, int mindim)
{
  float  *dx2=NULL, *dy2=NULL, *soln2=NULL;
  int    w2 = w/2, h2 = h/2;
  if (!Coarsest(w, h, mindim)) {
    dx2 = Allocate(w2, h2, dx_type);
    dy2 = Allocate(w2, h2, dy_type);
    soln2 = Allocate(w2, h2, soln_type);
```

```
    Restrict(dx2, dy2, w2, h2, dx, dy, NULL, NULL, soln, w, h);
    Zero(soln2, w2, h2);
    Ufmg(soln2, dx2, dy2, w2, h2, numit, mindim);
    ProlongAndAccumulate(soln, w, h, soln2, w2, h2, NULL, NULL);
  }
  /* perform V-cycle multigrid on fine grid */
  Umv(soln, dx, dy, w, h, numit, mindim);
}

/* V-cycle function for unweighted multigrid phase unwrapping. */
/* (Called recursively.)                                        */
void Umv(float *soln, float *dx, float *dy,
         int w, int h, int numit, int mindim)
{
  float *dx2=NULL, *dy2=NULL, *soln2=NULL;
  int    w2 = w/2, h2 = h/2;
  if (!Coarsest(w, h, mindim)) {
    Relax(soln, dx, dy, NULL, NULL, w, h, numit);
    dx2 = Allocate(w2, h2, dx_type);
    dy2 = Allocate(w2, h2, dy_type);
    soln2 = Allocate(w2, h2, soln_type);
    Restrict(dx2, dy2, w2, h2, dx, dy, NULL, NULL, soln, w, h);
    Zero(soln2, w2, h2);
    Umv(soln2, dx2, dy2, w2, h2, numit, mindim);
    ProlongAndAccumulate(soln, w, h, soln2, w2, h2, NULL, NULL);
    Relax(soln, dx, dy, NULL, NULL, w, h, numit);
  }
  else { /* coarsest */
    Relax(soln, dx, dy, NULL, NULL, w, h, 2*w*h);
  }
}
```

The header file is `ungrid.h`:

```
#ifndef __UNGRID
#define __UNGRID
void Ufmg(float *soln, float *dx, float *dy, int w, int h,
          int numit, int coarsest_dim);
void Umv(float *soln, float *dx, float *dy, int w, int h,
         int numit, int coarsest_dim);
#endif
```

The key multigrid operations of restriction and prolongation are in `gridops.c`. `ProlongAndAccumulate` performs the prolongation operations on the coarse grids and adds the prolonged results to the fine grids. `Restrict` performs the restriction operations from the fine to the coarse grids. Both functions are coded to accept (optional) weights so that the software may be re-used for the weighted multigrid algorithm of Section A.12. The file also contains the "housekeeping" functions `Zero` and `Coarsest` for array initialization and detection of the coarsest grid.

```
/*
 * gridops.c - prolongation and restriction operators for
 *             multigrid phase unwrapping.  Also contains
 *             functions for detecting coarsest array size
 *             and initializing an array to zero.
 */
#include <stdio.h>
#include <math.h>
#include "gridops.h"
#define MIN(x,y)  (((x) < (y)) ? (x) : (y))

/* Multigrid prolongation operator: weighted or unweighted. */
/* Resamples coarse grid values and ADDS to fine grid.      */
/* (Note: for unweighted case, pass null pointers for       */
/* coarse_dxwts and coarse_dywts.)                          */
void ProlongAndAccumulate(float *fine, int wf, int hf,
                          float *coarse, int wc, int hc,
                          float *coarse_dxwts, float *coarse_dywts)
{
  int    i, j, k, k1, k2, k3, k4, a, b, c;
  float  x1, x2, x3, x4, y1, y2, y3, y4, w1, w2, w3, w4, w;
  float  w1x, w2x, w3x, w4x, w1y, w2y, w3y, w4y;

  for (j=0, k=0; j<hc; j++) {
    for (i=0; i<wc; i++, k++) {
      a = 2*i;
      b = 2*j;
      c = b*wf + a;
      k1 = k;
      k2 = (i < wc - 1) ? k1 + 1 : k1;
      k3 = (j < hc - 1) ? k1 + wc : k1;
      k4 = (i < wc - 1) ? ((j < hc - 1) ? k1 + wc + 1 : k2) : k3;
      x1 = coarse[k1];
      x2 = coarse[k2];
      x3 = coarse[k3];
      x4 = coarse[k4];
      if (coarse_dxwts && coarse_dywts) {
        w1x = coarse_dxwts[k1];
        w2x = coarse_dxwts[k2];
```

```
      w3x = coarse_dxwts[k3];
      w4x = coarse_dxwts[k4];
      w1y = coarse_dywts[k1];
      w2y = coarse_dywts[k2];
      w3y = coarse_dywts[k3];
      w4y = coarse_dywts[k4];
      w1 = MIN(w1x, w1y);
      w2 = MIN(w2x, w2y);
      w3 = MIN(w3x, w3y);
      w4 = MIN(w4x, w4y);
      y1 = x1;
      w = w1 + w2;
      y2 = (w > 0.0) ? (w1*x1 + w2*x2)/w : 0.5*(x1 + x2);
      w = w1 + w3;
      y3 = (w > 0.0) ? (w1*x1 + w3*x3)/w : 0.5*(x1 + x3);
      w = w1 + w2 + w3 + w4;
      y4 = (w > 0.0) ? (w1*x1 + w2*x2 + w3*x3 + w4*x4)/w
                              : 0.25*(x1 + x2 + x3 + x4);
    }
    else {
      y1 = x1;                   y2 = 0.5*(x1 + x2);
      y3 = 0.5*(x1 + x3);        y4 = 0.25*(x1 + x2 + x3 + x4);
    }
    fine[c] += y1;
    fine[c + 1] += y2;
    fine[c + wf] += y3;
    fine[c + wf + 1] += y4;
    /* what if wf is odd? fill in extra columns */
    if (i==wc - 1) {
      b = 2*j;
      for (a=2*wc; a<wf; a++) {
        c = b*wf + a;
        fine[c] += y2;
        fine[c + wf] += y4;
      }
    }
    /* what if hf is odd? fill in extra rows */
    if (j==hc - 1) {
      a = 2*i;
      for (b=2*hc; b<hf; b++) {
        c = b*wf + a;
        fine[c] += y3;
        fine[c + 1] += y4;
      }
    }
```

```
      /* fill in lower right-hand corner */
      if (i==wc - 1 && j==hc - 1) {
        for (b=2*hc; b<hf; b++) {
          for (a=2*wc; a<wf; a++) {
            c = b*wf + a;
            fine[c] += y4;
          }
        }
      }
    }
  }
}

/* Multigrid restriction operator: weighted or unweighted. */
/* Restricts residuals of derivatives to coarser grid.     */
/* (Note: for unweighted case, pass null pointers for      */
/* dxwts and dywts.)                                        */
void Restrict(float *dx2, float *dy2, int wc, int hc,
              float *dx, float *dy, float *dxwts, float *dywts,
              float *soln, int wf, int hf)
{
  int     a, b, i, j, k;
  int     k1, k2, k3, k4, k5, k6, k7, k8, k9;
  int     k1x, k2x, k3x, k4x, k5x, k6x, k7x, k8x, k9x;
  int     k1y, k2y, k3y, k4y, k5y, k6y, k7y, k8y, k9y;
  float   scale;
  double  f1, f2, f3, f4, f5, f6, f7, f8, f9, wmult;
  double  w1, w2, w3, w4, w5, w6, w7, w8, w9;
  w1 = w2 = w3 = w4 = w5 = w6 = w7 = w8 = w9 = 1.0;

  /*  DX  */
  scale = (wf - 1.0)/(wc - 1.0);
  for (j=0; j<hc; j++) {
    b = 2*j;
    for (i=0; i<wc-1; i++) {   /* note: a < wf - 1 */
      a = 2*i;
      k = b*wf + a;
      /*  Indexes:    */
      /*  k1  k2  k3  */
      /*  k4  k5  k6  */
      /*  k7  k8  k9  */
      k5 = k;
      k4 = (a > 0) ? k5 - 1 : k5 + 1;
      k6 = (a < wf - 1) ? k5 + 1 : k5 - 1;
      k2 = (b > 0) ? k5 - wf : k5 + wf;
```

```
      k8 = (b < hf - 1) ? k5 + wf : k5 - wf;
      k1 = (a > 0) ? k2 - 1 : k2 + 1;
      k3 = (a < wf - 1) ? k2 + 1 : k2 - 1;
      k7 = (a > 0) ? k8 - 1 : k8 + 1;
      k9 = (a < wf - 1) ? k8 + 1 : k8 - 1;
      k1x = k2;         k4x = k5;         k7x = k8;
      k2x = k3;         k5x = k6;         k8x = k9;
      k3x = k3 + 1;     k6x = k6 + 1;     k9x = k9 + 1;
      if (dxwts && dywts) {
        /* weights: compute as MIN(k, k-1) */
        if ((w1 = dxwts[k1x]) > dxwts[k1]) w1 = dxwts[k1];
        if ((w2 = dxwts[k2x]) > dxwts[k2]) w2 = dxwts[k2];
        if ((w3 = dxwts[k3x]) > dxwts[k3]) w3 = dxwts[k3];
        if ((w4 = dxwts[k4x]) > dxwts[k4]) w4 = dxwts[k4];
        if ((w5 = dxwts[k5x]) > dxwts[k5]) w5 = dxwts[k5];
        if ((w6 = dxwts[k6x]) > dxwts[k6]) w6 = dxwts[k6];
        if ((w7 = dxwts[k7x]) > dxwts[k7]) w7 = dxwts[k7];
        if ((w8 = dxwts[k8x]) > dxwts[k8]) w8 = dxwts[k8];
        if ((w9 = dxwts[k9x]) > dxwts[k9]) w9 = dxwts[k9];
        wmult = 0.25*w5 + 0.125*(w2 + w8 + w4 + w6)
                          + 0.0625*(w1 + w3 + w7 + w9);
      }
      /* dx residuals */
      f1 = w1*(dx[k1] - (soln[k1x] - soln[k1]));
      f2 = w2*(dx[k2] - (soln[k2x] - soln[k2]));
      f3 = w3*(dx[k3] - (soln[k3x] - soln[k3]));
      f4 = w4*(dx[k4] - (soln[k4x] - soln[k4]));
      f5 = w5*(dx[k5] - (soln[k5x] - soln[k5]));
      f6 = w6*(dx[k6] - (soln[k6x] - soln[k6]));
      f7 = w7*(dx[k7] - (soln[k7x] - soln[k7]));
      f8 = w8*(dx[k8] - (soln[k8x] - soln[k8]));
      f9 = w9*(dx[k9] - (soln[k9x] - soln[k9]));
      if (dxwts && dywts) {
        if (wmult > 1.0e-6) dx2[j*wc + i] = scale*(0.25*f5
                 +  0.125*(f4 + f6 + f2 + f8)
                        +  0.0625*(f1 + f3 + f7 + f9))/wmult;
        else dx2[j*wc + i] = dx[k];
      }
      else {
        dx2[j*wc + i] = scale*(0.25*f5 + 0.125*(f4 + f6 + f2 + f8)
                                   +  0.0625*(f1 + f3 + f7 + f9));
      }
    }
  }
  /* correct dx at boundary */
```

```
for (j=0; j<hc; j++) {
  dx2[j*wc + (wc - 1)] = -dx2[j*wc + (wc - 2)];
}

/*  DY  */
scale = (hf - 1.0)/(hc - 1.0);
for (j=1; j<hc; j++) {   /* note: b > 0 */
  b = 2*j;
  for (i=0; i<wc; i++) {
    a = 2*i;
    k = b*wf + a;
    /*  Indexes:    */
    /*  k1  k2  k3  */
    /*  k4  k5  k6  */
    /*  k7  k8  k9  */
    k5 = k;
    k4 = (a > 0) ? k5 - 1 : k5 + 1;
    k6 = (a < wf - 1) ? k5 + 1 : k5 - 1;
    k2 = (b > 0) ? k5 - wf : k5 + wf;
    k8 = (b < hf - 1) ? k5 + wf : k5 - wf;
    k1 = (a > 0) ? k2 - 1 : k2 + 1;
    k3 = (a < wf - 1) ? k2 + 1 : k2 - 1;
    k7 = (a > 0) ? k8 - 1 : k8 + 1;
    k9 = (a < wf - 1) ? k8 + 1 : k8 - 1;
    k1y = k4;          k2y = k5;          k3y = k6;
    k4y = k7;          k5y = k8;          k6y = k9;
    k7y = k7 + wf;     k8y = k8 + wf;     k9y = k9 + wf;
    if (dxwts && dywts) {
      /* weights: compute as MIN(k, k-1) */
      if ((w1 = dywts[k1y]) > dywts[k1]) w1 = dywts[k1];
      if ((w2 = dywts[k2y]) > dywts[k2]) w2 = dywts[k2];
      if ((w3 = dywts[k3y]) > dywts[k3]) w3 = dywts[k3];
      if ((w4 = dywts[k4y]) > dywts[k4]) w4 = dywts[k4];
      if ((w5 = dywts[k5y]) > dywts[k5]) w5 = dywts[k5];
      if ((w6 = dywts[k6y]) > dywts[k6]) w6 = dywts[k6];
      if ((w7 = dywts[k7y]) > dywts[k7]) w7 = dywts[k7];
      if ((w8 = dywts[k8y]) > dywts[k8]) w8 = dywts[k8];
      if ((w9 = dywts[k9y]) > dywts[k9]) w9 = dywts[k9];
      wmult = 0.25*w5 + 0.125*(w2 + w8 + w4 + w6)
                       + 0.0625*(w1 + w3 + w7 + w9);
    }
    /* dy residuals */
    f1 = w1*(dy[k1] - (soln[k1y] - soln[k1]));
    f2 = w2*(dy[k2] - (soln[k2y] - soln[k2]));
    f3 = w3*(dy[k3] - (soln[k3y] - soln[k3]));
```

```
      f4 = w4*(dy[k4] - (soln[k4y] - soln[k4]));
      f5 = w5*(dy[k5] - (soln[k5y] - soln[k5]));
      f6 = w6*(dy[k6] - (soln[k6y] - soln[k6]));
      f7 = w7*(dy[k7] - (soln[k7y] - soln[k7]));
      f8 = w8*(dy[k8] - (soln[k8y] - soln[k8]));
      f9 = w9*(dy[k9] - (soln[k9y] - soln[k9]));
      if (dxwts && dywts) {
        if (wmult > 1.0e-6) dy2[j*wc + i] = scale*(0.25*f5
                + 0.125*(f4 + f6 + f2 + f8)
                      + 0.0625*(f1 + f3 + f7 + f9))/wmult;
        else dy2[j*wc + i] = scale*dy[k];
      }
      else {
        dy2[j*wc + i] = scale*(0.25*f5 + 0.125*(f4 + f6 + f2 + f8)
                                + 0.0625*(f1 + f3 + f7 + f9));
      }
    }
  }
  /* correct dy at boundary */
  for (i=0; i<wc; i++) {
    dy2[(hc - 1)*wc + i] = -dy2[(hc - 2)*wc + i];
  }
}

/* returns 1 if w or h is less than 3 */
int Coarsest(int w, int h, int mindim)
{
  if (mindim==0)
    return (w/2 < 3 || h/2 < 3);
  else
    return (w/2 < mindim || h/2 < mindim);
}

/* initialize array to zeroes */
void Zero(float *soln, int w, int h)
{
  int n=w*h, k;
  for (k=0; k<n; k++) {
    soln[k] = 0.0;
  }
}
```

The header file is gridops.h:

```
#ifndef __GRIDOPS
#define __GRIDOPS
```

```
void Restrict(float *dx2, float *dy2, int wc, int hc,
              float *dx, float *dy, float *dxwts,
              float *dywts, float *soln, int wf, int hf);
void ProlongAndAccumulate(float *fine, int wf, int hf,
                          float *coarse, int wc, int hc,
                          float *coarse_dxwts, float *coarse_dywts);
int Coarsest(int w, int h, int coarsest_dim);
void Zero(float *soln, int w, int h);
#endif
```

The dynamic memory management routines for the multigrid programs are found in `gridmem.c`. These functions allocate (`Allocate`) and free (`FreeAll`) the memory for the various pyramids of grids. Because the functions are called by the weighted multigrid algorithm of Section A.12, they also allocate and free the weighting arrays.

```
/*
 * gridmem.c - dynamic memory management functions
 *             for multigrid phase unwrapping
 */
#include <stdio.h>
#include <math.h>
#include "gridmem.h"

static int sizes[NUM_TYPES][MAX_ARRAYS];
static int num[NUM_TYPES];
static float *arrays[NUM_TYPES][MAX_ARRAYS];

/* allocate array: if already allocated, return pointer to it */
float *Allocate(int w, int h, ArrayType type)
{
  int size = w*h, i, j, k;
  k = (int) type;
  for (i=0; i<num[k]; i++) {
    if (sizes[k][i]==size)
      return arrays[k][i];
  }
  sizes[k][num[k]] = size;
  arrays[k][num[k]] = (float *) malloc(size*sizeof(float));
  if (!arrays[k][num[k]]) {
    printf("Error: cannot allocate memory (%d bytes)\n", size*4);
    return 0;  /* error */
  }
  return arrays[k][num[k]++];
}
```

```
/* free all allocated arrays */
void FreeAll()
{
  int i, j;
  for (j=0; j<NUM_TYPES; j++) {
    for (i=0; i<num[j]; i++) {
      free(arrays[j][i]);
    }
  }
}
```

The header file is gridmem.h:

```
#ifndef __GRIDMEM
#define __GRIDMEM
#define MAX_ARRAYS   100
#define NUM_TYPES      5
typedef enum {
  dx_type, dy_type, soln_type, dxwts_type, dywts_type
} ArrayType;
float *Allocate(int w, int h, ArrayType type);
void FreeAll();
#endif
```

The Gauss-Seidel iterations are performed by the function Relax in file relax.c. Because this function accepts (optional) weights, it is also called by the weighted multigrid functions of Section A.12.

```
/*
 * relax.c - function for solving weighted or unweighted
 *           version of Poisson equation (i.e., weighted
 *           or unweighted least-squares phase unwrapping)
 *           by means of red-black Gauss-Seidel relaxation
 */
#include <stdio.h>
#include <math.h>
#include "relax.h"

/*
 * Black-red Gauss-Seidel relaxation: assumes rho(i,j)
 *        = 0.25*(dx(i,j) - dx(i+1,j) + dy(i,j) - dy(i,j+1)
 * Thus, f(i,j)
 *    = 0.25*(f(i-1,j) + f(i+1,j) + f(i,j-1) + f(i,j+1)) + rho(i,j)
 * This version solves the weighted equation.  For the unweighted
 * equation, just pass null pointers for dxwts and dywts.
 */
```

```
void Relax(float *soln, float *dx, float *dy, float *dxwts,
           float *dywts, int w, int h, int numit)
{
  int i, j, k, n, ipass, istart;
  float  x1, x2, x3, x4, y, z, r;
  float  w1, w2, w3, w4, norm=0.0;
  double  diff, maxdiff, sum;

  if (w*h < 8*8 && numit < 2) numit = 2*(w + h);
  for (n=0; n<numit; n++) {                    /* iteration */
    for (ipass=0; ipass<2; ipass++) {   /* red and black sweeps */
      istart = ipass;
      sum = maxdiff = 0.0;
      for (j=0; j<h; j++) {                    /* row */
        for (i=istart; i<w; i+=2) {      /* column */
          k = j*w + i;
          /* weights */
          if (dxwts && dywts) {
            w1 = (i < w - 1) ? dxwts[k+1] : dxwts[k-1];
            if (w1 > dxwts[k]) w1 = dxwts[k];
            w1 = w1*w1;
            w2 = (i > 0) ? dxwts[k-1] : dxwts[k+1];
            if (w2 > dxwts[k]) w2 = dxwts[k];
            w2 = w2*w2;
            w3 = (j < h - 1) ? dywts[k+w] : dywts[k-w];
            if (w3 > dywts[k]) w3 = dywts[k];
            w3 = w3*w3;
            w4 = (j > 0) ? dywts[k-w] : dywts[k+w];
            if (w4 > dywts[k]) w4 = dywts[k];
            w4 = w4*w4;
            norm = w1 + w2 + w3 + w4;
          }
          /* surface points */
          x1 = (i < w - 1) ? soln[k+1] : soln[k-1];
          x2 = (i > 0) ? soln[k-1] : soln[k+1];
          x3 = (j < h - 1) ? soln[k+w] : soln[k-w];
          x4 = (j > 0) ? soln[k-w] : soln[k+w];
          /* derivatives */
          y = (i > 0) ? dx[k-1] : -dx[k];
          z = (j > 0) ? dy[k-w] : -dy[k];
          if (norm > 1.0e-6) {
            r = (w1*(x1 - dx[k]) + w2*(x2 + y)
                        + w3*(x3 - dy[k]) + w4*(x4 + z))/norm;
          }
          else {
```

```
          r = 0.25*(x1 - dx[k] + x2 + y + x3 - dy[k] + x4 + z);
        }
        diff = r - soln[k];
        if (diff < 0) diff = -diff;
        maxdiff = (maxdiff < diff) ? diff : maxdiff;
        sum += diff*diff;
        soln[k] = r;
      }
      istart = (istart) ? 0 : 1;
    }
  }
}
for (n=1; n<w; n*=2) printf("  ");
printf("Relax %d %d: rms = %lf\n", w, h, sqrt(sum/(w*h)));
}
```

The header file is relax.h:

```
#ifndef __RELAX
#define __RELAX
void Relax(float *soln, float *dx, float *dy, float *dxwts,
           float *dywts, int w, int h, int numit);
#endif
```

A.11 PCG ALGORITHM

This section contains the following files: mainpcg.c, pcg.c, and pcg.h.

We now present the program and functions for the PCG phase unwrapping algorithm of Section 5.4.3. The main program is in file mainpcg.c. The usage statement in the program listing below explains how to run the program. It is similar to those of the phase-unwrapping programs of the preceding sections. The maximum number of iterations (-iter) and the convergence tolerance (-converge) are optional parameters. If omitted, they default to the values 20 and 0.0, respectively, which will cause the PCG algorithm to iterate exactly 20 times. As discussed in Section 5.4.3, we recommend that the -thresh yes argument be specified, and that the resulting quality mask be fattened by one or two pixels (-fat 1 or -fat 2), because the PCG algorithm seems to work best with binary weights whose zero-weights have been thickened slightly. A typical command to execute the program would be

```
pcg -input mri.phase -format byte -output mri.surf -bmask
    mri.bmask -xsize 256 -ysize 256 -mode min_var -tsize 3
    -thresh yes -fat 2 -debug yes -iter 20
```

where pcg is the name of the executable file.

The program works as follows: The command line arguments are parsed by the functions `CommandLineParm`, `Keyword`, and `SetQualityMode`. The files are opened, the memory for the arrays is allocated, and the phase data, the quality map, and the (optional) border mask data are then read from the input files and processed. (The border mask, not to be confused with the quality map or mask, consists of pixels that should be ignored, such as the border pixels of the simulated IFSAR data shown in Figure 3.8b.) The border mask is fattened slightly by a call to the function `FattenMask`, the quality map may be thresholded to produce a quality mask, and the zero-weights of the quality mask may be fattened if the `-fat` argument is specified. There are no power-of-two restrictions on the array dimensions `xsize` and `ysize`, but if they are not of the form $2^n + 1$, then the arrays are embedded in larger arrays and zero padded. The dimensions of the larger arrays, `xsize_dct` and `ysize_dct`, are of the form $2^n + 1$. After additional arrays for the PCG algorithm are allocated, the phase Laplacian is computed by function `ComputeLaplacian`, and the function `PCGUnwrap` is called to perform the iterations of the PCG algorithm. The unwrapped results are extracted from the larger array in which they are embedded, made congruent to the wrapped input phase by the function `CongruentSoln`, and written to the output file.

```
/*
 * mainpcg.c -- phase unwrapping by means of PCG algorithm
 *
 * Source code files required:
 *      congruen.c          dct.c    dxdygrad.c      extract.c
 *       getqual.c         grad.c       histo.c      laplace.c
 *       mainpcg.c      maskfat.c         pcg.c     qualgrad.c
 *      qualpseu.c      qualvar.c     solncos.c         util.c
 */
#include <stdio.h>
#include <math.h>
#include "file.h"
#include "histo.h"
#include "pi.h"
#include "getqual.h"
#include "congruen.h"
#include "maskfat.h"
#include "util.h"
#include "extract.h"
#include "pcg.h"
#include "laplace.h"
#define BORDER    0x20
#define DEFAULT_NUM_ITER  20

main (int argc, char *argv[])
```

```
{
  int           i, j, k, n;
  FILE          *ifp, *ofp, *mfp=0, *qfp=0;
  float         *phase;     /* array */
  float         *soln;      /* array */
  float         *qual_map;  /* array */
  float         *rarray;  /* array */
  float         *parray;  /* array */
  float         *zarray;  /* array */
  unsigned char *bitflags;
  double        *xcos, *ycos;
  char          buffer[200], tempstr[200];
  char          infile[200], outfile[200];
  char          bmaskfile[200], qualfile[200];
  char          format[200], modekey[200];
  int           in_format, debug_flag;
  int           xsize, ysize;   /* dimensions of arrays */
  int           xsize_actual, ysize_actual;
  int           xsize_dct, ysize_dct;
  int           avoid_code, thresh_flag, fatten;
  int           tsize, num_iter=DEFAULT_NUM_ITER;
  double        rmin, rmax, rscale, epsi_con;
  double        one_over_twopi = 1.0/TWOPI;
  UnwrapMode    mode;
  char          use[] =      /* define usage statement */
   "Usage: prog-name -input file -format fkey -output file\n"
   "  -xsize x -ysize y [ -mode mkey -bmask file -corr file\n"
   "  -tsize size -debug yes/no -iter num -converge epsi\n"
   "  -thresh yes/no -fat n ]\n"
   "where 'fkey' is a keyword designating the input file type\n"
   "(key = complex8, complex4, float or byte), 'x' and 'y' are\n"
   "the dimensions of the file, bmask is an optional byte-file\n"
   "of masks for masking out undefined phase values, corr\n"
   "is an optional byte-file of cross-correlation values,\n"
   "tsize is the size of the square template for averaging the\n"
   "corr file or quality values (default = 1), and 'mkey' is\n"
   "a keyword designating the source of the quality values\n"
   "that guide the unwrapping path.  The value of mkey may be\n"
   "'min_grad' for Minimum Gradient unwrapping, 'min_var' for\n"
   "Minimum Variance unwrapping, 'max_corr' for Maximum\n"
   "Correlation unwrapping, or 'max_pseu' for Maximum\n"
   "Pseudocorrelation unwrapping.  All files are simple\n"
   "raster files, and the output file consists of floating\n"
   "point numbers that define the heights of the unwrapped\n"
   "surface.  If the 'debug' parm is 'yes', then intermediate\n"
```

```
  "byte-files are saved (quality map, etc.)  The maximum number\n"
  "of iterations is given by 'num' (default 20), and the con-\n"
  "vergence tolerance is given by 'epsi' (default = 0).  To\n"
  "apply an automatic threshold to the quality map to make\n"
  "a quality mask, the 'thresh' parm should be yes.  To\n"
  "thicken the quality mask, the 'fat' parm should be the\n"
  "number of pixels by which to thicken.\n";

printf("Phase Unwrapping by PCG algorithm\n");

/* GET COMMAND LINE PARAMETERS AND CHECK */
CommandLineParm(argc,argv, "-input", StringParm, infile, 1, use);
CommandLineParm(argc,argv, "-format", StringParm, format, 1, use);
CommandLineParm(argc,argv, "-output", StringParm, outfile, 1,use);
CommandLineParm(argc,argv, "-xsize", IntegerParm, &xsize, 1, use);
CommandLineParm(argc,argv, "-ysize", IntegerParm, &ysize, 1, use);
if (!CommandLineParm(argc, argv, "-mode", StringParm,
      modekey, 0, use)) strcpy(modekey, "none");
if (!CommandLineParm(argc, argv, "-bmask", StringParm,
      bmaskfile, 0, use)) strcpy(bmaskfile, "none");
if (!CommandLineParm(argc, argv, "-corr", StringParm,
      qualfile, 0, use)) strcpy(qualfile, "none");
if (!CommandLineParm(argc, argv, "-tsize", IntegerParm,
      &tsize, 0, use)) tsize = 1;
if (!CommandLineParm(argc, argv, "-debug", StringParm,
      tempstr, 0, use)) debug_flag = 0;
else debug_flag = Keyword(tempstr, "yes");
CommandLineParm(argc, argv, "-iter", IntegerParm,
                &num_iter, 0, use);
if (num_iter < 0) num_iter = 1;
if (!CommandLineParm(argc, argv, "-toler", DoubleParm,
      &epsi_con, 0, use)) epsi_con = 0.0;
if (!CommandLineParm(argc, argv, "-thresh", StringParm,
      tempstr, 0, use)) thresh_flag = 0;
else thresh_flag = Keyword(tempstr, "yes");
if (!CommandLineParm(argc, argv, "-fat", IntegerParm,
      &fatten, 0, use)) fatten = 0;

if (Keyword(format, "complex8"))  in_format = 0;
else if (Keyword(format, "complex4"))  in_format = 1;
else if (Keyword(format, "byte"))  in_format = 2;
else if (Keyword(format, "float"))  in_format = 3;
else {
  fprintf(stderr, "Unrecognized format: %s\n", format);
  exit(BAD_PARAMETER);
}
```

```
printf("Input file =  %s\n", infile);
printf("Input file type = %s\n", format);
printf("Output file =  %s\n", outfile);
printf("File dimensions = %dx%d (cols x rows).\n", xsize, ysize);

if (Keyword(bmaskfile, "none")) printf("No border mask file.\n");
else printf("Border mask file = %s\n", bmaskfile);
if (Keyword(qualfile, "none")) printf("No quality file.\n");
else printf("Correlation image file = %s\n", qualfile);

printf("Averaging template size = %d\n", tsize);
if (tsize < 0 || tsize > 30) {
  fprintf(stderr, "Illegal size: must be between 0 and 30\n");
  exit(BAD_PARAMETER);
}

printf("Quality mode = %s\n", modekey);
mode = SetQualityMode(modekey, qualfile, 0);
if (mode < 0) exit(BAD_PARAMETER);  /* error msg already printed */

/* Increase dimensions to power of two (plus one) */
xsize_actual = xsize;
ysize_actual = ysize;
for (xsize_dct = 1; xsize_dct + 1 < xsize_actual; xsize_dct*=2)
  ;
xsize_dct = xsize_dct + 1;
for (ysize_dct = 1; ysize_dct + 1 < ysize_actual; ysize_dct*=2)
  ;
ysize_dct = ysize_dct + 1;
if (xsize_dct != xsize_actual || ysize_dct != ysize_actual) {
  printf("Dim's increased from %dx%d to %dx%d for FFT/DCT's\n",
         xsize_actual, ysize_actual, xsize_dct, ysize_dct);
}

/*  OPEN FILES, ALLOCATE MEMORY   */
/* note: xsize_dct >= xsize;  ysize_dct >= ysize */
xsize = xsize_dct;
ysize = ysize_dct;
AllocateFloat(&phase, xsize*ysize, "phase data");
AllocateFloat(&soln, xsize*ysize, "scratch data");
AllocateFloat(&qual_map, xsize*ysize, "quality map");
AllocateByte(&bitflags, xsize*ysize, "bitflags array");
OpenFile(&ifp, infile, "r");
OpenFile(&ofp, outfile, "w");
if (!Keyword(bmaskfile, "none")) OpenFile(&mfp, bmaskfile, "r");
```

```
if (mode==corr_coeffs) OpenFile(&qfp, qualfile, "r");

/*  READ AND PROCESS DATA  */
xsize = xsize_actual;
ysize = ysize_actual;
printf("Reading input data...\n");
GetPhase(in_format, ifp, infile, phase, xsize, ysize);

if (qfp) {
  printf("Reading quality data...\n");
  /* borrow the bitflags array temporarily */
  ReadByte(qfp, bitflags, xsize*ysize, qualfile);
  /* process data and store in quality map array */
  AverageByteToFloat(bitflags, qual_map, tsize, xsize, ysize);
}

/* border mask data */
printf("Processing border mask data...\n");
if (mfp) {
  ReadByte(mfp, bitflags, xsize*ysize, bmaskfile);
}
else {
  for (k=0; k<xsize*ysize; k++)
    bitflags[k] = 255;
}
for (k=0; k<xsize*ysize; k++) {
  bitflags[k] = (!bitflags[k]) ? BORDER : 0;
}
if (mfp) FattenMask(bitflags, BORDER, 1, xsize, ysize);

GetQualityMap(mode, qual_map, phase, bitflags, BORDER,
              tsize, xsize, ysize);

if (thresh_flag) {
  HistoAndThresh(qual_map, xsize, ysize, 0.0, 0, 0.0,
                 bitflags, BORDER);
  if (fatten > 0)
    FattenQual(qual_map, fatten, xsize, ysize);
}
/* Set border (masked) pixels to 0-wts and free bitflags */
for (k=0; k<xsize*ysize; k++) {
  if (bitflags[k] & BORDER) qual_map[k] = 0.0;
}
free(bitflags);   /* bitflags array no longer needed */
```

```
if (debug_flag) {
  char filename[300];
  sprintf(filename, "%s.qual", outfile);
  SaveFloatToImage(qual_map, "quality", filename, xsize, ysize,
                   0, 0, 0);
}

/* embed arrays in possibly larger FFT/DCT arrays */
for (j=ysize_dct-1; j>=0; j--) {
  for (i=xsize_dct-1; i>=0; i--) {
    if (i<xsize_actual && j<ysize_actual)
      phase[j*xsize_dct + i] = phase[j*xsize_actual + i];
    else phase[j*xsize_dct + i] = 0.0;
  }
}
if (qual_map) {
  for (j=ysize_dct-1; j>=0; j--) {
    for (i=xsize_dct-1; i>=0; i--) {
      if (i<xsize_actual && j<ysize_actual)
        qual_map[j*xsize_dct + i] = qual_map[j*xsize_actual + i];
      else
        qual_map[j*xsize_dct + i] = 0.0;
    }
  }
}

/* Set dimensions to DCT dimensions */
xsize = xsize_dct;
ysize = ysize_dct;
/* Allocate more memory */
AllocateFloat(&rarray, xsize*ysize, "r array data");
AllocateFloat(&zarray, xsize*ysize, "z array data");
AllocateFloat(&parray, xsize*ysize, "p array data");

/*  UNWRAP  */
printf("Unwrapping...\n");
for (k=0; k<xsize*ysize; k++)  soln[k] = 0.0;
ComputeLaplacian(phase, rarray, qual_map, NULL, xsize, ysize, 1);
PCGUnwrap(rarray, zarray, parray, soln, qual_map, NULL,
          xsize, ysize, num_iter, epsi_con);
printf("\nFinished\n");
PrintMinAndMax(xsize, ysize, soln, "solution");

/* restore dimensions to input sizes */
xsize = xsize_actual;
ysize = ysize_actual;
```

```
  /* extract results from enlarged arrays */
  for (j=0; j<ysize; j++) {
    for (i=0; i<xsize; i++) {
      soln[j*xsize + i] = soln[j*xsize_dct + i];
    }
  }
  for (j=0; j<ysize; j++) {
    for (i=0; i<xsize; i++) {
      qual_map[j*xsize + i] = qual_map[j*xsize_dct + i];
    }
  }
  for (j=0; j<ysize; j++) {
    for (i=0; i<xsize; i++) {
      phase[j*xsize + i] = phase[j*xsize_dct + i];
    }
  }
  /* make result congruent to input phase */
  CongruentSoln(soln, phase, qual_map, xsize, ysize);
  /* scale and save result */
  for (k=0; k<xsize*ysize; k++) soln[k] *= TWOPI;
  WriteFloat(ofp, soln, xsize*ysize, outfile);
  printf("Wrote congruent surface to file %s\n", outfile);
  free(qual_map);
  free(phase);
  free(soln);
  free(rarray);
  free(parray);
  free(zarray);
}
```

The bulk of the PCG algorithm is found in file `pcg.c`. The function `PCGUnwrap` controls the main loop, while the operations within the loop are performed by `PCGIterate`. The iterations continue until either the convergence tolerance (`epsi_con`) or the maximum number of iterations (`max_iter`) is achieved.

```
/*
 *  pcg.c -- function for Preconditioned Conjugate Gradient solution
 *           of weighted least-squares phase unwrapping problem
 */
#include <stdio.h>
#include <math.h>
#include "solncos.h"
#include "pcg.h"
#include "util.h"
#define SIMIN(x,y) (((x)*(x) < (y)*(y)) ? (x)*(x) : (y)*(y))
```

```
/* Main function of PCG algorithm for phase unwrapping.  If dywts */
/* is null, then the dx and dy weights are calculated from dxwts  */
void PCGUnwrap(float *rarray, float *zarray, float *parray,
               float *soln, float *dxwts, float *dywts, int xsize,
               int ysize, int max_iter, double epsi_con)
{
  int      i, j, k, iloop;
  double   sum, alpha, beta, beta_prev, epsi;
  double   *xcos, *ycos;
  for (k=0, sum=0.0; k<xsize*ysize; k++)
    sum += rarray[k]*rarray[k];
  sum = sqrt(sum/(xsize*ysize));
  AllocateDouble(&xcos, xsize, "cosine terms");
  AllocateDouble(&ycos, ysize, "cosine terms");
  for (iloop=0; iloop < max_iter; iloop++) {
    PCGIterate(rarray, zarray, parray, soln, dxwts, dywts,
               xsize, ysize, xcos, ycos, iloop, sum, &alpha,
               &beta, &beta_prev, &epsi);
    if (epsi < epsi_con) {
      printf("Breaking out of main loop (due to convergence)\n");
      break;
    }
  }
  free(xcos);
  free(ycos);
}

/* Main function within iterative loop of PCG algorithm. */
void PCGIterate(float *rarray, float *zarray, float *parray,
                float *soln, float *dxwts, float *dywts, int xsize,
                int ysize, double *xcos, double *ycos, int iloop,
                double sum0, double *alpha, double *beta,
                double *beta_prev, double *epsi)
{
  int      i, j, k;
  int      k1, k2, k3, k4;
  double   sum, w1, w2, w3, w4, btemp, delta, avg, scale;
  float    *wts;

  scale = 1.0/(xsize*ysize);
  /* remove constant bias from rarray */
  for (k=0, avg=0.0; k<xsize*ysize; k++)  avg += rarray[k];
  avg *= scale;
  for (k=0; k<xsize*ysize; k++)  rarray[k] -= avg;
  /* compute cosine transform solution of Laplacian in rarray */
```

```
for (k=0; k<xsize*ysize; k++)  {
  zarray[k] = rarray[k];
}
DirectSolnByCosineTransform(zarray, xsize, ysize, xcos, ycos);
/* calculate beta and parray */
for (k=0, *beta=0.0; k<xsize*ysize; k++) {
  *beta += rarray[k]*zarray[k];
}
printf("beta = %lf\n", *beta);
if (iloop == 0) {
  for (k=0; k<xsize*ysize; k++) {
    parray[k] = zarray[k];
  }
}
else {
  btemp = (*beta)/(*beta_prev);
  for (k=0; k<xsize*ysize; k++) {
    parray[k] = zarray[k] + btemp*parray[k];
  }
}
/* remove constant bias from parray */
for (k=0, avg=0.0; k<xsize*ysize; k++)  avg += parray[k];
avg *= scale;
for (k=0; k<xsize*ysize; k++)  parray[k] -= avg;
*beta_prev = *beta;
/* calculate Qp */
for (j=0; j<ysize; j++) {
  for (i=0; i<xsize; i++) {
    k = j*xsize + i;
    k1 = (i < xsize - 1) ? k + 1 : k - 1;
    k2 = (i > 0) ? k - 1 : k + 1;
    k3 = (j < ysize - 1) ? k + xsize : k - xsize;
    k4 = (j > 0) ? k - xsize : k + xsize;
    if (dxwts==NULL && dywts==NULL) {    /* unweighted */
      w1 = w2 = w3 = w4 = 1.0;
    }
    else if (dxwts==NULL || dywts==NULL) {
      /* one set of weights */
      wts = (dxwts) ? dxwts : dywts;
      w1 = SIMIN(wts[k], wts[k1]);
      w2 = SIMIN(wts[k], wts[k2]);
      w3 = SIMIN(wts[k], wts[k3]);
      w4 = SIMIN(wts[k], wts[k4]);
    }
    else {     /* two sets of weights */
```

```
        w1 = dxwts[k];
        w2 = (i>0) ? dxwts[k-1] : dxwts[k];
        w3 = dywts[k];
        w4 = (j>0) ? dywts[k-xsize] : dywts[k];
      }
      zarray[k] = (w1 + w2 + w3 + w4)*parray[k]
                   - (w1*parray[k1] + w2*parray[k2]
                         + w3*parray[k3] + w4*parray[k4]);
    }
  }
  /* calculate alpha */
  for (k=0, *alpha=0.0; k<xsize*ysize; k++) {
    *alpha += zarray[k]*parray[k];
  }
  *alpha = *beta/(*alpha);
  printf("alpha = %lf\n", *alpha);
  /* update rarray */
  for (k=0; k<xsize*ysize; k++) {
    rarray[k] -= (*alpha)*zarray[k];
  }
  /* update parray */
  for (k=0; k<xsize*ysize; k++) {
    soln[k] += (*alpha)*parray[k];
  }
  /* remove constant bias from soln */
  for (k=0, avg=0.0; k<xsize*ysize; k++)  avg += soln[k];
  avg *= scale;
  for (k=0; k<xsize*ysize; k++)  soln[k] -= avg;
  /* compute epsi and delta */
  for (k=0, sum=0.0; k<xsize*ysize; k++) {
    sum += rarray[k]*rarray[k];
  }
  *epsi = sqrt(sum/(xsize*ysize))/sum0;
  for (k=0, sum=0.0; k<xsize*ysize; k++) {
    sum += (*alpha)*(*alpha)*parray[k]*parray[k];
  }
  delta = sqrt(sum/(xsize*ysize));
  printf("ITER %d: EPSI %lf DELTA %lf\n", iloop, *epsi, delta);
}
```

The header file is pcg.h:

```
#ifndef __PCG
#define __PCG
void PCGUnwrap(float *rarray, float *zarray, float *parray,
```

```
               float *soln, float *dxwts, float *dywts, int xsize,
               int ysize, int max_iter, double epsi_con);
void PCGIterate(float *rarray, float *zarray, float *parray,
                float *soln, float *dxwts, float *dywts, int xsize,
                int ysize, double *xcos, double *ycos, int iloop,
                double sum0, double *alpha, double *beta,
                double *beta_prev, double *epsi);
#endif
```

A.12 WEIGHTED MULTIGRID ALGORITHM

This section contains the following files: `mainfmg.c`, `fmg.c`, `fmg.h`, `grid.c`, and `grid.h`.

We can now present the software for the weighted multigrid algorithm of Section 5.4.4. The main program is in file `mainfmg.c`. The usage is similar to that of the PCG algorithm of Section A.11. The maximum number of multigrid cycles (`-cycles`) and the number of Gauss-Seidel iterations (`-iter`) can be specified. If omitted, they each default to the value two. We recommend that the `-thresh yes` argument be specified, because the algorithm seems to work best with binary weights. In general, there seems to be no need to thicken the zero-weights. A typical command to execute the software would be

```
fmg -input mri.phase -format byte -output mri.surf -bmask
    mri.bmask -xsize 256 -ysize 256 -mode min_var -tsize 3
    -thresh yes -fat 0 -debug yes -cycles 2 -iter 2
```

where `fmg` is the name of the executable file.

The program works as follows: The command line arguments are parsed by the functions `CommandLineParm`, `Keyword`, and `SetQualityMode`. The files are opened, the memory for the arrays is allocated, and the phase data, the quality map, and the (optional) border mask data are then read from the input files and processed. The border mask is fattened slightly by a call to the function `FattenMask` and the quality map may be thresholded to produce a quality mask. (The border mask, not to be confused with the quality map or mask, consists of pixels that should be ignored, such as the border pixels of the simulated IFSAR data shown in Figure 3.8b.) There is no power-of-two restriction on the array dimensions. After additional arrays for the phase derivatives are allocated, the function `MultigridUnwrap` is called to execute the multigrid algorithm. After the unwrapped solution is made congruent to the wrapped input phase by the function `CongruentSoln`, the results are written to the output file.

```
/*
 * mainfmg.c -- phase unwrapping by means of multigrid algorithm
 *
 * Source code files required:
 *     congruen.c    dxdygrad.c     extract.c         fmg.c
 *      getqual.c        grad.c        grid.c     gridmem.c
 *      gridops.c       histo.c     mainfmg.c     maskfat.c
 *     qualgrad.c    qualpseu.c     qualvar.c       relax.c
 *         util.c
 */
#include <stdio.h>
#include <math.h>
#include "file.h"
#include "histo.h"
#include "pi.h"
#include "getqual.h"
#include "congruen.h"
#include "maskfat.h"
#include "util.h"
#include "extract.h"
#include "fmg.h"
#define BORDER             0x20
#define DEFAULT_NUM_ITER    2
#define DEFAULT_NUM_CYCLES  2

main (int argc, char *argv[])
{
  int            i, j, k, n;
  FILE           *ifp, *ofp, *mfp=0, *qfp=0;
  float          *phase;     /* array */
  float          *soln;      /* array */
  float          *qual_map;  /* array */
  float          *dx;        /* array */
  float          *dy;        /* array */
  unsigned char  *bitflags;
  char           buffer[200], tempstr[200];
  char           infile[200], outfile[200];
  char           bmaskfile[200], qualfile[200];
  char           format[200], modekey[200];
  int            in_format, debug_flag;
  int            xsize, ysize;   /* dimensions of arrays */
  int            avoid_code, thresh_flag, fatten, tsize;
  int            num_iter=DEFAULT_NUM_ITER;
  int            num_cycles=DEFAULT_NUM_CYCLES;
  double         rmin, rmax, rscale, one_over_twopi = 1.0/TWOPI;
```

```
UnwrapMode      mode;
char            use[] =   /* define usage statement */
  "Usage: prog-name -input file -format fkey -output file\n"
  "  -xsize x -ysize y [ -mode mkey -bmask file -corr file\n"
  "  -tsize size -debug yes/no -cycles numc -iter num\n"
  "  -thresh yes/no -fat n ]\n"
  "where 'fkey' is a keyword designating the input file type\n"
  "(key = complex8, complex4, float or byte), 'x' and 'y' are\n"
  "the dimensions of the file, bmask is an optional byte-file\n"
  "of masks for masking out undefined phase values, corr\n"
  "is an optional byte-file of cross-correlation values,\n"
  "tsize is the size of the square template for averaging the\n"
  "corr file or quality values (default = 1), and 'mkey' is\n"
  "a keyword designating the source of the quality values\n"
  "that guide the unwrapping path.  The value of mkey may be\n"
  "'min_grad' for Minimum Gradient unwrapping, 'min_var' for\n"
  "Minimum Variance unwrapping, 'max_corr' for Maximum\n"
  "Correlation unwrapping, or 'max_pseu' for Maximum\n"
  "Pseudocorrelation unwrapping.  All files are simple\n"
  "raster files, and the output file consists of floating\n"
  "point numbers that define the heights of the unwrapped\n"
  "surface.  If the 'debug' parm is 'yes', then intermediate\n"
  "byte-files are saved (quality map, etc.)  The number of\n"
  "multigrid cycles is given by 'numc'.  The number of Gauss-\n"
  "Seidel iterations is given by 'num'.  To apply an auto-\n"
  "matic threshold to the quality map in order to produce\n"
  "a quality mask, the 'thresh' parm should be yes.  To\n"
  "thicken the quality mask, the 'fat' parm should be the\n"
  "number of pixels by which to thicken.\n";

printf("Phase Unwrapping by Weighted Multigrid Algorithm\n");

/* GET COMMAND LINE PARAMETERS AND CHECK */
CommandLineParm(argc,argv, "-input", StringParm, infile, 1, use);
CommandLineParm(argc,argv, "-format", StringParm, format, 1, use);
CommandLineParm(argc,argv, "-output", StringParm, outfile, 1,use);
CommandLineParm(argc,argv, "-xsize", IntegerParm, &xsize, 1, use);
CommandLineParm(argc,argv, "-ysize", IntegerParm, &ysize, 1, use);
if (!CommandLineParm(argc, argv, "-mode", StringParm, modekey,
      0, use)) strcpy(modekey, "none");
if (!CommandLineParm(argc, argv, "-bmask", StringParm, bmaskfile,
      0, use)) strcpy(bmaskfile, "none");
if (!CommandLineParm(argc, argv, "-corr", StringParm, qualfile,
      0, use)) strcpy(qualfile, "none");
if (!CommandLineParm(argc, argv, "-tsize", IntegerParm, &tsize,
```

```
      0, use)) tsize = 1;
if (!CommandLineParm(argc, argv, "-debug", StringParm, tempstr,
      0, use)) debug_flag = 0;
else debug_flag = Keyword(tempstr, "yes");
CommandLineParm(argc, argv, "-iter", IntegerParm, &num_iter,
                0, use);
if (num_iter < 0) num_iter = 1;
CommandLineParm(argc, argv, "-cycles", IntegerParm, &num_cycles,
                0, use);
if (num_cycles < 0) num_cycles = 1;
if (!CommandLineParm(argc, argv, "-thresh", StringParm, tempstr,
      0, use)) thresh_flag = 0;
else thresh_flag = Keyword(tempstr, "yes");
if (!CommandLineParm(argc, argv, "-fat", IntegerParm, &fatten,
      0, use)) fatten = 0;

if (Keyword(format, "complex8"))  in_format = 0;
else if (Keyword(format, "complex4"))  in_format = 1;
else if (Keyword(format, "byte"))  in_format = 2;
else if (Keyword(format, "float"))  in_format = 3;
else {
  fprintf(stderr, "Unrecognized format: %s\n", format);
  exit(BAD_PARAMETER);
}

printf("Input file =  %s\n", infile);
printf("Input file type = %s\n", format);
printf("Output file =  %s\n", outfile);
printf("File dimensions = %dx%d (cols x rows).\n", xsize, ysize);

if (Keyword(bmaskfile, "none")) printf("No border mask file.\n");
else printf("Border mask file = %s\n", bmaskfile);
if (Keyword(qualfile, "none")) printf("No quality file.\n");
else printf("Correlation image file = %s\n", qualfile);

printf("Averaging template size = %d\n", tsize);
if (tsize < 0 || tsize > 30) {
  fprintf(stderr, "Illegal size: must be between 0 and 30\n");
  exit(BAD_PARAMETER);
}

printf("Quality mode = %s\n", modekey);
mode = SetQualityMode(modekey, qualfile, 0);
if (mode < 0) exit(BAD_PARAMETER);  /* error msg already printed */
```

```
/*  OPEN FILES, ALLOCATE MEMORY   */
OpenFile(&ifp, infile, "r");
OpenFile(&ofp, outfile, "w");
if (!Keyword(bmaskfile, "none"))
  OpenFile(&mfp, bmaskfile, "r");
if (mode==corr_coeffs)
  OpenFile(&qfp, qualfile, "r");
AllocateFloat(&phase, xsize*ysize, "phase data");
AllocateFloat(&soln, xsize*ysize, "scratch data");
AllocateByte(&bitflags, xsize*ysize, "bitflags array");
AllocateFloat(&qual_map, xsize*ysize, "quality map");

/*  READ AND PROCESS DATA  */
printf("Reading phase data...\n");
GetPhase(in_format, ifp, infile, phase, xsize, ysize);

if (qfp) {
  printf("Reading quality data...\n");
  /* borrow the bitflags array temporarily */
  ReadByte(qfp, bitflags, xsize*ysize, qualfile);
  /* process data and store in quality map array */
  AverageByteToFloat(bitflags, qual_map, tsize, xsize, ysize);
}

/* border mask data */
printf("Processing border mask data...\n");
if (mfp) {
  ReadByte(mfp, bitflags, xsize*ysize, bmaskfile);
}
else {
  for (k=0; k<xsize*ysize; k++)
    bitflags[k] = 255;
}
for (k=0; k<xsize*ysize; k++) {
  bitflags[k] = (!bitflags[k]) ? BORDER : 0;
}
if (mfp) FattenMask(bitflags, BORDER, 1, xsize, ysize);

GetQualityMap(mode, qual_map, phase, bitflags, BORDER,
              tsize, xsize, ysize);

if (thresh_flag) {
  HistoAndThresh(qual_map, xsize, ysize, 0.0, 0, 0.0,
                 bitflags, BORDER);
  if (fatten > 0)
```

```
    FattenQual(qual_map, fatten, xsize, ysize);
  }

  for (k=0; k<xsize*ysize; k++) {
    if (bitflags[k] & BORDER) qual_map[k] = 0.0;
  }
  free(bitflags);

  if (debug_flag) {
    char filename[300];
    sprintf(filename, "%s.qual", outfile);
    SaveFloatToImage(qual_map, "quality", filename, xsize, ysize,
                     0, 0, 0);
  }

  /*  UNWRAP  */
  printf("Unwrapping...\n");
  AllocateFloat(&dx, xsize*ysize, "dx derivs");
  AllocateFloat(&dy, ysize*ysize, "dy derivs");
  DxPhaseGradient(phase, dx, xsize, ysize);
  DyPhaseGradient(phase, dy, xsize, ysize);
  for (k=0; k<xsize*ysize; k++)  soln[k] = 0.0;
  MultigridUnwrap(soln, dx, dy, qual_map, NULL, xsize, ysize,
                  num_cycles, num_iter);
  printf("\nFinished\n");
  PrintMinAndMax(xsize, ysize, soln, "solution");

  /* make result congruent to wrapped input phase */
  CongruentSoln(soln, phase, qual_map, xsize, ysize);
  /*  scale and save result  */
  for (k=0; k<xsize*ysize; k++) soln[k] *= TWOPI;
  printf("Saving unwrapped congruent surface to %s\n", outfile);
  WriteFloat(ofp, soln, xsize*ysize, outfile);
  free(dx);
  free(dy);
  free(phase);
  free(qual_map);
  free(soln);
}
```

The function `MultigridUnwrap` in file `fmg.c` controls the multigrid cycle iterations and calls the function that executes the main operations.

```
/*
 *  fmg.c -- function for multigrid solution of weighted
```

```
 *          least-squares phase-unwrapping problem
 */
#include <stdio.h>
#include <math.h>
#include "fmg.h"
#include "grid.h"
#include "dxdygrad.h"

/*  Call the functions for performing the multigrid phase-  */
/*  unwrapping algorithm.  If dywts is a null pointer, then */
/*  dxwts are copied into an array for dywts.               */
void MultigridUnwrap(float *soln, float *dx, float *dy, float *dxwts,
  float *dywts, int xsize, int ysize, int num_cycles, int num_iter)
{
  int  n, dywts_was_null=0, coarsest_dim=3;
  if (dywts==NULL) {
    dywts_was_null = 1;
    AllocateDouble(&dywts, xsize*ysize, "dy wts");
    for (n=0; n<xsize*ysize; n++) dywts[n] = dxwts[n];
  }
  for (n=0; n<num_cycles; n++) {
    printf("\nFMG CYCLE %d\n", n+1);
    FullMultigridVcycle(soln, dx, dy, dxwts, dywts,
                        xsize, ysize, num_iter, coarsest_dim);
  }
  if (dywts_was_null) free(dywts);
}
```

The header file is `fmg.h`:

```
#ifndef __FMG
#define __FMG
void MultigridUnwrap(float *soln, float *dx, float *dy,
                     float *dxwts, float *dywts, int xsize,
                     int ysize, int num_cycles, int num_iter);
#endif
```

The functions in `grid.c` perform the main multigrid operations. (The prolongation and restriction operations are performed by the functions in the file `ungrid.c` of Section A.10.) `FullMultigridVcycle` is a recursive function that performs the weighted full multigrid algorithm, while `Vcycle` recursively performs the weighted V-cycle algorithm. The functions `RestrictDxwts` and `RestrictDywts` restrict the weights from the fine to the coarse grids.

```
/*
 * grid.c - functions for (weighted) multigrid phase unwrapping
 */
#include <stdio.h>
#include <math.h>
#include "grid.h"
#include "gridops.h"
#include "gridmem.h"
#include "relax.h"

/* Main function for weighted multigrid (called recursively) */
void FullMultigridVcycle(float *soln, float *dx, float *dy,
   float *dxwts, float *dywts, int w, int h, int numit, int mindim)
{
  float  *dx2=NULL, *dy2=NULL, *soln2=NULL;
  float  *dxwts2=NULL, *dywts2=NULL;
  int    w2 = w/2, h2 = h/2;
  if (!Coarsest(w, h, mindim)) {
    dxwts2 = Allocate(w2, h2, dxwts_type);
    dywts2 = Allocate(w2, h2, dywts_type);
    dx2 = Allocate(w2, h2, dx_type);
    dy2 = Allocate(w2, h2, dy_type);
    soln2 = Allocate(w2, h2, soln_type);
    RestrictDxwts(dxwts2, w2, h2, dxwts, w, h);
    RestrictDywts(dywts2, w2, h2, dywts, w, h);
    Restrict(dx2, dy2, w2, h2, dx, dy, dxwts, dywts, soln, w, h);
    Zero(soln2, w2, h2);
    FullMultigridVcycle(soln2, dx2, dy2, dxwts2, dywts2, w2, h2,
                        numit, mindim);
    ProlongAndAccumulate(soln, w, h, soln2, w2, h2, dxwts2, dywts2);
  }
  /* perform V-cycle multigrid on fine grid */
  Vcycle(soln, dx, dy, dxwts, dywts, w, h, numit, mindim);
}

/* V-cycle multigrid algorithm (called recursively) */
void Vcycle(float *soln, float *dx, float *dy, float *dxwts,
            float *dywts, int w, int h, int numit, int mindim)
{
  float *dx2=NULL, *dy2=NULL, *soln2=NULL;
  float *dxwts2=NULL, *dywts2=NULL;
  int    w2 = w/2, h2 = h/2;
  if (!Coarsest(w, h, mindim)) {
    Relax(soln, dx, dy, dxwts, dywts, w, h, numit);
    dxwts2 = Allocate(w2, h2, dxwts_type);
```

```
    dywts2 = Allocate(w2, h2, dywts_type);
    dx2 = Allocate(w2, h2, dx_type);
    dy2 = Allocate(w2, h2, dy_type);
    soln2 = Allocate(w2, h2, soln_type);
    RestrictDxwts(dxwts2, w2, h2, dxwts, w, h);
    RestrictDywts(dywts2, w2, h2, dywts, w, h);
    Restrict(dx2, dy2, w2, h2, dx, dy, dxwts, dywts, soln, w, h);
    Zero(soln2, w2, h2);
    Vcycle(soln2, dx2, dy2, dxwts2, dywts2, w2, h2, numit, mindim);
    ProlongAndAccumulate(soln, w, h, soln2, w2, h2, dxwts2, dywts2);
    Relax(soln, dx, dy, dxwts, dywts, w, h, numit);
  }
  else { /* coarsest */
    Relax(soln, dx, dy, dxwts, dywts, w, h, 2*w*h);
  }
}

/* multigrid restriction operation for dx weights */
void RestrictDxwts(float *coarse, int wc, int hc,
                   float *fine, int wf, int hf)
{
  int     a, b, i, j, k, m, n;
  int     k1, k2, k3, k4, k5, k6, k7, k8, k9, k10, k20;
  float   *f, *c;
  f = fine;          c = coarse;
  /*  Indexes:     */
  /*  k1   k2    k3  */
  /*  k4   k5=k  k6  */
  /*  k7   k8    k9  */
  for (j=0; j<hc; j++) {
    b = 2*j;
    for (i=0; i<wc; i++) {
      a = 2*i;
      k = b*wf + a;
      k5 = k;
      k6 = (a < wf - 1) ? k5 + 1 : k5 - 1;
      k4 = (a > 0) ? k5 - 1 : k5 + 1;
      if (b < hf - 1) {
        k7 = k4 + wf;   k8 = k5 + wf;    k9 = k6 + wf;
      }
      else {
        k7 = k4 - wf;   k8 = k5 - wf;    k9 = k6 - wf;
      }
      if (b > 0) {
        k1 = k4 - wf;   k2 = k5 - wf;   k3 = k6 - wf;
```

```
      }
      else {
        k1 = k4 + wf;   k2 = k5 + wf;   k3 = k6 + wf;
      }
      if (f[k5]==0.0 && f[k9]==0.0) c[j*wc + i] = 0.0;
      else if (f[k6]==0.0 && f[k8]==0.0) c[j*wc + i] = 0.0;
      else if (f[k6]==0.0 && f[k9]==0.0) c[j*wc + i] = 0.0;
      else if (f[k5]==0.0 && f[k8]==0.0) c[j*wc + i] = 0.0;
      else if ((i==0 || c[j*wc + i - 1] != 0.0) &&
        ((f[k4]==0.0 && f[k8]==0.0)
          || (f[k5]==0.0 && f[k7]==0.0)))
                                      c[j*wc + i] = 0.0;
      else c[j*wc + i] = 0.25*(f[k5] + f[k6] + f[k8] + f[k9]);
    }
  }
}

/* multigrid restriction operation for dy weights */
void RestrictDywts(float *coarse, int wc, int hc,
                   float *fine, int wf, int hf)
{
  int     a, b, i, j, k, m, n;
  int     k1, k2, k3, k4, k5, k6, k7, k8, k9, k10, k20;
  float  *f, *c;
  f = fine;           c = coarse;
  /*  Indexes:     */
  /*  k1   k2    k3  */
  /*  k4   k5=k  k6  */
  /*  k7   k8    k9  */
  for (j=0; j<hc; j++) {
    b = 2*j;
    for (i=0; i<wc; i++) {
      a = 2*i;
      k = b*wf + a;
      k5 = k;
      k6 = (a < wf - 1) ? k5 + 1 : k5 - 1;
      k4 = (a > 0) ? k5 - 1 : k5 + 1;
      if (b < hf - 1) {
        k7 = k4 + wf;    k8 = k5 + wf;    k9 = k6 + wf;
      }
      else {
        k7 = k4 - wf;    k8 = k5 - wf;    k9 = k6 - wf;
      }
      if (b > 0) {
        k1 = k4 - wf;    k2 = k5 - wf;    k3 = k6 - wf;
```

```
      }
      else {
         k1 = k4 + wf;      k2 = k5 + wf;      k3 = k6 + wf;
      }
      if (f[k5]==0.0 && f[k9]==0.0) c[j*wc + i] = 0.0;
      else if (f[k6]==0.0 && f[k8]==0.0) c[j*wc + i] = 0.0;
      else if (f[k8]==0.0 && f[k9]==0.0) c[j*wc + i] = 0.0;
      else if (f[k5]==0.0 && f[k6]==0.0) c[j*wc + i] = 0.0;
      else if ((j==0 || c[(j - 1)*wc + i] != 0.0) &&
          ((f[k2]==0.0 && f[k6]==0.0)
            ||  (f[k3]==0.0 && f[k5]==0.0)))
                                        c[j*wc + i] = 0.0;
      else c[j*wc + i] = 0.25*(f[k5] + f[k6] + f[k8] + f[k9]);
    }
  }
}
```

The header file is `grid.h`:

```
#ifndef __GRID
#define __GRID
void FullMultigridVcycle(float *soln, float *dx, float *dy,
                         float *dxwts, float *dywts, int w, int h,
                         int numit, int coarsest_dim);
void Vcycle(float *soln, float *dx, float *dy, float *dxwts,
            float *dywts, int w, int h, int numit, int coarsest_dim);
void RestrictDxwts(float *coarse, int wc, int hc, float *fine,
                   int wf, int hf);
void RestrictDywts(float *coarse, int wc, int hc, float *fine,
                   int wf, int hf);
#endif
```

A.13 L^p-NORM ALGORITHM

The following files are contained in this section: `mainlpno.c`, `lpnorm.c`, and `lpnorm.h`.

We can now present the software for the minimum L^p-norm algorithm[1] of Section 5.5. The main program is in file `mainlpno.c`. The usage is very similar to that of the PCG algorithm of Section A.11. The maximum number of main iterations `-iter`, the maximum number of PCG iterations `-pcg_iter` and the normalization parameter `-e0` can be specified. If omitted, these parameters

[1] The software is for the $p = 0$ case. It can be modified easily to work for any p by changing the definition of the weights in function `ComputeDerivWts` in file `laplace.c` in Section A.1.

default to the values 20, 20, and 0.001, respectively. (The normalization parameter is in units of radians/2π.) As in the case of the PCG algorithm, we recommend that the `-thresh yes` argument be specified, since the algorithm seems to work best with binary weights. A typical command to execute the software would be

```
lpn -input mri.phase -format byte -output mri.surf -bmask
    mri.bmask -xsize 256 -ysize 256 -mode min_var -tsize 3
    -thresh yes -fat 0 -pcg_iter 20 -iter 20 -e0 0.001
```

where `lpn` is the name of the executable file.

The program works as follows: The command line arguments are parsed by the functions `CommandLineParm`, `Keyword` and `SetQualityMode`. The files are opened, the memory for the arrays is allocated, and the phase data, the quality map, and the (optional) border mask data are then read from the input files and processed. (The border mask, not to be confused with the quality map or mask, consists of pixels that should be ignored, such as the border pixels of the simulated IFSAR data shown in Figure 3.8b.) The border mask is thickened slightly by a call to the function `FattenMask`, the quality map may be thresholded to produce a quality mask, and the zero-weights of the quality mask may be fattened if the `-fat` argument is specified. If the array dimensions `xsize` and `ysize` are not of the form $2^n + 1$, then the arrays are embedded in larger arrays and zero padded. The dimensions of the larger arrays, `xsize_dct` and `ysize_dct`, are of the form $2^n + 1$. After additional arrays are allocated for the PCG arrays and the data-dependent weights, the phase Laplacian is computed by function `ComputeLaplacian`, and the function `LpNormUnwrap` is called to execute the L^p-norm algorithm. The unwrapped solution is then extracted from the larger array in which it was embedded and written to the output file.

```
/*
 * mainlpno.c -- phase unwrapping by means of min. Lp norm algorithm
 *
 * Source code files required:
 *     congruen.c         dct.c    dxdygrad.c      extract.c
 *      getqual.c        grad.c       histo.c      laplace.c
 *       lpnorm.c    mainlpno.c     maskfat.c          pcg.c
 *     qualgrad.c    qualpseu.c     qualvar.c       raster.c
 *     residues.c     solncos.c        util.c
 */
#include <stdio.h>
#include <math.h>
#include "file.h"
#include "histo.h"
#include "pi.h"
```

```
#include "getqual.h"
#include "congruen.h"
#include "maskfat.h"
#include "util.h"
#include "extract.h"
#include "getqual.h"
#include "lpnorm.h"
#define BORDER  0x20
#define DEFAULT_NUM_ITER  10
#define DEFAULT_PCG_ITER  20
#define DEFAULT_E0        0.001

main (int argc, char *argv[])
{
  int            i, j, k, n;
  FILE           *ifp, *ofp, *mfp=0, *qfp=0;
  float          *phase;     /* array */
  float          *soln;      /* array */
  float          *qual_map;  /* array */
  float          *rarray;    /* array */
  float          *zarray;    /* array */
  float          *parray;    /* array */
  float          *dxwts;     /* array */
  float          *dywts;     /* array */
  unsigned char  *bitflags;
  double         *xcos, *ycos;
  char           buffer[200], tempstr[200];
  char           infile[200], outfile[200];
  char           bmaskfile[200], qualfile[200];
  char           format[200], modekey[200];
  int            in_format, debug_flag;
  int            xsize, ysize;   /* dimensions of arrays */
  int            xsize_actual, ysize_actual, xsize_dct, ysize_dct;
  int            avoid_code, thresh_flag, fatten, tsize;
  int            num_iter=DEFAULT_NUM_ITER;
  int            pcg_iter=DEFAULT_PCG_ITER;
  double         rmin, rmax, rscale, e0=DEFAULT_E0;
  double         one_over_twopi = 1.0/TWOPI;
  UnwrapMode     mode;
  char           use[] =     /* define usage statement */
    "Usage: prog-name -input file -format fkey -output file\n"
    "  -xsize x -ysize y [ -mode mkey -bmask file -corr file\n"
    "  -tsize size -debug yes/no -iter num -pcg_iter nump\n"
    "  -e0 e0val -thresh yes/no -fat n ]\n"
    "where 'fkey' is a keyword designating the input file type\n"
```

```
    "(key = complex8, complex4, float or byte), 'x' and 'y' are\n"
    "the dimensions of the file, bmask is an optional byte-file\n"
    "for masking out undefined phase, corr is an optional byte-\n"
    "file of cross-correlation values, tsize is the size of the\n"
    "square template for averaging the corr file or quality\n"
    "values (default = 1), and 'mkey' is a keyword designating\n"
    "the type of quality map.  The value of mkey may be\n"
    "'min_grad' for Minimum Gradient, 'min_var' for Minimum\n"
    "Variance, 'max_corr' for Maximum Correlation, 'max_pseu' for\n"
    "Maximum Pseudocorrelation, or 'none' for no quality map.\n"
    "All files are simple raster files, and the output file\n"
    "is a raster file of elevation values of the unwrapped\n"
    "surface.  If the 'debug' parm is 'yes', then intermediate\n"
    "byte-files are saved.  The maximum number of iterations is\n"
    "'num', and the number of PCG iterations is 'nump'.  The\n"
    "normalization parameter is 'e0'.  To apply an automatic\n"
    "threshold to the quality map, the 'thresh' parm should be\n"
    "yes.  To thicken the quality mask, the 'fat' parm should be\n"
    "the number of pixels by which to fatten it.\n";

printf("Phase Unwrapping by Minimum Lp Norm Algorithm\n");

/* GET COMMAND LINE PARAMETERS AND CHECK */
CommandLineParm(argc,argv, "-input", StringParm, infile, 1, use);
CommandLineParm(argc,argv, "-format", StringParm, format, 1, use);
CommandLineParm(argc,argv, "-output", StringParm, outfile, 1,use);
CommandLineParm(argc,argv, "-xsize", IntegerParm, &xsize, 1, use);
CommandLineParm(argc,argv, "-ysize", IntegerParm, &ysize, 1, use);
if (!CommandLineParm(argc, argv, "-mode", StringParm,
      modekey, 0, use)) strcpy(modekey, "none");
if (!CommandLineParm(argc, argv, "-corr", StringParm,
      qualfile, 0, use)) strcpy(qualfile, "none");
if (!CommandLineParm(argc, argv, "-tsize", IntegerParm,
      &tsize, 0, use)) tsize = 1;
if (!CommandLineParm(argc, argv, "-bmask", StringParm, bmaskfile,
      0, use)) strcpy(bmaskfile, "none");
if (!CommandLineParm(argc, argv, "-debug", StringParm, tempstr,
      0, use)) debug_flag = 0;
else debug_flag = Keyword(tempstr, "yes");
if (!CommandLineParm(argc, argv, "-thresh", StringParm,
      tempstr, 0, use)) thresh_flag = 0;
else thresh_flag = Keyword(tempstr, "yes");
if (!CommandLineParm(argc, argv, "-fat", IntegerParm,
      &fatten, 0, use)) fatten = 0;
CommandLineParm(argc, argv, "-iter", IntegerParm, &num_iter,
```

```
                0, use);
if (num_iter < 0) num_iter = 1;
CommandLineParm(argc, argv, "-pcg_iter", IntegerParm, &pcg_iter,
                0, use);
if (pcg_iter < 0) pcg_iter = 1;
CommandLineParm(argc, argv, "-e0", DoubleParm, &e0, 0, use);

if (Keyword(format, "complex8"))  in_format = 0;
else if (Keyword(format, "complex4"))  in_format = 1;
else if (Keyword(format, "byte"))  in_format = 2;
else if (Keyword(format, "float"))  in_format = 3;
else {
  fprintf(stderr, "Unrecognized format: %s\n", format);
  exit(BAD_PARAMETER);
}

printf("Input file =  %s\n", infile);
printf("Input file type = %s\n", format);
printf("Output file =  %s\n", outfile);
printf("File dimensions = %dx%d (cols x rows).\n", xsize, ysize);
if (Keyword(bmaskfile, "none")) printf("No border mask file.\n");
else printf("Border mask file = %s\n", bmaskfile);
printf("Quality mode = %s\n", modekey);
if (modekey==none) {
  printf("No quality map.\n");
  strcpy(qualfile, "none");
}
else {
  if (Keyword(qualfile, "none")) printf("No correlation file.\n");
  else printf("Correlation image file = %s\n", qualfile);
  printf("Averaging template size = %d\n", tsize);
  if (tsize < 0 || tsize > 30) {
    fprintf(stderr, "Illegal size: must be between 0 and 30\n");
    exit(BAD_PARAMETER);
  }
}
mode = SetQualityMode(modekey, qualfile, 1);
if (mode < 0) exit(BAD_PARAMETER);  /* error msg already printed */

/* Increase dimensions to power of two (plus one) */
xsize_actual = xsize;
ysize_actual = ysize;
for (xsize_dct = 1; xsize_dct+1 < xsize_actual; xsize_dct*=2)
  ;
```

```
xsize_dct+=1;
for (ysize_dct = 1; ysize_dct+1 < ysize_actual; ysize_dct*=2)
  ;
ysize_dct+=1;
if (xsize_dct != xsize_actual || ysize_dct != ysize_actual) {
  printf("Dimensions increase from %dx%d to %dx%d for FFT's\n",
         xsize_actual, ysize_actual, xsize_dct, ysize_dct);
}

/*  OPEN FILES, ALLOCATE MEMORY   */
/* note: xsize_dct >= xsize;  ysize_dct >= ysize */
xsize = xsize_dct;
ysize = ysize_dct;
AllocateFloat(&phase, xsize*ysize, "phase data");
AllocateFloat(&soln, xsize*ysize, "scratch data");
AllocateFloat(&qual_map, xsize*ysize, "quality map");
AllocateByte(&bitflags, xsize*ysize, "bitflags array");
OpenFile(&ifp, infile, "r");
OpenFile(&ofp, outfile, "w");
if (!Keyword(bmaskfile, "none")) OpenFile(&mfp, bmaskfile, "r");
if (mode==corr_coeffs) OpenFile(&qfp, qualfile, "r");

/*  READ AND PROCESS DATA  */
xsize = xsize_actual;
ysize = ysize_actual;
printf("Reading input data...\n");
GetPhase(in_format, ifp, infile, phase, xsize, ysize);

if (qfp) {
  printf("Reading quality data...\n");
  /* borrow the bitflags array temporarily */
  ReadByte(qfp, bitflags, xsize*ysize, qualfile);
  /* process data and store in quality map array */
  AverageByteToFloat(bitflags, qual_map, tsize, xsize, ysize);
}

/* border mask data */
printf("Processing border mask data...\n");
if (mfp) {
  ReadByte(mfp, bitflags, xsize*ysize, bmaskfile);
}
else {
  for (k=0; k<xsize*ysize; k++)
    bitflags[k] = 255;
}
```

```
  for (k=0; k<xsize*ysize; k++) {
    bitflags[k] = (!bitflags[k]) ? BORDER : 0;
  }
  if (mfp) FattenMask(bitflags, BORDER, 1, xsize, ysize);

  GetQualityMap(mode, qual_map, phase, bitflags, BORDER,
                tsize, xsize, ysize);
  if (thresh_flag) {
    HistoAndThresh(qual_map, xsize, ysize, 0.0, 0, 0.0,
                   bitflags, BORDER);
    if (fatten > 0)
      FattenQual(qual_map, fatten, xsize, ysize);
  }
  /* Set border (masked) pixels to 0-wts  */
  for (k=0; k<xsize*ysize; k++) {
    if (bitflags[k] & BORDER) qual_map[k] = 0.0;
  }

  if (debug_flag) {
    char filename[300];
    sprintf(filename, "%s.qual", outfile);
    SaveFloatToImage(qual_map, "quality", filename, xsize, ysize,
                     0, 0, 0);
  }

  /* embed arrays in possibly larger FFT/DCT arrays */
  for (j=ysize_dct-1; j>=0; j--) {
    for (i=xsize_dct-1; i>=0; i--) {
      if (i<xsize_actual && j<ysize_actual)
        phase[j*xsize_dct + i] = phase[j*xsize_actual + i];
      else phase[j*xsize_dct + i] = 0.0;
    }
  }
  for (j=ysize_dct-1; j>=0; j--) {
    for (i=xsize_dct-1; i>=0; i--) {
      if (i<xsize_actual && j<ysize_actual)
        bitflags[j*xsize_dct + i] = bitflags[j*xsize_actual + i];
      else bitflags[j*xsize_dct + i] = BORDER;
    }
  }
  if (qual_map) {
    for (j=ysize_dct-1; j>=0; j--) {
      for (i=xsize_dct-1; i>=0; i--) {
        if (i<xsize_actual && j<ysize_actual)
```

```
          qual_map[j*xsize_dct + i] = qual_map[j*xsize_actual + i];
        else
          qual_map[j*xsize_dct + i] = 0.0;
      }
    }
  }

  /* Set dimensions to DCT dimensions */
  xsize = xsize_dct;
  ysize = ysize_dct;
  /* Allocate more memory */
  AllocateFloat(&rarray, xsize*ysize, "r array data");
  AllocateFloat(&zarray, xsize*ysize, "z array data");
  AllocateFloat(&parray, xsize*ysize, "p array data");
  AllocateFloat(&dxwts, xsize*ysize, "dx weights");
  AllocateFloat(&dywts, xsize*ysize, "dy weights");

  /*  UNWRAP  */
  for (k=0; k<xsize*ysize; k++) soln[k] = 0.0;
  printf("Unwrapping...\n");
  LpNormUnwrap(soln, phase, dxwts, dywts, bitflags, qual_map,
    rarray, zarray, parray, num_iter, pcg_iter, e0, xsize, ysize);
  printf("\nFinished\n");
  PrintMinAndMax(xsize, ysize, soln, "solution");

  /* restore dimensions to input sizes and save solution*/
  xsize = xsize_actual;
  ysize = ysize_actual;
  for (j=0; j<ysize; j++) {
    for (i=0; i<xsize; i++) {
      soln[j*xsize + i] = TWOPI*soln[j*xsize_dct + i];
    }
  }
  /* save solution */
  printf("Saving unwrapped surface to file '%s'\n", outfile);
  WriteFloat(ofp, soln, xsize*ysize, outfile);
  free(rarray);
  free(zarray);
  free(parray);
  free(soln);
  free(qual_map);
  free(bitflags);
  free(dxwts);
  free(dywts);
}
```

The function `LpNormUnwrap` controls the main loop of the L^p-norm algorithm and is listed below in file `lpnorm.c`. The function iterates until the phase residual becomes residue-free or the maximum number of iterations is achieved. Within each iteration the function `PCGUnwrap` is called to execute the PCG phase-unwrapping algorithm on the residual phase. After the main loop iterations are completed, one of two steps is performed. If the residual (computed by function `ResidualPhase`) is residue-free, then it is unwrapped with a simple phase-unwrapping algorithm (`RasterUnwrap`) and added to the intermediate solution. If, on the other hand, the residual contains residues, the solution is made congruent to the wrapped input phase.

```
/*
 *  lpnorm.c -- functions for minimum Lp-norm phase unwrapping
 */
#include <stdio.h>
#include <math.h>
#include "pcg.h"
#include "raster.h"
#include "residues.h"
#include "laplace.h"
#include "lpnorm.h"
#include "util.h"
#include "congruen.h"
#define POS_RES  0x01
#define NEG_RES  0x02
#define BORDER   0x20

/* Main iteration of minimum Lp-norm phase-unwrapping algorithm */
void LpNormUnwrap(float *soln, float *phase, float *dxwts,
             float *dywts, unsigned char *bitflags, float *qual_map,
             float *rarray, float *zarray, float *parray, int iter,
             int pcg_iter, double e0, int xsize, int ysize)
{
  int  i, j, k, n;
  float  *residual;
  residual = rarray;  /* borrow rarray to compute the residual */
  ResidualPhase(residual, phase, soln, xsize, ysize);
  n = Residues(residual, bitflags, POS_RES, NEG_RES,
               BORDER, xsize, ysize);
  for (k=0; k<iter && n>0; k++) {
    printf("\nIter %d: %d residues\n", k+1, n);
    ComputeDerivWts(phase, soln, dxwts, dywts, qual_map, e0,
                    xsize, ysize);
    ComputeLaplacian(phase, rarray, dxwts, dywts, xsize, ysize, 1);
    /* borrow zarray temporarily to compute Laplacian of soln */
```

```
    ComputeLaplacian(soln, zarray, dxwts, dywts, xsize, ysize, 0);
    for (i=0; i<xsize*ysize; i++) rarray[i] -= zarray[i];
    PCGUnwrap(rarray, zarray, parray, soln, dxwts, dywts,
              xsize, ysize, pcg_iter, 0.0);
    ResidualPhase(residual, phase, soln, xsize, ysize);
    n = Residues(residual, bitflags, POS_RES, NEG_RES,
                 BORDER, xsize, ysize);
  }
  printf("%d residues after %d iter.  Processing residual...\n",
         n, k);
  /* see if the entire nonmasked array is residue-free */
  n = Residues(residual, NULL, POS_RES, NEG_RES,
               BORDER, xsize, ysize);
  if (n==0) {
    /* Unwrap residual and add to solution */
    /* (This could be improved by unwrapping only the non-masked */
    /* pixels, and removing the nonmasked residue test above)    */
    RasterUnwrap(residual, zarray, xsize, ysize); /* borrow zarray */
    for (k=0; k<xsize*ysize; k++)  soln[k] += zarray[k];
  }
  else {
    /* Make solution congruent to wrapped input phase */
    CongruentSoln(soln, phase, NULL, xsize, ysize);
  }
}


/* Compute the residual phase, i.e., the wrapped difference */
/* between the (partially) unwrapped solution "soln" and     */
/* and the wrapped input phase "phase".                      */
void ResidualPhase(float *resid, float *phase, float *soln,
                   int xsize, int ysize)
{
  int    n, k;
  double r;
  for (k=0; k<xsize*ysize; k++) {
    r = phase[k] - soln[k];
    if (r < 0) r += (2 - ((int)r));  /* make r positive */
    r -= (int)r;  /* get fractional part of r */
    if (r > 0.5) r -= 1.0;  /* wrap to range (-0.5, 0.5) */
    resid[k] = r;
  }
}
```

The header file is `lpnorm.h`:

```
#ifndef __LPNORM
#define __LPNORM
void LpNormUnwrap(float *soln, float *phase, float *dxwts,
         float *dywts, unsigned char *bitflags, float *qual_map,
         float *rarray, float *zarray, float *parray, int iter,
         int pcg_iter, double e0, int xsize, int ysize);
void RasterUnwrap(float *phase, float *soln, int xsize, int ysize);
void ResidualPhase(float *resid, float *phase, float *soln,
                   int xsize, int ysize);
#endif
```

APPENDIX B

SOFTWARE INDEX

In this appendix we list the files that make up the software of Appendix A and the sections in which they are located, followed by a list of the constants, macro names, data types, and functions and the files in which they are defined. We begin with the list of files. They are listed in alphabetical order along with their section numbers and brief descriptions of their contents.

FILES

SECTION	FILE	DESCRIPTION
A.2	`brcut.c`	functions for placing branch cuts in arrays
A.2	`brcut.h`	header file for `brcut.c`
A.4	`congruen.c`	functions for congruence operation
A.4	`congruen.h`	header file for `congruen.c`
A.4	`dct.c`	two-dimensional fast cosine transform
A.4	`dct.h`	header file for `dct.c`
A.2	`dipole.c`	function for finding dipole residues
A.2	`dipole.h`	header file for `dipole.c`
A.2	`dxdygrad.c`	functions for computing phase gradients
A.1	`dxdygrad.h`	header file for `dxdygrad.c`
A.1	`extract.c`	functions for extracting phase data
A.1	`extract.h`	header file for `extract.c`
A.1	`file.h`	error codes
A.9	`flynn.c`	functions for Flynn's algorithm
A.9	`flynn.h`	header file for `flynn.c`
A.12	`fmg.c`	Multigrid phase-unwrapping algorithm
A.12	`fmg.h`	header file for `fmg.c`
A.3	`getqual.c`	functions for generating quality maps
A.3	`getqual.h`	header file for `getqual.c`

A.6	gold.c	Goldstein's algorithm for branch cuts
A.6	gold.h	header file for gold.c
A.2	grad.c	function for computing phase gradient
A.2	grad.h	header file for grad.c
A.12	grid.c	functions for multigrid phase unwrapping
A.12	grid.h	header file for grid.c
A.10	gridmem.c	memory management functions for multigrid
A.10	gridmem.h	header file for gridmem.c
A.10	gridops.c	multigrid operations
A.10	gridops.h	header file for gridops.c
A.3	histo.c	functions for histograms and thresholding
A.3	histo.h	header file for histo.c
A.4	laplace.c	functions for Laplacians, residual phase
A.4	laplace.h	header file for laplace.c
A.2	list.c	functions for managing pixels to unwrap
A.2	list.h	header file for list.c
A.13	lpnorm.c	minimum L^p-norm-unwrapping algorithm
A.13	lpnorm.h	header file for lpnorm.c
A.5	maindiff.c	main program for comparing two surfaces
A.9	mainflyn.c	main program for Flynn's algorithm
A.12	mainfmg.c	main program for multigrid algorithm
A.6	maingold.c	main program for Goldstein's algorithm
A.5	mainjump.c	main program for measuring discontinuities
A.13	mainlpno.c	main program for L^p-norm algorithm
A.8	mainmcut.c	main program for mask cut algorithm
A.11	mainpcg.c	main program for PCG algorithm
A.7	mainqual.c	main program for quality-guided algorithm
A.10	mainunmg.c	main program for unweighted multigrid
A.10	mainunwt.c	main program for unweighted DCT algorithm
A.8	maskcut.c	function for mask cut algorithm
A.8	maskcut.h	header file for maskcut.c
A.3	maskfat.c	functions for thickening masks
A.3	maskfat.h	header file for maskfat.c
A.3	maskthin.c	functions for thinning a mask
A.3	maskthin.h	header file for maskthin.c
A.2	path.c	function for unwrapping around branch cuts
A.2	path.h	header file for path.c
A.11	pcg.c	functions for PCG algorithm
A.11	pcg.h	header file for pcg.c
A.1	pi.h	definitions of π and 2π
A.3	qualgrad.c	functions for maximum phase gradient maps
A.3	qualgrad.h	header file for qualgrad.c
A.7	quality.c	function for quality-guided algorithm
A.7	quality.h	header file for quality.c
A.3	qualpseu.c	functions for pseudocorrelation maps

We now list the items that are defined in the header files, namely the constants, function macros and data types. They are listed below in alphabetical order along with the section numbers and the header files in which they are defined. The page numbers are also listed.

CONSTANTS, MACROS, DATA TYPES

We now list the functions. The main programs (i.e., the contents of the source files whose names are prefixed with `main` such as `mainpcg.c`) are not listed. The functions are listed alphabetically along with the section numbers and the files in which they are defined. The page numbers of the functions are also listed.

FUNCTIONS

APPENDIX C

SOFTWARE ftp DOWNLOAD

All of the software listed in Appendix A is available for downloading on the Wiley ftp server. The data files for the phase unwrapping examples of Chapter 3, including the wrapped phase data, the correlation maps, and the "ground truth" surfaces, are posted on the server as well. The server can be accessed through a web browser or ftp program. To access the software and data, connect online at ftp://ftp.wiley.com/public/sci_tech_med/ and go to the phase_unwrapping directory. There you will find directions on how to download the software and data.

These software and data should not be considered public domain. They are made available to purchasers of this book. Please refer to the license statement and disclaimer of warranty in Appendix A.

INDEX

Lightning Source UK Ltd.
Milton Keynes UK
UKOW05n1541010414

229225UK00001B/82/P

9 780471 249351